# Sustainability and the Rights of Nature in Practise

# Social–Environmental Sustainability

*Series Editor*

## Chris Maser

## Published Titles

**Decision-Making for a Sustainable Environment:**
A Systemic Approach
*Chris Maser*

**Biosequestration and Ecological Diversity:**
Mitigating and Adapting to Climate Change and Environmental Degradation
*Wayne A. White*

**Insects and Sustainability of Ecosystem Services**
*Timothy D. Schowalter*

**Land-Use Planning for Sustainable Development, Second Edition**
*Jane Silberstein and Chris Maser*

**Interactions of Land, Ocean and Humans:**
A Global Perspective
*Chris Maser*

**Sustainability and the Rights of Nature:**
An Introduction
*Cameron La Follette and Chris Maser*

**Fundamentals of Practical Environmentalism**
*Mark B. Weldon*

**Economics and Ecology:**
United for a Sustainable World
*Charles R. Beaton and Chris Maser*

**Resolving Environmental Conflicts:**
Principles and Concepts, Third Edition
*Chris Maser and Lynette de Silva*

**Sustainability and the Rights of Nature in Practise**
*Cameron La Follette and Chris Maser*

For more information on this series, please visit: https://www.crcpress.com/Social-Environmental-Sustainability/book-series/CRCSOCENVSUS

# Sustainability and the Rights of Nature in Practise


Edited by
Cameron La Follette
Chris Maser



CRC Press
Taylor & Francis Group
Boca Raton London New York

CRC Press is an imprint of the
Taylor & Francis Group, an **informa** business

Front cover photo of the Whanganui River of New Zealand © Lynne L. Reid, 2017.

CRC Press
Taylor & Francis Group
6000 Broken Sound Parkway NW, Suite 300
Boca Raton, FL 33487-2742



Printed on acid-free paper

International Standard Book Number-13: 978-1-138-58451-8 (Hardback)

DOI: 10.1201/9780429505959



**Library of Congress Cataloging-in-Publication Data**

Names: La Follette, Cameron, author. | Maser, Chris, author.
Title: Sustainability and the rights of nature in practise / Chris Maser, Cameron La Follette.
Description: Boca Raton : CRC Press 2019. | Summary: "This book provides practitioners from diverse cultures around the world the opportunity to describe their own projects, successes, and challenges in moving toward a legal personhood for Nature. It includes contributions from Nepal, New Zealand, U.S. Native American cultures, and Scotland, amongst others, by practitioners working on projects that can be integrated into a Rights of Nature framework. The authors also tackle required changes to shift the paradigm such as thinking of Nature in a sacred manner, questioning Nature's rights versus human rights, conceptualization of restoration, and the removal of a large-scale energy infrastructure"-- Provided by publisher.
Identifiers: LCCN 2019021497 | ISBN 9781138584518 (hardback) | ISBN 9780429505959 (ebook)
Subjects: LCSH: Environmental law, International. | Rights of nature. | Sustainable development--Law and legislation. | Nature conservation--Law and legislation. | Environmental protection--International cooperation. | Environmentalism--International cooperation. | Sustainability--International cooperation. | Nature--Effect of human beings on.
Classification: LCC K3585 .L34 2019 | DDC 344.04/6--dc23
LC record available at https://lccn.loc.gov/2019021497

**Visit the Taylor & Francis Web site at**
**http://www.taylorandfrancis.com**

**and the CRC Press Web site at**
**http://www.crcpress.com**

*We dedicate this book to all those working to bring true sustainability, and a respectful relationship with Nature, to fruition for the benefit of all generations*

# Contents

Series Preface ....... ix
Preface ....... xi
Acknowledgements ....... xiii
Editors ....... xv
Contributors ....... xvii

## Section I Concepts and Overviews

1. **Introduction: Rights of Nature, Sacred Lands and Sustainability in the Western Tradition** ....... 3
   *Cameron La Follette*

2. **Defending the Tree of Life: The Ethical Justification for the Rights of Nature in a Theory of Justice** ....... 13
   *Kathryn Anne Gwiazdon*

3. **Rights of Nature: Myth, Films, Laws and the Future** ....... 39
   *Eugen Cadaru*

4. **Nature's Rights in Permaculture** ....... 59
   *W. D. Scott Pittman*

## Section II The Struggle for Sustainability and the Rights of Nature

5. **Kiribati and Climate Change** ....... 79
   *His Eminence Anote Tong*

6. **'When God Put Daylight on Earth We Had One Voice' Kwakwaka'wakw Perspectives on Sustainability and the Rights of Nature** ....... 89
   *Douglas Deur, Kim Recalma-Clutesi and Clan Chief Kwaxsistalla Adam Dick*

7. **Environmental Sustainability: The Case of Bhutan** ....... 113
   *Dechen Lham*

8. **The Restoration of the Caledonian Forest and the Rights of Nature** ....... 141
   *Alan Watson Featherstone*

9. **The Significance of the Stewardship Ethic of the Indigenous Peoples of Nigeria's Niger Delta Region on Biodiversity Conservation** ....... 167
   *Ngozi F. Unuigbe*

**10. German Energiewende: A Way to Sustainable Societies?** ........ 191
*Michael W. Schröter and Dani Fössl*

**11. Seasonally Flooded Savannas of South America: Sustainability and the Cattle-Wildlife Mosaic** ........ 205
*Almira Hoogesteijn, José Luis Febles and Rafael Hoogesteijn*

**12. Ocean Rights: The Baltic Sea and World Ocean Health** ........ 237
*Michelle Bender*

## Section III Rights of Nature in the Law

**13. A River Is Born: New Zealand Confers Legal Personhood on the Whanganui River to Protect It and Its Native People** ........ 259
*The Honorable Christopher Finlayson*

**14. The Rights of Nature in Ecuador: An Overview of the New Environmental Paradigm** ........ 279
*Hugo Echeverria and Francisco José Bustamante Romo Leroux*

**15. The Godavari Marble Case and Rights of Nature Discourse in Nepal** ........ 295
*Jony Mainaly*

**16. Nature's Rights: Why the European Union Needs a Paradigm Shift in Law to Achieve Its 2050 Vision** ........ 311
*Mumta Ito*

**17. Nature's Rights through Lawmaking in the United States** ........ 331
*Lindsey Schromen-Wawrin and Michelle Amelia Newman*

**18. The Experiment with Rights of Nature in India** ........ 365
*Kelly D. Alley and Tarini Mehta*

**19. Caring for Country and Rights of Nature in Australia: A Conversation between Earth Jurisprudence and Aboriginal Law and Ethics** ........ 385
*Mary Graham and Michelle Maloney*

**20. Conclusion: Nature's Laws of Reciprocity** ........ 401
*Chris Maser*

**Index** ........ 405

# *Series Preface*

In reading this book, keep in mind that Nature functions perfectly without any human 'help' or intervention, which means Nature's value is entirely intrinsic and self reinforcing. But the majority of people feel that Nature, as a resource, is wasted if it is not used for their benefit – hence the concept of 'conversion potential'. In other words, how can a particular natural resource be converted into a commercial product for human use, thereby giving it economic value?

Here, a basic principle is that community programs must be founded on local requirements and cultural values in balance with those of the broader world, which includes understanding and acknowledging environmental issues, long-term biophysical trends, and their social–environmental ramifications. After all, social–environmental sustainability is a common relationship between people and the natural environment, based on the constraints of the underlying biophysical principles that maintain the lands and waters in a sustainably productive state. Simply put, as we honour our relationship with Nature in a sustainable way, we honour all generations. As we abuse Nature by overexploitation, we abuse all generations.

Maintaining a respectful relationship with Nature – by placing its right to flourish foremost in human economics – forms a critical, worldwide nexus between the social–environmental sustainability of people in the present and those of the future. With respect to every culture worldwide, it is imperative that we take personal responsibility for our words, deeds, decisions, actions and their consequences, because the first step toward social–environmental sustainability begins with the respect and the quality of the care we give ourselves. We must then extend that respect and care to our families, friends, neighbours and Nature, because community sustainability is the foundation of every nation. This said, the degree of mutual caring, cooperation and long-term sustainability of the landscape characterises a community and reflects the psychophysical health of its citizenry.

And it is our humility and consent to the Rights of Nature, by prioritising and repairing our relationship with Nature, that determines the legacy we leave – one that in today's world either progressively liberates or progressively impoverishes all generations. The choice of how we, the adults of the world, behave is ours – either with psychological maturity and sacred humility or self-indulgence and monetary arrogance. Fortunately, our human consciousness is elevating as the Rights of Nature paradigm continues to spread in countries throughout the world – illustrated in part by the courageous articles in this book.

Finally, this CRC series of books on the various facets of social–environmental sustainability is a forum wherein those who dare to seek harmony and wholeness can struggle to integrate disciplines and balance the material world with the spiritual, the scientific with the social, and in so doing expose their vulnerabilities, human frailties and hope, as well as their visions for a good-quality, sustainable future.

As the title of this book implies, the human component of the world is critically important – but often an overlooked or blatantly ignored dimension of social–environmental sustainability. Yet, it is the integrity of the relationships among the diverse elements of any system that both defines it through its functional processes and, in obeying the Right of Nature, confers global social–environmental sustainability to all generations of life on Earth.

**Chris Maser**
*Series Editor*

# *Preface*

In 2017, Chris Maser and I published *Sustainability and the Rights of Nature: An Introduction*, to survey current efforts worldwide to reorient human resource use to Nature's ecological limits. We also sketched how resource use in the United States would have to change to fit within such a model.[1] We focused on the so-called 'Rights of Nature' paradigm, because it is the most comprehensive effort thus far to create a legal framework that cultures can adapt to fit their own circumstances.

After the initial book, we realised the importance of hearing other voices worldwide on the struggle to create true sustainability by respecting Nature's right to flourish. Only by on-the-ground examples can we discover what people in varying environments and cultures are doing to protect landscapes, restrain human overuse, restore degraded areas and create and enforce respectful relationships with the hills and valleys, forests and seashores, they call home. One of the most difficult decisions we had was locating efforts that really focused on Nature's right to flourish, whether directly or indirectly. We considered whether the effort – local or on a broad scale – was *systemic*, rather than symptomatic. We have tried to avoid including papers focusing on a symptomatic approach to the despoliation of Nature, because it is narrower, human-focused and fragmented, and does not seek to heal the overall human relationship to Nature.

In this book, we aim to show how various cultures of the world are grappling with the need to mend this relationship. Some of the case studies focus more on sustainability; others focus explicitly on creating the legal framework for a Rights of Nature-based form of ecological governance. It is important to recognise that all cultures have rich histories of interaction with their local environments and therefore have their own traditions to draw from when grappling with the excesses of industrialisation, whether it took place rapidly and recently, over decades or even centuries (as in England, where industrialisation began). Thus, it is most important to have reports from people living in those cultures about how they are facing the need for systemic change.

There are three overview papers: one looking at cinema and its effects on people's understanding of Nature's intelligence, another exploring the ethical implications of ecological governance and a third describing Permaculture, which provides a thoughtful model for living sustainably worldwide. Following these, the papers explore environmental degradation and powerful, creative responses to the problem from across the globe, ranging from Bhutan to Germany, Canada to Venezuela, Scotland to Nigeria, Nepal to New Zealand and Australia, Ecuador to Kiribati and India, the European Union to the United States. The principal point to make is that there are exciting frontiers ahead: people wrestling creatively with the urgency and focus appropriate to the warning signals emanating from the world's ecosystems. This is how the relationship that has been too long ignored is repaired; in the process, human cultures are revitalised.

My role in this book has been to collaborate with many talented, devoted people worldwide to bring their struggles, challenges and philosophies before readers. I am deeply grateful to them all for taking the time from busy lives to write these case studies and work with me through the long editorial process. I also extend my gratitude to all my colleagues in the environmental and conservation fields worldwide: their courage and farsightedness, often in very difficult situations, are an inspiration always.

All the opinions I have stated in this book are personal to me only, and do not reflect the opinions of any employers past or present.

**Cameron La Follette**

## Note

1. La Follette, Cameron and Maser, Chris. 2017. *Sustainability and the Rights of Nature: An Introduction.* CRC Press, Boca Raton, FL.

# Acknowledgements

We, the editors, gratefully acknowledge the many people without whom this book would never have come to pass. First and foremost, we extend our profoundest gratitude to the authors, who contributed their knowledge, dedication and experience to writing for this volume. They have taken time out of exceedingly busy lives, while engaged in valuable work, to describe what is happening, or what could be done, to protect Nature, and thus human cultures, for all generations to come. It has been our great privilege to work with these far-sighted individuals from around the world.

We also acknowledge the dedicated staff at CRC Press, without whom this volume, and our earlier book, *Sustainability and the Rights of Nature: An Introduction*, would never have been completed. In particular, we offer special thanks our editor, Irma Shagla Britton, who is always standing by with advice, guidance and wise counsel. Without Marsha Hecht's wonderful attention to detail as project editor, and Arun Kumar's guidance as project manager, this book would not have seen the light of day.

In addition, I (Chris) offer sincere thanks to my wife, Zane, for her understanding and patience while I worked on the manuscript. I am also grateful for a second chance to work with Cameron La Follette, whom I have found to be a superb coauthor. Finally, I (Cameron) gratefully acknowledge Chris Maser's tremendous and unflagging support that led to my working on both these books on the exciting new paradigm of Rights of Nature.

# Editors

**Cameron La Follette** has a law degree from Columbia University School of Law, a Master's in Psychology from New York University, and a Bachelor's in Journalism from the University of Oregon. Her initial environmental activism (1978–1982) was with Oregon non-profit organisations that focused on preserving ancient forests on Federal public lands managed by the U.S. Forest Service and U.S. Bureau of Land Management to protect salmon habitat, clean drinking water and forest ecosystems. She served on the Salem, OR, Planning Commission for 3 years (2002–2005), applying the City of Salem's land use and zoning ordinances to many situations ranging from residential housing to industrial and commercial properties. Since 2010, she has been Executive Director of an environmental and land use nonprofit that focuses on protecting the natural resources of the Oregon coast, working with residents to oppose ill-advised land use projects and helping maintain liveable coastal communities. She is the coauthor, with Chris Maser, of *Sustainability and the Rights of Nature: An Introduction*, published by CRC Press in 2017. She was the lead author on 'Oregon's Manila Galleon', a special issue of *Oregon Historical Quarterly* published in June 2018, and has also written many articles on Oregon coastal history for the online Oregon Encyclopedia.

**Chris Maser** spent over 25 years as a research scientist in natural history and ecology in forest, shrub steppe, subarctic, desert, coastal and agricultural settings. Trained primarily as a vertebrate zoologist, he was a research mammalogist in Nubia, Egypt (1963–1964), with the Yale University Peabody Museum Prehistoric Expedition and a research mammalogist in Nepal (1966–1967), where he participated in a study of tick-borne diseases for the U.S. Naval Medical Research Unit #3 based in Cairo, Egypt. He conducted a 3-year (1970–1973) ecological survey of the Oregon Coast for the University of Puget Sound, Tacoma, Washington. He was a research ecologist with the U.S. Bureau of Land Management for 13 years – the first 7 (1974–1981) studying the biophysical relationships in rangelands in southeastern Oregon and the last 6 (1982–1987) studying old-growth forests in western Oregon. He also spent a year as a landscape ecologist with the U.S. Environmental Protection Agency (1990–1991).

He is an independent author as well as an international lecturer and facilitator in resolving environmental conflicts, vision statements and sustainable community development. He is also an international consultant in forest ecology and sustainable forestry practises. He has written or edited over 290 publications, including 43 books.

# Contributors

**Kelly D. Alley** is Alma Holladay Professor at Auburn University, Auburn, AL.

**Michelle Bender** is the Ocean Rights Manager, Earth Law Center, Executive Committee, Global Alliance for the Rights of Nature, and IUCN World Commission on Environmental Law.

**Francisco José Bustamante Romo Leroux** is an Environmental Law Professor at Universidad Internacional Sek del Ecuador.

**Eugen Cadaru** is an independent researcher. He has a PhD in Cinematography and Media from the National University of Theatre and Film in Bucharest, Hungary, and a Bachelor's degree in Legal Sciences from the University of Bucharest.

**Douglas Deur** is an Associate Research Professor in the Department of Anthropology, Portland State University, and an Adjunct Professor of Environmental Studies at the University of Victoria (BC), Canada.

**Hugo Echeverria** is a Professor at the Colegio de Jurisprudencia, Universidad San Francisco de Quito, Ecuador.

**Alan Watson Featherstone** is the Founder of the award-winning Scottish charity, Trees for Life.

**José Luis Febles** is Research Assistant, Human Ecology Department, Cinvestav, Merida Unit, Mexico.

**The Honorable Christopher Finlayson** is from Wellington, New Zealand, where he graduated from Victoria University with degrees in Latin and Law. He served as a Minister of the New Zealand Parliament from 2005 to 2019 representing Rongotai. During his tenure in Parliament, he served as Attorney General, Cabinet Minister for Treaty of Waitangi Negotiations, Minister for Culture and Heritage and Associate Minister of Maori Affairs. As Treaty Minister, Mr. Finlayson signed a record 59 deeds of settlement with indigenous tribes. He also represented New Zealand at the International Court of Justice on Whaling and the United Nations Security Council on Terrorism. Mr. Finlayson currently practices commercial litigation based in Auckland, New Zealand..

**Dani Fössl** is Advisor for Environmental Affairs at the Representation of Rhineland-Pfalz at the Federal and European Unions.

**Mary Graham** is Adjunct Associate Professor at the University of Queensland, and a Kombumerri First Nations person.

**Kathryn Anne Gwiazdon** is the Executive Director, Center for Environmental Ethics and Law; Founding Member, Ecological Law and Governance Association; and Expert, United Nations Harmony with Nature.

**Almira Hoogesteijn** is Senior Researcher, Human Ecology Department, Cinvestav, Merida Unit, Mexico.

**Rafael Hoogesteijn** is Conflict Program Director, Jaguar Program, Panthera Brazil, Cuiaba, MT, Brazil.

**Mumta Ito** is an attorney and Founder and President of Nature's Rights, a Scottish charity organization.

**Clan Chief Kwaxsistalla Adam Dick** was the clan chief (Oqwa'mey) of the Kawadillikalla Clan of the Dzawatainuk Tribe of the Kwakwaka'wakw Nation in British Columbia, Canada.

**Dechen Lham,** a resident of Thimphu, Bhutan, is currently a PhD candidate at the University of Zurich, Switzerland.

**Jony Mainaly** is a lecturer at National Law College, Tribhuvan University, Nepal.

**Michelle Maloney** is the Co-Founder and National Convenor of the Australian Earth Laws Alliance and Adjunct Senior Fellow at the Law Futures Centre, Griffith University.

**Tarini Mehta** is a student at Pace University Law School, New York, NY.

**Michelle Amelia Newman** is an attorney with the Natural Resources Defense Council.

**W. D. Scott Pittman** is the Owner/Managing Director of Environmental Design Concepts, Inc., Founder and director of the Permaculture Drylands Institute, and President and Co-founder, with Dr. Bill Mollison, of the Permaculture Institute, established in 1997.

**Kim Recalma-Clutesi** is a Kwakwaka'wakw/Pentlatch writer, film director, leader, and activist, based on the Qualicum Indian Reserve in British Columbia, Canada.

**Lindsey Schromen-Wawrin** is an attorney in private practice in Port Angeles, WA (Klallam (nəxʷsƛ̕áy̕əm̕) Territory), and currently serves on the Port Angeles City Council.

**Michael W. Schröter** is Advisor for Environmental Affairs at the Representation of the Free Hanseatic City of Bremen at the Federal and European Unions.

**His Eminence Anote Tong** was three-term President of the Republic of Kiribati (2003–2016), two-time nominee for the Nobel Peace Prize and winner of the Sunhak Peace Prize. He is a world-renowned leader in the battles against climate change and for ocean conservation.

**Ngozi F. Unuigbe** is Environmental Policy Research Fellow, United Nations University – Institute for Natural Resources in Africa, Accra, Ghana; and an Associate Professor, University of Benin, Nigeria.

# Section I

# Concepts and Overviews

# 1

# *Introduction: Rights of Nature, Sacred Lands and Sustainability in the Western Tradition*

**Cameron La Follette**

**CONTENTS**

Rights of Nature: Beginnings ..... 3
Rights of Nature and Sacred Lands Tradition in Western Culture ..... 6
Sustainability Projects in the United States ..... 8
Conclusion ..... 11
Notes ..... 11

The question continues of how best to repair humanity's relationship with Nature, which has been reduced to a 'relationship' allowing human use of all and any 'natural resources' since the advent of widespread industrialisation. That the lack of respect for natural processes has led to severe and worsening environmental consequences, ranging from desertification and widespread air, water and soil pollution to climate change, is no longer disputed.[1]

## Rights of Nature: Beginnings

Recognising that human activity has major implications for the Earth's ability to support human life in the future, governments, thinkers and advocates worldwide are searching for new paradigms, or refashioning traditional paradigms, that can restore the relationship of mutual respect. Some new ideas encompass the world; others focus on the local community or region. This is exactly what is needed, since human relationship to the environment is always local. In order to change current practises, it is necessary to change technologies, expectations and philosophies. This would seem to be a tall order, but it does not all have to be done at once. The most immediate need is simply to curb human overuse of the natural environment, so that ecosystems can once again flourish according to the ecological processes that create and maintain them. Deeper changes in human cultures, expectations and needs can evolve more slowly, as long as humans show restraint meantime in impinging on the natural environment.

It is, however, essential to approach the massive problems *systemically*. Symptomatic fixes of local ecological devastation – partial, and human focused – are commonplace worldwide, and tend to be popular, because they can focus narrowly and usually do not require deep-seated cultural or economic change. But they also maintain the focus on human needs rather than on what Nature needs to flourish. Most of these efforts, if carried to their

DOI: 10.1201/9780429505959-1

conclusion, would provide minimal natural function, just enough to keep the mostly depleted ecosystem viable for maximum human use. Nature is not allowed to flourish, its resilience is greatly diminished and the fabric of the local ecosystems – and, frequently, many species – trembles on the edge of extinction. Treating Nature this way also severely diminishes the richness of human communities.

There are many well-thought-out proposals that, if implemented, would nevertheless provide only symptomatic change in practise. Recent studies of the need to change laws to more closely model the resilience and fluctuations of ecological systems are a case in point. It has been noted that legal systems covering natural resources law, especially in the United States, tend to assume the environment is globally stable and functioning in a steady state. The law is slow to innovate, as its structures favour incremental adjustments. Thus, human relationships with Nature are regulated based on linear patterns of stability, rather than fluctuating patterns of ecological resilience. Laws need to become more adaptive, in order to help maintain both ecological resilience and the adaptive capacity of social systems – all of which depend on the environment – to become and remain more sustainable.[2]

These are important considerations. But making legal systems more supple and flexible without a systemic commitment to placing Nature's flourishing first will only lead to more efforts at management of natural systems, more striving to create loopholes by those who seek to plunder natural resources and more unstable societies that cannot maintain their own resilience within, and as a part of, Nature's resilience. Ironically, the current network of laws regulating human use of such vast ecosystems as coastal shorelines, or millions of acres of forest, at times provides the means for so much public scrutiny under different legal rubrics that inappropriate developments are ultimately defeated. That is not an argument for creating webs of overlapping and ineffective laws, but it is a reminder that the commitment to fitting human societies inside Nature's flourishing, so they contribute to both Nature's resilience and their own simultaneously, must reside at the head of American – and international – natural resources law.

Different parts of the world show varying levels of ecological trauma. Some cultures remain respectful of living within Nature's limits, and as a result the ecosystems are fairly healthy. Bhutan is a good example of this. Other regions show the lack of environmental concern that seems essential for maximum industrialisation. For example, widespread soil pollution in China, including about 19% of farmland, indicates the results of untrammelled use of pesticides and fertilisers, as well as the results of high industrialisation activity leading to inorganic heavy metal pollution such as cadmium, mercury and arsenic.[3]

'Rights of Nature' is a shorthand term for a multipronged effort in many regions to deepen sustainability practises. The concept focuses on restoring *relationship* to human interaction with Nature, so that humans do not thoughtlessly plunder Nature as if it were merely an inanimate storehouse of resources. Arising out of an international conference in 2010 in Bolivia, the international Rights of Nature movement fused ideas and philosophies from around the world towards the common goal of placing respect for Nature's right to flourish at the forefront of human activities.

Though descriptive, the term gives the impression of a primarily legal struggle, as rights are guaranteed in law and enforced via the courts. But although some Rights of Nature controversies have ended up in court, especially in the United States, the effort to forge profoundly sustainable societies that respect Nature's right to flourish above human use is not primarily a legal matter. The law only sets the framework and acts as an enforcer.

There are many ways of conceptualising and undertaking the work to build a society where Nature flourishes first. Often the efforts do not go by the name of 'Rights of Nature', but are clearly directed towards a systemic vision of sustainability. They all require, among other things, that Nature be understood as a living entity, capable of entering into relationship with humans. If there is a true relationship, then humans have the responsibilities that inhere in any relationship: to respect the other party profoundly, and to care for their fate, health and future. This is all the more essential when Nature provides *all* that humans require for sustenance, from air to food and water. But Nature's rights are inherent, rather than of human making. Humans recognise them and enter into respectful relationship with Nature in order to live sustainably, but do not create its rights. Humans only enforce their side of the relationship, to make sure that Nature flourishes. Understanding this is an important first step in grappling with the concept of the Rights of Nature.

The contemporary focus on the form of ecological governance nicknamed 'Rights of Nature' has its roots in the 1972 United States Supreme Court case *Sierra Club v. Morton*. In this case, the Sierra Club, an environmental group, tried to establish its legal standing (i.e. that it had sufficient connection to, and harm from, the proposed action to support its participation in the case) to oppose in court the Disney Corporation's proposal to build a large ski resort on leased National Forest Land in Mineral King Valley, in California. The Sierra Club lost this case, though it was ultimately successful in stopping the proposed resort.

However, Justice William O. Douglas wrote a dissent in which he argued trees, valleys and natural objects ought to have standing on their own to sue, as they would be the ones most damaged by a resort.[4] Just before the case was heard by the Supreme Court, Christopher Stone, a University of Southern California law professor, published a law review article entitled 'Should Trees Have Standing?' which amassed legal theories and case evidence to bolster an argument that natural entities can indeed have standing in the law to bring suit against proposed human actions that will harm them.[5]

These ideas met with incredulity among some in the American legal profession, as exemplified by the scornful ditty by John M. Naff Jr. published in the *Journal of the American Bar Association* in 1972:

> If Justice Douglas has his way –
> O come not that dreadful day –
> We'll be sued by lakes and hills
> Seeking a redress of ills.
> Great mountain peaks of name prestigious
> Will suddenly become litigious.
> Our brooks will babble in the courts,
> Seeking damages for torts.
> How can I rest beneath a tree
> If it may soon be suing me?
> Or enjoy the playful porpoise
> While it's seeking habeas corpus?
> Every beast within his paws
> Will clutch an order to show cause.
> The courts, besieged on every hand,
> Will crowd with suits by chunks of land.
> Ah! But vengeance will be sweet
> Since this must be a two-way street.
> I'll promptly sue my neighbor's tree
> For shedding all its leaves on me.[6]

In the ensuing decades, it became clear that the American legal system contains mechanisms that approach providing Nature legal protection, often via a guardianship model, or the ability to sue by proxy.[7] However, neither the United States, nor any of its individual states, provide an overarching legal and constitutional mandate requiring deep sustainability, much less granting Nature the right to flourish and making that right prior to all human rights. The problem remains: How best to implement an ecological governance-oriented system? There are reams of unanswered questions about definition of terms, fairness to humans whose activities deplete Nature's functioning, legal structures that minimise cheating, how to protect Nature's flourishing when 'Nature' is not a static entity, what human activities contribute to flourishing versus diminishing of Nature, the scope of a Rights of Nature form of governance – a watershed? an ecosystem? the entire world? – and many others. Finding answers to these questions is even more important as various cultures and countries integrate a Rights of Nature, or legal personhood, concept into their governing frameworks.

But however a legal and political system that prioritises Nature's flourishing might be designed, implemented and enforced, all Rights of Nature systems are rooted in an understanding that Nature and its many components, ranging from rivers to deer, are sacred, and Nature is a living entity.

## Rights of Nature and Sacred Lands Tradition in Western Culture

One frequently sees articles and books to the effect that the Western tradition (out of which industrial capitalism arose) has no place for the concept that Nature is sacred, and therefore must be treated respectfully, as the other member of a critical relationship. But this is untrue. All nonindustrial cultures have an understanding of Earth sacredness that is fundamental to their worldview, in the West as elsewhere. Though the constellation of Earth sacredness understanding has become attenuated in Western thought historically, especially subsequent to industrialisation, there are many strands showing that Europeans, in common with all other peoples, understood the sacredness of the Earth and lived in intimate communion with their environment. Western thought therefore has a rich trove of sacred understanding from which Westerners can draw in reinvigorating a sacred-Earth philosophy from within their own traditions. Here are a few of many examples.

France and Spain are home to some of the most ancient, richest and most complex cave art in the world. Well-known European Palaeolithic cave art includes the stunning animal paintings of Chauvet Cave in France, which contains paintings of bison, horses and other animals estimated at 35,000 years old, with many others stemming from the later Palaeolithic as well; Lascaux, also in France, with paintings thought to be about 17,000 years old; and Altamira Cave in Spain, where some of the animal paintings are thought to be 20,000 years old or more, and some of the other symbols to be more than 35,000 years old.[8] There are many other caves in these two countries that also contain extensive Palaeolithic art depicting animals such as bison, aurochs and horses, and human hands as well as abstract designs; most of the caves were in use, at least intermittently, for thousands of years.[9]

There have been many interpretations of the paintings, usually found on the ceilings and walls of cave chambers twisting deep into the hillsides in utter darkness. Perhaps the most comprehensive theory was proposed by Jean Clottes and David Lewis-Williams, and subsequently explored in much greater depth by Clottes, postulating that the cave art resulted from a shamanic belief system that utilised deep trance states for vision-seeking. The visions were painted in the caves as part of the community's shamanic religion, perhaps as an aspect of initiations, other important community rituals and wisdom-seeking from the sacred powers found in and beyond these often vast and labyrinthine cave systems.[10]

Looking at sacred understanding of the local landscape from the other side of Europe, there is extensive evidence throughout rural Greece for longstanding – reaching back to the archaic era – worship of nymphs, the local deities of rivers, streams, springs and watersheds, including mountains, meadows and trees (dryads and hamadryads). There were also nymphs of the sea, the Okeanides. The earliest literary mentions of nymphs, praising their sacredness and beauty, occur in Homer's *Iliad* and *Odyssey*, and subsequently in the Homeric Hymns, as well as the poet Hesiod. Greek lyric and choral poetry also described nymphs, as did the later tragedies and comedies emanating mainly from urban Athens. Nymphs were frequently worshipped in caves, especially those that sheltered a spring. In the caves, votive offerings, carvings and inscribed stones, as well as altars, have commonly been found. The earliest archaeological evidence for nymph worship comes from the sixth or seventh century BCE, though clearly it stretched back at least to the eighth century, during the Homeric period. The earliest datable objects showing representations of nymphs are coins and black-figure vases. By the fifth and fourth centuries, nymph worship was at its height throughout Greece, and there are many representations of nymphs in the archaeological record.[11]

The Celtic lands, especially Ireland and Scotland, provide more examples of locally based relationship with the sacred Earth and its deities. Perhaps the two most comprehensive windows into the dense network of belief in local land-based deities, use of healing charms, agricultural prayers and ceremonies and lifecycle rituals, is to be found in John Grigorson Campbell's *The Gaelic Otherworld*, comprising his two great works on the Scottish Highlands and Islands, and Alexander Carmichael's *Carmina Gadelica*, a compendium of Scottish prayers and incantations from the same regions. Both men collected their materials in the nineteenth century, and both works show in great detail how the rural people of the Highlands and Islands lived for centuries with their natural environment and its supernatural beings.[12]

Additionally, the early Nature poetry and epigrams of Ireland especially (but also Wales and Scotland) are renowned for their sensitive and intimate portrayal of the natural environment, embellished with much praise and rich descriptions of beauty.[13] Irish literature also provides glimpses of the intimately local understanding of the land by its residents – from early times into the mediaeval era – who named every feature of their local home regions. They had intricate, vital tales about the hills, rivers and valleys that memorialised the community's mythic interactions with the places they lived in.[14]

England also – different culturally from the Celtic lands due to Anglo-Saxon immigration from the mid-fifth to early seventh centuries, and the eleventh-century Norman Conquest as well – also has a strongly relationship-based outlook to Nature in its past. This can best be seen by examination of the archaeological record and the motifs of surviving Old English poetry, where it is clear that an animistic outlook, characterised by an understanding of vital essence in Nature and its species, was commonplace. For example, Anglo-Saxons designed metal boar crests or cheek guards on ancient helmets; *Beowulf*, the well-known Old English epic (c. 700–1000 CE), contains descriptions of boar protection for warriors. Boar images on war gear clearly were not merely aesthetic; they were, instead, effective protectors of the warriors who wore them – spirit helpers and guardians. There are many other references to a worldview anchored in an understanding of sacred, animated Nature, including soul journeys, shamanic exile and animal spirit helpers, in Anglo-Saxon society, as we see from epigrams, healing charms and oral poetry.[15]

These examples could be multiplied many times over, were there room for it. However, the central point is that the European and Anglo-American traditions contain as rich and intimate an understanding of land sacredness as any other tradition in the world. The theory of 'human exceptionalism', an anthropocentric philosophy that places humans at the centre of the cosmos, also a longstanding strain in European thought, gained malignant prominence with the rise of industrialisation in the eighteenth century, which began principally in England. Industrialisation swiftly became a tool worldwide to leap boundaries constraining production, whether limits of Nature or limits of human community. 'Nature' became merely a repository of resources available for human use, with no abiding spiritual power, sacredness or life, thus stripping it of any relationship with humans; unsurprisingly, Nature was often portrayed in increasingly abstract terms.[16]

German sociologist Max Weber (1864–1920) famously referred to this process of the modern era's emphasis on rationality, thus eliminating the Earth's sacredness and magic, as 'the disenchantment of the world'. Though he himself was interested in magic, nature mysticism and healing and the occult, he sought to understand the secularising trends he saw expanding throughout Western culture. How was it that Nature, originally understood to be filled with supernatural power ('Mana') and then with spirits, deities and other supernatural beings, ended up being considered merely a lifeless hoard of resources for human use? In several works, he traced the levels of increasing abstraction and disenchantment that the Western tradition undertook to arrive at this conclusion.[17]

However, the intimately sacred relationship between Nature and human communities did not disappear in the West. It has survived in rural pockets all over Western Europe and the United States. The elaboration and celebration of this relationship frequently found refuge in poetry. Especially in England and Anglo-America, poets became the shamans of the society, reminding people continuously of Nature's beauty, mystery and power and the necessity of human relationship with Nature. It is clear that an understanding of Nature's sacredness and living presence has remained vital in the Western tradition, albeit often pushed to the margins.[18] Now, however, it is much more widely understood that symptomatic efforts to grapple with industrial pollution and severe ecological damage by perfecting technological solutions are not going to solve the problems; only a widespread cultural metamorphosis, with emphasis on systemic solutions, will. This will require both restraint in human use and undertaking regional restoration projects, large and small, to restore Nature's ability to flourish.[19]

## Sustainability Projects in the United States

The United States is a very influential country in the worldwide effort to repair humanity's relationship with Nature. Not because the United States is a leader in this field – it is not – but because of its vast economic and political influence worldwide. Thus, it seems appropriate to sketch a few important sustainability and restoration projects underway in the United States to illumine the local work being undertaken in many communities to repair relationships. These examples only spotlight a few of the thousands of projects spearheaded by communities, working and acting locally, across the United States.

This necessitates defining the overused term 'sustainability', which has been stretched to fit so many situations that it has nearly lost all meaning. As corporate sleight-of-hand, 'sustainability' is often merely another word for a sophisticated form of 'greenwashing', by which companies parade environmental images that mask environmentally damaging activities. But four inviolable concepts underlie the true sustainability at the core of the

Rights of Nature concept: (a) true sustainability prohibits mitigation or substitution of any kind for political or monetary gain; (b) sustainable projects create sustainable levels of human use and do not maintain current levels of overconsumption; (c) sustainable use shrinks the human footprint on Earth and does not expand it; (d) true sustainability fluctuates with climate, biological cycles, human population, economic needs and other factors, but always maintains Nature's integrity.[20]

With this definition in mind, the survey of some important American projects focusing on systemic restoration, and repairing ecosystem resilience, can begin.

One of the most impressive undertakings in the United States is the American Prairie Reserve Project. Its mission is 'to create the largest nature reserve in the continental United States, a refuge for people and wildlife preserved forever as part of America's heritage'.[21] It is working to assemble a multimillion-acre nature reserve in Montana's grasslands, part of the Great Plains ecosystem. American Prairie Reserve is aiming to amass more than 3 million acres of public and private land into the reserve, purchasing private lands to connect fragmented public lands. Since temperate grasslands like the Great Plains are the least protected ecosystem in the world, there are few opportunities to create a large-scale conservation project, and Montana, in the Western United States, is one of the places where it can be successful. In this sparsely populated region, it is possible to create an ecosystem-scale reserve that can provide a resilient native landscape.

Currently the Reserve stands at more than 350,000 acres. The Reserve must not only purchase or lease the needed lands, it must actively restore and conserve the properties and labour to restore populations of severely depleted native wildlife ranging from American bison to prairie dogs, cougar, foxes and wolves. There are many challenges, centring on conflicts between people and large wildlife populations, community resistance to the return of predators such as wolves and cougars and social intolerance for prairie dogs, who were once poisoned on a vast scale.[22]

There are several highly important and systemic efforts to reintroduce keystone species – those that have a major effect on an ecosystem's functioning and integrity – throughout the United States, including bison, wolves and sea otters. The problem in all three cases is primarily one of public education, since Americans are not now used to living adjacent to large, or even medium-sized, wildlife populations. Conflicts and opposing philosophies have made wildlife reintroductions contentious and often less successful at the outset than would otherwise have been the case.

At least 30 million bison once roamed the Great Plains of North America. But habitat loss, overhunting and human conflict have meant that bison have dwindled away in the United States. There are now only three free-ranging bison herds in the United States: about 3,500 in Yellowstone National Park, and two smaller herds in southern Utah. In addition to the American Prairie Reserve's work, conservation organisations are helping return bison to Charles Russell National Wildlife Refuge and tribal lands in Montana.[23]

Wolves were once ubiquitous across much of the United States, but most wolf populations were hunted, trapped and poisoned to extinction or very close to it by settlers and other rural populations living in wolf habitats. Mexican grey wolves, a subspecies native to the southwestern United States and Mexico, came even closer to extinction when the last few known wild ones were captured in 1980 for a captive breeding programme. Between 1982 and 1998, the breeding programme successfully brought the population back from the brink, and the wolves were reintroduced to their native habitat. As of 2017 there were at least 114 Mexican grey wolves in the wild. Still critically endangered, the populations are slowly increasing.[24]

Reintroduction of grey wolves to Yellowstone National Park began in 1995, after decades of battle between opponents and proponents. The principal biological problem was that wolves, as keystone predators, are essential to controlling elk populations, which otherwise become too large; with wolf absence, elk populations in the Yellowstone area ballooned. Return of the wolves stabilised the elk population. The grey wolf has been delisted from the Endangered Species List in the Northern Rocky Mountains region, which includes Yellowstone. The federal government aims for a long-term average of about 1,000 wolves in this region – much smaller than historically, but at least a viable nucleus population. Wolves have begun expanding into Oregon and northern California as well.[25]

Sea otters historically lived along the rim of the Pacific Ocean from Japan to the Aleutian Islands, the Alaskan coast and on down the northwest coast as far south as Baja California. Historically, the worldwide sea otter population may have been as large as 300,000 animals. Intensive commercial fur hunting in the eighteenth and nineteenth centuries, mainly undertaken by Russians, British and Americans, nearly extirpated the sea otter from its entire range. By 1911, when it was finally protected under the International Fur Seal Treaty, there were only about 2,000 sea otters remaining, principally in Alaska. Ending commercial hunting allowed the sea otter population to rebound in Alaska (with some local translocations to reintroduce them to former habitat), but areas south of Alaska, including British Columbia, remained depleted.

A remnant population discovered in California in 1938 has slowly expanded to repopulate a small portion of the original California range, with much conservation help from nonprofit organisations engaged in monitoring, fostering abandoned pups and public education.[26] Restoring populations in Washington required a translocation of Alaskan otters in 1970–1971. The population, now numbering more than 1,000 individuals, is concentrated in one area of the Washington coast.[27] A similar translocation in Oregon in 1970–1971 failed. Little if any monitoring occurred, so the reasons are unclear, but the otters appear to have left the translocation sites, perhaps to return to Alaska. The Oregon coast remains without otters, except for a trickle of strays, probably from Washington.[28] In all areas outside Alaska, and some regions of British Columbia where sea otters have once again populated their traditional habitats, numbers are low, and only a fraction of the original habitat once again supports this critically important species.

A network of conservation organisations along the Pacific coast is working, individually and collectively, to increase sea otter presence in the coastal regions they once called home. Otters are extremely important keystone species, being in large measure responsible for the growth of healthy kelp forests, which in turn support a rich array of marine life.[29]

Turning from landscape-scale conservation and the urgency of returning keystone species to their habitats, it is important to shine a brief spotlight on another major concern to the Rights of Nature and sustainability mosaic: the health of rivers. The struggle to keep rivers healthy or return them to health is of major concern to people everywhere. The United States, though not a world leader in giving rivers the right to flourish as legal persons – as New Zealand is, for example – nevertheless is the world leader in one area of river sustainability: the removal of dams. Dam removal is significant because it is costly, always includes significant river restoration activity in addition to actual dam removal and also clearly marks a communitywide shift in perspective about the value of a river. Rather than being harnessed for human use, it becomes more valuable as a free water system, providing a healthy riverine and watershed ecosystem for humans as well as wildlife.

In the twentieth century, the United States led the world in damming rivers for diverse purposes, ranging from hydropower to irrigation and flood control. At least 90,000 dams greater than 6 feet high block the nation's waterways great and small, and there are many thousand smaller ones, equally damaging to local ecosystems, blocking fish migration and riverine function. But a nascent dam removal movement intensified in the 1990s; as of 2018,

American Rivers estimates some 1,480 dams have been breached or removed nationwide.[30] The two largest dam removals in the world to date, completed in 2014, were Elwha and Glines Dams in Washington State. Glines Dam was more than 200 feet high and Elwha Dam over 100 feet high. Removing them restored the Elwha River in its 45-mile sprint from the Olympic Mountains to the Strait of Juan de Fuca. Before the dams were built in 1914 and 1927, the Elwha was a very rich salmon river. It took nearly 20 years for the political consensus to coalesce for dam removal, and there were many unknowns, as no one had attempted to remove dams as massive as these two, behind which were trapped some 27 million cubic yards of sediment. In the event, both dams were removed slowly, so the sediment was released incrementally. The freed river created some 70 acres of new beach and riverside estuarine habitat at its mouth with the sediment that flowed downstream. The Elwha has rebounded dramatically in natural habitats and in its once-legendary salmon populations.[31]

## Conclusion

It has become a cliché to say that the world is at a historic turning point in stewarding the health of the world's ecosystems in the face of crises that may be undermining the Earth's ability to support human life. But humans do not pour vast resources of time, money and self-sacrifice to restore and repair ecosystems out of fear; those are acts of love. We have a relationship with Nature to repair, a relationship more vital than any words can describe. Rights of Nature is fundamentally a means to relationship; it is neither a luxury nor an impediment to human progress. It is, instead, at the core of human life, as it always has been. The labour, creativity, communal willpower and self-sacrifice necessary to do it are there. In a thousand ways, worldwide, the respectful relationship with Nature, so brutally torn apart, is being repaired.

## Notes

1. United Nations Environment Programme. 2018. Global Environment Outlook (GEO-6): Regional Assessments. https://www.un.org/sustainabledevelopment/blog/2016/05/rate-of-environmental-damage-increasing-across-planet-but-still-time-to-reverse-worst-impacts/ (accessed 22 February 2019).
2. Garmestani, A.S. and Allen, C.R. (eds). 2014. *Social-Ecological Resilience and Law.* New York, NY: Columbia University Press.
3. Chen, R., de Sherbinin, A., Ye, C. and Shi, G. 2014. China's soil pollution: Farms on the frontline. *Science*, Vol. 344, Issue 6185, p. 691. DOI: 10.1126/science.344.6185.691-a (accessed 16 January 2019).
4. Stone, C. 2010. *Should Trees Have Standing? Law, Morality and the Environment.* Third Edition. New York, NY: Oxford University Press.
5. Ibid.
6. Naff, J.M. Jr. 1972. Reflections on the Dissent of Douglas, J., in *Sierra Club v. Morton.* 58 *ABA Journal*, Vol. 820, 1972.
7. Stone, C. 2010. *Should Trees Have Standing? Law, Morality and the Environment.* Third Edition. New York, NY: Oxford University Press.
8. (a) Lawson, A.J. 2012. *Painted Caves: Palaeolithic Rock Art in Western Europe.* London: Oxford University Press; (b) Pike, A.W.G., Hoffman, D.L. Garvia-Diez, M., Pettitt, P.B., Alcolea, J.,

De Balbin, R., González-Sainz, C., de las Heras, C., Lasheras, J.A., Montes, R., Zilhao, J. 2012. U-Series Dating of Paleaolithic Art in 11 Caves in Spain. *Science*, Vol. 336, Issue 6087, pp. 1409–1413. DOI: 10.1126/science.1219957

9. Lawson, A.J. 2012. *Painted Caves: Palaeolithic Rock Art in Western Europe*. London: Oxford University Press.
10. (a) Clottes, J. and Lewis-Williams, D. 1998. *The Shamans of Prehistory: Trance and Magic in the Painted Caves*. New York, NY: Harry N. Abrams; (b) Clottes, J., Martin, O.Y. and Martin, R.D. (translators). 2016. *What Is Palaeolithic Art: Cave Painting and the Dawn of Human Creativity*. Chicago, IL: University of Chicago Press.
11. Larson, J. 2001. *Greek Nymphs: Myth, Cult, Lore*. New York, NY: Oxford University Press.
12. (a) Campbell, J.G., Black, R. (eds). 2005. *The Gaelic Otherworld*. Edinburgh, Scotland: Birlinn Limited; (b) Carmichael, A. 1992, 1994. *Carmina Gadelica: Hymns and Incantations*. New York, NY: Lindsfarne Press, and Edinburgh, Scotland: Floris Books.
13. Jackson, K.H. 1951, 1971, 1986. *A Celtic Miscellany: Translations from the Celtic Literatures*. New York: Dorset Press.
14. Dooley, A. and Roe, H. (translators). 1999, 2008. *Tales of the Elders of Ireland: A New Translation of Acallam na Senórach*. London: Oxford University Press.
15. Glosecki, S.O. 1989. *Shamanism and Old English Poetry*. New York, NY and London: Garland Publishing.
16. See, e.g. (a) Sessions, G. (ed). 1995. *Deep Ecology for the 21st Century: Readings on the Philosophy and Practice of the New Environmentalism*. New York, NY and London: Shambala; (b) Sale, K. 1996. *Rebels Against the Future: The Luddites and Their War on the Industrial Revolution*. Boston, MA: Addison-Wesley Publishing Company.
17. Josephson-Storm, J.A. 2017. *The Myth of Disenchantment: Magic, Modernity and the Birth of the Human Sciences*. Chicago, IL: University of Chicago Press.
18. Preece, R. 1999. *Animals and Nature*. Vancouver, BC, Canada: University of British Columbia Press.
19. (a) Ibid.; (b) La Follette, C. and Maser, M. 2017. *Sustainability and the Rights of Nature: An Introduction*. Boca Raton, FL: CRC Press.
20. La Follette, C. and Maser, M. 2017. *Sustainability and the Rights of Nature: An Introduction*. Boca Raton, FL: CRC Press.
21. American Prairie Reserve: https://www.americanprairie.org/mission-and-values (accessed 21 January 2019).
22. American Prairie Reserve: https://www.americanprairie.org/news-blog/not-yet-yellowstone-wildlife-restoration (accessed 21 January 2019).
23. National Wildlife Federation: https://www.nwf.org/Our-Work/Wildlife-Conservation/Bison (accessed 21 January 2019).
24. U.S. Fish and Wildlife Service: https://www.fws.gov/news/ShowNews.cfm?ref=2017-u.s.-mexican-wolf-population-survey-completed-&_ID=36277&Source=iframe (accessed 21 January 2019).
25. U.S. Fish and Wildlife Service: https://www.fws.gov/mountain-prairie/es/grayWolf.php (accessed 23 January 2019).
26. The Otter Project, http://www.otterproject.org/about-sea-otters/natural-history/ (accessed 22 January 2019).
27. U.S. Fish and Wildlife Service, https://www.fws.gov/wafwo/articles.cfm?id=149489662 (accessed 22 January 2019).
28. McLeish, T. 2018. *Return of the Sea Otter*. Seattle, WA: Sasquatch Books.
29. Ibid.
30. American Rivers: https://figshare.com/articles/American_Rivers_Dam_Removal_Database/5234068 (accessed 22 January 2019).
31. (a) Nijhuis, M. World's Largest Dam Removal Unleashes U.S. River after Century of Electric Production. August 27, 2014. *National Geographic*, https://news.nationalgeographic.com/news/2014/08/140826-elwha-river-dam-removal-salmon-science-olympic/ (accessed 22 January 2019); (b) Mapes, L.V. 2013. *Elwha: A River Reborn*. Seattle, WA: The Mountaineers Books.

# 2

## *Defending the Tree of Life: The Ethical Justification for the Rights of Nature in a Theory of Justice*

**Kathryn Anne Gwiazdon**

**CONTENTS**

Introduction ... 13
Extending the Ethical Roots of Justice to the Roots Themselves ... 16
A Worldview: Living Ethics for a Living Earth ... 18
A Methodology: A Rooted Cosmopolitanism Approach to Law and Life ... 19
The Motivations: Love Matters ... 23
Foundational Principles: Healing through *Ubuntu* and Reconciliation Ecology ... 24
Rights of Nature Is a Natural Evolution of Justice ... 25
*Sapere aude*!: An Eternal Inquiry towards Perpetual Peace ... 27
Wrath and Reason Are Just Responses to Grave Injustices ... 29
Conclusion ... 31
Notes ... 32

### Introduction

The Rights of Nature movement advocates for the development, incorporation and defence of the inherent rights of Nature in legal mechanisms around the world. Its most basic premise is to give a voice to Nature in the legal system and, through appointed custodians or the general public, bring more, better and stronger claims of harm, demands for justice or calls for protection. Rights of Nature is gaining momentum in local, national and international instruments around the world, is found in a growing number of laws and constitutions, is being defended and argued in courts and civil society and nongovernmental organizations are picking up its banner around the world. Yet despite these advancements, it faces major philosophical, political and scientific challenges to its incorporation and implementation in law and governance.

For example, Rights of Nature is diametrically opposed to anthropocentricism and seeks ecocentric, Earth-centred recognition and rights, yet within the human-created, anthropocentric constructs of law and governance. Also, similar to human rights discourse, it aims to honour the inherent rights of life by giving Nature rights in law, yet is unclear with the definition and prioritization of rights, as well as its scale and scope in application, whether law-based or science-based. For example, if 'Mother Earth is a living being', and all the Earth has inherent rights, do rights extend to living viruses and microscopic particles in the air, soil and oceans?[1] And if all life on Earth is interrelated, how can it be possible to defend 'the right [of Mother Earth] to regenerate its bio-capacity and to continue its vital cycles and processes *free from human disruptions*'. (emphasis added)?[2] After all, the breath of every living thing alters every living thing.

DOI: 10.1201/9780429505959-2

In addition, we must also consider whether a rights-based approach is an effective approach to protecting life and the foundations of life. Human rights laws exist in nearly every nation, and even within criminally enforceable global governance institutions, yet grave injustices continue to occur every day. Are rights and standing enough, if there is no civil or political will? Or if power corrupts that civil and political will? If legal rights do not translate to adequate protection, we must explore why they do not. What forces prevent the consideration and administration of justice for life on Earth?

There are also interrelated questions of the science of the Rights of Nature concept. For example, where does an ecosystem end? Where does the influence of a river stop? How can we measure harms to a mountainside in order to make the mountain whole again? Where, in time, does 'whole' begin? When and how are species considered indigenous, or invasive? And *should* we even attempt to measure the immeasurable? Also, how do we, with humility, 'think like a mountain' or give proper voice to an endangered wolf that would make it louder than any calls for human economic development (acknowledging that the framing for that question is often politically and purposefully deceptive)?[3] And what if that wolf has been genetically cloned or modified to allow for its continued existence – does that change its rights or value?

Rights of Nature is a major, expansive undertaking – I think much larger than what most of its advocates even properly recognize – because it seeks to give rights to life, to life processes and to contributors to life processes (animate and inanimate). Who defines what life is? Who decides the priorities of justice or protection to it, whether invasive or indigenous, created by humans or Nature? And who sets the proper scale and scope? Life on Earth owes its life to the cosmos itself – where do the boundaries of the Rights of Nature end? Even from a most basic understanding of what Rights of Nature aims to do, it is also clear that Rights of Nature must be informed by several disciplines, from philosophy to jurisprudence, with numerous managerial (e.g. what is possible?), linguistic, political, scientific, economic and cultural considerations. The depth of its complexities, however, does not minimise the importance of its principles in ethics and justice, and ethics and justice can help us explore these complex questions.

Whatever the answers to these questions ultimately resolve to be, it is crucial to understand that Rights of Nature seeks to give rights to life *in defence of life*. And it is this 'defence of life' that gives the Rights of Nature concept its power and potential. This article seeks to show that although Rights of Nature faces major challenges – indeed, all efforts in environmental or life protection do – its values seek to defend life on Earth, and those are values that must be defended for the future of life on Earth. Rights of Nature is also based in ethical principles already found throughout our justice system, just not yet extended to include all life. Therefore, it is not the Rights of Nature concept that needs to fit within a broken system, but rather a broken system which needs to change to better protect the values that Rights of Nature, and many of its sister movements, advance.[4]

And let there be no doubt, ours is a broken system. We cannot deny the data; the charts show our trajectories. Our current system of law and governance, from human rights to Nature's rights to everything in between, is not protecting life on Earth. We are in a conservation crisis because we are in a governance crisis, and the underlying cause of each is a values crisis. Those in power, those making the decisions and those framing the arguments are not advancing values that protect life on Earth. The unbridled quest for power is creating vast inequality around the world. Citizens and states are hypercompetitive, hyperindividualised and increasingly turning to saviours and strongmen who foster partisanship and separateness. The rise of the fascist, nationalist state will leave no rights to humans – or Nature. And neoliberalism as the ruling governing rationality separates, utilitises, monetises and marketises every living and nonliving thing on Earth, often with the blessing of the mainstream environmental movement.[5]

This article offers ethical engagement as a methodology for the inclusion of Rights of Nature in the theory, and practise, of justice. Ethics, as the domain of inquiry into what we consider right or wrong, is the foundation for the rule of law. And even though ethical engagement is complex and cumbersome, which is most likely why it is so often avoided by policy-makers, it is critical for understanding our current conservation and governance crises and for finding a better path forward. It can help us highlight and understand the values that harm life, and the values that support life, on this truly interdependent, interrelational, infinitesimal 'pale, blue dot'.[6] And once those values are highlighted and understood, they can help us confront injustice and create better laws, policies and institutions for life on Earth, as well as help motivate the prerequisite civil and political will.

To begin, the article will highlight the ethical underpinnings of the Rights of Nature movement. Guided by the work of the Biosphere Ethics Initiative, it will review the principles advanced in the Universal Declaration of Rights of Mother Earth and the Statutes of the Global Alliance for the Rights of Nature, the leading international network that advances the implementation and adoption of Rights of Nature.[7] The Biosphere Ethics Initiative is an international soft law program led by the U.S.-based nongovernmental organisation the Center for Environmental Ethics and Law, and has been adopted by the International Union for the Conservation of Nature (IUCN), the world's oldest and largest international environmental organisation.[8] Through its *Relato* Methodology for ethical engagement, based in rooted cosmopolitanism, it bears witness to, highlights and advances ethics in action by local communities, as well as state and global institutions, and is informed and guided in theory and practise by noted philosophers and jurists.

Rio State *Relato*. Karla Monteiro Matos describing the challenges and successes of members of the community recycling collectives in Mesquita, Brazil. (Photograph by Kathryn Gwiazdon.)

Words found consistently throughout this article are relation, relational, relationship, interrelated and so on. This is no accident. The rights of Nature are, ultimately, relational rights, advocating for relational freedom and relational justice.[9] As such, I propose Rights of Nature consider *ubuntu* as a foundational underlying ethical and legal principle to advance their work. *Ubuntu* is an ethic of interrelatedness that originates from several indigenous cultures of southern Africa. It roughly translates to 'I am because we are': my humanity exists because of your and our shared humanity, and advances principles of care, equity and reconciliation.[10] *Ubuntu* should be extended beyond our human relationships to the entire community of life, and can be done so without changing a word: 'I am because *we* are'.

Throughout the analysis, I will also highlight some of the major challenges that Rights of Nature must confront in order to advance its values; namely, the imbalance of power; the increasingly statist, or nationalist, state and neoliberalism. It will conclude that the values advanced by Rights of Nature proponents are the values that advance life on Earth, and through a process of broadening our notions of ethics and justice to the entire community of life, humanity can heal itself, can heal our relationships with one another and with the natural world and can heal our legal and governance systems.

## Extending the Ethical Roots of Justice to the Roots Themselves

> Even idle reflection may conclude that every legal code is capable of improvement, for it is possible to imagine what is most glorious, exalted, and beautiful is being more glorious, exalted, and beautiful still. But a large and ancient tree puts out more and more branches without thereby becoming a new tree; yet it would be foolish to refuse to plant a tree just because it might produce new branches.[11]
>
> **Georg Wilhelm Friedrich Hegel (1770–1831)**

Wendell Phillips (1811–1884) was an American lawyer and social reformer dedicated to the abolition of slavery. Confronting this grave inhumanity, he argued, 'the first duty of society is justice'.[12] The very purpose of the justice system is to provide the rules and institutions for governing sustainable and stable human societies, and inasmuch as possible, for preventing cruelty and great harm. To the Greek philosopher Epicurus (341–270 BC), 'Justice never is anything in itself, but in the dealings of men with one another in any place whatever at any time it is a kind of compact not to harm or be harmed'.[13] Justice requires that, if there is a harm, those responsible are held to account, and that the recipient of the harm is made whole again. For democratic nations, this pursuit for justice, for what it means to be a human,[14] is done primarily through the ethical and legal principles of liberty, equity and solidarity.

If justice asks, 'What is just?', ethical engagement is what guides our answer. The law, its rules and institutions, reflects what we believe is right or wrong, when responsibility attaches and which vehicles we believe will see those determinations through. Ethics is the *domain of inquiry* into what is considered right or wrong, when responsibility attaches and which vehicles would best suit that progress toward rightness. From a democratic perspective, laws and public policies are ethically justified and legitimate to the extent that they emerge from the reasonable deliberation of free and equal citizens who will be significantly affected by them.[15]

The inquiry into justice is eternal. The noted German philosopher Georg Wilhelm Friedrich Hegel knew that 'there is a constant need for new legal determinations'.[16] He saw it essential to understand that 'the very nature of the finite material entails an infinite progression ... it is therefore mistaken to demand that a legal code should be comprehensive in the sense of absolutely complete and incapable of any further determinations ...'[17] Justice can never be perfect or complete; it is alive, evolving, progressing. As such, and as Amartya Sen, the noted philosopher and author of *The Idea of Justice*, argues, 'We need *justitia*, not *justitium*'.[18]

Ethics is the foundation of justice and jurisprudence, on what is considered fair, on what is considered right or wrong, on what will be included and implemented in our rule of law. Indeed, ethical underpinnings are in every decision we make, or don't make, every law we draft, adopt, define or implement – or choose not to draft, fail to adopt, pervert when defining or never actually implement. Ethical underpinnings are in all of our decisions, and sometimes often and even purposefully hidden. A helpful tool is to identify our ultimate aim so that we can identify the values that support that ultimate aim. For an ethical inquiry into incorporating Rights of Nature in our justice system, we need to ask: What are the values that sustain life on Earth that allow life to flourish, and do our national and global law and governance structures advance those values? The Honourable Albert Sachs is a former Justice of the Constitutional Court of South Africa. As an anti-apartheid activist, he saw great injustices and great suffering due to the actions of man. He was instrumental in drafting the new constitution that set the path for a new South Africa, and, as a Justice, for actualising those values in law and life; he believed 'We have a constitutional duty to defend deep core values which are part of emerging world jurisprudence'.[19]

Ethics is practical, is actual and actionable, is life and advances the cause of life. Just as justice is a 'life-enhancing motivator', (i.e. do not harm), so is ethics.[20] Albert Schweitzer, noted philosopher, theologian, medical doctor and Nobel Peace Prize recipient, believed that, 'Ethics is nothing other than reverence for life. Reverence for life affords me my fundamental principle of morality, namely, that good consists in maintaining, assisting and enhancing life, and to destroy, to harm or to hinder life is evil'.[21] If reverence is deep respect for someone or something, how do human societies show respect? We honour it, we do not harm it; and if we have the power, we protect it. Rights of Nature asks: Do humans revere life? And if we do, how do we actualise that reverence? Sachs argued that the 'very purpose of entrenching rights is to provide a framework of constitutional *regard* ...' (emphasis added).[22] Sachs was referring to a regard for humans, but the principle remains. If we revere, regard, respect something, how can we translate that into action?

Rights are generally understood as 'that which is morally correct, just, or honorable', 'the cause of truth or justice' or, as afforded in law, 'a moral or legal entitlement to have or obtain something or to act in a certain way' or 'something to which one has a just claim'.[23] Rights of Nature would limit human activity; indeed, that would be their very purpose. However, to afford rights to Nature would be a liberating act of restriction; as Hegel noted, 'Because our social life is in harmony with our individuality, the duties of ethical life do not limit our freedom but actualize it. Such duties do not restrict us, but liberate us'.[24] Rights of Nature uses the law to protect Nature from humanity, and by extension, protect humanity from humanity. It is the covenants that constrain us, that sustain us. J. Ronald Engel is a philosopher and global ethicist who has focused much of his life's work on foundational covenantal relationships,

> We are, and will always be, limited creatures alongside other limited creatures, in a limited world. Yet we have the unique obligation as humans to self-impose the limits that are required to live sustainably and abundantly in this world. There is no escaping this reality. We are therefore in constant need to remind ourselves that we are not 'limitless animals,' but are created to live democratically, by self-rule, by self-imposed principles of mutual respect, care, and entrustment, or we cease to be human.[25]

And law is a crucial tool for self-imposing those limits, for putting ethics into action. Hegel is foundational to seeing ethics as action: 'The ethical is not abstract like the good, but it is intensely actual'.[26] In his *Elements of the Philosophy of Right*, he argues, 'Ethical life (or *Sittlichkeit*) is the idea of freedom as the living good which has its knowledge and volition in self-consciousness, and its actuality through self-conscious *action*'.[27] (emphasis added). And not only is it 'eternal justice', but a rational system of social institutions.[28] We are seeking an eternal inquiry into an eternal question of justice in order to cultivate real ethical action. How can we cultivate the spirit, the laws, the institutions, not to simply sustain a society, but to sustain life for all societies on Earth? And does Rights of Nature provide us with the values to cultivate such a spirit?

## A Worldview: Living Ethics for a Living Earth

The Declaration for the Rights of Mother Earth (the Declaration) was adopted in 2010 at the World People's Conference on Climate Change and the Rights of Mother Earth in Cochabamba, Bolivia.[29] It is the guiding document to the world's largest network that advances Rights of Nature, the Global Alliance for the Rights of Nature. It is structured with a Preamble and four articles that outline its understanding of Mother Earth, the inherent rights of Mother Earth, the obligations of humans and definitions. It encompasses the values of a movement to protect life on Earth, yet it also identifies some of its challenges, such as capitalism; predatory behaviour; toxic or radioactive waste and nuclear, chemical and biological weapons.

The Declaration opens with the foundational consideration that 'we are all part of Mother Earth, an indivisible, living community of interrelated and interdependent beings with a common destiny; gratefully acknowledging that Mother Earth is the source of life, nourishment and learning and provides everything we need to live well…'[30] The idea of Earth as a living entity, and humanity as part of the whole, is a core component of indigenous cultures and has been accepted widely in Western society.

Craig Kauffman is one of the leading voices for the practical incorporation of Rights of Nature in law and governance.[31] He and his colleagues explore many of the questions introduced above, including how to construct Rights of Nature norms and how to measure ecosystems. In his and Pamela Martin's upcoming article 'When Rivers Have Rights: Case Comparisons of New Zealand, Colombia, and India', they look to indigenous leaders and traditions to find a better, more accurate understanding of the relationships between humans and Nature. They note Bruce Kendall Goldtooth, a member of the Navajo Nation and the Executive Director of the Indigenous Environmental Network, who advances an indigenous worldview that 'sees humans as part of nature, an integrated whole in which the component parts have a "harmonious, awake, loving, and intelligent relationship with all other aspects of creation"'.[32] They also cite a beautiful quote from Casey Camp Horinek, the leader of the Ponca Nation of Oklahoma, while at the International Rights of Nature Tribunal in Quito, Ecuador in 2014,

> If you drank the water this morning or liquids, if you ate of the hooded nations or the four legged; if you breathe; if your body became warm from the fires of the earth, then you must recognize and understand that there is no separation between humans and Earth and all that are relatives of Earth and the cosmos, because you live in relation with her as a result of being one with her and there is no separation.[33]

Kauffman and Martin note, 'This harmony between humans and nature is the basis of *sumak kawsay*, or wellbeing', an ancient Quichua principle that translates to '"good living" in harmony with our communities, ourselves, and most importantly, our living, breathing environment'.[34] It is expressed in the Ecuadoran Constitution (2008), as 'Nature or Pachamama, where life is reproduced and exists, has the right to exist, persist, maintain and regenerate its vital cycles, structure, functions and its processes in evolution'.[35] They highlight that this principle 'is also reflected in the Iroquois (or Haudenosaunee) normative framework for living called the Good Mind. Other Lakota and Dakota nations refer to this harmonious relationship as *Mitakuye Owasin*, "All My Relations"'.[36]

The interconnectedness of all life is also well understood in Western science and philosophy. Aldo Leopold (1887–1948) was a forester, a conservationist, a philosopher, and is considered by many to be 'the father of wildlife ecology'.[37] In his renowned work, *A Sand County Almanac,* he defines ecology as 'the natural processes by which land and living things upon it maintain their existence'.[38] He believes that 'Men are only fellow-voyagers with other creatures in the odyssey of evolution'.[39]

Alexander von Humboldt (1769–1859) was a naturalist, a philosopher, a man of the particular, a man of the cosmos. His observations of life on Earth, united under several disciplines, were noted and embraced by great statesmen, such as Thomas Jefferson, and great philosophers, such as Henry David Thoreau. Humboldt, too, saw Nature alive and found harmony in diversity, 'animated by one breath – from pole to pole, one life is poured on rocks, plants, animals, and even into the swelling breast of man, but that breath came from earth itself…'[40] He also helped Thoreau develop his own understandings of Nature: 'Humboldt allowed Thoreau to weave together science and imagination, the particular and the whole, the factual with the wonderful'.[41] Earth was not dead, but 'lives and grows', it is 'living poetry…not a fossil earth, but a living specimen'. To Humboldt and Thoreau, 'Detail led to the unified whole'.[42] The methodology of this concept – the particular and the whole, the detail in the unified whole – will be explained when discussing rooted cosmopolitanism below.

In *Walden,* Henry David Thoreau professed, 'The Earth is all alive and covered with papillae. The largest pond is as sensitive to atmospheric changes as the globule of mercury in its tube'.[43] The Declaration acknowledges this same understanding of a living, interdependent community in its Preamble, noted above.[44] And Hegel believed, 'The sun, moon, mountains, rivers and all natural objects around us are. They have, in relation to consciousness, the authority not only of being in the first place, but also having a particular nature which the consciousness acknowledges, and by which it is guided in its behavior towards them, its dealings with them, and its use of them'.[45]

The Declaration's 'Inherent Rights of Mother Earth' mirror several well-understood and widely adopted human rights, including the right to life, to be respected, to regenerate and to maintain its integrity, as well as several human rights currently being advanced in domestic and international environmental law, such as the right to water and clear air and the right to be free from contamination. In the spirit of interrelatedness, and evolution, it also affords the right to each being's place and role in the harmonious functioning of Mother Earth. Many of these same principles are also found in the work of the Biosphere Ethics Initiative.

## A Methodology: A Rooted Cosmopolitanism Approach to Law and Life

Similar to the Declaration for Rights of Nature proponents, the guiding document of the Biosphere Ethics Initiative is the Evolving Biosphere Ethic.[46] The work of the Initiative began in 2004 in Bangkok, Thailand at the 3rd IUCN World Conservation Congress. The adoption of IUCN Resolution 3.020, *Drafting a Code of Ethics for Biodiversity Conservation*, was advanced by the IUCN Comité français and reaffirmed IUCN's 'commitment to an ethical view of nature conservation, based on respect for the diversity of life, as well as the cultural diversity of peoples'.[47] The work to develop this code was tasked to the Ethics Specialist Group of the IUCN World Commission on Environmental Law, where it continues today as the Biosphere Ethics Initiative. Although its origins are with IUCN, and the work continues to be advanced within IUCN as now-adopted policy, the work is currently led by the U.S.-based nongovernmental organisation the Center for Environmental Ethics and Law.[48]

The Evolving Biosphere Ethic is the evolving, textual embodiment of the peoples, institutions and work behind the Initiative. This includes numerous developmental meetings at local and global meetings, but most importantly, it includes the results of the *Relatos*. As noted above, *Relatos* are the heart of the Initiative. *Relato* is a Portuguese word meaning 'to relate', and its meetings are to do just that: create a space for ethical engagement and relational experiences. They are meetings of local leaders with global experts in an effort to highlight and advance ethical action in law and governance, and they are most often precipitated as a response to some particular harm to a community. Since 2007, there have been six *Relatos*: the Chicago Wilderness *Relato* in Chicago, Illinois; the South African National Parks (SANParks) *Relato* in Cape Town, South Africa; the *Relato* of the Indiana Dunes Region at the southern border of Lake Michigan in the United States; the Jordan *Relato* in sites throughout Jordan; the Rio State *Relato* in Nova Friburgo, Gramacho, and Rio de Janeiro, Brazil and the Gangjeong Village *Relato* in Jeju, Republic of Korea.[49]

South African National Parks *Relato*. The *Relato* group meeting with the women-led poverty relief construction teams at Slangkop Tented Camps, Table Mountain National Park. (Photograph by Kathryn Gwiazdon.)

The *Relato* Methodology is a rooted cosmopolitan approach that values diversity, yet seeks commonality; that is experiential, yet is unpacked by global scholars; that is place-based, yet understands our place to be within our communities and within the cosmos. Kwame Anthony Appiah explores rooted cosmopolitanism in *Cosmopolitanism* and *The Ethics of Identity*.[50] It is the idea that we can be informed and rooted by our local experiences, without losing sight of our global place, of our global relationships. Appiah's own personal history helped develop his ideas: 'Our community was Asante, was Ghana, was Africa, but it was also… England, the Methodist Church, the Third World: and, in [my father's] final words of love and guidance… it was also all humanity'.[51] Everything that we are and that we value is determined by the relationships around us, from the most personal and intimate to the broadest and most public.

Hegel also explored cosmopolitanism, and saw that we do not lose individuality by seeking uniformity in diversity, but rather we are all better protected: 'Particular interests should certainly not be set aside, let alone suppressed; on the contrary, they should be harmonized with the universal, so that both they themselves and the universal are protected'.[52] He later wrote an addendum that, 'Everything depends on the unity of the universal and the particular within the state'.[53] He also believed that our relationships must be truth-based, 'individuals must be conscious of this harmony. There is no freedom at all in a society whose members "identify" themselves with it only because they are victims of deception, illusion, or ideology'.[54] The Evolving Biosphere Ethic notes, 'People from all backgrounds and ages carry wisdom that is important for our common future, and we must build trust among generations, cultures, and professions'.[55] And trust-building requires truth-telling, 'The media has a key ethical role in environmental education, impact, empowerment and the truth of what they report'.[56]

Rights of Nature, the Evolving Biosphere Ethic and those who identify with their principles acknowledge a foundational truth: humanity, human development and life development (or evolution) is utterly dependent on Nature, on the diversity of all life and on our relationships. The Evolving Biosphere Ethic notes, 'There is strength in diversity; differences in language, belief, and practice, or bio-cultural diversity, are necessary for biological evolution, support scientific knowledge, and nurture solutions to our shared problems', as well as understanding the importance of our diverse spaces to all life, 'Both animals and humans have sacred spaces in nature that are vital to their lives and healthy development' and 'Play in and around nature is vital to the healthy development of children and adults'. [57]

In recognition of our constant learning, it also incorporates the precautionary principle into our own understanding of life, 'We have a responsibility to act with precaution when faced with scientific uncertainty'.[58] If we aim to give a voice to Nature, acting even in the face of incomplete knowledge, embracing humility will be a crucial consideration. As Leopold notes, 'The wild things that live on my farm are reluctant to tell me, in so many words, how much of my township is included within their daily or nightly beat. I am curious about this because it gives me the ratio between the size of their universe and the size of mine, and it conveniently begs the much more important question, who is more thoroughly acquainted with the world in which he lives?'[59]

The Evolving Biosphere Ethic also addresses key ethical considerations concerning humanity's impact on evolution: 'It is dangerous for the future of life to commodify life, which includes placing a price on nature, appropriating life processes, and synthetically creating new life forms and introducing them into the biosphere; and understands that scientific knowledge is not value-neutral, in its research or application'.[60] We must ask ourselves who funds the research, who uses or has access to the results, and who silences the results?

Rights of Nature embraces care for local places and spaces, as well as care for the wider, often unknown, community of life. Immanuel Kant (1724–1804) is one of the world's most influential philosophers. He investigated human progress, civilisation and morality, and his writings are foundational to human rights discourse. He, too, explored cosmopolitanism, within an aim of perpetual peace, particularly as it relates to strangers. If Rights of Nature expects to expand the circle of care of humans outside of humanity, a better understanding of our relationships with strangers may be helpful. Kant believed that through hospitality, as opposed to inhospitable conduct, strangers can enter into peaceful mutual relations (which may later turn into law and policy), 'thus bringing the human race nearer and nearer to a cosmopolitan constitution'.[61] He wrote that 'The peoples of the earth have thus entered in varying degrees into a universal community, and it has developed to the point where a violation of rights in one part of the world is felt everywhere. The idea of a cosmopolitan right is therefore not fantastic or overstrained; it is a necessary complement to the unwritten code of political and international right, transforming into a universal right of humanity. Only under this condition can we flatter ourselves that we are continually advancing towards a perpetual peace'.[62]

Rights of Nature is alive, and it is experiential. Known as one of Germany's greatest poets, Johann Wolfgang von Goethe saw that 'behind variety was uniformity'.[63] In our attempts to find unity in complex diversity, we draft and launch statements, declarations and covenants – sometimes as enforceable law, sometimes as soft law aimed to guide law and policy-making. The principles are beautiful, but to be universal, they are often anonymous; abstract; dissociated from local, particular experiences and set in stone once launched – closing off the continued process, engagement and ownership of the text of the documents. It seems we have 'a propensity, which is natural to all men, but which philosophers in particular are apt to cultivate with a particular fondness, as the great means of displaying their ingenuity, to account for all appearances in as few principles as possible'.[64]

The living Nature, based in real experiences, of the Evolving Biosphere Ethic is an attempt to address these seemingly lifeless documents. It aims to live and evolve, in honour of the real people and real events that it learns from. And it draws equally from awe and the actual: 'The well-springs of imagination lay less in logic than in the realm of human experience – the realm in which law ultimately operated and had meaning'.[65] Rights of Nature gives rights to real rivers, with real names, in real places, with real importance around the world, while also advancing global legal and ethical norms. It actualises experience, making all life part of that experience, and with the respect and reverence life and life-givers are owed. Schweitzer believed that, 'The fundamental fact of human awareness is this: "I am life that wills to live in the midst of life that wills to live." A thinking man feels compelled to approach all life with the same reverence he has for his own. Thus, all life becomes part of his own experience'.[66] Leopold believed, 'There is value in any experience that reminds us of our dependency on the soil-plant-animal-man food chain, and of the fundamental organization of the biota'.[67] Rights of Nature is rooted in that reality.

The Evolving Biosphere Ethic is a result of and a reflection of our ongoing ethical inquiry for conservation and justice for life on Earth, and it incorporates several key considerations that Rights of Nature advocates could find helpful:

1. What Is Your Fair Share: What is a fair share of each individual's use of natural resources that does not compromise the existence of other life?
2. How to Channel Resources: How can we channel the resources required to mitigate climate change and redress vast inequalities in the world?
3. Short-Term Arguments versus Long-Term Engagement: How do we counter short-term economic arguments against biodiversity conservation with the long-term thinking required in ethical inquiry?

4. The Individual in the Community: How do we support the common good alongside individual ambition, within a society dominated by private property rights and those seeking excessive power?
5. Protecting the Whole: How can we sustain the integrity of the biosphere without much greater equality in economic opportunities and incomes?
6. Ethical Institutions: What kind of domestic and international law systems do we need to empower institutions to respect, support and implement the Biosphere Ethics Initiative?
7. Balancing Rights: How can we ethically balance the rights of nature and the rights of humans?
8. What Is Your Population Responsibility: What obligations should societies and individuals assume in recognition of the pressures that our growing population places on one another and the biosphere?
9. Empowerment during Distress: How can we empower communities in ecological, military and/or economic distress to promote biodiversity conservation?

## The Motivations: Love Matters

> Ecological solidarity between humans and nature, with the obligation to respect and the compassion of love, is the basis for genuine care of living beings, places and people: love for the beauty and gift of the natural world with all of its living diversity; love for our places and our homes; and love for the people of today and tomorrow.[68]

The four loves that are expressed in the first principle of the Evolving Biosphere Ethic have an important role in the advancement of Rights of Nature. Biophilia, the love of living things; sociophilia, the love of one another; locaphilia, the love of place and yes, we even need a bit of egophilia, the love of self – for if we do not love ourselves, how can we expect to love another? There is a beautiful inscription, often attributed to Plato, at the entrance to the U.S. Department of Justice in Washington D.C., 'Justice in the life and conduct of the State is possible only as first it resides in the hearts and souls of the citizens'.[69] We cannot underestimate the power and potential of what stirs the heart.

Great harm is occurring, the awe and wonder of Nature are under assault and human emotions are stirring – and governments should take heed. Martha Nussbaum, a distinguished professor of law and philosophy at the University of Chicago, believes that liberal societies should politically cultivate emotion for justice and equality. In her profound work, *Political Emotions: Why Love Matters for Justice*, she argued that nations have two tasks before them: first, to 'engender and sustain strong commitment to worthy projects that require effort and sacrifice', such as environmental protection or other projects that 'get people to think larger thoughts and re-commit to the common good', and second, to 'keep at bay forces that lurk in all societies and, ultimately, in all of us: tendencies to protect the fragile self by denigrating and subordinating others'.[70] To Nussbaum, love is at the core of all emotions that 'sustain a decent society'. 'Love', she writes, 'is what gives respect for humanity its life… [and] it is needed all the more urgently in real, imperfect societies that aspire for justice'.[71]

But Nussbaum understands, as noted above, the importance of real, personal experiences to law and life, 'Real people are sometimes moved by the love of just principles presented ... abstractly; but the human mind is quirky and particularistic, more easily able to conceive a strong attachment if these high principles are connected to a particular set of perceptions, memories, and symbols that have deep roots in the personality and in people's sense of their own history'.[72] In a similar vein, Bruce Jennings, a noted political scientist and philosopher, believes that emotion and reason can build a better way forward, 'We must find a way to integrate and synthesize enchantment (imagination and wonder) and enlightenment (reason and control) as complementary and mutually essential for a new ecological social contract and worldview'.[73] He explores relational love, self-love and society in Jean Jacques Rousseau's *Discourse on Inequality*:

> ... *amour de soi-même* is a natural sentiment which inclines every animal to watch over its own preservation, and which ... produces humanity and virtue. [It is...] always good and always in conformity with order ... [and is] contented when true needs are satisfied. *Amour proper* is ... artificial and born in society, which drives each individual to have a greater esteem for himself than anyone else, inspires in men all the harm they do to one another ... [It] makes comparisons, is never content and never could be ...[74]

*Amour de soi-même* is literally 'love' (amour) 'of' (de) 'self-' (soi) 'even' (même), but *soi-même* together is 'one-self', and is understood together as self-love; *amour propre* is literally 'love' (amour) and 'own' (propre), and is understood as ego and pride. The Rights of Nature philosophy advances relational love to better our global society, in direct confrontation of harmful hyperindividualism. But *amour de soi-même* doesn't stop there, it also produces better humans: in others, we can see ourselves. And Rights of Nature should strongly embrace that it does the same. Not only is Rights of Nature a natural evolution of justice, as discussed below, it is a natural evolution of the goodness of humanity.

## Foundational Principles: Healing through *Ubuntu* and Reconciliation Ecology

The ethical and legal principles of *ubuntu* also see diversity as fundamental for understanding our common humanity. I first learned about *ubuntu* at the South African National Parks *Relato* that took place with several governmental and nongovernmental partners throughout Cape Town and the surrounding area. It is a relational ethic from tribes across Southern Africa and has been explored and advanced in great depth by some of the world's most thoughtful political and spiritual minds, such as Archbishop Desmond Tutu and Nelson Mandela. As noted above, *ubuntu* is an ethic of interdependence. It has several interpretations, but generally can be understood as, 'I am because we are,' or in other words, only through others are we able to see our own humanity. Therefore, only through particularities, only through our diversity, can we truly be.

It has also been argued that *ubuntu* can be extended to all of our living relationships, as well, in the truest sense of interdependence and interrelatedness. Motlatsi Khosi, a lecturer in African philosophy at the University of South Africa, argues that '[*ubuntu*] also relates to the way people interact with both the natural and metaphysical worlds, the latter consisting of unseen elements such as ancestors and God ... It should be about how to make sense of the connection with the universe, animals, nature and humans'.[75]

*Ubuntu* is related to another principle, ecological reconciliation, also learned at the South African National Parks *Relato.* Ecological reconciliation is confronting and addressing the harms we have done to humans and nature in the name of humans and nature.

> Ecological reconciliation is necessary for a sustainable future: we must confront the truth of the past, the harms to humans and nature in the name of humans and/or nature, and reconcile ourselves with those impacts; we must be able to forgive and move forward, without blame or guilt.[76]

Reconciliation is the healing aspect of *ubuntu,* as most notably seen in the Truth and Reconciliation Commission of post-Apartheid South Africa. As explored in further detail below, perhaps Rights of Nature, through principles such as *ubuntu* and reconciliation, can help us begin to heal our humanity, by righting our relationship with Nature.

Similar to Rights of Nature, *ubuntu* is used as a legal principle to heal broken relationships. In *Afri-Forum and Another vs. Malema and others,* the hate speech trial of Julius Malema, Judge Colin Lamont decided, '*Ubuntu* is recognized as being an important source of law within the context of strained or broken relationships amongst individuals or communities and as an aid for providing remedies which contribute towards more mutually acceptable remedies for the parties in such cases'.[77] Justice Sachs often advanced the constitutional values of *ubuntu* in his own decisions. He views *ubuntu* as restorative justice, with its key elements being encounter, reparation, reintegration and participation: 'Reparation focuses on repairing the harm that has been done rather than doling out punishment. Reintegration into the community depends upon the achievement of mutual respect for and mutual commitment to one another'.[78] We need to heal, and we need to respect, we need to recommit to our relationships with nature – and Rights of Nature can help us do that.

## Rights of Nature Is a Natural Evolution of Justice

> There is no easy walk to freedom anywhere.[79]
>
> **President Nelson Mandela (1918–2013)**

Nature alive, ethics alive, justice alive – and a world dying. Species extinctions are rising; biodiversity is collapsing; oceans are acidifying; humans are overpopulating, hurting and migrating – and each of these crises is due to unbridled, or wrongfully bridled (or wrongfully governed), human activity. It is not a stretch of the imagination to extend foundational notions of justice – respect life, do not denigrate, do not subordinate, do not harm (and if you harm, try to make whole again) – to the natural world; after all, they would be founded in the same ethical principles, motivated by the same values. Kant believed, 'This homage which every state pays (in words at least) to the concept of right proves that man possesses a greater moral capacity, still dormant at present, to overcome eventually the evil principle within him (for he cannot deny that it exists), and to hope that others will do the same'.[80]

The foundational aims of the Biosphere Ethics Initiative and Rights of Nature are rooted in life and justice: to protect life, to 'keep nature alive and flourishing in the biosphere', and to promote just governance, just economies and just development and consumption.

> We must always strive to attain social, economic, and environmental justice and support these and other universal human rights, for poverty and environmental and human health are inextricably linked and cannot be solved separately.[81]

Gangleong Village *Relato.* The ancient way of life of the indigenous villagers of Gangleong, who have lived in harmony with Nature for centuries, was threatened due to development of a military base. (Photograph by Kathryn Gwiazdon.)

We cannot separate our conservation crises from our governance crises. Indeed, we cannot separate any of our life crises. We cannot address the protection of wild areas without addressing the protection of the poor; we cannot address the rights of a river without addressing the rights of refugees; we cannot address the mass-scale loss of biodiversity, the acidification of our oceans and the warming of our climate without addressing the failures in ethics and justice of nation-states and global governance bodies. We cannot address how to heal our relationships with Earth without addressing militarisation, war and peace. To fail to connect the harms being done to humans by humans, and the harms being done to nature by humans, harms us all. After all, 'we are *all* part of Mother Earth, an indivisible, living community of interrelated and interdependent beings *with a common destiny*' (emphasis added).[82] This shared destiny also includes shared responsibilities, 'every individual, every sector, every community, and every state has a shared responsibility to protect our future, dependent upon what harm they have caused and what good they are capable of doing'.[83]

The foundational worldview, aims and ethical principles, as advanced by the Biosphere Ethics Initiative, are those advanced by Rights of Nature. The Declaration promotes democracy, participation, learning and communication. It advances harmony, peace and precaution. In principle, there is nothing in this document that is unfamiliar to the laws and institutions of modern-day liberal democracies. It advances scientific truths, human enlightenment and development and rights already afforded to other living beings, humans. Understanding that no human rights can exist without the life processes that support humanity, it is an honest document. But is it an actionable document? Can a theory of justice, which would require implementation by a governing nation, and possibly a global governance institution, be extended to all life on Earth? If, as Hegel suggests, 'The state is the actuality of the ethical Idea', it should not be too difficult to extend what many believe to be the ethics that should guide human behaviour to the ethics that should guide the rules and institutions that guide human behaviour.[84] The state has a 'pivotal importance in setting life conditions for all on a basis of equal respect, and as the largest unit ... that is decently accountable to people's voices and capable of expressing their desire to give themselves laws of their own choosing'.[85]

## *Sapere aude*!: An Eternal Inquiry towards Perpetual Peace

Following the principles of humility and recognising our own human capacity, Sen argues that we need to look at what is possible, as the Transcendentalists did, but also be realisation-focused.[86] Doing this 'makes it easier to understand the importance of the prevention of manifest injustice in the world, rather than seeking the perfectly just. The subject of justice is not merely about trying to achieve – or dream of achieving – some perfectly just society or social arrangements, but about preventing manifestly severe injustices'.[87] This helps society address grave injustices while continuing to explore our never-ending inquiry into what is just. When humanity calls for justice, it is not a call for perfect justice, but rather a call for 'the elimination of some outrageously unjust arrangements'.[88] We know a society with slavery is unjust, but we also know that ending slavery will not create a perfectly just society. So, even if some believe that Rights of Nature is not appropriate or ready for inclusion into our legal systems because it does not yet have all of the answers, Sen warns us not to destroy the world in search of perfect justice, '[citing Ferdinand I] Let justice be done, though the world perish'.[89] Therefore, we must always be learning, always be striving for better, continually improving toward 'perpetual peace', as Kant expects of a rational human society.[90]

Leopold believed that 'all ethics so far evolved rest upon a single premise: that the individual is a member of a community of interdependent parts'.[91] And so it was only natural that any ethic would extend respect and care to the entire community of life, 'The land ethic simply enlarges the boundaries of the community to include soils, waters, plants, and animals, or collectively: the land'.[92] In doing so, 'the role of *Homo sapiens* [is changed] from conqueror of the land-community to plain member and citizen of it. It implies respect for his fellow-members, and also respect for the community as such'.[93] He saw 'The extension of ethics [to the land as] an evolutionary possibility and an ecological necessity'.[94]

Rights of Nature embraces the land ethic by extending notions of justice to the entire community of life. Even Kant set his context of human rights within the whole, 'the whole of humanity (universorum), united in Earthly society and distributed in national groups'.[95] To include, or even consider inclusion of, Rights of Nature within our justice systems is merely evidence of the evolution of human progress and jurisprudence. Following Sen's guidance that it is easier to know injustice than perfect justice, nearly everything we have been taught in law and philosophy about what is good and right is against arbitrary harm, unjust war and those who seek to oppress, destroy or enslave those who are weak, vulnerable or have no voice.

If we acknowledge that Nature is alive, yet we have no respect for Nature, no acknowledgement of our place within it, is Nature nothing more than life bent to our will? Is Nature life, but life without liberty – to grow, to develop, to nurture others, to be free 'from the intrusive interference of others?'[96] If this is true, then is Nature some lower rung on the hierarchy of life, even though it is the source of all life? Is Nature no more than a slave to humanity, 'subject to the inconstant, uncertain, unknown, arbitrary will of another man'?[97] Are humans nothing more than unjust conquerors in a constant state of war?[98] Or perhaps we are mere despots, imperialists or colonialists with the ultimate dominion over another, to take away life, whenever we please?[99] Are we truly incapable of saying that there is no common bond between the entire community of life, to deny justice to all life on Earth? Democracy moved us away from ancient regimes based on hierarchy and subordination; international law has unequivocally and universally denounced slavery and colonialism.[100] Indeed, 'All oppression creates a state of war'.[101] Let us extend those same notions of liberty and equality that we afford to humankind, to all life, let us 'focus on a peace treaty to end this unwinnable war'.[102]

Jennings explored relational liberty in his most recent book, *Ecological Governance: Toward a New Social Contract with the Earth*. He wanted to address the void in liberal theories of freedom 'of the web of interdependencies – cultural meaningful roles, styles, and self-identities, shared values, rituals, and practices'.[103] Indeed, everything that makes life, life. He defines relational liberty as 'freedom in and through relationships of interdependency ... [it is] freedom through transactions and relationships of interdependency with others that exemplify justice (parity of social membership, voice, and participation) and solidarity (mutuality of civic concern and respect)'.[104] Jennings's notions of relational liberty are actually very much in line with the ethic of *ubuntu*, as described above. He continues, 'The essence of the philosophical strategy ... is to internalize the common good into the individual good, to read the "we" into each "I". And ... to internalize the freedom and well-being of all (both human and nonhuman creatures and systems of life) into the freedom and well-being of each'.[105] And he also ultimately connects this liberty with justice,

> At heart of the concept of liberty is the notion that to be free is to be in certain kinds of relationship with others. Human freedom in the relational sense cannot be understood in isolation from either the social or the ecological biotic webs within which the individual resides and are the preconditions of human survival, development, and nourishing. Relational freedom cannot exist within the context of unjust structures of power, wealth, social opportunity, health, and psychological integrity.[106]

Seeking to give nonhuman entities rights within the construct we use to govern ourselves, and how we exist alongside other humans, friend and stranger, is a natural evolution of justice and liberty, and it is also evidence to our own enlightenment. To Kant, 'Enlightenment is man's emergence from his self-imposed immaturity. Immaturity is the inability to use one's understanding without guidance from another'.[107] Our capacity to care for the welfare of others is a value that our governments should embrace. Nussbaum, in citing Auguste Comte's philosophy on human religion,[108] not only understood that 'human conduct, including the operations of sympathy and love, can be understood in a lawlike way ...' but that 'a broader sympathy is more advanced, more mature, than the narrow sympathy with family and kin by which most people are animated'.[109]

Kant believed that 'This immaturity is self-imposed when its cause lies not in lack of understanding, but in lack of resolve and courage to use it without guidance from another. *Sapere aude*! [Dare to know!] Have the courage to use your own understanding! That is the motto of enlightenment'.[110] Rights of Nature is an extension of Comte's view that 'Sympathy extends ever outward, until the unity of the entire species, past and future as well as present, becomes its object...'[111] Let us have the courage to advocate for this understanding of life in law: we are all connected, we are all dependent upon one another. Care and empathy, a yearning to protect the vulnerable and voiceless, are values that protect life. Even if we do not yet know *how* to translate these values towards other species and life processes into law, we know that the current way we treat Nature is against human nature.

## Wrath and Reason Are Just Responses to Grave Injustices

> Against eternal injustice, man must assert justice, and to protest against the universe of grief, he must create happiness.
>
> **Albert Camus (1913–1960)**

In advancing a theory for justice for life, and to heal life, we cannot underestimate the power behind witnessing injustice. Violent, racial apartheid motivated Justice Sachs's constitutional development on democracy, liberty, equality and justice for all; and philosophers note the crucial importance of healing to their own discipline, 'there is no use in philosophy, unless it casts out the suffering of the soul'.[112] Philosophy helps us bring structure and 'reach to reflections on values and priorities', but it also helps us confront 'the denials, subjugations and humiliations which humans suffer across the world'.[113] And as noted above, the way we treat Nature is a reflection of own humanity, our limitations and our capabilities.

French philosopher Michel Foucault believed it was the duty to hold states accountable for the suffering they cause, 'Because they claim to be concerned with the welfare of whole societies, governments arrogate to themselves the right to pass off as mere abstract profit or loss the human unhappiness that their decisions provoke or their negligence permits. It is a duty of an international citizenship to always bring the testimony of people's suffering to the eyes and ears of governments, sufferings for which it's untrue that they are not responsible. The suffering of men must never be a mere silent residue of policy. It grounds an absolute right to stand up and speak to those who hold power'.[114] Foucault saw power not as a thing, but as relations 'such that one can direct the behavior of another or determine the behavior of another'. [115] In national and international governance, power is the nation-state over human individuals and over all life, and 'Power relations always leave open the possibility for resistance'.[116]

Sen explores power relations through Gautama Buddha's teachings. Buddha argued that 'we have responsibility to animals precisely because of the asymmetry between us, not because of any asymmetry that takes us to the need for cooperation'.[117] If we are enormously more powerful than another, such as the relationship between a mother and child, we have some responsibility towards the other that connects exactly with this asymmetry of power.[118] This asymmetry can be applied between individuals, between individuals and states, between states, between state and global governance and between humans and the natural world.

Through the rule of law, the state governs human behaviour, and at all levels of political society. Not only do the power-holders within states decide the laws and the institutions to guide and correct human behaviour, they also frame the arguments before us. With an almost religious fanaticism, instrumental and utilitarian arguments dominate modern day policy-making. Global climate change ethicist Donald Brown believes that instrumental rationality is the greatest challenge to ethical thinking and acting.[119] Instrumental rationality is operationalised throughout law and governance as decision-making based on what is most efficient or cost beneficial, absent of the values involved. Kant also saw the harm of instrumental rationality and made the connection between morality and governance. He believed that political moralists 'fashion their morality to suit their own advantage as a statesman'.[120] They 'wield power on Earth…' by 'advocating might instead of right' and '[making their] principles subordinate to [their] end'.[121]

Often through fear and deceit, the powerful are placing the protection of life against some of the most powerful forces on Earth: Nature versus nationalism, or even, in the case of climate change refugees, conservation versus terrorism and national and human security. Why? Why do they fund massive climate change denial campaigns or purposefully foster tensions between environmental protection and economic development, those 'frivolous environmental regulations' versus the 'vital protection of American jobs'? Society – the good of society – cannot exist without the foundations of life. But perhaps the good of society – justice – is not what those seeking excessive power, those corrupted by money or ego, really desire.

Archbishop Tutu believed that, 'Harmony, friendliness, community are great goods. Social harmony is for us the *summum bonum* – the greatest good'.[122] If social harmony is the greatest good, then knowingly creating, or not trying to prevent, social disharmony is the ultimate harm.[123] Hegel may have termed this an 'infinite injury'.[124] In his chapter on *The Ecological Conscience*, Leopold understood that 'Conservation is a state of harmony between man and land'.[125] When the foundations of life are harmed, our harmony with Nature is harmed – all life is harmed. Indeed, Nature is the epitome of the vulnerable, marginalised 'other' in human society, and Rights of Nature demands justice for the harms made against her. Sachs understood the link between the treatment of vulnerable populations, in particular, and the welfare of the state, 'Our society as a whole is demeaned when state action intensifies rather than mitigates [the poor's] marginalization'.[126] And Hegel believed that, 'The poor in civil society are victims of not of some natural misfortune, but of a social wrong'.[127] Our treatment of the natural world is a social wrong that must be righted, and in doing so, all life will be better protected. As the Evolving Biosphere Ethic states, 'If we protect the vulnerable people, places, species, and societies, we protect everyone and everything'.[128]

If, as Sen believes, justice is an exploration into what it means to be human, then an exploration into injustice could be, 'how can one human being do this to another [human being]?'[129] This became the dominant theme of the Truth and Reconciliation Commission, its creation a direct response to witnessing grave injustices. Sachs saw that 'What was at stake was an affirmation of the values of our society. It was not just a question of calculating losses and gains, of counterposing the advantages and disadvantages of certain types of state conduct. It was a question of what kind of people we were. What were we about? What kind of country did we live in? Did we have shame, and if so, what were the things that brought it upon us?'[130] Perhaps the next International Tribunal on the Rights of Nature can lead us through a truth and reconciliation process. How can humans do this to life on Earth? What is at stake is the affirmation of the values of our society.

For Sachs, he turned the laws of apartheid South Africa, which constituted 'a barricade of injustice' to freedom, into 'a primary instrument for accomplishing peaceful revolution'.[131] He said that the process of making new laws not only helped his country heal, but also himself.[132] But to do so, they had to have a serious reflection into gross injustices, into evil behaviour, how it is condoned and how it spreads and through what institutional mechanisms and what culture.[133] For apartheid South Africa, 'the enemy was not a people or a population, but a system of injustice'.[134] Rights of Nature is a response to a system of injustice and can be a part of a larger process that allows us to heal our humanity, our relationships with life on Earth and life itself.

Similar to Nussbaum, Sen saw 'reasoning and feeling as deeply interrelated activities' and emphasised the important role for 'instinctive revulsion to cruelty and to insensitive behavior'.[135] That it is a good thing to emphasise 'caring about the miseries and happiness of others'.[136] In studying Mary Wollstonecraft's writings on *A Vindication of the Rights of Women*, he noted that resistance to injustice 'draws on both indignation and argument'. Wollestonecraft was experiencing subjugation, and the anger and exasperation was clear in her work, 'reason calls for this respect, and loudly demands JUSTICE for one half of the human race'.[137] Applying this to Rights of Nature, in no system of justice would great harm be allowed to occur without accountability. In no system of justice would great harm to life – and a harm that exponentially creates even more harm to more life – be allowed. As humans that respect the principles of justice, we have a right to be angry before such inconsistent justice. As caring and empathetic creatures, we have a right to be angry before such harm. Perhaps it is time to combine 'wrath and reasoning';[138] perhaps 'a little rebellion now and then is a good thing, and as necessary in the political world as storms in the physical'.[139]

The Rights of Nature philosophy is a call to end a patent injustice: the harm, caused by humanity, against life on Earth. And it advances principles already ingrained in citizens of democratic nations. Rights of Nature extends civil rights leader Martin Luther King, Jr.'s (1929–1968) understanding that 'Injustice anywhere is a threat to justice everywhere'.[140] It advances principles of inherent rights of life and liberty in living beings, as found in the U.S. Declaration of Independence, 'We hold these truths to be self-evident, that all men are created equal, that they are endowed by their Creator with certain unalienable Rights, that among these are Life, Liberty and the pursuit of Happiness…'[141] When U.S. President Abraham Lincoln (1809–1865) made 'these truths' the centrepiece of his Gettysburg Address, he was responding to a grave injustice, calling for an end to slavery, and an actualisation of these principles in the U.S. Constitution.[142] And Walt Whitman (1819–1892), in his elegy for President Lincoln, 'When Lilacs Last in the Dooryard Bloom'd', used natural images together with images of human activity to invoke liberty and equality of all Americans.[143] As Nussbaum noted, '…emotions directed at the geographical features of a nation are ways of channeling emotions towards its key commitments – inclusiveness, equality, the relief of misery, the end of slavery'.[144] A wild river, a soaring eagle. Life, liberty and happiness are certainly not unique to humanity, and in word and deed, in symbolism or actualism, they are certainly not separated from the natural world.

## Conclusion

> When a father asked him for advice about the best way of educating his son in ethical matters, a Pythagorean replied, 'Make him the citizen of a state with good laws.'[145]
>
> **Hegel**

Life is relational. Justice is relational. Liberty is relational. Rights are relational. What makes us happy, and what makes us human, is relational. And certainly, harm is relational. And this is why it takes a relational, local, global, diverse, unifying response in law and governance to better protect life on Earth.

We are imperfect humans working within imperfect human constructs in our eternal strive toward eternal justice. We need to acknowledge this truth, act with humility and act courageously. Humans may be capable of destroying all life on Earth, but all life on Earth does not exist to support us. The foundations of human life may not be the same as the foundations of all life, and even if our actions destroy ourselves, it is possible that Earth will continue, heal and flourish without us. We are not the centre of the universe – or the Earth. It is time our governance systems reflect that simple truth.

What we do know with certainty is that we can always strive to be better than we are today, better than what our current legal system allows or limits. This is not only evidenced by the crises our current governance systems have created, but also as a fundamental understanding of human development. We can always do better; indeed, to survive, we must. Extending rights beyond ourselves is a direct response to our crises, a direct response to injustice and an embrace of the natural evolution of justice and the natural evolution of what it means to be human.

Rights of Nature is an attempt to right our relationship with Nature. We are more than pirates and robbers of life; if life is to be taken away or harmed, just reasons must be given, and unjust reasons, and violators, be held to account. We must heal ourselves, and to do that, we must reconcile ourselves with how we treat the larger community of life. Sachs understood that there was great potential in reconciliation, 'The nation wishing to understand and deal with its past… is asking much larger questions: how could it happen, what was it like for all concerned, how can you spot the warning signs, and how can it be prevented from occurring again?' [146] And it all begins with serious inquiry, serious dialogue, 'Dialogue is the foundation of repair. The dignity that goes with dialogue is the basis for achieving common citizenship. It is the equality of voice that marks a decisive start, the beginning of a sense of shared morality and responsibility'.[147]

Rights of Nature is an extension of values that already exist at the foundation of our human societies – at the foundation of justice – and rationality and progress demand that we address the inquiry of their inclusion with seriousness. Only then may we find that, 'the disagreements that exist may be removed through reasoning, helped by questioning established prejudices, vested interests, and unexamined preconceptions' and, ultimately, that there is room in the broad theory of justice for Rights of Nature.[148]

In closing, I leave you with words from Aldo Leopold, a sort of elegy to the passenger pigeon,

> On a Monument to the Pigeon… There will always be pigeons in books and museums, but these are effigies and images, dead to all hardships and to all delights. Book-pigeons cannot dive out of a cloud to make the deer run for cover, or clap their wings in thunderous applause of mast-laden woods. Book-pigeons cannot breakfast on new-mown wheat in Minnesota, cannot dine on blueberries in Canada. They know no urge of the seasons; they feel no kiss of the sun, no lash of wind and weather. They live forever by not living at all.[149]

Let us not give eternal life to species by noting their extinction in books; let us not find perpetual peace in the grave.[150] Extending rights to Nature extends life to us all.

## Notes

1. Global Alliance for the Rights of Nature. 'Universal Declaration of the Rights of Mother Earth.' www.therightsofnature.org/universal-declaration. (accessed 6 January 2019).
2. Ibid.

3. Leopold, A. 1949. *A Sand County Almanac and Sketches Here and There.* Oxford: Oxford University Press.
4. See generally, the UN Harmony with Nature program, www.harmonywithnatureun.org; the Earth Democracy project of the Ethics Specialist Group of the IUCN World Commission on Environmental Law, www.iucn.org/commissions/world-commission-environmental-law/our-work/ethics; the Ecological Law and Governance Association, www.elga.world; and The Hague Principles for a Universal Declaration on Responsibilities for Human Rights and Earth Trusteeship, http://www.earthtrusteeship.world/the-hague-principles-for-a-universal-declaration-on-human-responsibilities-and-earth-trusteeship/.
5. This is explored more thoroughly in Gwiazdon, Kathryn. 2019. 'From Stardust to Sacred Sands: Protecting Life on Earth through a Human Story of Ethics, Care, and the Cosmos'. Forthcoming in *The Crisis in Global Ethics and the Future of the Earth Charter: Essays in Honor of J. Ronald Engel.* Edited by Peter Burdon et al. Cheltenham: Edward Elgar.

   For her discussion on the danger of the environmental movement's embrace of neoliberalism, Gwiazdon cites Monbiot, George. 15 April 2016. 'Neoliberalism – The Ideology at the Root of All Our Problems'. *The Guardian.* Accessed 21 January 2019. https://www.theguardian.com/books/2016/apr/15/neoliberalism-ideology-problem-george-monbiot. 'Neoliberalism sees competition as the defining characteristic of human relations. It redefines citizens as consumers, whose democratic choices are best exercised by buying and selling… Inequality is recast as virtuous: a reward for utility and a generator of wealth, which trickles down to enrich everyone'. See also: Monbiot, George. 15 May 2018. 'The UK government wants to put a price on nature – but that will destroy it'. *The Guardian.* (Accessed 21 January 2019). https://www.theguardian.com/commentisfree/2018/may/15/price-natural-world-destruction-natural-capital. 'In reality, natural wealth and human-made capital are neither comparable nor interchangeable. If the soil is washed off the land, we cannot grow crops on a bed of derivatives'. And 'still more deluded is the expectation that we can defend the living world through the mindset that's destroying it. The notions that nature exists to serve us; that its value consists of the instrumental benefits we can extract; that this value can be measured in cash terms; and that what can't be measured does not matter, have proved lethal to the rest of life on Earth'.

   Gwiazdon also cites Klein, Naomi. 2015. *This Changes Everything: Capitalism vs. the Climate.* New York: Simon & Schuster. 'Instead of calling for bans on harmful chemicals, or pursuing legal recourse, [environmental organizations] are "pro-business, non-confrontational, and ready to polish even the most tarnished corporate logos." In efforts to be taken seriously "in circles where seriousness is equated with toeing the market line," they are adopting the frameworks and the language of the markets.' A complete excerpt from the Gwiazdon article,

   > This pro-market behavior not only neutralizes environmental organizations, but it is taking away the language of an entire movement. Powerful terms like "nature," "preservation," "protection," "conservation," and even what it means to be an "environmental" organization, are being made impotent. Even a tree is no longer a tree, but a "carbon sink" for a cap and trade mechanism that allows for polluting industries to continue with business as usual, so long as they can purchase carbon credits a thousand miles away. And if a politically weak community lives in that "carbon sink"? A community that lives in harmony with nature, that should be embraced instead of erased? Simply push them aside, business must continue.
   >
   > "Natural capital," "cap and trade" mechanisms, "ecosystem services,": the language of the new environmental movement embraces market ideology. It is turning the interconnected web of life into something that is disconnected, divided, and labeled with a price tag. And it is causing harm. Monbiot believes that "the natural capital agenda is the definitive expression of our disengagement from the living world. First, we lose our wildlife and natural wonders. Then we lose our connections with what remains of life on Earth. Then we lose the words that described what we once knew. Then we call it capital and give it a price. This approach is morally wrong, intellectually vacuous, emotionally alienating and self-defeating.

6. Sagan, C. 1985. *Pale Blue Dot: A Vision of the Human Future in Space.* New York: Ballantine Books.
7. For the Biosphere Ethics Initiative, see www.environmentalethicsandlaw.org; and for the Global Alliance for the Rights of Nature, see www.therightsofnature.org.
8. See Key IUCN Resolutions at www.environmentalethicsandlaw.org/ethics-at-iucn.
9. I am not sure what other discourses or trains of thought may have adopted this phrase, 'relational rights,' but for purposes of this article, it means rights owed to a being or entity due to – but not dependent upon – their integral and reciprocal relationships to other right-holders. Therefore, RON does not owe its inherent rights to humanity, nor does humanity owe its inherent rights to nature, but they are or should be legally recognized because of their relationships to each another.
10. Tutu, A. D. 1999. *No Future Without Forgiveness.* New York: An Image Book, Doubleday.
11. Hegel, G. W. F. 1991. *Elements of the Philosophy of Right.* Edited by Allen W. Wood. Translated by H. B. Nisbet. Cambridge: Cambridge University Press.
12. Note: This quote is often improperly cited to Alexander Hamilton. Philipps, Wendell. 1894. *Speeches, Lectures, and Letters. Volume 1.* Boston: Lee and Shepard. He continued, 'A prosperous inequity', says Jeremy Taylor, 'is the most unprofitable condition in the world.' And, 'It is not Northern laws or officers they fear, but Northern conscience'. For a thorough exploration of the theory of justice, see also John Rawls, 'Justice is the first virtue of social institutions' in Rawls, J. 1971. *The Theory of Justice.* Cambridge: Belknap Press of Cambridge University.
13. Epicurus. 1926. *Epicurus, the Extant Remains.* Translated by Cyril Bailey. Oxford: Clarendon Press.
14. Sen, A. 2009. *The Idea of Justice.* Cambridge: The Belknap Press of Cambridge University Press.
15. Jennings, B. 2016. *Ecological Governance: Towards a New Social Contract with the Earth.* Morgantown: West Virginia University Press.
16. Hegel, G. W. F. 1991. *Elements of the Philosophy of Right.* Edited by Allen W. Wood. Translated by H. B. Nisbet. Cambridge: Cambridge University Press.
17. Ibid.
18. Sen, A. 2009. *The Idea of Justice.* Cambridge: The Belknap Press of Cambridge University Press.
19. Sachs, J. A. 2009. *The Strange Alchemy of Life and Love.* Oxford: Oxford University Press.
20. Sen, A. 2009. *The Idea of Justice.* Cambridge: The Belknap Press of Cambridge University Press.
21. Schweitzer, A. 1923. *Civilization and Ethics.* London: A. & C. Black.
22. Sachs, J. A. 2009. *The Strange Alchemy of Life and Love.* Oxford: Oxford University Press.
23. Definitions of 'right' taken from https://en.oxforddictionaries.com/definition/right and https://www.merriam-webster.com/dictionary/right.
24. Hegel, G. W. F. 1991. *Elements of the Philosophy of Right.* Edited by Allen W. Wood. Translated by H. B. Nisbet. Cambridge: Cambridge University Press.
25. Bosselmann, K. J., Engel, R. and Taylor, P. 2008. *Governance for Sustainability.* Gland: IUCN.
26. Hegel, G. W. F. 1991. *Elements of the Philosophy of Right.* Edited by Allen W. Wood. Translated by H. B. Nisbet. Cambridge: Cambridge University Press.
27. Ibid.
28. Ibid. Ethical life has been represented to nations as eternal justice. §145; Hegel's name for a rational system of social institutions is 'ethical life' (*Sittlichkeit*) §144–145.
29. Global Alliance for the Rights of Nature. 'Universal Declaration of the Rights of Mother Earth'. www.therightsofnature.org/universal-declaration. (accessed 6 January 2019).
30. Ibid.
31. Kauffman, C. and Martin, P. 'When Rivers Have Rights: Case Comparisons of New Zealand, Colombia, and India'. Forthcoming in *Law & Policy* (available upon request).
32. Ibid.
33. Ibid.
34. Pachamama Alliance. 'Ancient Teachings of Indigenous Peoples'. https://www.pachamama.org/sumak-kawsay. (Accessed 6 January 2019).

35. Global Alliance for the Rights of Nature. 'Rights of Nature Articles in Ecuador's Constitution'. https://therightsofnature.org/wp-content/uploads/pdfs/Rights-for-Nature-Articles-in-Ecuadors-Constitution.pdf. (Accessed 6 January 2019).
36. Ibid.
37. The Aldo Leopold Foundation. 'Aldo Leopold'. https://www.aldoleopold.org/about/aldo-leopold/. (Accessed 6 January 2019).
38. Leopold, A. 1949. *A Sand County Almanac and Sketches Here and There.* Oxford: Oxford University Press.
39. Ibid.
40. Wulf, A. 2015. *The Invention of Nature: The Adventures of Alexander von Humboldt, The Lost Hero of Science.* London: John Murray.
41. Ibid.
42. Ibid.
43. Thoreau, H. D. 1995. *Walden*. United States of America: Houghton Mifflin Harcourt.
44. Declaration.
45. Hegel, G. W. F. 1991. *Elements of the Philosophy of Right*. Edited by Allen W. Wood. Translated by H. B. Nisbet. Cambridge: Cambridge University Press.
46. Center for Environmental Ethics and Law. 'The Evolving Biosphere Ethic'. https://environmentalethicsandlaw.org/the-evolving-biosphere-ethic. (Accessed 6 January 2019).
47. 'RES 3.020 Drafting a code of ethics for biodiversity conservation'. IUCN. Accessed 6 January 2019. http://cmsdata.iucn.org/downloads/wcc3_res_020.pdf.
48. See IUCN Council Decisions C/74/18 and C/75 (2010) on the Biosphere Ethics Initiative and IUCN WCC Resolution 004 (2012), Establishment of the Ethics Mechanism, available at 'Ethics at IUCN'. *Center for Environmental Ethics and Law.* Accessed 6 January 2019. https://environmentalethicsandlaw.org/ethics-at-iucn.
49. Center for Environmental Ethics and Law. 'The *Relato* Methodology'. https://environmentalethicsandlaw.org/relato methodology. (Accessed 6 January 2019).
50. (a) Appiah, K. A. 2006. *Cosmopolitanism*. New York: W.W. Norton; (b) Appiah, K. A. 2007. *The Ethics of Identity.* Princeton: Princeton University Press.
51. Ibid.
52. Hegel, G. W. F. 1991. *Elements of the Philosophy of Right*. Edited by Allen W. Wood. Translated by H. B. Nisbet. Cambridge: Cambridge University Press.
53. Ibid.
54. Ibid.
55. Center for Environmental Ethics and Law. 'The Evolving Biosphere Ethic'. https://environmentalethicsandlaw.org/the-evolving-biosphere-ethic. (Accessed 6 January 2019).
56. Ibid.
57. Ibid.
58. Ibid.
59. Leopold, A. 1949. *A Sand County Almanac and Sketches Here and There.* Oxford: Oxford University Press.
60. Center for Environmental Ethics and Law. 'The Evolving Biosphere Ethic'. https://environmentalethicsandlaw.org/the-evolving-biosphere-ethic. (Accessed 6 January 2019).
61. Kant, I. 2009. *An Answer to the Question: What Is Enlightenment?* Translated by H. B. Nisbet. London: Penguin Books.
62. Ibid.
63. Wulf, A. 2015. *The Invention of Nature: The Adventures of Alexander von Humboldt, The Lost Hero of Science.* London: John Murray.
64. Sen, A. 2009. *The Idea of Justice*. Cambridge: The Belknap Press of Cambridge University Press.
65. Sachs, J. A. 2009. *The Strange Alchemy of Life and Love.* Oxford: Oxford University Press.
66. Meyer, M. 2002. *Reverence for Life: The Ethics of Albert Schweitzer for the Twenty-First Century.* Syracuse: Syracuse University Press.

67. Leopold, A. 1949. *A Sand County Almanac and Sketches Here and There*. Oxford: Oxford University Press.
68. Center for Environmental Ethics and Law. 'The Evolving Biosphere Ethic'. https://environmentalethicsandlaw.org/the-evolving-biosphere-ethic. (Accessed 6 January 2019).
69. Author unknown, often attributed to Plato but unverified. Inscription over the 10th Street entrance of the U.S. Department of Justice, Washington, D.C.
70. Nussbaum, M. C. 2013. *Political Emotions: Why Love Matters for Justice*. Cambridge: The Belknap Press of Harvard University Press.
71. Ibid.
72. Ibid.
73. Jennings, B. 2016. *Ecological Governance: Towards a New Social Contract with the Earth*. Morgantown: West Virginia University Press.
74. Ibid.
75. Chibba, S. 2013. '*Ubuntu* Is about Relationships'. Brand South Africa. September 19. Accessed 6 January 2019. https://www.brandsouthafrica.com/people-culture/people/ubuntu-is-about-relationships.
76. Center for Environmental Ethics and Law. 'The Evolving Biosphere Ethic'. https://environmentalethicsandlaw.org/the-evolving-biosphere-ethic. (Accessed 6 January 2019).
77. *Afri-Forum and Another vs. Malema and others*. 2011. 23 (The Equality Court of South Africa).
78. Sachs, J. A. 2009. *The Strange Alchemy of Life and Love*. Oxford: Oxford University Press.
79. Mandela cited Jawaharlal Nehru in his 1952 address to the African National Congress from Robben Island. 'Presidential Address by Nelson R. Mandela to the ANC (Transvaal) Congress 21 September 1953'. South Africa History Online. Accessed 6 January 2019. https://www.sahistory.org.za/archive/no-easy-walk-freedom-presidential-address-nelson-r-mandela-anc-transvaal-congress-21-septemb.
80. Kant, I. 2009. *An Answer to the Question: What is Enlightenment?* Translated by H. B. Nisbet. London: Penguin Books.
81. Center for Environmental Ethics and Law. 'The Evolving Biosphere Ethic'. https://environmentalethicsandlaw.org/the-evolving-biosphere-ethic. (Accessed 6 January 2019).
82. Global Alliance for the Rights of Nature. 'Universal Declaration of the Rights of Mother Earth'. www.therightsofnature.org/universal-declaration. (accessed 6 January 2019).
83. Center for Environmental Ethics and Law. 'The Evolving Biosphere Ethic'. https://environmentalethicsandlaw.org/the-evolving-biosphere-ethic. (Accessed 6 January 2019).
84. Hegel, G. W. F. 1991. *Elements of the Philosophy of Right*. Edited by Allen W. Wood. Translated by H. B. Nisbet. Cambridge: Cambridge University Press.
85. Nussbaum, M. C. 2013. *Political Emotions: Why Love Matters for Justice*. Cambridge: The Belknap Press of Harvard University Press.
86. Sen, A. 2009. *The Idea of Justice*. Cambridge: The Belknap Press of Cambridge University Press.
87. Ibid.
88. Ibid.
89. Ibid.
90. Kant, I. 2009. *An Answer to the Question: What Is Enlightenment?* Translated by H. B. Nisbet. London: Penguin Books.
91. Leopold, A. 1949. *A Sand County Almanac and Sketches Here and There*. Oxford: Oxford University Press.
92. Ibid.
93. Ibid.
94. Ibid.
95. Kant, I. 2009. *An Answer to the Question: What Is Enlightenment?* Translated by H. B. Nisbet. London: Penguin Books.
96. Sen, A. 2009. *The Idea of Justice*. Cambridge: The Belknap Press of Cambridge University Press.

97. Locke, J. 1980. *Second Treatise of Government*. Edited by C. B. Macpherson. Indianapolis: Hackett Publishing Company. 'a liberty ... not to be subject to the inconstant, uncertain, unknown, arbitrary will of another man: as freedom of nature is, to be under no other restraint but the law of nature'.
98. Ibid. Locke states, 'Slavery is the state of war continued, between a lawful conqueror and a captive'.
99. Ibid. Locke states, 'Despotical power is an absolute, arbitrary power one man has over another, to take away his life, whenever he pleases'.
100. Nussbaum, M. C. 2013. *Political Emotions: Why Love Matters for Justice*. Cambridge: The Belknap Press of Harvard University Press.
101. De Beauvoir, S. 2010. *The Second Sex*. Translated by Constance Borde and Sheila Malovany-Chevallier. New York: Alfred A. Knopf.
102. Jennings, B. 2016. *Ecological Governance: Towards a New Social Contract with the Earth*. Morgantown: West Virginia University Press.
103. Ibid.
104. Ibid.
105. Ibid.
106. Ibid.
107. Kant, I. 2009. *An Answer to the Question: What Is Enlightenment?* Translated by H. B. Nisbet. London: Penguin Books.
108. Nussbaum, M. C. 2013. *Political Emotions: Why Love Matters for Justice*. Cambridge: The Belknap Press of Harvard University Press.
109. Ibid.
110. Kant, I. 2009. *An Answer to the Question: What Is Enlightenment?* Translated by H. B. Nisbet. London: Penguin Books.
111. Nussbaum, M. C. 2013. *Political Emotions: Why Love Matters for Justice*. Cambridge: The Belknap Press of Harvard University Press.
112. Ibid.
113. Sen, A. 2009. *The Idea of Justice*. Cambridge: The Belknap Press of Cambridge University Press.
114. Foucault, M. 2002. *Power Vol. III, The Essential Works of Foucault 1954–1984*. Edited by James Faubion. London: Allen Lane Publishing.
115. Foucault, M. 2007. *The Politics of Truth*. Boston: MIT Press.
116. Foucault, M. 2004. *Society Must Be Defended: Lectures at the Collège de France, 1975–76*. London: Penguin.
117. Sen, A. 2009. *The Idea of Justice*. Cambridge: The Belknap Press of Cambridge University Press.
118. Ibid.
119. Gwiazdon, K. *State Global Responsibility for Environmental Crises: The Ethical and Legal Implications of a State's Failure to Protect Human Rights*. Forthcoming publication, edited by L. Westra, et al. For the June 2019, 27th gathering of the Global Ecological Integrity Group (available upon request).
120. Kant, I. 2009. *An Answer to the Question: What Is Enlightenment?* Translated by H. B. Nisbet. London: Penguin Books.
121. Ibid.
122. Tutu, A. D. 1999. *No Future Without Forgiveness*. New York: An Image Book, Doubleday.
123. Gwiazdon, K. *State Global Responsibility for Environmental Crises: The Ethical and Legal Implications of a State's Failure to Protect Human Rights*. Forthcoming publication, edited by L. Westra, et al. For the June 2019, 27th gathering of the Global Ecological Integrity Group (available upon request).
124. Hegel, G. W. F. 1991. *Elements of the Philosophy of Right*. Edited by Allen W. Wood. Translated by H. B. Nisbet. Cambridge: Cambridge University Press.
125. Leopold, A. 1949. *A Sand County Almanac and Sketches Here and There*. Oxford: Oxford University Press.
126. Sachs, J. A. 2009. *The Strange Alchemy of Life and Love*. Oxford: Oxford University Press.
127. Hegel, G. W. F. 1991. *Elements of the Philosophy of Right*. Edited by Allen W. Wood. Translated by H. B. Nisbet. Cambridge: Cambridge University Press.

128. Center for Environmental Ethics and Law. 'The Evolving Biosphere Ethic.' https://environmentalethicsandlaw.org/the-evolving-biosphere-ethic. (Accessed 6 January 2019).
129. Sachs, J. A. 2009. *The Strange Alchemy of Life and Love.* Oxford: Oxford University Press.
130. Ibid.
131. Ibid.
132. Ibid.
133. Ibid.
134. Ibid.
135. Sen, A. 2009. *The Idea of Justice.* Cambridge: The Belknap Press of Cambridge University Press.
136. Ibid.
137. Ibid.
138. Ibid.
139. Jefferson, P. T. 1999. *Jefferson: Political Writings.* Edited by J. Appleby and T. Ball. Cambridge: Cambridge University Press.
140. King Jr., M. L. 1963. *Letter from the Birmingham Jail.* Accessed 6 January 2019. Available at https://www.africa.upenn.edu/Articles_Gen/Letter_Birmingham.html.
141. U.S. Declaration of Independence. 1776. Accessed 6 January 2019. Available at https://www.archives.gov/founding-docs/declaration-transcript.
142. Lincoln, President Abraham. 1863. Gettysburg Address. Accessed 6 January 2019. Available at http://rmc.library.cornell.edu/gettysburg/good_cause/transcript.htm.
143. Nussbaum, M. C. 2013. *Political Emotions: Why Love Matters for Justice.* Cambridge: The Belknap Press of Harvard University Press.
144. Ibid.
145. Hegel, G. W. F. 1991. *Elements of the Philosophy of Right.* Edited by Allen W. Wood. Translated by H. B. Nisbet. Cambridge: Cambridge University Press.
146. Sachs, J. A. 2009. *The Strange Alchemy of Life and Love.* Oxford: Oxford University Press.
147. Ibid.
148. Sen, A. 2009. *The Idea of Justice.* Cambridge: The Belknap Press of Cambridge University Press.
149. Leopold, A. 1949. *A Sand County Almanac and Sketches Here and There.* Oxford: Oxford University Press.
150. Kant, I. 2009. *An Answer to the Question: What Is Enlightenment?* Translated by H. B. Nisbet. London: Penguin Books.

# 3

# *Rights of Nature: Myth, Films, Laws and the Future*

**Eugen Cadaru**

**CONTENTS**

Nature's Intelligence: A Continuous Presence in Popular Culture ..... 39
Nature's Intelligence Portrayed in Cinema ..... 40
First Stop: Animation ..... 41
Next Station: Commercial Films Designed for Older Children ..... 43
Further On: Commercial Films Intended for Adults ..... 45
Last Stop: Auteur Cinema ..... 48
The Impact of Animism on the Current Collective Mentality ..... 50
Animism in Law ..... 51
Frontiers of Tomorrow ..... 53
Conclusion ..... 54
Notes ..... 56

## Nature's Intelligence: A Continuous Presence in Popular Culture

Granting Nature rights to flourish, which can be legally defended in court, is an expanding new movement worldwide. The concept of granting rights to Nature presupposes that Nature – and parts of Nature, such as mountains, trees, coral reefs and entire ecosystems – is a being and has personhood. Entities are granted rights in human legal systems because they are beings. But in what way is Nature a being? Does Nature have intelligence, which humans can recognise and live with in relationship? There are no definitive answers yet from the halls of modern science. But human beings have lived with the knowledge of Nature's intelligence for millennia, knowledge reflected in their religious traditions, myths and stories. Modern Western culture is no different, but since officially Nature is considered to be just a bundle of resources for human use, the exploration of Nature's consciousness has taken another form: through literature and cinema. Here we will explore how contemporary cinema has, and continues to, explore and deepen our understanding of Nature's intelligence.

According to the Cambridge online dictionary, animism is 'the belief that all natural things, such as plants, animal, rocks and thunder have spirits and can influence human events'.[1] Even the official Western anthropocentric culture of the last two millennia has understood this perspective as the need of 'primitive or uneducated' people to explain and deal with the surrounding natural world, which appears both mysterious and dangerous. Not so long ago, this view of the universe was a predominant philosophy all over the world. In other words, over a long period of time, animism was an essential way of understanding the intelligence of Nature in a nonmaterialist framework.

The first instances of this belief in European culture can be found, most likely, in Celtic, Greek and Roman religious traditions. Greek spirit beings – such as nymphs, dryads,

DOI: 10.1201/9780429505959-3

hamadryades, nereids, oceanides, satyrs and Gaea – personalised trees, rivers, seas, other aspects of nature and planet Earth herself. Roman 'genius loci', such as Silvanus, Faunus and Flora, were followed during the European Middle Ages by other spiritual nature beings such as the Green Man, Puck, gnomes, leshi, nixie/neck, rusalka, vodyanoy and many others. These nature spirits or elementals could be found throughout Europe, among all nationalities from Scandinavia to the Balkans and from the Atlantic Ocean to the Ural Mountains.

Legends and tales about spirits of Nature can also be found in all non-European cultures. Tribal populations worldwide frequently revere nature deities as one of their main manifestations of spiritual activity, as did civilisations such as the Aztecs, Inca and Maya, and Middle Eastern and Asian cultures. Wherever human societies flourish, religious activity and tradition about nature deities has appeared.

Reverence for trees is a good example. It is said that *Siddhartha Gautama,* a former Prince from ancient Nepal who gave up his luxurious life in order to become an ascetic, in his continuous quest for the truth, gained illumination and became Buddha when he was meditating under a fig tree. Sacred trees are also common in Hinduism and Shinto and they are greatly revered in the belief that nature spirits inhabit them. In the Jewish tradition, trees are a key element in the Garden of Eden: the history of mankind begins from the Tree of the knowledge of good and evil. The ancient Greeks understood dryads and hamadryades to be spiritual beings that live in trees and forests; hamadryades in particular were bonded to a particular tree. Any harm done to a tree directly affected the hamadryad. Moreover, in pre-Christian Norse mythology, Yggdrasil is the World-tree that stands at the centre of the cosmos, upholding the universe and connecting the nine worlds to one another. In Christian iconography there is a famous image representing Jesus Christ as a *Tree of Life,* with the 12 Apostles as twigs growing from His main trunk. In Islamic culture, the motif of the *Tree of Immortality* can be found both in Quran and in Muslim art. Tribal cultures – such as African, Native American, Aboriginal Australian, Polynesian, Micronesian, Papuan and Maori – developed their own specific sacred traditions on the importance of trees.

To conclude, we notice that all over the world and through all times, trees and many other beings from rivers to animals have been conceived by popular, mythic and religious understanding as being connected with some kind of fundamental universal energy, as alive and intelligent. Does Nature have a spirit? Or are animals and plants intelligent? Is our Planet Earth a sensitive being? These questions are as old as human civilisation, but there has been no final and indisputable answer that satisfies contemporary scientific questioning. So the debate continues throughout the Western world. There are many comprehensive overviews of this fascinating global phenomenon.[2]

All in all, as we have already seen, cultural ideas pass through the ages, wearing as many 'clothes' (ways of expression) as humanity has invented, including oral tales and songs, drawings and paintings, stories and studies. Therefore, coming to the present, it seems appropriate to survey how the main art of the twentieth century – cinema – mirrored and depicted Nature intelligence. It sheds light on metaphysics, culture and the evolution of society.

## Nature's Intelligence Portrayed in Cinema

Even in the contemporary modern age, some scientific understanding[3] seems to confirm that Nature is alive and has conscious spirit, as folk traditions and many religions have always held. Starting with the first means of expression, such as oral tales, songs and simple drawings, then continuing with classical art and science, human cultures transported this thesaurus of ideas to the twentieth century, though often only as a backdrop. Then cinema, a new invented art, took over the concept of Nature's intelligence and brought it back to the mainstream, as it once was in humankind's ancient traditions. Cinema, in its attempt to offer new entertainment, took full advantage of humanity's treasury of ideas. Among these was the fascinating concept that Nature is intelligent, and Earth or/and its elements possess consciousness. At the intersection of this idea and the syncretism of film, some major art masterpieces appeared.

Beginning with a short survey of the animated movie industry, we will see how the expression of animism in designed-for-children films is shaping the future. At the next level we observe how Nature's intelligence is highlighted in commercial blockbusters like *Avatar, Star Wars, Dances with Wolves, Planet of the Apes* and *King Kong,* in which the main focus is on the emotion of amazement produced by the animist approach to reality. Metaphorically and subtly, similar concepts were introduced in more intellectual films as well, especially those in the so-called 'auteur cinema' genre, where the film director exercises strong subjective control over the work, as would the author of a novel. Auteur cinema is frequently used to describe film directors who bring uniquely recognisable styles or thematic focus to their productions. Such films include *Solaris, Twin Peaks* and *L'ours (The Bear)* which present Nature's intelligence as something quite ordinary, a common – but often neglected – feature of the surrounding universe.

Together, these films reflect the animistic idea in all its components. Some, such as *Solaris* and *Avatar,* make viewers familiar with the concept of Earth intelligence. Others explore animal consciousness, notably *Two Brothers, War Horse* and the *Life of Pi.* Yet others examine the sensitivity of plants, especially *A Monster Calls, Lord of the Rings* and *Pan's Labyrinth.* Finally, some films, such as *Princess Mononoke,* survey concepts of nature spirits.

In this analysis we will also see how some of these wonderful pieces of film art – such as *Stalker, Lord of the Rings, The Chronicles of Narnia* and *Jungle Book* – have their roots in equally celebrated works of literary art from different cultures, proving once more that popular and classic literatures are one of the most important vectors of animist ideas in the modern world.

## First Stop: Animation

One might think that discussing cartoons in this study is inappropriate, as animated films presumably contain naive ideas designed for children. Actually, the opposite is true: animated films and all their subtly presented ideas provide the unconscious setting for every adult who had access to animation during his/her childhood. For this reason, the films of the animation industry are crucially important for the evolution of ideas about Nature's intelligence and animal intelligence.

Possibly the most common depiction of animated films and television series for children portrays animals or other elements of the natural world possessing human abilities: speaking, reasoning, feeling deep emotions and revealing characteristics such as kindness or wickedness. It would be rather difficult to find an animation work without these elements. Through animation people are accustomed, from their earliest years, to the basic idea that all natural entities possess consciousness in the same way that people do. Even later on, when children are taught that such ideas are 'just for the kids' and are not true in the real world, this adult teaching only has a superficial effect, creating adulthood patterns for social behaviour. But it cannot change the basic deeply embedded notion from childhood: that elements of natural world have intelligence and that Nature spirits exist.

As the number of children's animation films is nearly endless, I mention only a few of the best known.

First, many films show pets whose adventures happen in an urban context, introducing the idea that our immediate animal partners are more intelligent than we are accustomed to considering. For example, the *Tom and Jerry* cartoon series, created in 1940 by William Hanna and Joseph Barbera for Metro-Goldwyn-Mayer, features a house cat and mouse, and other animal characters in endless slapstick but violent (though bloodless) comedy. The usual plot line focuses on Tom's efforts to capture Jerry the mouse, which he rarely does due to the mouse's cunning and cleverness. *The Aristocats*, an American animated romantic musical comedy produced in 1970 by Walt Disney Animation Studios, tells the story of an aristocratic mother cat and her three kittens, who live with a wealthy opera diva. She decides to give her fortune first to the cats, and then to her butler. The plot follows the adventures of the aristocratic cats and the alley cat who befriends them when the butler abandons them in the countryside in order to inherit the fortune more quickly. Finally, *The Secret Life of Pets*, a 2016 Illumination Entertainment computer-animated comedy film, chronicles the daily lives, tribulations and jockeying among of a group of pets living with their owners in a Manhattan apartment.

Then there are films that depict an entire world of deeply sensitive and rational beings far away in the wilderness. Perhaps the best known is *Bambi*, a 1942 Walt Disney Animation Studios production based on the book *Bambi, a Life in the Woods* by Felix Salten. This classic movie depicts the life of Bambi the fawn, who will one day become the Prince of the Forest like his father, guarding the woodland creatures from hunters. Bambi's mother is killed by hunters, and he himself must confront them before becoming the new Prince. *Jungle Book*, a 1967 Disney animated feature film based on the book of the same name by Rudyard Kipling, follows Mowgli, an abandoned human child raised by wolves in the jungles of India. His friends Bagheera the panther and Baloo the bear seek to return him to the human village for his own safety, before the arrival of Shere Khan, a man-eating Bengal tiger. But Mowgli does not want to leave the jungle, and has many adventures there before finally accepting life back in the village. Disney released a live-action remake in 1994, and an animated sequel in 2003. Another popular Disney animated film is *Brother Bear*, produced in 2003 by the Feature Animation Studio, followed by a sequel in 2006. This film explores the aftermath when Kenai, an Inuit boy, kills a bear in revenge for his brother's death in a battle he himself provoked. The Spirits change Kenai into a bear as punishment, and he must learn the meaning of brotherhood before being allowed to become human again.

Related to these are films portraying the concept that animals in all ages and all parts of the world were intelligent. *Ice Age*, an animated comedy drama produced in 2002 and released by 20th Century Fox, stars prehistoric animals during the ice age and their troubles as they begin migrating south to escape the increasingly harsh winters. Much of the plot explores the sabre-toothed tigers' desire for revenge against humans, who had killed half their pack, when they find a human baby. The movie was a box office success and was followed by four popular sequels. Similarly, *The Good Dinosaur*, a 2015 animated film by Pixar Animation Studios, features dinosaurs and humans in a world where dinosaurs never went extinct. An orphaned cave boy is adopted by dinosaurs, who protect him until finally they give him to a human family, so he can be with his own kind.

Classic animated films like *Alice in Wonderland,* created in 1951 by Walt Disney Animation Studios, and based on the book *Alice in Wonderland* by Lewis Carroll, or *The Wizard of Oz,* a 1982 Japanese movie (also released in the United States) directed by Fumihiko Takayama and based on the book *The Wonderful Wizard of Oz* by L. Frank Baum, try to convince us that, in parallel with the common world that we all perceive and feel, there is another (better?) one. In *The Wizard of Oz,* Dorothy, who lives on a farm in Kansas, is lifted by a tornado into the magical realm of Oz, and has many adventures there with sensitive, intelligent animals and other beings, ranging from the Cowardly Lion and the Hungry Tiger to the Scarecrow and the Tin Woodman. *Alice in Wonderland* follows the adventures of a girl who falls through a rabbit hole into a wondrous world of astonishing animals such as the Cheshire Cat, who vanishes until only his smile remains, the clever Mad Hatter and the wise Caterpillar. These movies, and the books that inspired them, beckon us to cross the border into Wonderland or the Land of Oz, where we will find a range of clever animals living in kingdoms where animals, humans and other kinds of beings share intelligence and even magical ability.

Finally, there is another class of animated stories that show how the roles between nonhumans and humans are interchangeable, presenting once more the idea that spirit/intelligence is the ultimate reality, which can take different material forms. Many of these are cinematic versions of classic fairy tales. For example, *The Little Mermaid,* a 1989 Walt Disney Animation Studios film, is based on the fairy tale with the same name by Hans Christian Andersen, which Disney followed up in 2018 with a 'live action' movie, both on the theme of a mermaid trapped among humans, and her trials living with and hoping to escape from, life among human beings. *Beauty and the Beast* is a 1991 Walt Disney Animation Studios film based on a classic French fairy tale by Jean-Marie Leprince de Beaumont. The Beast, a prince transformed into a monster as a punishment for arrogance, must learn to love Belle, the young woman he imprisons in his castle, and earn her love in return, or else the enchantress's curse will keep the Beast a monster forever. Finally, *Pinocchio* is a 1940 Walt Disney Animation Studios film based on the novel *The Adventures of Pinocchio* by Carlo Collodi. It is now considered one of the greatest animation films ever made. The plot explores the efforts of Pinocchio, a wooden puppet brought to life by the Blue Fairy, to learn how to become brave, truthful and unselfish, and thus become a real boy. His many adventures include being led astray by a fox and a cat, and rescuing his puppet-maker father from the belly of a giant whale. His conscience is Jiminy Cricket, a real cricket who helps Pinocchio through all his tribulations.

## Next Station: Commercial Films Designed for Older Children

Working as an extension of the animated movies industry, this category of films intended for older children reshape the main themes of Nature and animal intelligence, to make them entertaining for a more mature level of understanding.

Here again, we frequently find parallel worlds where animals speak and act exactly like humans, and gifted children are able to easily interact with them. A widely known example is *The Chronicles of Narnia* film series, consisting (so far) of three motion pictures between 2005 and 2010, produced by Walt Disney Pictures and based on a series of novels with the same name by C.S. Lewis. The films, like the novel, centre on the adventures of four siblings in the magical land of Narnia. Aslan, a wise lion who is the true king of Narnia, guides the children through many adventures and struggles against evil in Narnia.

Another recent and highly popular example is the *Harry Potter* series of eight motion pictures, produced between 2001 and 2011 by Warner Bros. Pictures, and based on a series of novels with the same name by J.K. Rowling. Five of the films are among the highest-grossing films of all time. The film series focuses on Harry Potter, a young magician-in-training at Hogwarts School of Witchcraft and Wizardry, and his many struggles against the evil magician Lord Voldemort, who wishes to destroy him. Harry and his classmates have constant interactions with mortal and magical animals and other magical beings, ranging from gnomes to a phoenix.

There are also films featuring mythological worlds similar to ours, but where trees and other nonhuman beings are almost completely anthropomorphised. Perhaps the best known of these is *The Lord of the Rings* series of three motion pictures produced between 2001 and 2003 by WingNut Films, and based on the fantasy novels with the same name by J.R.R. Tolkien. Tolkien's Middle Earth features wizards, sentient trees, talking eagles, beautiful elves, trolls, hobbits, dwarves and dragons, all highly intelligent and highly motivated in their fight against the evil of Sauron and his minions, who are destroying Middle Earth.

Baggins Residence, Hobbiton, New Zealand. (Photograph by Pseudopanax/Wikimedia https://commons.wikimedia.org/wiki/File:Baggins_residence_%27Bag_End%27.jpg.)

Another example is the *Guardians of the Galaxy* series, consisting of two motion pictures in 2014 and 2017 produced by Marvel Studios, and based on a comic book series with the same name published by Marvel Comics. These films include a genetically modified raccoon and a tree-like humanoid who sacrifices himself to save the group of companions who are the guardians of the galaxy. The hero, Peter Quill, turns out to be only half human, his father a member of an unknown species.

Films featuring mythological creatures who are perfectly intelligent and fit to enter into close friendship with humans find their rightful place under this section also. A good example is *The Water Horse-Legend of the Deep* of 2007, produced by Revolution Studios and based on a novel with the same name by Dick King-Smith. It tells the story of a lonely boy who nurtures a mysterious egg until it hatches. It turns out to be a Celtic water-horse, of which there can be only one in the world at a time, and which becomes the Loch Ness Monster, and remains the boy's friend through various trials.

Another film with similar features is *Pete's Dragon*, produced in 2016 by Walt Disney Pictures, based on an unpublished short story of the same name written by the famous Hollywood writer Seton I. Miller. In this story, Pete loses his mother and father in an accident, and is befriended by a dragon. After various adventures in which Pete and the dragon protect each other, they separate, because (as the dragon says firmly) Pete will always be in danger from other humans with a dragon around. Eventually, the dragon is reunited with his fellows, and Pete is adopted into a family.

Close to the end of this very minimal list of feature movies (picked up, again, from an entire ocean of similar titles), a special realistic film needs mention: *The Fox and the Child/Le renard et l'enfant* produced in 2007 by BonnePioche and Canal +. This feature was made by Luc Jacquet, the director who won the Oscar for Best Documentary Feature in 2005 with his *March of the Penguins*. The movie emphasises the sensitivity of the animal world (embodied by a fox) as a key element that could help humans understand the vastness and beauty of Nature. The film follows the adventures of a 10-year old girl who sees a fox in the eastern Jura Mountains of France. By her patient observation and interaction, the two become accustomed to one another, and the fox eventually lets her see its den, as well as helps her escape a wolf pack. At the end of the film, the fox, visiting the girl and feeling trapped in her house, escapes by breaking the glass as it jumps through the window. This leads the girl to realise she cannot keep a wild creature as a pet.

To conclude this section, we note that straddling the border between the designed-for-children-shows and those intended for adult entertainment is *The Muppet Show*, a unique television and cinema feature, that premiered as a television series between 1976–1981 in the United Kingdom and was also made into feature films in 1979 and 2011. In all these movies, the mixture between human and intelligent animal characters is so well staged that the difference between the two apparently separate categories of beings is almost erased. The series is a music hall-style song-and-dance variety show starring Kermit the Frog, other Muppeteers and a human guest star in episodes of slapstick comedy and humorous parody.

## Further On: Commercial Films Intended for Adults

In this category of feature movies that are accessible both to teenagers and adults, the artistic discourse tends to become more intellectual, the spectacular visions of Nature framed by serious inquiries on the relationship between humans and the natural environment.

First, several films focus on the relation between man and his closest biological relative: the ape. From the multitude of films treating this theme, we especially note the classic *Tarzan* franchise, a series of films based on the novel *Tarzan of the Apes* and its 23 sequels by Edgar Rice Burroughs. There have been many major Tarzan motion pictures between 1918 and 2016, beginning in the era of silent films and continuing unabated into movies with soundtracks, beginning in 1929. All the Tarzan films present the idea that the ape's sensitivity and intelligence are perfectly compatible with the human. The plot revolves around the newborn son of an earl and countess marooned in West Africa. Both die when the child is about a year old, and the apes adopt him in the area where he is found and raise him in ignorance of his human heritage. Eventually Tarzan finds his parents' cabin and undergoes various struggles in relation to discovering his human heritage and how (or whether) to become a part of Western civilisation.

The same idea can be found in another film, *Instinct*, produced in 1999 by Touchstone Pictures, and inspired by the novel *Ishmael* by Daniel Quinn. In this film, the plot focuses on an anthropologist, missing for years and discovered to be living with wild gorillas in Africa. Convicted of killing several park rangers, he is sent to prison. There a young psychiatrist eventually learns that the anthropologist had been accepted by a gorilla family, which he was seeking to defend when rangers began shooting them. Horrified by the violence of a guard in the prison, the anthropologist escapes and goes back to Africa. Both films suggest that life in the 'wild' society of manlike animals could be preferable to a life spent in a cruel human civilisation.

The clash between these two worlds is explored by the *King Kong* franchise, which ballooned into a series of eight motion pictures between 1933 and 2017, with at least one more on the horizon, as well as novels and theatre adaptations. American filmmaker Merian C. Cooper created the concept and the King Kong figure. The movies centre around Kong, a giant gorilla who lives on Skull Island in the Indian Ocean, alongside other gigantic animals such as dinosaurs. In the original motion picture, an American film crew captures Kong and transports him to New York to be exhibited. In the various plots of all the subsequent films – which include stories about Kong's partner and son, as well as confrontations and violence with humans both in the West and back on his home island – King Kong retains his fundamental character of semihuman intelligence and great physical strength. Central to the plot of all the movies is hostility between a sensitive beast and cruel, cynical humans.

This concept is taken to its logical conclusion in the *Planet of the Apes* series of films, consisting of nine motion pictures between 1968 and 2017 produced by 20th Century Fox, and based on the novel *La Planete des Singes* by Pierre Boulle. There was also a popular television franchise. The initial 1968 film was immensely popular and one of that year's top money-makers in North America.

The plot centres on an American astronaut who travels to a planet where apes rule primitive humans, and focuses on his clashes with the ape rulers and orang-utan scientists. The final, iconic scene shows the astronaut coming upon a ruined Statue of Liberty, and realising he is on Earth – not a strange planet. The movie introduced a strange and very disturbing hypothesis: a world dominated by apes that have became more intelligent than humans.

If *Tarzan* and *Instinct* gave us a hint of how an emotional connection between humans and anthropoid animals could work, other films explore the interaction between humans and other living creatures, domestic or wild. We will mention here only two such titles. The first one, *War Horse*, a 2011 film by DreamWorks Pictures, directed by Steven Spielberg and based on a novel of the same name by Michael Morpurgo, describes the level of mutual attachment that can be reached between a young man and his horse, even through the hellish events of World War I.

The second one, *Dances with Wolves*, was produced in 1990 by Majestic Film International, and based on a novel of the same name by Michael Blake. The plot revolves a Union Army officer at the time of the Civil War, who requests a posting to the American frontier so he can experience it before it is gone. Alone, he mans an Army outpost, befriends the Sioux people and a wolf. Marrying into the tribe, he abandons the fort and lives as a Sioux until captured by the Army as a deserter. Army soldiers kill the wolf when he tries to follow. The movie focuses in part on the way an animal sacrifices its life as a price paid to support his human friend. But more than this, the wolf is the 'messenger' of a pure wild Nature that expects the hero's arrival, being at the same time the embodiment of the call coming from the depths of the hero's heart.

Speaking of how nature communicates with the human spirit, we mention here two other movies, both a little on the dark side, though exploring ideals of Nature's wisdom. The first is *A Monster Calls*, a 2016 film created by Participant Media and based on a novel of the same name by Patrick Ness. A young boy, Conor, must accept his mother's terminal cancer. The Monster, a huge yew tree-like being, tells Conor three wisdom stories, while also obligating the boy to then tell his own nightmare of his mother's pending death. The Monster embodies the voice of wisdom, supporting the boy in wrestling with the most difficult moment of his life.

Another important movie explores how the Spirit of Nature can show many other faces. This is the 2006 dark fantasy film *Pan's Labyrinth* of 2006, written, directed and produced by Guillermo del Toro, in which the subtle and commonly imperceptible world manifests in ordinary life by the presence of an insect, a mandrake root or a faun. Ofelia's story takes place in Spain in 1944 in the early Franco period, and the movie fuses the 'real' world with the mythical world in an overgrown labyrinth, where the girl meets magical creatures, especially the half-human faun, who leads her through challenges in the labyrinth. The movie reveals that the king of the underworld built the labyrinth as a portal for the return of his dead daughter's spirit, which turns out to be Ofelia.

We now touch upon *Star Wars*, which consists of 11 motion pictures between 1977 and 2018, based on a concept created by George Lucas. The movies created many spin-offs of books, television shows, video games and theme parks. Together, these rank as the third highest-grossing media franchise of all time. The films themselves are considered the second-highest-grossing film series in history. *Star Wars* is set in a distant galaxy, where many alien species, often humanoid, co-exist, and space travel between planets is frequent. The series chronicles the rise and fall of different governments throughout the galaxy, which is bound together by 'the Force', an energy field created by all living beings. In these films many nonhuman creatures of high intelligence think, speak and behave like humans.

The centrepiece of contemporary commercial film art in which Nature's intelligence is presented under both the primitive and the scientific aegis is the blockbuster *Avatar*, released in 2009. It was written, directed and produced by James Cameron. The film was highly successful, becoming the highest-grossing film of all time, the first film to gross more than $2 billion, and the best-selling film of 2010 in the United States. Four sequels are planned. The plot explores a future time when humans, having depleted Earth's natural resources, are colonising Pandora, a habitable moon in the Alpha Centauri star system, in order to mine unobtanium, an important superconductor. The mining colony threatens the Na'vi, an indigenous humanlike species that lives in harmony with Nature and worships a mother goddess. The film explores the clashes, efforts at understanding and ultimate expulsion of humans (except for a favoured few) from Pandora.

The movie explores both sides of what the Spirit of Nature could mean. On one side, we are confronted with the view of the Na'vi natives of the Omaticaya clan, who speak with the spirits of the animals they hunt and kill, and also connect themselves with the soul of universe by entwining parts of their bodies with branches of the trees, especially the sacred Tree of Souls. On the other hand, we hear a human scientist (with a name that defines her character), Dr. Grace Augustine, explain how the roots of Pandora's giant trees are interconnected, generating a huge biological neural network which can be accessed by the local population in order to upload and download data from its structure. In this way, we are already primed with the explanation for the final scene in which the entire planet (we recognise here the ancient Gaea myth of a living planet), being called to take action against the human invaders, sends all her fighting resources – that is, all the wildlife – to stand against the aggression. In other words, the film leads us to understand how the 'primitive' Omaticaya approach to Nature is proved correct by advanced scientific research.

## Last Stop: Auteur Cinema

We begin investigating this class of films with a magnificent piece of art that, exactly like *Avatar*, focuses entirely on the topic of Nature's intelligence: *Princess Mononoke*. This 1997 Japanese animated fantasy film, written and directed by Hayao Miyazaki, though a commercial product, is deeply imprinted in all its components by its creator's personality. Though Miyazaki frequently introduced references to animism in most of his works, in this masterpiece he created an outstanding transcultural artistic discourse on cutting-edge ideas such as the spirit of Nature, as well as human responsibility with respect to the life of the planet, including animals and spirit beings. The plot revolves around Prince Ashitaka, who is wounded by a boar god as it seeks to harm his village. The prince must seek a cure in the western lands the deity came from, and there meets the Great Forest Spirit and many *kodama* (tree spirits). He intervenes in the struggle over Iron town, built by clear-cutting forests and focused on mining iron, which has led to conflict with local forest deities. After an epic struggle, Iron town is rebuilt along lines more harmonious with Nature. Those who see the film will have the opportunity to experience, besides the conflict between protectors of nature and those who irrationally exploit it, an astonishing representation of intelligent entities: huge white talking wolves, numerous *kodama* and the Great Forest Spirit, represented during the day by a deerlike animal and by night as a giant night-walker.

Grizzly Bear in Yellowstone National Park, USA. (Photograph by U.S. Fish and Wildlife Service.)

Another breathtaking picture that carries the unmistakable mark of its director is *The Bear/L'óurs*, a1988 film directed by Jean-Jacques Annaud, adapted from the novel *The Grizzly King* by James Oliver Curwood. This movie portrays one of the kindest and most sensitive approaches to the natural world in the history of cinema. It narrates the story of an orphan bear cub who struggles to survive, carrying his childish dreams throughout a wild mountain environment where bear hunters are looking for their prey. The film is narrated from the cub's point of view as he seeks to bond with a male grizzly after his mother is killed in a rockslide. Ultimately the grizzly decides not to kill one of the hunters when he is defenceless, and the hunters in turn decide not to kill the grizzly. The film thus provides an amazing lesson of humanity for viewers.

We should also note that Jean-Jacques Annaud made a similar film, *Two Brothers/Deux Freres*, in 2004, produced by Pathe. This movie follows the epic effort of two Cambodian tiger cubs in the 1920s who, being separated from childhood – one remaining in the wild until captured as a pet for a Khmer prince, and the other trapped as an exhibit in a circus – struggle to reunite with each other and find peace in the forest again, finally also reuniting with their mother.

Two other films merit a mention on this theme of Nature's intelligence. In *Chocolat*, a film produced in 2000 by Miramax Films, directed by Lasse Hallstrom, and based on the novel *Chocolat* by Joanne Harris, viewers are provided with a very subtle personification of the North Wind, who acts like a wise adviser for the heroine, leading her to a quiet French town, where she opens a chocolaterie and slowly inspires the villagers to experience more joy and emotional sharing. Then, in the *Life of Pi*, a 2012 feature by Fox 2000 Pictures, directed by Ang Lee, and based on the novel of the same name by Yann Martel, we witness an interchange of roles and position between humans and animals. The plot explores the consequences of an Indian family deciding to emigrate to Canada and there sell the animals of their zoo. The ship founders, and the son, Pi, is the only survivor, along with several of the zoo animals. The ending narrative twist leaves viewers with a strong suggestion that souls are the same, regardless of the material form they take from one moment to another.

Finally, let us look at three films that at first glance have little in common: (a) *Twin Peaks*, two television series in 1990 and 2017, and a film in 1992, directed by David Lynch, and created by Mark Frost and David Lynch; (b) *Stalker*, a 1979 Soviet art film directed by Andrei Tarkovsky, based on the novel *Roadside Picnic* by Boris & Arkady Strugatsky and (c) *Solaris*, 1972 a Soviet science fiction film also directed by Andrei Tarkovsky, based on the novel of the same name by Stanislaw Lem. Despite surface differences, all three films stage, in similar ways, primarily at the suggestive level, the *Gaea* myth that Earth is a living, intelligent being.

*Twin Peaks* revolves around the murder of a teenager in a fictional town, and the many unsettling and terrifying events that occur as a result, including possession, strange disappearances, and angelic visions. But the opening sequence from each episode of *Twin Peaks* shows a majestic waterfall, with fog floating over its waters or over the dark green forests, that seems to cover the world. Furthermore, there are also many images of serene natural beauty sprinkled through the film, leading viewers towards an intense feeling of nature's identity and personality as a being, rather than an object.

*Stalker* describes an expedition by the Stalker to take two clients to a mysteriously restricted and sentient 'Zone', which contains a room able to fulfil a person's innermost true desires – which may not be their conscious ones. The three travel through many uncanny areas filled with the debris of modern society to reach the Zone, but end up not entering the room. In *Stalker*, Nature is portrayed with full green intensity, and its seemingly endless paths transform it from an object into a being that guides the humans who enter its realm. In both films, the Gaea presence communicates with gifted humans in many different ways. For example, in *Twin Peaks* we see a woman receiving messages from a thinking log and in *Stalker* a mysterious big black dog guards the hero during its metaphysical sleep.

*Solaris* takes place aboard a space station orbiting the oceanic planet Solaris. The two remaining scientists of a three-man crew are all experiencing separate and mysterious emotional crises, visions and recreations of the dead, as does the psychologist sent to evaluate the unknown phenomenon that has affected the others. Ultimately the movie reveals that Solaris itself is causing the apparitions, showing how a planet can behave as an intelligent entity, confronting attackers (who test her by using destructive x-rays) with the worst shadows of their own consciences. The subtlety of the images transforms what ostensibly appears to be science into the philosophy of a sentient and deeply sensitive planet.

## The Impact of Animism on the Current Collective Mentality

The previous two sections explored how a very old, but very vigorous, idea – animism – flourished through the centuries and became a major theme in cinema, the most popular art form of the present. We surveyed a number of splendid films, most of them from the Western tradition, which presented an alternative perspective on reality from present-day materialism. These films explore the hypothesis that animals and plants could be rational and sensitive. They invite us to see how the world could look if seen through their eyes, and learn how communication between us and the other elements of nature could work. At the end of this marathon of movies (most of them based on literary works), any filmgoer could ask themselves at least two questions: Are we, the humans, the only holders of intelligence on this planet? Should we treat Nature as something other than a commodity to be exploited for our benefit?

During the past century, and continuing to the present, some scientific approaches explored and even confirmed, at least partially, the old theory asserted by animism. A simple search on the internet reveals many articles, some of them published in the most prestigious scientific journals or magazines, such as *National Geographic*, the BBC, *Scientific American* and *New Scientist*, to name a few. For example, some researchers studying plants from varying perspectives now maintain that plants can perceive, feel emotions of others, intuit, learn from experience, react to different stimuli, adapt their behaviour, give signals, communicate and coordinate between themselves as a community and defend themselves by chemical mechanisms in response to attacks. All of these abilities and many similar others drove part of the scientific community to the conclusion that, even though plants do not have a neuronal system and a brain, they still are able to analyse, calculate and solve problems. But does this mean they have conscience? The answer to that question could change the perspective of materialism, which holds plants to have only a low, vegetal life. A positive approach to this dilemma is given by a plant neurobiologist, Professor Stefano Mancuso, and a journalist, Alessandra Viola, in their book *Brilliant Green: the Surprising History and Science of Plant Intelligence.*[4]

As we might expect, many in the scientific community have not embraced this approach. But there are increasing numbers of books exploring similar themes, such as the well-known forester and author Peter Wohlleben's work *The Hidden Life of Trees*[5] or *The Intelligence of the Substance,*[6] written by Dumitru Constantin Dulcan, a renowned Romanian neurologist and psychiatrist specialising in the philosophy of science. According to Dulcan, the universe is caused and driven by an Intelligence that can be found at all levels, in each of its components. A similar idea (also known as *panpsychism*) is sustained by the theory of 'quantum Animism' of the American physicist Nick Herbert, according to whom mind permeates the world at every level and every natural system has an inner life, a conscious centre, from which it directs and observes its action.[7]

Whether such a vision could be confirmed by science remains to be seen. But irrespective of the answer, a new trend focusing on greater respect for Nature has appeared worldwide, especially in the last 20 years, and is continuously growing. It appears the world has been lifted into a new ecological conscience, as activities devoted to this new culture paradigm can be seen almost everywhere.

There are many articles and debates in mass media, both online and offline; there are also scores of specialised television documentary channels focusing on nature and wildlife. In addition, there are campaigns targeted to protect a wide variety of species, as well as many other conservation actions. Politicians, decision-makers, and opinion leaders like Leonardo DiCaprio and Prince Charles now frequently take strong positions in favour of ecological causes. Governments have advanced from political discussion and conferences to the signing of international treaties meant to maintain the natural balance of the environment. Governments worldwide have also designed grants and policies encouraging large-scale restoration projects and creating environmentally focused education materials. Many corporations have also contributed to social responsibility by investing financial resources in large eco-targeted projects.

Finally, we note how urban people all over the world continue to focus attention on Nature, trying to spend as much time as possible in natural areas. The phenomenon of 'forest bathing',[8] called *shinrin-yoku* in Japanese, which arose out of Japanese medical philosophy, is a good example. This practise simply requires people to wander through wooded areas, attuned with all senses to the forest's presence. City dwellers are also seeking to make urban areas as natural as possible by enlarging parks and other urban natural areas. There also appears to be a trend of people leaving high-tech cities to spend their lives in small rural communities surrounded by Nature.[9]

Thus, it certainly appears that a major change in the collective consciousness has already taken place with respect to the environment. We are witnessing a worldwide trend of people reorienting to Nature. However, it is difficult to ascertain whether this transformation occurred due to the ancient philosophy of animism renewing its popularity in modern times, or to the generalised concern that potential and looming ecological malfunction could endanger the continuation of human life on this planet.

## Animism in Law

All the above-mentioned examples of eco-activities clearly show that people are very concerned about how Nature should be treated. This new reality is undoubtedly a huge step away from the classic view that dominated Western culture during the previous centuries, according to which Nature was seen primarily as mere resources to be put to human use – or wasted if not used. But how comprehensive will this shift in mentality be? Will this change include a radical new vision with regard to the ontological status of Nature within the universe and the civilisation we created? Will animist theory, so often portrayed in films for children and adults, prevail? It is premature to provide a final answer, but it is possible to portray the progression of change. The best way to do this is to survey a crucial indicator of every civilisation: the law. Perhaps surprisingly, the concept of animism, or an animist philosophy, is already producing concrete consequences in several countries, both creating new laws and changing existing ones. This is a tremendous change in legal theory and, ultimately, in people's lives.

The first country to adopt animist philosophy into its laws was Switzerland (a country which belongs to the core of Western civilisation), whose Constitution[10] in Article 120(2) states: 'The Confederation shall legislate on the use of reproductive and genetic material from animals, plants, and other organisms. In doing so, it shall take into account the dignity of living beings as well as the safety of human beings, animals and the environment, and shall protect the genetic diversity of animal and plant species'. This language acknowledges that living beings – a term which should include animals, plants and other organisms – have a right to dignity. And we all know that only a 'person' can be granted the attribute of dignity. Nevertheless, what exactly this right means, and what obligations it places on human beings to respect the dignity of other beings, is not yet clear. Since promulgating the language in 2000, Switzerland has not enacted any primary or secondary legislation to clarify how this principle might work in practise.

In 2008, Ecuador became the first country in the world to clearly state, in article 10 of its new Constitution, 'Nature shall be the subject of rights'. Via this single phrase – stating for the first time that Nature has the same legal capacity as a human person does – a new horizon for human civilisation has emerged. For the first time, nonhuman entities became equal in law with humans, having legal standing to defend their rights. In addition, an entire section of the Constitution (Chapter 7, entitled 'Rights of Nature') enumerates the rights granted to Nature. Legal provisions recognise the rights of ecosystems to exist and flourish, give people the authority to petition on behalf of Nature and require the Government to protect Nature and remedy violations of these rights.[11]

The government of New Zealand took the next significant step in this worldwide change. Following extensive debate and negotiation with the indigenous Maori people, New Zealand in 2013 and 2017 granted legal rights to Te Urewera National Park and Whanganui River. Under the implementing legislation, these entities have all the rights, powers, duties and liabilities of a legal person, including the right to bring legal actions (via guardians) to protect their interests. Significant debates and Rights of Nature statutes, court rulings and legal arguments continue to expand worldwide, in Bolivia – which in 2010 enacted the first comprehensive Rights of Nature statute – the United States, Australia, Scotland, Venezuela, India, Nepal, Bhutan and Colombia, among others.[12]

Debate has also begun in the European Union institutions about adoption of a similar Directive. The project's philosophy argues as follows: human beings are a part of the natural environment and totally dependent on it for existence; experience shows that legal frameworks based on regulating human uses of Nature, where Nature is considered merely an object under the law, have proven unable to protect the natural environment and achieve sustainability; therefore, there must be a fundamental change in the perceived relationship between humankind and the rest of the natural world. This means the formal recognition of Nature as a living being, having intrinsic value in itself for which humans have a duty of care. The core of this nonanthropocentric relationship with Nature is securing the rights of the natural environment itself in human institutions. Since humans are entirely dependent upon Nature for all their needs, only such a respectful relationship, thoroughly implemented, has a chance to secure the survival of humans on Earth.

## Frontiers of Tomorrow

Over the centuries, people in many cultures preserved and transferred an animist philosophy, frequently embedded in religion, from one generation to the other, up to the present time. In contemporary Western culture, literature and cinema have reinvigorated it and returned it to the spotlight, to the point that some countries have mandated legal provisions granting rights to Nature – sometimes ecosystems, sometimes specific rivers, mountains or regions. The multistate European Union is debating a similar Directive. It is therefore legitimate to ask if we are witnessing a complex cultural phenomenon in which a fusion of ancient tradition and popular culture create a new frontier for the functioning of human civilisations. And if so, what might it look like?

Scrutinising the ideas put forth in recent cinema, as well as actions taken by nation-states to secure Nature's rights to flourish, there are many possible scenarios. There could be an International Charter of Nature's Rights. Humans could build a global eco-technological civilisation that equally respects human rights and Nature's rights. Advanced digital devices and intelligent robots might be used to monitor every tree and river in order to protect Nature from any human damage, with global eco-security deemed more important than the sovereignty of nation-states. For example, there could be a Global Environmental Agency that would act with the authority of an Environmental Planetary Law based on a Rights of Nature concept.

It is always risky to assert how the future will look, but it is clear that an economic system based on the ruthless exploitation of Nature will shortly face complete bankruptcy. It has already passed over the red line, entered the unsustainable zone and threatens worldwide ecosystemic collapse. Humans have no other planet to move to; therefore, humanity will have to treat the human right to a healthy environment as a vital need. In other words, being faced with an urgent and global risk, we will be obliged to provide an urgent and global answer. The question is whether limiting human consumption to the Earth's regenerative capacity would solve the predicament, or whether there needs to be an entirely new paradigm of relationship between humans and Nature. Creating a new paradigm would allow humans to enter relationship with a world that is complex, integrative and holistic. This would require humans to shift their relationship with Nature from an object to a partner, with a respectful Rights of Nature at the centre of culture and philosophy.

Currently, the culture of Nature's Rights is just beginning; thus, we should not expect spectacular results overnight, but should move ahead step by step. With the range of concepts that already discussed in mind, we could imagine the strategy for implementing Rights of Nature as a square. At each corner a different action would be needed.

In the first corner would be integrative research, encompassing many fields ranging from biology and physics to ecology, psychology, ethics and philosophy. Multidisciplinary institutes would perform in-depth analyses of natural processes, including a focus on their subtle aspects. These are almost entirely ignored by contemporary science, but well known to the people that explore these ancient and once-common sides of relationship with Nature. Such concepts (the geometry of Nature, and relationship with the conscious energy/elemental spirits of Nature, are found in shamanically oriented religions worldwide). Studies must focus on them – even though these themes are currently unpopular with scientific investigation – as one of the basic roles of science is to investigate all hypotheses, explore the nature of reality and expand the knowledge of it.

At the second corner would be public communication about Rights of Nature. Global companies focused on science popularisation, such as *National Geographic*, and *New Scientist*, could both inform the public about the results of the above-mentioned research and also inspire people to think more deeply about this subject and to take action.

The third corner would consist of education programmes targeted especially to young people. Partnerships for action on the environment need to be multiplied all over the world, encouraging children to connect with Nature, plant and adopt trees, take care of streams and explore the environment as a close friend that guarantees their survival.

The last corner of the square would consist of public policies enacted into laws, enforced and respected, to guarantee that no abuse of Nature occurs again. Human abuse of Nature, and the danger that arises from it, is especially well dramatised by the plight of the world's forests. Nowadays trees are cut, frequently unsustainably, for two main purposes: to provide more agricultural land or to manufacture more objects ranging from houses to furniture and paper. This occurs daily around the world, despite all scientific evidence describing the benefits trees provide humans: they provide the air we breathe, they are the home of many other species ranging from birds to insects, they create stable ecosystems, they support the soil and thus prevent landslides and nutrient depletion, as well as providing spiritual health benefits to humans. In addition, more advanced theories, supported by studies,[13] assert concepts that many people are aware of at intuitive level: trees are conscious beings.

After decades of debate, many of the world's nations seem to understand – at least at a theoretical level – the risks that come from excessive logging, and adopted legal provisions to curb it. Unfortunately, this has not ended the problem; often, national authorities are not able to guarantee the protection of forests, as recent mapping shows.[14] In some instances, the laws protecting forests are strong enough to succeed when economic interests put pressure on them; in other cases, public authorities are overwhelmed by illegal logging; and, upon occasion, corrupt officials even provide direct or indirect support for it.

A better paradigm to treating forests as goods managed by different entities, ranging from corporations to governments, would be to designate forests as rights-holders, whose existence is interlinked with human life. This approach, transposed into binding legal provisions, would greatly improve the status of forests around the world. Such a project should become the Rights of Nature movement's main target for the immediate future, a first step to provide rights to the entire biosphere.

## Conclusion

Nature's rights could become the most important aspect of human rights: protecting and living in relationship with Nature, guaranteeing her ability to flourish so that humans might also flourish. But there are even deeper reasons than techniques of survival for humans to return to greater focus on Nature. Humans are deeply attuned to Nature at every level of being, even in ways that seem too commonplace for notice. For example, high-technology cell phones frequently have pre-installed images of Nature. Why? According to recent research referred to in *New Scientist*,[15] contemplation of Nature does more than relax people. It triggers feelings of awe, which dissolve one's very sense of self and lessen inner suffering. Awe can change people's entire range of emotion and behaviour, leading to sensations of wonder, joy, humbleness, curiosity and tranquillity. This is the human response to the sublime: a growing sense of communion with one's surroundings, in which feelings of separateness dissolve and happiness increases. People begin to feel more ability to explore and be creative, to be compassionate and generous.

Cinema has explored many aspects of the human attachment to Nature, and the long-stifled (in the Western tradition) understanding of Nature's intelligence. Perhaps cinema can further explore the road back to a more respectful relationship. One might, for example, begin by viewing the 2009 film *The Road*, directed by John Hillcoat, portraying the struggles of a man and his son to stay alive in an ecologically collapsed, postapocalyptic United States peopled mainly by scattered refugees and roaming gangs seeking to survive in a devastated landscape. This film gives viewers a realistic scenario of how bleak the world could be after a disaster of human-made horrors. The movie was filmed in abandoned and decayed postindustrial landscapes, especially in the Pittsburgh, Pennsylvania (USA) region. That in itself should be cause for alarm to moviegoers, as well as cause to begin exploring alternatives to the massively disrespectful industrial economy.

Steel Mills and Staircase in Hazelwood, Pittsburgh, PA, in January, 1940. (Photo by Jack Delano. U.S. Prints and Photographs Division, Library of Congress.)

Following a film that portrays so much ugliness and bleak ecological despair, perhaps Rights of Nature will seem like a highly reasonable path to pursue, to finding our way back to a respectful relationship with Nature. Cinema, due to its ability to explore the imagination's horizons and to expand the frontiers of the possible, can help greatly in this essential transition.

People only act on what they can conceive of, and here cinema may continue to play an outsize role in exploring conflicts, options and solutions to a new world fashioned in a respectful relationship with Nature.

First, the cinema can fulfil an expanded role in helping children explore and assimilate concepts of a living Earth and maintaining respect for both the seen and unseen aspects of Nature. Films can and should go beyond the landscape of merely portraying intelligent and clever animals, and should explore the subtlety of surrounding world in ways children can approach easily. Films such as *Two Brothers* and *Princess Mononoke* have already succeeded in this goal, and hopefully other films will continue the trend.

Creative films for adults will be equally important; they can excel in exploring themes such as human interaction with Nature's elementals, Nature's unknown powers, and similar concepts. *Pan's Labyrinth* is a good example of one film that has already fulfilled this task. In addition, these films should enquire into more difficult topics, such as whether our current ultra-technological civilisation could make humans less arrogant, perhaps stepping into a new Ecological Age driven by an enhanced, ecologically oriented technology.

Of course, these are concepts for the future, most of them hovering now at the frontiers of the imagination, but cinema has the tools to explore possible pathways to accomplishing them. In addition, new documentaries, beyond disseminating important results of scientific research, could portray the lives of the actual ecological heroes, inspiring people to take action to protect their own future.

All in all, the rich treasury of cinematic ideas has full capacity to enhance for the future, as it already has in the past, needed discourse and debate over animist philosophy and Nature's intelligence. Animist philosophies lie at the heart of mythology, religion and spirituality, and can become the seed from which we turn back to a forgotten communion with Nature, and explore new frontiers of Nature's intelligence.

## Notes

1. *https://dictionary.cambridge.org/dictionary/english/animism* (Accessed November 2018).
2. A few examples: *Introduction to a Science of Mythology* (Jung, C.G. and Kerenyi, C.; Routledge and Kegan Paul, 1951); *Myth and Reality* (Eliade, M.; Harper & Row, 1963); *Myth and Meaning: Cracking the Code of Culture* (Levi-Strauss, C.; Penguin Random House 1995); *New Larousse Encyclopaedia of Mythology* (1987, Crescent Books), *A Dictionary of World Mythology* (1997, Oxford University Press), *The Ultimate Encyclopaedia of Mythology* (1999, Anness Publishing).
3. See, for example, the theory of the Mother tree developed by Professor Suzanne Simard, the works and vision of the American scientist David Bohm (1917–1992), the works and vision of the American scientist Fritjof Capra:

   Gorzelak, M., Asay, A.K., Pickles, B.J. and Simard, S.W. 2015. Inter-Plant Communication through Mycorrhizal Networks Mediates Complex Adaptive Behavior in Plant Communities. *AOB Plants* 7. https://www.researchgate.net/publication/276412552_Inter-plant_communication_through_mycorrhizal_networks_mediates_complex_adaptive_behaviour_in_plant_communities (Accessed 23 December 2018).

   Simard, S.W. 2018. Mycorrhizal Networks Facilitate Tree Communication, in *Memory and Learning in Plants*, pp. 191–213. Baluska, F., Gagliano, M. and Witzany, G, eds. Springer International Publishing.

   Bohm, D. 1980, 2002. *Wholeness and the Implicate Order*. Routledge.

   Capra, F. 1997. *The Web of Life*. Anchor.

   Capra, F. 2010. *The Tao of Physics: An Exploration of the Parallels between Modern Physics and Eastern Mysticism*. Shambala. 5th Edition.
4. Stefano, M. and Viola, A. 2015. Brilliant Green: the Surprising History and Science of Plant Intelligence. Island Press.

5. Peter, W. 2016. *The Hidden Life of Trees.* Greystone Books.
6. Dulcan, D. C. 2009. *Inteligența Materiei/The Intelligence of the Substance.* Eikon.
7. Herbert, N. 1985. *Quantum Reality.* Doubleday.
8. Qing, L. 2018. *Forest Bathing.* Penguin Random House.
9. https://www.telegraph.co.uk/property/uk/great-exodus-many-people-moving-cities-going/
10. https://www.admin.ch/opc/en/classified-compilation/19995395/index.html (Accessed November 2018).
11. Ecuadorian Constitution. (Title II, Chapter 7). [Ecuadoran] Constitution, full language: http://pdba.georgetown.edu/Constitutions/Ecuador/english08.html
12. La Follette, C. and Maser, C. 2017. *Sustainability and the Rights of Nature: An Introduction.* CRC Press, Boca Raton, FL.
13. (a) McFarlane, R. The Secrets of the Wood Wide Web at NewYorker.com (Accessed 10 January 2019); (b) Simard, S. and Wohleben, P. Intelligent Trees (www.intelligent-trees.com) (Accessed 10 January 2019).
14. World Bank's interactive map shows where deforestation is severe at www.geospatialworld.net (Accessed 10 January 2019).
15. Marchant, J. 2017. Awesome Awe. *New Scientist* (Issue 3136/29 July 2017).

# 4

# *Nature's Rights in Permaculture*

**W. D. Scott Pittman**

**CONTENTS**

Permaculture: The Beginnings ..... 59
Permaculture Ethics ..... 63
  Care of the Earth ..... 64
  Care of People ..... 66
  Set Limits to Consumption and Reproduction and Redistribute Surplus ..... 66
Principles of Nature ..... 66
  Observe and Interact ..... 67
  Catch and Store Energy ..... 68
  Obtain a Yield ..... 69
  Apply Self-Regulation and Accept Feedback ..... 69
  Use and Value Renewable Resources and Services ..... 69
  Produce No Waste ..... 69
  Design from Pattern to Details ..... 70
  Integrate Rather Than Segregate ..... 72
  Use Small and Slow Solutions ..... 72
  Use and Value Diversity ..... 72
  Use Edges and Value the Marginal ..... 73
  Creatively Use and Respond to Change ..... 73
Permaculture and the Rights of Nature ..... 73
Conclusion ..... 74
Notes ..... 75

## Permaculture: The Beginnings

Permaculture is a design system for human habitats that is based on the fundamental principles of Nature. The principles of Nature have evolved over millennia to support the life within a total ecosystem, thereby maintaining a sustainable community of the living things. Permaculture is also designed to maintain human resilience by mimicking Nature's model, and re-establish that humans are a part of Nature rather than owners of Nature.

For mankind to survive, we must understand how to follow the laws of Nature and to respect that Nature has the same rights of survival as its many complex parts. One of those parts, a small part, is humans, though our influence on all natural systems has been catastrophically huge. Just as humanity has evolved legal systems to protect its rights, we, in permaculture, are advocates of Nature having the same rights.

DOI: 10.1201/9780429505959-4

In order to clearly explain what 'designing like Nature' looks like, we teach about all of the systems involved in our living situation, such as architecture, horticulture, agriculture, water management and soil regeneration, which we call visible structures. Permaculture describes legal, financial and social systems as invisible structures. By comparing natural systems with human systems we are able to show that Nature is much better at long-term sustainability than humans are.

A cursory glance at the current industrial food system reveals an incredibly inefficient system of agriculture that requires a huge investment in chemical fertilisers and pesticides, as compared to the system evolved by a natural forest that requires none of these manufactured inputs and is able to produce more abundantly and efficiently. This is equally true of other human systems, including creating habitat, water use, waste generation, food storage, access, energy and community.

Our predecessors, the indigenous inhabitants of the many ecosystems of the planet, had a clear understanding of the laws of nature since their rules of behaviour took into account those natural laws. Swidden (slash-and-burn) agriculture is an historic system of moving a community, complete with agriculture and hunting, from one area to another depending on the amount of time needed for the ecosystem to recover from human use. In the Central American tropics it was, typically, 7 to 25 years minimum from leaving a swidden site before returning to that site. There has been little research to establish the time needed to recover full soil fertility and plant/animal recovery following this slash-and-burn technique. My experience in Equatorial Guinea is that once settled agriculture was introduced by the Spanish colonists, the villages stayed in place, since the indigenous population provided labour for the colonists, and the gardens moved further and further away from the gardeners. Some villagers walked a full day to reach their gardens, and then a whole day to return. This makes it very uneconomical to maintain the swidden tradition.

The laws of these indigenous societies were a direct reflection of the natural laws of their environment. Different tribal members were in charge of various roles in the community such as healers, hunters and gatherers, and these individuals communicated directly with nature spirits: for example, the shaman with plant spirits, the hunter with animal spirits. For thousands of years humanity lived in a direct relationship with the natural world around them, and evolved a set of sanctions (laws) reflecting their admiration for, and a belief in, direct kinship with all aspects of their world. The separation from Nature brought on by settled agriculture, and therefore settled humans, began to loosen the human connection to the natural world.

Permaculture derived a lot of its ethics and principles from a study of aboriginal societies. We have also taken building techniques, gardening tactics and a deep reverence for observing and mimicking Nature, including human society, from our indigenous forebears, and incorporated this knowledge into the Permaculture curriculum.

To this day it is the aboriginal people who principally carry the wisdom of living in cooperation with Nature. It should be no surprise that Jair Bolsonaro, the recently elected (2018) President of Brazil, is removing protection of tribal lands for exploitation of raw materials. The United States and Canada are similarly targeting tribal lands in the race for more fossil fuel mining and pipelines. In the book *Thy Will Be Done* by Gerard Colby and Charlotte Bennett, the authors relate the cynical collaboration of the missionaries of the Summer Language Institute and Standard Oil to convert the tribes of the Amazon to Christianity in order to exploit the oil reserves on their traditional lands.[1] Certainly this partnership knew that they had to overcome these tribal people's belief in Nature's Rights.

Permaculture was founded in Australia during the international foment of the 1960s. Founders Bill Mollison and David Holmgren were both involved in academia: Holmgren as a student at Tasmanian College of Advanced Design in Hobart, and Mollison as a lecturer and tutor at the University of Tasmania in Hobart. Both Mollison and Holmgren were students of ecology, as well as environmental activists.

The combination of deforestation and agricultural salting of the soil were the driving forces igniting development of a system of design that would improve both forestry and agriculture. Many concerned people in Tasmania had already given up on mass activist demonstrations. It was obvious the government was supportive of mainstream farmers and timber industries and that they were willing to wage a war of clubs, tear gas and water cannons to prove that support. This led activists to long meetings and gatherings looking for new, less confrontational ways to save forests and agricultural land. Ultimately, they arrived at the idea of creating alternative solutions. Bill Mollison came up with the portmanteau name 'permaculture', a contraction of the two words 'permanent' and 'agriculture'.

Looking at the problems of soil salting and clear-cutting forests from a more holistic vantage point, it became obvious these problems were conjoined. Salting resulted from poor irrigation practises in arid lands, and deforestation was responsible for flooding and soil loss higher in the watersheds, as well as in regions with higher water tables. Clearing the land of trees increased surface heat, which drove the capillary action that drew up water from the higher water table, and left the dissolved salt on the surface of the soil. Over time this salt accumulates in the soil profile and causes the soil to collapse. This increased salting leads to loss of more trees from poisoning, as well as desiccation of the rhyzosphere (root zone of plants). The rhyzosphere is extremely susceptible to salt, and it causes the tender feeder roots of plants to desiccate and die. The only solution to soil salting is to flood it with fresh water to dissolve the salt and flush it from the soil.

An example of extreme salting can be found in Sonora, Mexico, where arid land was irrigated and used to farm hybrid wheat varieties developed by Norman Borlaug (1914–2009). Borlaug is considered the initiator of the 'Green Revolution' that changed agriculture worldwide, by promoting an input-intensive combination of inorganic fertilisers, pesticides and hybrids or genetic crossbreeding to increase crop yields.[2] The heavy reliance on petroleum-based fertilisers increased salting and the further demise of soil microorganisms.

A considerable amount of salt is introduced to the soil by irrigating from saline ground water. Over time this salt causes soil collapse, as the pore space of soil cannot resist the desiccating effects of salt. Once the soil collapses it is virtually impossible to regain the pore space in the crumb structure of soil where many of the microorganisms live, and where much nutrient exchange happens. This is a common problem of dry-lands, and humans are creating more arid lands through the increasing practise of damming waterways and introducing irrigation canals.[3] Water is then no longer able to spread and hydrate the total landscape.

The problem of soil salting and deforestation has continued worldwide, to the present, in an attempt to satisfy industrial capitalism's insatiable need to create a profit from the abusive use of the Earth's resources. After more than 30 years of teaching and practising permaculture, I find the biggest roadblock to restoring true planetary sustainability is the incredible avarice found worldwide. This can only be justified if one perceives the Earth as feed-stock for an insatiable corporate regime. This was the problem we faced in the early days of Permaculture, and remains the primary problem to shifting monocropping and commodity agricultural practises to a long-term sustainable model.

More and more tropical forests worldwide are being cut to turn into cattle pastures and soy farms. The monocropping of oil palm, *Elaeis guineensis*, now covers over 127 million hectares of the Earth's surface and is used primarily as a gasoline additive, though also as a food additive and in chemical products.[4] Over time these concentrations of single-species crops remove all the nutrients from the soil and require water to be imported from rivers and streams in order to survive. The uplands, being stripped of water for irrigation, become less and less capable of maintaining the diversity of flora and fauna that completes its nutrient cycle.

It is obvious that humans are part of a whole system, but just a small part. Humans are ambulatory and self-reflecting, but that does not mean that we are masters of all we survey. Looking at the Earth's systems and trying to find the human place in that total ecology led to Permaculture morphing into 'permanent culture' rather than 'permanent agriculture'. This change, and the ethics of Permaculture, were serious objects of debate at the second and third International Permaculture Convergence, a biannual meeting alternating between developed and developing nations. The definition was 'officially' changed in 1988 by Bill Mollison, in his publication *Permaculture: A Designer's Manual*.[5] As it turns out, Permaculture thinkers needed to look far beyond the immediate problems of toxic agriculture and clear-cutting to find the human place in the system that sustains all living beings. Human ownership of anything is purely temporary; a magnificent redwood, for example, is much closer to immortality than are we.

Mollison had already spent from 1954 until 1963 in the Tasmanian forest as a wildlife observer for the Commonwealth Scientific and Industrial Research Organization (CSIRO), and had developed some of his design ideas based on the interactions he observed in the forest. Holmgren brought design skills and ecology background to the collaboration. Together they wrote *Permaculture One: A Perennial Agricultural System for Human Settlements*,[6] published in 1978, and the rest is history.

Mollison's work with the aboriginal cultures of Australia provided another perspective for looking at Nature's rights among aboriginal people worldwide. The Australian aborigines had lived for thousands of years with little impact on the biophysical integrity of the land they occupied. In most cases indigenous people did not see themselves as separate from Nature but rather a part of, and a product of, Nature.

Permaculture is based on the knowledge that Nature has rights. Mollison stated in his lectures that a tree should have equal rights as humans. His time in the Tasmanian forest convinced him that the ecosystems he was observing could be replicated in the human environment. Based on this, Mollison and Holmgren developed the Principles of Nature, and Holmgren in 2002 published *Permaculture: Principles and Pathways Beyond Sustainability*.[7] In this book, Holmgren integrated a quarter century of teaching the twelve simple design principles that lie at the heart of Permaculture. Holmgren first described the primary principles found in Nature that direct and organise its growth and functions. These then become the guiding principles for designing human homesteads and settlements. I will use the principles outlined by Holmgren to illustrate how both Nature and humans could share the Earth in nondestructive ways.

To convey what Permaculture is, I will use a small homestead as an example. One has to do some initial basic research on the legal structures governing property ownership and land use restrictions. It is also necessary to figure out the finances of land ownership and cost of building. These are discussed in the last section of the Permaculture design course. Then one must design where things should go to provide the best outcomes and the least disturbance. Here Permaculture teaches basic building techniques using the least need to buy materials. This is often mud brick (adobe), straw bale or cob structures powered by photovoltaics and heated by the sun. How humans feed themselves is always of concern, and Permaculture focuses on building soils, insect control and basic design of the garden space, which is organic and eliminates use of fertilisers or biocides. Permaculture also provides guidelines for raising livestock, rainwater harvesting and creating an Earth-friendly livelihood. Finally, Permaculture focuses on building and living in community, whether it is in a commune or any other kind of community.

Permaculture teachings cover everything a human should know about to live a regenerative lifestyle, available in a two-week certification course. Everything in the course arises directly from the principles and ethics of Permaculture, which in turn derive from observation and respect for Nature and natural processes.

## Permaculture Ethics

There are three ethics in Permaculture, and they are the touchstone of all design decisions. It should be clear that Permaculture is a 'design' system and not merely a gardening or agricultural system. Permaculture was the first discipline I encountered that had an ethical base, except for the Hippocratic Oath that guides the conduct of physicians. Ethics was what drew me to Permaculture.

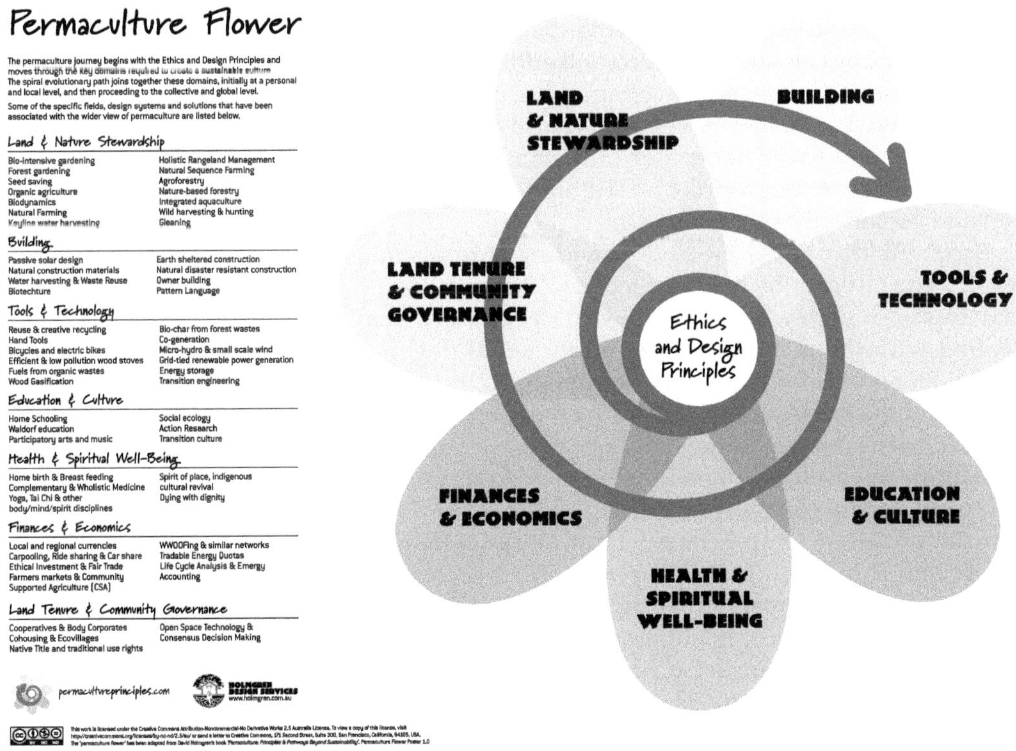


The Permaculture Flower shows the spiral flow through all the items in the design process, beginning with ethics and principles. (Graphic courtesy of PermaculturePrinciples.com.)

### Care of the Earth

This ethic encompasses the other two, which are listed to give emphasis and more specificity to the overall ethic. I had a lot of difficulty with this first ethic because it is such a sweeping statement, and humans know so little about how the Earth functions. Embracing this ethic could stop anyone from doing anything for fear of creating disaster out of ignorance. This is where the principle 'use small and slow solutions' provides guidance. The difficulty in these times is that humans have wreaked such destruction on the Earth it is hard to know if we are past the tipping point in creating permanent, ecological disequilibrium to ecosystems so they cannot support life in the manner that has allowed humans to flourish for millennia.

If humans are to care for the Earth, we have to totally restructure our cultural norms. The current economic system encourages pillage of the resources of the whole planet for both private profit and mass consumerism. There are no cultural/ethical brakes strong enough to stop a system run amuck. Our educational institutions were designed, and still operate as, a training ground to provide the producers and consumers to an industrial society, an era that is fast disappearing into history, as it cannot be sustained on a finite resource base.

In the discipline of Permaculture, we seek to understand how to return to a cooperative relationship with natural systems and how to shed unnatural lifestyles and return to a life that is regenerative and nonviolent towards Nature. Though we look to Nature for our answers, we find that there is less and less of Nature to learn from. It seems a little absurd that Ecuador and Bolivia are returning rights to Nature that were never surrendered, but were taken by force by Spaniards during the colonial era, as happened similarly in North America under English and French colonisation. The indigenous people around the world were and are greatly changed by colonisation, often severed from their own cultures that were based on a deep belief of care for Nature and living within Nature's limits. Much of the knowledge and social behaviour of Permaculture has come from our interaction with the native peoples of the world, who still understand that they are connected to Nature in a deep and profoundly spiritual way.

I like to think that it was the understanding of indigenous wisdom that led Arne Naess to develop the philosophy of Deep Ecology.[8] Naess coined the term 'deep ecology' in 1972, and contrasted it with what he called shallow ecology. Shallow ecology espouses a policy of dealing with the symptoms of environmental decline, rather than the causes of the decline. It focuses on the difference of humans in the environment, rather than humans as an element of the environment. Only by giving equality to all living beings with the same rights as humans will we understand our place in the natural world. The deep ecologists I know educate themselves by spending time in wild, natural places – much as an indigenous person would engage in a vision quest.

Mollison criticised Deep Ecology as 'all philosophy and no action'. I disagree with his criticism, and think that a philosophy of deep ecology runs throughout Permaculture. It was a profound understanding of Nature that led Naess to pinpoint the lack of true understanding of how deeply humans are affected by their self-imposed separation. I also was greatly moved by Dolores LaChapelle,[8] whose writings about Nature inspired my own exploration into Deep Ecology. Many in the Deep Ecology movement were teachers, as was Bill Mollison; Deep Ecology was first articulated in 1971, just a few years before the early Permaculture publications.

There was much action in the environmental movement during the 1960s and 1970s. But in spite of marches, Greenpeace's efforts to curb whaling via direct confrontation with whaling ships, and the more radical tree-spiking some activist groups undertook to prevent old-growth forests from being clear-cut, the environment was literally losing ground. One of the motivations for the development of Permaculture was to offer people incentives to stop the desecration of the natural world. The thought was that being *for* something, rather than merely against many environmentally destructive activities, we might gain some traction toward respecting forests, waterways, grasslands, oceans and each other while beginning the long process of restoring what has been lost.

Mollison started the not-for-profit Permaculture Institute of Australia and taught the first Permaculture certification class in the early eighties. In 1981 he won the international Right Livelihood Award, sometimes referred to as the alternative Nobel Prize, which is given to honour and support those offering answers to urgent contemporary challenges facing humanity, such as sustainable development and environmental protection. Mollison then took his teaching on the road. In the early 1990s Russia awarded him with the Vavilov Gold Medal for contributions to sustainable agriculture, as well as inducting him into the Russian Academy of Agricultural Science. There are now tens of thousands of graduates of Permaculture's intensive design course in Earth and People Care.

The General Sherman Tree in Sequoia National Park is some 2,000 years old. It is the largest known living single-stem tree on Earth. One has to be awestruck by the wisdom of Nature viewing this tree. (Photograph by Neal Parish/Wikimedia. https://commons.wikimedia.org/wiki/File:General_Sherman_2426497682.jpg)

### Care of People

The second Permaculture ethic is a special branch of Earth Care. This ethic focuses widely on social behaviour, education and governance. The goal is to create an environment that heals racism, sexism, poverty, fear and hunger and provide a step-by-step guide to that end. There are many techniques provided in the Permaculture literature for gardening, animal husbandry, farming, forestry and other activities that provide health, comfort and a caring community. But these techniques are a small part of Permaculture; the major thrust of Permaculture is teaching a way to look at the Earth and all it contains as a whole system, thus becoming an insider rather than an outsider of the planet's life-sustaining processes. The Permaculture principles gleaned from natural ecosystems are all an attempt to teach people how to interact with the Earth, and realise humans are an integral part of it.

### Set Limits to Consumption and Reproduction and Redistribute Surplus

The third ethic addresses the problem that humankind has developed an insatiable consumer appetite. It does not seem to matter so much what people purchase as what the purchase adds to one's status in society and among peers. A growing wave of narcissism is spreading worldwide, greatly enlarged by advertising that impinges on personal space through media that is ubiquitous and very compelling. The escalating consumptive hunger is largely responsible for the amount of waste that has stuffed landfill sites, oceans and, through incineration, the air. Food waste alone accounts for 33% of waste worldwide,[9] equivalent to 68 billion US dollars in the industrialised world. The current worldwide figure for total waste is 2.12 billion tons per year.[10] The alarming growth of the world's population has created an epidemic of consumerism and waste. This third ethic is sometimes called the 'recycle ethic' or the 'seventh generation ethic', meaning that if humans could limit consumption to that which can be cycled back into Care of the Earth and People societies would go a long way towards limiting the destructive activity that is looting Earth's natural resources and polluting our air, water and soils.

The ethics of Permaculture attempt to guide designs toward creating communities, homesteads and urban dwellings that encourage Care of the Earth, Care of People and a Reduction of Consumption and, through education, population reduction. In order to accomplish this, Permaculture looks to natural systems for guidance. The principles of Nature inform our choices.

## Principles of Nature

I have chosen the 12 principles outlined by David Holmgren in his book *Permaculture Principles and Pathways Beyond Sustainability*[11] to look at more closely. These 12 principles encompass the multitude of principles that Mollison lists, while attempting to consolidate both his and those of many other ecologists, particularly H.T. Odum and his brother E.C. Odum.[12]

Holmgren directs his principles towards the human designer, rather than as first appearing in Nature and then being adapted by humans. Without these original principles to guide us, humans have slowly lost the natural wisdom derived from the ancient participation in natural systems, though it has taken some 10,000 years (since settled agriculture) to replace this wisdom with the current materialism.

Derrick Jensen in his book *The Myth of Human Supremacy* describes a traditional, deeply troubling ideal of Nature's inherent order: 'One of the most harmful notions of Western Civilization – and one of the most foundational – is that of the Great Chain of Being, or Latin scala naturae (which literally means 'ladder or stairway of nature'), closely related to the divine right of kings. It is a hierarchy of perfection, with God at the top, then angels, then kings, then priests, then men, then women, then mammals, then birds, and so on, through plants, then precious gems, then other rocks, then sand'.[13] This is a profoundly body-hating notion, as, according to those who articulated the hierarchy, those at the top – the perfect – are pure spirit; and those at the bottom – the imperfect, the corrupt – are pure matter. Both men and women live in a battleground of spirit and body, with men tending to represent mind/spirit/better/perfected, and women tending to represent body/life/death/corruption/imperfection. In this construct, humans are the centre of attention, with those above humans being bodiless and perfected, and those below being fully embodied, imperfect and increasingly mindless.

Of course, within each category there are subcategories. So civilised man is more perfected than 'primitive' man, who is barely removed from animals. One sees this hierarchy everywhere within Western culture, except that as the West secularised, God and the angels were removed from the hierarchy, leaving civilised men at the top. In the Great Chain of Being, those further up in the hierarchy can exploit those below, as they have less sentience, or at least are less perfected.

The Great Chain of Being has long been used to rationalise whatever hierarchies those in power wish to rationalise. It is a very versatile tool. The Great Chain of Being also underlies the modern belief that the world consists of resources to be exploited by humans. Traditional indigenous peoples do not believe in this hierarchy; instead, they believe the world consists of other beings with whom we should enter into respectful relationship. This is one reason these other cultures have often been sustainable.

The Western perception of evolution is infected with this belief in the Great Chain of Being, as so often people, including scientists, think and write and act as though evolution aims to fashion more and more perfect creatures, leading eventually to that most perfect creature: the human being.

Holmgren tailored his principles to help secularised people return to living inside, rather than separate from, the web of life. Taken together, they restore the ability to participate in the relationships of the world with all its living things. By teaching these principles along with its ethics, Permaculture creates a map to return to a more humble, less hubristic, view of the world. Is soon becomes obvious that the Western world especially has been asleep in a web of hierarchy, and the consequences drive people to view the world as a human kingdom rather than a broader web of relationships. The Permaculture principles can lead people to a deeper understanding of Nature.

## Observe and Interact

I always encourage my students to spend at least a year observing and interacting with the proposed design environment before acting. Thankfully, there are good records of weather dating back many years. There are also sophisticated maps that accurately describe most terrain. The most critical instrument for the designer is his own sensorium of sight, smell, sound, taste and touch with which to observe the site. Once people have walked the site and accumulated all of the information the senses can provide, the process of design can begin. The deeper the observation, the more profound is the understanding of the site.

An initial site interaction should be devoid of all judgment, and should be akin to the inquiry of a 3- or 4-year-old discovering the world. The educated person is addicted to the gold star of perfection, so is always looking for reward at the end of the task rather than being open to information. This is the first practise of observing and interacting. Action will come after there is a deeper understanding of where and when the student is in the landscape. The process of walking and observing the land is the fundamental interaction. Mollison called it 'walking Zazen'. Zazen literally means a seated meditation. The goal of this meditation to just to observe without judgment, letting thoughts, words, ideas and images pass by without any attempt at understanding or getting involved in them.

Later one may write down these observations and thoughts and attempt to understand the language of the landscape that will guide the design. As a person walks through a forest, for example, they have to train themselves to seek the patterns that constitute all creation. The movement and pattern of streams from the mountain ridge to the estuary is a branching pattern that is replicated over and over in Nature: trees branch, our lungs and kidneys branch. The order of that branching is coherent with the amount of fluids or gasses being transported at each order of branching. Finding those natural patterns and understanding their function is critical to designing systems that function like a forest, a meadow or any other natural ecosystem. It is also the path to regeneration.

A critical part of this observation exercise is to notice the connections throughout the system. 'Whole system' thinking is crucial. It is, for example, much like the hunter who spots his prey not by looking for the prey, but looking for anomalies in the surrounding environment.

## Catch and Store Energy

This principle is very obvious; in natural systems, storing energy is the key to growth, health and regeneration. The sun is the overriding provider of energy to all life on Earth. Trees use this energy to create sugars, to move sap through their bodies and, through pheromones, to communicate with themselves and other species. A squirrel is the perfect representative of this principle; squirrels work diligently all winter gathering seeds, in the form of nuts, to carry them through the winter. Humans use the sun to generate electricity via solar panels, and use other energies such as the wind to pump water from wells, or water wheels to grind grain and many other tasks.

Permaculture frequently turns to trees to understand this principle most deeply. Trees catch the wind in their canopies, and the swaying back and forth of the trunk creates heat from the friction between epithelial cells in the wood. That same tree uses its leaves to convert the sun's rays to sugars, by combining carbon dioxide with water via its chlorophyll. Mollison used the tree form to describe many of the patterns in nature, such as branching. Another common pattern in Nature is annidation or stacking. Many species of plants and animals occupy the same space in a forest, from tubers and roots to grasses, fungi, forbs, shrubs, small trees, vining plants and the canopy trees, as well as emergent trees. All together there are up to nine layers of plants in a healthy Amazon primary rain forest, for example. This does not even include all the insects, reptiles, amphibians and mammals that also occupy that same space.

### Obtain a Yield

Without a yield, there is no possibility of sustainability. Everything in Nature obtains a yield; it may be fruit, leaves, growth, manure or any number of other things. But it is necessary to produce more than one consumes in order to continue living. This is a lesson that modern humans cannot seem to master. We are great at consuming, but terrible at saving; therefore, we have managed to reduce global forests by 80%, according to the Global Reforest Resources Assessment.[14] In 2016 alone humans cut down, burned or otherwise destroyed 73.4 million acres of forests, according to data from the University of Maryland.[15] If we look beyond trees we quickly realise that the loss in habitat, diversity, water retention, shade and rainfall is enormous. This is a primary reason we are seeing massive extinction at all levels of life.

Humans have reduced the wealth of the Earth in all areas, including mineral, soil, biota and atmosphere. In exchange, we have gained more and more glamour, gadgets and opportunities for consumption, at the peril of being unable to continue life.

### Apply Self-Regulation and Accept Feedback

Self-regulation is obvious in natural ecologies. It is extremely rare to find an obese deer; they eat only what they need, and if not they accept the feedback of being a meal for predators. Plants have developed various strategies to avoid inbreeding and crowding, and are self-regulating through evolutionary feedback. Many cultures worldwide are not currently embracing this principle, as the West especially has developed an industrial society that is no longer accepting self-regulation or feedback. The outcome is overpopulation, ecological destruction and rampant obesity, all of which are the outcome of ignoring feedback.

Illness is an excellent feedback mechanism. Most of the illnesses we suffer in an affluent social environment are the result of ignoring the feedback that our food is no longer providing nourishment. Diseases such as diabetes, high cholesterol, heart attacks, depression and others are the result of bad diet and social conditions fostered by industrial food and miserable work conditions.

### Use and Value Renewable Resources and Services

It is a miracle that rainforests can exist and thrive on soils that have had most of the nutrients washed out by inundation rainfall.[16] The whole forest depends on the constant falling of leaves, mosses, insect carapaces and manure. These renewable resources make the whole dynamic system work in the absence of much soil nutrient. This is a key metaphor for understanding this Principle.

Prior to World War 1, the majority of American farmers raised livestock that not only provided eggs, milk, meat and other goods but provided the manure that enhanced the soil's fertility for cropping. With the discovery of the process of removing nitrogen from natural gas, this all changed, and over time farmers switched to petrochemical fertilisers. Unfortunately, artificial inorganic fertilisers don't provide the necessary nutrients for maintaining soil health, and in many regions worldwide there has been a rapid deterioration of soil microorganisms, leading to soil infertility, salt accumulation, soil chemistry imbalances and erosion.[17]

### Produce No Waste

If one truly observes how natural systems operate, it is clear there is no waste. Everything in a forest, meadow or steppe is utilised by some segment of the ecosystem. Not only is there no waste, but the waste yield from the ecosystem impedes the overland flow of water and topsoil from the landscape, further enriching and maintaining its future productivity.

As noted above, the human landscape is awash in wasted by-products; oceans, beaches and waterways littered with plastic and other trash attest to the profligate use of raw materials. In the not-too-distant past, containers, lids and other objects were made of durable materials, which were returned to the manufacturer or reused at home.

The idea of composting has seen a resurgence in many cities and towns. Compost, generally speaking, is a mixture of decayed or decaying organic material used to fertilise soil. It can include things such as leaves, grass clippings, vegetables and other food scraps, which are placed together in a bin and decomposed via the action of aerobic and anaerobic bacteria, fungi and other organisms. Composting is one of many ways to recycle our waste stream into a valuable additive to our soil and plants. Some people use worms to recycle kitchen scraps, by building a vermiculture habitat and putting the scraps in that container to be turned from food waste to fertiliser by the worms.

Virtually everything that Nature produces is recycled within a short period of time, and it is recycled into the living tissue of its inhabitants. Humans, on the other hand, have developed materials that do not recycle and, to the contrary, cause harm to their environment, or if they are recyclable they are incinerated or buried in landfills polluting our ground water or the air we breathe. We have littered our water, our soil, our atmosphere and many hapless creatures with litter that is either physically harmful like plastic, or poisonous to living things.

## Design from Pattern to Details

The patterns found in Nature give us an enormous amount of information on how to place things in the landscape so that they function like an ecosystem. Teasing out the patterns and their interactions in natural systems is a lifelong study that only gets richer as one looks deeper. One of the most ubiquitous megapatterns in Nature is the spiral.

The Ekman spiral is a phenomenon of wind and oceanic flow and, for the most part, an invisible pattern. Tree windbreaks that were planted to protect crops during the Great Depression utilised the Ekman spiral form to throw the wind over the cultivated crop so that, as a secondary benefit, the field received much of the particulate matter that is carried in the wind. The wind is forced up and over the trees, and as it comes back to the ground and slows, the soil and other material carried in the wind falls out. A 20-foot-high windbreak will shift the wind up to 200 feet downwind.

It is easy to observe spirals at night in the Milky Way and other constellations, and in the hurricane and tornado. The arrangement of leaves on a plant stem as well as the phylotaxy of the branches themselves shows a spiral pattern, often in the Fibonacci sequence (a numerical sequence in which each number is the sum of the previous two, on to infinity). Rivers create spirals by meandering, and are thereby oxygenated. Meanders throw the water from one bank to the other. The water flowing along the bank is thus slowed by friction, and creates a velocity differential in the river, causing it to turn over (spiral) as it flows downstream. This allows a greater percentage of the water to interface with the oxygen-filled surface atmosphere; meanwhile, higher in the terrain, the faster-flowing streams form plunge pools that create a more oxygen-rich environment and habitat for oxygen-loving fish such as trout. The downstream rivers have less oxygen available, so the fish that evolved in this environment are slower and larger, like catfish and carp.

There are also the Von Kármán vortices (long linear chains of spiral eddies that form when fluid flow is disturbed by an object) that are created by waters, winds or other fluids flowing around islands or other solid objects. Downstream of the obstruction, two large spiralling arms appear in a pattern that recurs several times, and then dissipates.

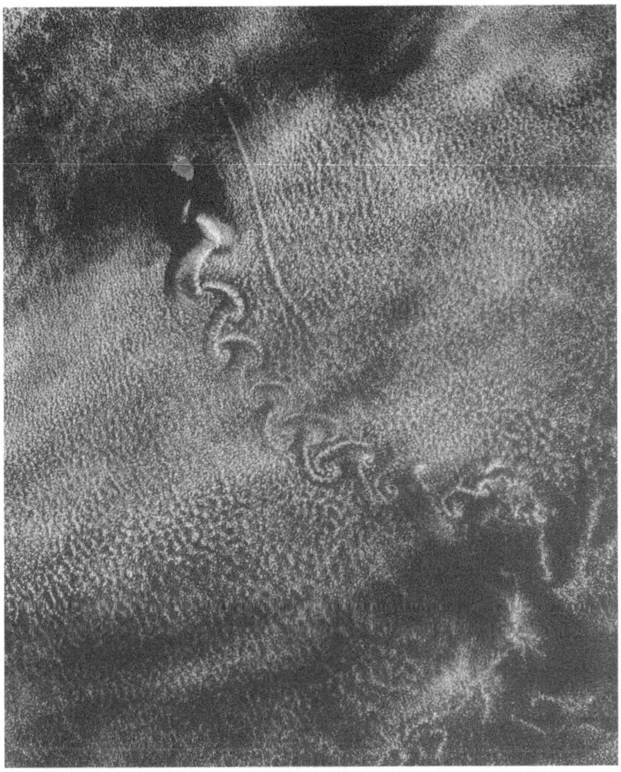

Von Kármán vortices in clouds downwind of Isla Socorro. (Photograph by Jeff Schmaltz, U.S. National Aeronautics and Space Administration [NASA] on August 19, 2013. https://earthobservatory.nasa.gov/images/81883/a-swirl-of-clouds-over-the-pacific)

The many temporal patterns, such as the diurnal and seasonal changes, shape local ecosystems. Rainfall and snow patterns create vegetative patterns in forests and meadows. The 'scatter pattern' can be viewed from the air when flying over dry lands where resources are sparse; there the vegetation is scattered according to availability of limited resources.

There is a surprisingly limited number of these natural patterns, but they shape a very dynamic, yet balanced and resilient, Nature that has evolved to sustain this planet. We, as humans, can replicate those patterns to create a more regenerative future. Christopher Alexander and colleagues in 1977 wrote a groundbreaking book, *A Pattern Language*, which examines human settlement and urban patterns of human activity.[18] This classic tome from the Center for Environmental Structure provides a practical language and virtual toolkit for planning and building based on the patterns of Nature, ranging from designing a single room to an entire community. It comprehensively examines how we can create space to allow for sensible transportation, housing, industry and other needs based on the nature of things rather than artificial design constructs.

## Integrate Rather Than Segregate

One of the outstanding patterns in nature is the integration of multiple species to provide for the needs of all. This is the opposite of the human pattern of segregation of plants to maximise profits, as in monocropping. In the contemporary industrial agriculture paradigm, we spend an inordinate amount of time and labour eliminating any encroaching plant or animal.

Permaculture has followed the wisdom of J. Russell Smith, who wrote *Tree Crops: A Permanent Agriculture*.[19] This pioneering book surveys the best ways to use trees for food, soil conservation and sustainable agriculture. The model used by Permaculture is called 'Food Forestry'. Food forestry is an attempt to replicate forest architecture using food, fibre and nutrient-producing plants to create a perennial agriculture. This is combined with developing plant guilds that place plants together to provide protection from predation and to interact with each other to provide fertility, rather than using petrochemical fertilisers. This template for urban ecological gardening is a growing methodology that is being adopted by city parks and home gardeners in urban neighbourhoods. The late Toby Hemenway describes this in his books *Gaia's Garden* and *The Permaculture City*.[20]

## Use Small and Slow Solutions

A focus on using small and slow solutions is an attempt to temper the urge for creating human habitat and community by making the houses, needed objects and travel faster and bigger – thus destroying the very community the designers attempt to create. My attempt to get clients to spend, at a minimum, four seasons to avoid making any big mistakes that will have to be redone or replaced is a prime example of this principle.

A leisurely walk in the woods tells us all we need to know about incremental growth over time. Unfortunately, humans now have access to the wonders of fuel and equipment to multiply their personal energy, which leads to the temptation to build a house that takes decades of work to pay for and enjoy. Rather than spending life energy paying off a 30-year mortgage, we could slowly create our own living space, and it undoubtedly would be smaller than what the mortgage would have provided. The tiny house movement is an admirable attempt to build human habitat using less materials and using our own labour. Some of the houses are mobile, so they are easily moved as work or adventure calls for.

## Use and Value Diversity

The Earth is losing some 150–200 species per day of plants, insects, birds and mammals, as a result of human activities.[21] This is many times the 'natural' or 'background' extinction rate! Daily there are further alarming reports from around the world of excessive loss of many kinds of insects, including pollinating insects such as honeybees.[22] We are additionally seeing a loss of diversity in food plants that has been caused and then accelerated by industrial agriculture, which selects for seed varieties that produce poundage rather than quality. There is also a trend toward corporate ownership of seeds and seed patenting, both of which shrink the genetic diversity of critically important food plants worldwide.[23] Natural forests constantly diversify to meet new environmental stressors or natural disasters, but managed plantations do not. In diversity there is insurance for a sustainable future, but in a monoculture there is little resilience with which to weather disaster.

### Use Edges and Value the Marginal

Where two different ecosystems meet, such as forest and meadow, one finds the highest diversity, because both the species of the forest and of the meadow dwell at this edge, as well as the unique edge species that prefer the edge environment. Raptors prefer this edge where the hunting and visual access provide both food and protection. Many human settlements are also placed at the meeting (edge) of ocean and shore, or forest and river.

Fences, pathways and other human constructs create edge and thereby opportunity for more diversity and higher productivity. Agronomists first noticed the 'edge effect' as they were trying to predict the yield of a particular field. They realised that the rates they predicted did not match the actual yield. After some time they came to the realisation that if one measured at the edge of the field, the yields were much higher than if they measured in the middle of the field. Once this was taken into account their predictions were much more likely to match reality.[24]

Whenever practical I include edge in my designs, whether it is city lots or thousand-acre properties. By doing this I can increase both diversity and yield, with a much healthier environment as a bonus. The edge effect does not just occur on the horizontal plane, but also the vertical. Where the soil and water meet the atmosphere is where most life occurs. This is the realm of the fungi that break down organic material, turning it into richer soil and mushrooms.[25]

### Creatively Use and Respond to Change

Humans are the major change-makers throughout this world, without much creativity, and our response to unwanted change is often wrongheaded. As a result, we are facing massive problems via climate change, and seem like deer frozen in oncoming headlights in our response. Natural systems are often slow in response to change, and their adjustment is based on evolutionary processes. For example, in the Rocky Mountains the response to an avalanche that scours the hillsides of vegetation is often the arrival of the aspen grove to re-establish stability of the slope and provide fertility for future change, as if in anticipation.

Many of the changes in human life are social or occupational, very often seen as crises rather than an opportunity to try new and different ways of being. In my case it was grasshoppers that suddenly appeared in my field and my garden, and were in the process of destroying my plants. I decided that I would add Guinea fowl to my menagerie as grasshopper predators. They are fierce predators, and don't scratch up plants like chickens do; with their help, I managed to save my garden.

## Permaculture and the Rights of Nature

This paper has provided only a very truncated and basic description of the principles that guide our actions in the natural landscape as well as the human-made landscape. But, hopefully, this summary gives a sense of how embedded Permaculture is in natural systems, focusing on mimicking Nature in our own environment. Nature should have its rights to flourish defended in human legal systems. Humans have evolved as part of a living system that itself evolved over millions of years to establish equilibrium among all its elements. For example, trees have developed a way to support themselves by forming alliances with other species. When one looks at a forest – and really observes it – one sees a community of individuals that cooperates for the benefit of the whole system.

The mycelial network that reveals itself through its fruit, the mushroom, creates a nutrition-sharing field that connects almost everything in the forest.[26] This field is capable of communication with other species and operating as an emergency store of water and nutrient in times of drought. With its mushroom fruit, the network lures animals whose presence enriches the soil while they burrow and divot the soil searching for fruit. They provide opportunity for water retention, and help spread the mushroom spores they have gathered on their fur throughout the forest. The vegetative world gives back, through their roots, the sugars that they have created out of sunlight via the chlorophyll in their leaves. The leaf fall is processed over time by the mycelium as food and is returned to the soil as nutrient. Both the mycelium and the root structure of plants provide habitat for the teeming microbiological life that also contributes back in the form of elemental nutrients. There is no waste in this system, and any competition is usually caused by human or natural disturbance.

These living natural systems are capable of teaching humans much about cooperation and living in community, and they are worth far more than their value as commodities. It is comparable to burning books in a library to heat the rooms. The commodification of natural resources is not only creating species loss and diminishing the resilience of the natural world, but also destroying the human knowledge base. I also think that we are impoverishing the human spirit. Permaculture has a zoning system to guide practitioners where to place elements in a system. There are five zones, based on the amount of time one spends in each zone during a year. Zone one is the household and the immediate area surrounding that is visited daily or more often. Zone five is left wild, free of human tinkering. This wild zone is the library of understanding how Nature works and what it can teach those who observe and interact.

Not only have we become estranged from Nature, we have become estranged from each other, which has led to all manner of pathological expressions. There is lingering racism, a heritage from slavery; sexism and misogyny left over from another form of slavery; constant warfare; competition for resources; lack of social programmes for the less fortunate and an incredible misallocation of wealth, especially in the United States, where 1% of the population owns more wealth than the bottom 90%. Looking worldwide, the richest 85 people own as much wealth as the entire bottom half of humanity.[27]

For this reason, Permaculture chose to model its education and efforts on a system that is more egalitarian and cooperative. The model is wild Nature. Permaculture drew all of its principles from observations of Nature. First and foremost Nature is cooperative; every living thing in Nature cooperates and maintains its domain sustainably. The result of this interlocking system has been thousands of years without the kind of outbreaks and traumas seen recently in the human community.

Ultimately, humans are going to have to learn to live more cooperatively in a much simpler lifestyle, inside the relationship with Nature, or we will create a world that is forced into competition and violence to survive. The lesson seems to be to work towards less materialism and narcissism, encapsulated in the saying 'small is beautiful'.[28]

## Conclusion

I have been watching the leaves fall from our fruit trees these past two weeks as ambient temperature drops, and marvel at the slow but steady excision of leaves as the plants send out the hormone signal that it is time to hibernate for another winter season. There is a beautiful carpet of golden leaves beneath a large apricot that lies like a magic carpet at the foot of the tree. There is enormous hubris involved in the question of whether Nature should be granted rights in human legal systems. For too long, the human attitude towards Nature has been one primarily of use and abuse. Since humans are the only ones who can convey these abrogated rights of Nature and enforce them against our own depredations, we need to work hard to free the whole planet from our modern tendency to treat Nature as a storeroom of commodities and nothing more. Permaculture, based as it is on the inherent patterns that create and maintain Nature's resilience, can help humans live more collaboratively inside Nature's constraints, and fashion a lifeway in which Nature's right to flourish is both a given and a ground from which the community springs.

## Notes


1. Colby, G. and Dennett, C. 1995. *Thy Will Be Done: The Conquest of the Amazon; Nelson Rockefeller and Evangelism in the Age of Oil*, Harper Collins.
2. Shiva, V. 1991. *The Violence of the Green Revolution: Third World Agriculture, Ecology and Politics*, Zed Books.
3. United Nations Interagency Taskforce Decade for Deserts and the Fight Against Desertification. *Awareness, Raising Communication and Education Unit (ARCE)*, various reports and documents. United Nations Convention to Combat Desertification (UNCCD) Secretariat, Bonn, Germany. http://www.un.org/en/events/desertification_decade/contact.shtml (accessed 22 February 2019).
4. Rainforest Rescue. 'Palm Oil: Deforestation for Everyday Products'. https://www.rainforest-rescue.org/topics/palm-oil#start (accessed 25 February 2019).
5. Mollison, B. 1988. *Permaculture – A Designer's Manual*, Tagari Publications.
6. Holmgren, D. and Mollison, B. 1978. *Permaculture One: A Perennial Agricultural System for Human Settlement*. Tagari Publications.
7. Holmgren, D. 2002. *Permaculture – Principles and Pathways Beyond Sustainability*. Holmgren Design Services.
8. Naess, A. and Drengson, A. 2008. *The Ecology of Wisdom*. Counterpoint.
9. LaChapelle, D. 1978. *Earth Wisdom*. Guild of Tutors Press.
10. Steff, K. 2017. 'The World Is Drowning in Ever-Growing Mounds of Garbage'. The Washington Post, November 21, 2017. https://www.washingtonpost.com/world/africa/the-world-is-drowning-in-ever-growing-mounds-of-garbage/2017/11/21/cf22e4bd-17a4-473c-89f8-873d48f968cd_story.html?hpid=hp_rhp-top-table-high_lagos-globalwaste%3Ahomepage%2Fstory&utm_term=.622c9e808fbf (accessed 22 February 2019).
11. Holmgren, D. 2002. *Permaculture Principles and Pathways beyond Sustainability*. Holmgren Design Services.
12. Odum, E. P. and Barrett, G. W. 2004. *Fundamentals of Ecology*, 5th edition. Cengage Learning.
13. Jensen, D. 2016. *The Myth of Human Supremacy*, Seven Stories Press.
14. McDickers, K., Jonsson, O., Pena, L., Marklund, L. et al. 2015. *Global Forest Resource Assessment*, 2nd edition. United National Food and Agriculture Organization.
15. Plumer, B. 2018. 'Tropical Forests Suffered Near-Record Tree Losses in 2017'. *New York Times*, 27 June 2018. https://www.nytimes.com/2018/06/27/climate/tropical-trees-deforestation.html (accessed 22 February 2019).
16. Kawa, N. C. 2016. *Amazonia in the Anthropocene: People, Soils, Plants, Forests*. University of Texas Press.

17. (a) Gilani, N. 2019. 'The Effects of Synthetic Fertilizers'. SFGate Home Guides. https://homeguides.sfgate.com/effects-synthetic-fertilizers-45466.html (accessed 25 February 2019); (b) Thompson, D. 2019. 'The Disadvantages of Inorganic Fertilizer'. SFGate Home Guides. https://homeguides.sfgate.com/disadvantages-inorganic-fertilizer-64756.html (accessed 25 February 2019).
18. Alexander, C., Ishikawa, S., Silverstein, M. et al. 1977. *A Pattern Language: Towns, Buildings, Construction*. Center for Environmental Structure. Oxford University Press, New York.
19. Smith, R. J. 1953. *Tree Crops, A Permanent Agriculture*. The Devin-Adair Company, New York.
20. (a) Hemenway, T. 2001. *Gaia's Garden: A Guide to Home-Scale Permaculture*. Chelsea Green Publishing, White River Junction, VT; (b) Hemenway, T. 2015. *The Permaculture* City: Regenerative Design for Urban, Suburban and Town Resilience. Chelsea Green Publishing, White River Junction, VT.
21. (a) Tomek, V. 2006. 'The Impairment of the Ecosphere: The Current Status and Outlook'. Ontario Consultants on Religious Tolerance. http://www.religioustolerance.org/tomek08.htm (accessed 25 February 2019); (b) Gordon, I., Calatayud, P.-A., Le Gall, P. and Garnery, L. 2019. 'We Are Losing the "Little Things That Run the World"'. UN Environment Programme Foresight Brief 011, January 2019. https://environmentlive.unep.org/foresight (accessed 25 February 2019); (c) Tilman, D., Clark, M., Kimmel, K. et al. 2017. 'Future Threats to Biodiversity and Pathways to Their Prevention'. *Nature* 546, 73–81 (June 1, 2017). https://www.nature.com/articles/nature22900 (accessed 25 February 2019).
22. Gordon, I., Calatayud, P.-A., Le Gall, P. and Garnery, L. 2019. 'We Are Losing the "Little Things That Run the World"'. UN Environment Programme Foresight Brief 011, January 2019. https://environmentlive.unep.org/foresight (accessed 25 February 2019);
23. (a) Croptrust.org. 'Why Is Crop Diversity Important?' https://www.croptrust.org/our-mission/crop-diversity-why-it-matters/why-crop-diversity-important/ (accessed 25 February 2019); (b) Seed Freedom, https://seedfreedom.info/ (accessed 25 February 2019).
24. Gaffarizadeh, M., Preclac, F. G. and Gruse, R. M. 1994. 'Grain Yield Response of Corn, Soybean and Oats Grown in a Strip Intercropping System'. *American Journal of Alternative Agriculture* 19(4).
25. Stamets, P. 2005. *Mycelium Running: How Mushrooms Can Help Save the World*. Ten Speed Press, Berkeley, CA.
26. Ibid.
27. (a) Kristof, N. 2014. 'An Idiot's Guide to Inequality'. *New York Times*, July 23, 2014. https://www.nytimes.com/2014/07/24/opinion/nicholas-kristof-idiots-guide-to-inequality-piketty-capital.html (accessed February 27, 2019); (b) Fuentes-Nieva, R. and Galasso, N. 2014. 'Working for the Few: Political Capture and Economic Inequality'. 178 Oxfam Briefing Paper, January 20, 2014. https://www.oxfam.org/sites/www.oxfam.org/files/bp-working-for-few-political-capture-economic-inequality-200114-summ-en.pdf (accessed 27 February 2019).
28. Schumacher, E. F. 1973. *Small Is Beautiful: Economics as if People Mattered*. Harper Perennial Reprint Edition, 2010.

# Section II

# The Struggle for Sustainability and the Rights of Nature

# 5

# *Kiribati and Climate Change*

**His Eminence Anote Tong**

**CONTENTS**

The Island Nation of Kiribati .......... 79
The Crisis of Climate Change .......... 80
Kiribati's Response to Climate Change .......... 82
Future Uncertainties for National Sovereignty .......... 85
Conclusion .......... 86
Notes .......... 86

## The Island Nation of Kiribati

The Republic of Kiribati, formerly known as the Gilbert Islands and a former colony of the United Kingdom, comprises three groups of 33 atoll islands scattered astride both the Equator and the International Date Line. It has the distinction of being the only country to occupy the four corners of the globe. Inclusive of its 200-mile (322-kilometre) Exclusive Economic Zone, it covers 1.3 million square miles (3.5 million square kilometres) of the centre of the Pacific Ocean, in contrast to its total landmass of just over 800 square kilometres (309 square miles). Atoll islands are coral islands that form atop submerged seamounts, typically aligned as a string according to the movement of the tectonic plate where they are located and the hotspots beneath the plate, which burn through the plate to form the seamounts. The islands are narrow strips of coral sand surrounding a lagoon, and are low lying, on average rising no more than 2 metres (6.5 feet) above sea level.

The people of Kiribati are of Micronesian ethnicity and, according to traceable history, have occupied these islands for around 3,000 years. The total population is about 120,000 people, spread over 20 of the inhabited islands, with over half the population now concentrated on a small portion of the main island of Tarawa. One of the most historic and fiercest conflicts of World War II was the Battle of Tarawa, fought on this tiny strip of island between the occupying Japanese forces and the U.S. Marines in 1943.[1] Christmas Island, in the eastern group of Kiribati, was also the site of the hydrogen bomb atmospheric tests completed by the United Kingdom in the late 1950s.[2]

As a former colony of the United Kingdom, Kiribati has modelled its political system on the Westminster Parliamentary system but with modifications to make it a republic with close similarities to the U.S. presidential system. Since attaining independence from Britain in 1979, Kiribati has enjoyed unparalleled political stability, and, in spite of the comparatively low per-capita gross domestic product, the people have enjoyed a typically idyllic island community lifestyle without the usual rush and pressure of so-called industrialised modern society. Extreme poverty is not a part of life in Kiribati.

DOI: 10.1201/9780429505959-5

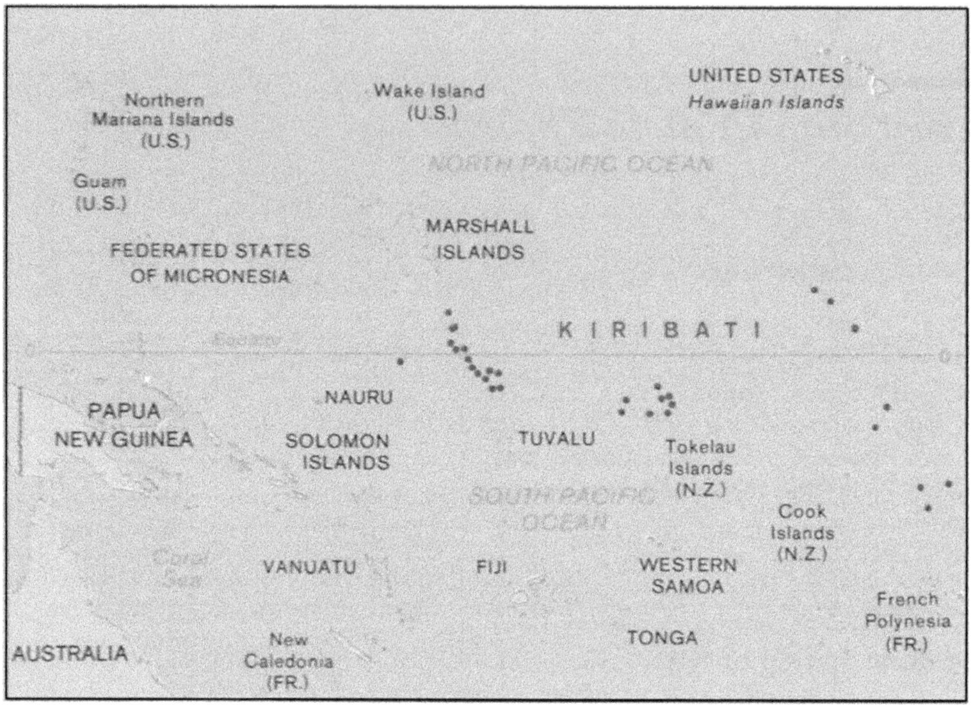


Map of Kiribati in relation to other Pacific States. (Map courtesy of United States Congress, 1989.)

As a society, the people of Kiribati remain relatively isolated from international events and social media. They are mostly unaware of, and therefore do not care much for, what goes on in the rest of the world, let alone what the implications are likely to be on their life or their future. In a typically traditional existence, most do not worry much beyond the next day, and as a nation and a people do not really plan much beyond daily existence, let alone the next generation.

Being located on the Equator, the islands have always enjoyed the relative tranquillity of the doldrums, where cyclones often form but travel either southwards or to the north, but are not supposed to remain in the doldrums due to the rotation of the Earth.[3] Wind speeds rarely exceed 40 mph, but on those rare occasions in the past when they have, severe damage often results due to the extreme vulnerability of the atoll islands.

## The Crisis of Climate Change

For the most vulnerable countries on the frontline, climate change poses an existential challenge. The Fourth Assessment Report of the Intergovernmental Panel on Climate Change (IPCC), released in 2007, was for the first time quite unequivocal in its projections of the rise in sea level, other impacts of the ongoing rise in global temperatures and the dire implications for the most vulnerable countries.[4] The report also for the first time went significantly further than previously in resolving the ongoing controversy within the scientific community on the causes of the rise in global temperatures.

Subsequent reports of the IPCC, including the most recent one of October 2018, expand on the full implications for the global community if the 1.5 Centigrade-degree target temperature reduction is not met by 2030. These later reports have not only reaffirmed the predictions of the Fourth Assessment Report but have concluded that they were too conservative, and that it is more likely that the tipping points to serious and irreversible climate change will be reached much sooner than earlier predicted.[5] Other independent reports by academic and research institutions which are being released and posted on social media virtually on a daily basis also indicate that both polar ice caps are melting much faster than earlier reported. In 2018, during the northern summer, extreme heat and unprecedented bush fire disasters caused severe damage to property and loss of life in the U.S. state of California.[6] In January 2019 Australia issued public warnings due to the severe heat wave sweeping through the country.[7] The frequency and intensity of storms in different regions of the world, including in the Southern and Central Pacific, have also become a common occurrence.

The question is: What has been the experience for the most vulnerable island countries in the Pacific Region to the impacts of climate change, in particular for Kiribati? Due to the inherent fragility of atoll island systems, extreme weather events have always been a threat to the stability of the island ecosystems, but on the rare occasions in past decades when there have been extra high tides and stronger-than-usual winds, the islands have been able to recover, and life gradually returned to normal. 'Have you noticed a rise in sea level?' is the question frequently asked by foreign media. And the honest answer has been, 'No'. However, what has been happening over the years clearly indicates that changes have been taking place.

Makin, Kiribati. The islands of Kiribati are low-lying coral atolls, which are highly vulnerable to climate change. (Photograph by Anna Szramkowska.)

In one of the most populated islands, a whole village community had to relocate its entire population to a different part of the island, with the church and the community hall being the only remaining structures of what was once a thriving community. Most of the communities on the more isolated islets are also undergoing severe challenges from erosion and salt water intrusion into the freshwater ponds and aquifers, which destroys food crops and the community's source of potable water. Following complaints raised by members of Parliament for government assistance to provide protective seawalls, some of the islands were subjected to thorough surveys to ascertain the extent of the damage and to determine the assistance needed. It was concluded that most of the damages were beyond the government's capacity to repair and that assistance could only be provided to protect public and community infrastructure.[8]

Even several of the repair works undertaken at huge expense have not been entirely successful in providing protection from constant weathering.[9] In 2015, Cyclone Pam, which devastated the islands of Vanuatu, veered northwards to severely flood and destroy homes in the islands of Tuvalu and on Tamana and Arorae, the two southernmost islands in Kiribati. Cyclone Pam also flooded the rest of Kiribati to a lesser extent, but enough to cause unprecedented damage to decades-old food trees and the freshwater lens.[10] In December of 2018 swells generated by a storm system north of the islands of Fiji again caused severe damage to homes on the same two islands, and lesser damage and flooding to the rest of the islands in the Gilberts group. Most recently, during the January 2019 full moon spring tide, flooding and minor damages were again reported and evident on the more vulnerable coastal communities on the main island of South Tarawa.[11] Similar impacts would have also been experienced on the outer islands, although these have not yet been reported.

The predictions of the 2007 Intergovernmental Panel on Climate Change Fourth Assessment Report were quite clear that the projected rise in sea level within this century will be well above the current elevation of most of the islands in Kiribati, Tuvalu and the Marshall Islands, even if carbon dioxide emission levels were to be reduced to zero. That report has since been reviewed by the IPCC itself, as well as other independent research institutes, as being too conservative.[12] So, taking the increasing intensity and frequency of weather events now being experienced in Kiribati on the ground, together with the scientific projections, it is quite clear that the low-lying atoll island countries face serious challenges to their future existence. However, against these findings is the reported result of research based on satellite imagery by Paul Kench of Auckland University, suggesting that, contrary to intuitive analysis, these islands have over the past four decades grown in area and are likely to continue to remain above the rising sea level.[13]

## Kiribati's Response to Climate Change

These contradictory messages have considerably dampened the urgency for action, and also provided climate deniers with a counterargument to the call for action on climate change. In Kiribati, as is also the case for other island countries facing the same challenge, it has not been easy for people to accept the possibility that their home islands will no longer be able to continue to provide a homeland for future generations. As predominantly Christian societies, there is also a belief espoused by church leaders, and endorsed by the government, that God will not allow such a catastrophic flood to happen again after the covenant following the biblical Great Flood in Noah's time. But no matter how emotionally unacceptable it may be, the brutal reality is that unless the islands can continue to remain above the rising seas and increasing intensity of the storms, and unless they can remain habitable either through natural processes or radical adaptation initiatives, future generations in countries on the frontline of the climate change impacts will have no option but to relocate.

Based on the scientific information from the IPCC and the observations of what is happening in different parts of the world, but especially in the polar and glacial regions, it is clear that global temperatures have been rising – and will continue to rise, and along with this temperature increase, we will see sea level rise and increasingly higher-energy storms. Given these scenarios, including the contrary findings that land is growing, and religious beliefs, what policy options are available for these countries to pursue? Since the Auckland University report, subsequent studies indicate that climate change will definitely affect the habitability of the islands whether they are eroded, recover their size or even grow over time.[14]

Up to 2016, the Kiribati government's strategy had been to undertake whatever adaptive measures were necessary not only to maintain the sovereignty of the nation, but also to acknowledge the inevitability that whatever adaptive measures were undertaken, they might not be adequate to accommodate the current or future population size. Some will have to relocate, either because they have to or because they choose to do so. In acknowledging the possibility of relocation for some of the population, the policy advocates that any migration must be planned.[15] Those who do migrate must be able to do so with dignity, as people with skill and qualifications, to ensure that they do not migrate as climate refugees and bear all the stigma that usually goes with it.[16] The experience of the European countries over the last few years in dealing with the massive migration from North Africa is a good lesson to learn and not repeat. The policy of migration with dignity is a proactive response to this reality/challenge, and advocates a programme of education and up-skilling in-country, to ensure that those who migrate possess the skills and have the capacity to integrate easily into their new societies, so it is as painless for them as it is for the receiving community.

The loss of one's ancestral homeland is overwhelmingly unacceptable. It has far-reaching implications, including the indignity of loss of identity, the emptiness from the severance of spiritual and ancestral connections to the land, loss of culture and tradition. It is emotionally unacceptable. It would be a huge step into nothingness, which frightens most people. In the Pacific, the possibility that one day relocation may be the only option available was initially rejected outright by many of the leaders and people of the countries in question. To this very day in Kiribati this possibility continues to be denied by a large (but rapidly declining) sector of the population, including at the leadership level. This desperate position is supported by a misinterpretation of ostensibly countermanding research, and a misplaced reliance on what constitutes God's promise of protection. Such a position is also convenient, in that it relieves those facing this challenge, as well as those responsible for addressing it, from the burden of dealing with such an impossible task. However, the increasing frequency of damage from unprecedented weather events is beginning to bring closer to people the true magnitude of the challenge ahead, and that burying one's head in the sand may not be the solution.

Based on the IPCC predictions and other independent research, it is unlikely that Kiribati can continue to survive as a viable sovereign nation into the decades ahead without significant and radical adaptation measures.[17] And even if such measures are undertaken, it is extremely unlikely that sufficient resources could ever be mobilised to build climate resilience for all of the islands. It was on the basis of this possible scenario that a study was commissioned in 2016, with assistance from the United Arab Emirates and undertaken by the same Dutch engineers (Van Oord) who built Palm Island in Dubai, to explore options for raising parts of the main island of Tarawa in a test pilot project.[18] The intention was that, if proven successful, the model would be replicated in the other atoll island nations in the Pacific as part of the initiative by the Coalition of Atoll Island Nations on Climate Change, whose membership includes The Marshall Islands, Tuvalu, Tokelau and Kiribati.

It was also the genesis of what later evolved into the Pacific Rising Project, which is a proactive collective response to the challenge by those countries on the frontline of the climate challenge.[19] Other initiatives included studies undertaken by the government of the Republic of Korea to explore alternative sources of water from the ocean depths and energy using ocean thermal energy conversion technology,[20] and the more radical concept of floating islands as future homes for displaced peoples, which is being explored by Seasteading Institute and Blue Frontiers, in partnership with the government of French Polynesia.[21] It seems that the current (2019) Kiribati government has not followed up on any of these initiatives, perhaps due to its different view of, and response to, the challenge posed by climate change.

The concept of Pacific Rising was premised on seeking options for ensuring that whatever happens, the island nations will be able to build the degree of climate resilience needed to ensure that they do not lose their homeland entirely, in order to retain their sovereignty and all that would otherwise be lost. Tuvalu had already, with assistance from the Government of New Zealand, undertaken a project to fill in the so-called 'borrow pits' dug following World War II to provide material for the construction of the existing airport runway. The result was not only that the pits were restored, but also that parts of the island were raised well above the predicted rise in sea level due to climate change. All this was done at a cost of NZ $12 million.[22]

This experience suggests that with properly designed retaining walls, provided the resources can be mobilised, it is possible to raise the level of the islands above the rising seas and increasing intensity of the storms. As indicated earlier, in Kiribati over 50% of the population is concentrated on a small part of South Tarawa, and many of them are subjected to the periodic flooding and weathering events that are increasing in intensity and frequency. Therefore, building the retaining walls needed to raise and significantly extend the island would not only provide the needed protection, but also provide the space desperately needed to accommodate the highly congested areas around the island.

Mexico Village, Makin, Kiribati. (Photograph by Anna Szramkowska.)

With progressive incremental scaling, and with proper planning, such a structure could be extended to accommodate the entire current and future population sizes, which can be guaranteed greater security in the face of intensifying climate change impacts. Needless to say, a great deal of detail will need more careful study, and much depends on a number of variables. The idea of selecting just one or two islands to raise will also be highly emotional, and therefore extremely political. But, as a nation, Kiribati must never lose sight of the larger picture, which is its survival as a people.

## Future Uncertainties for National Sovereignty

Two questions remain to be addressed: (a) How will the resources needed to undertake such an initiative be mobilised? and (b) Will it all be worthwhile?

There is no doubt that undertaking or even thinking of doing something like raising an island is an acknowledgment of the impending catastrophe, and together with the enormity of the work envisaged would overwhelm the mental capacity of leaders normally used to dealing with small-scale projects. The resources needed for such a venture would be in the hundreds of millions, even billions, of dollars, a staggering amount well beyond the capacity of any of the countries involved. Whether the industrialised countries, which have benefited from the cause of the damage now facing countries on the frontline, should foot the bill has been an ongoing debate. The establishment in 2010 of the Green Climate Fund, under the auspices of the United Nations Framework Convention on Climate Change, is a reluctant and arms-length acknowledgement of that liability by the industrialised nations.[23]

But should the island nations rely entirely on international resources to deliver the needed solutions in time? Experience so far has demonstrated that administration of the Green Climate Fund remains bureaucratically sluggish, and contributions remain short of commitments, while demand for resources is growing. Under the circumstances, it would seem prudent for countries like Kiribati, whose needs are more urgent, to explore sources in addition to those available from the international community.

Kiribati already has sovereign funds approaching AU$1 billion built up over the last 40 years, and has a large fishery resource from which it currently derives annual revenue of around AU$200 from foreign fishing access licence fees. However, foreign fishing access fees represent a very small proportion (around 10% of the landed value of the catch at the wharf) of a multibillion-dollar industry. Therefore, if Kiribati and other island countries in similar situations were able to capture a higher rate of return (through enhanced participation in the industry) on the final value of the abundant fish resources in their exclusive economic zones, they would be able to generate the resources needed to build the climate resilience needed. In addition, deep seabed mining for manganese nodules, the long-awaited source of wealth for those countries in possession of it, now seems to be getting underway.[24] A number of countries in the region have already issued exploratory mining licences within and beyond their Exclusive Economic Zones and some, including Kiribati, are very interested in actual mining.[25] Therefore, the idea of countries leveraging future earnings from resources within the Exclusive Economic Zones against the issuing of Blue/Tuna Bonds,[26] which is being actively discussed as a potential source of funding in building climate resilience, is indeed a realistic option under the circumstances.

The option of maintaining the integrity of their homeland and retaining their sovereignty has always been a matter of first choice for all the island nations, hence the strong resistance to the possibility of massive relocation. For many of the people, but especially the elderly, the prospect of having to move somewhere else is extremely intimidating and traumatic. In several cases where elderly people have migrated under the Pacific Access Category Scheme to join their children who are in New Zealand, the majority have invariably chosen to return home, where conditions are not as comfortable as in New Zealand.[27] So there will always be a sector of the population who could never assimilate well into a different society, and would suffer disproportionately if their special circumstances are not accommodated.

## Conclusion

To look at the other side, the possibility of being able to accommodate the entire population in a single raised landmass also presents a host of new opportunities. The cost savings from the rationalisation of services currently provided to the widely dispersed island communities, including health, education and transportation, would result in significant reductions in the national budget allocations for these services, while at the same time presenting opportunities for improving efficiency in their delivery. Since very little, if any, revenue is derived from any activities on the outer islands, there would be no loss of revenue. Therefore, with the potential for significant increases in revenue from higher rates of return from marine resources, the opportunities for building a relatively well-off society are indeed possible.

## Notes

1. See (a) Alexander, J. H. 1995. *Utmost Savagery: The Three Days of Tarawa*. Naval Institute Press; (b) Graham, M. B. 1998. *Mantle of Heroism: Tarawa and the Struggle for the Gilberts, November 1943*. Presidio Press; (c) Wukovitz, J. 2007. *One Square Mile of Hell: The Battle for Tarawa*. NAL Trade.
2. Arnold, L. and Smith, M. 2006. *Britain, Australia and the Bomb: The Nuclear Tests and Their Aftermath*. Basingstoke: Palgrave McMillan.
3. See (a) Linacre, E. and Geerts, B. 1998. Movement of the South Pacific Convergence Zone. Available at: http://www-das.uwyo.edu/~geerts/cwx/notes/chap12/spcz.html (Accessed 7 March 2019). (b) Braddock, K. L., Kaplan, A. and Gouriou, Y. et al. 2006. Tracking the Extent of the South Pacific Convergence Zone since the Early 1600s. *American Geophysical Union*. First published May 3, 2006. Free access: Available at: https://doi.org/10.1029/2005GC001115 (Accessed 7 March 2019).
4. Intergovernmental Panel on Climate Change. 2007. *Climate Change 2007: Synthesis Report. Contribution of Working Groups I, II and III to the Fourth Assessment Report of the Intergovernmental Panel on Climate Change*, Core Writing Team, Pachauri, R. K and Reisinger, A. (eds.). IPCC, Geneva, Switzerland, 104 pp.
5. Intergovernmental Panel on Climate Change. 2018. Summary for Policymakers. In: *Global Warming of 1.5°C*. An IPCC Special Report on the impacts of global warming of 1.5°C above pre-industrial levels and related global greenhouse gas emission pathways, in the context of strengthening the global response to the threat of climate change, sustainable development, and efforts to eradicate poverty [Masson-Delmotte, V., Zhai, P., Pörtner, H. O., Roberts, D., Skea, J., Shukla, P. R., Pirani, A., Moufouma-Okia, W., Péan, C., Pidcock, R., Connors, S., Matthews, J. B. R., Chen, Y., Zhou, X., Gomis, M. I., Lonnoy, E., Maycock, T., Tignor, M., Waterfield, T., (eds.)]. World Meteorological Organization, Geneva, Switzerland, 32 pp.

6. Kasler, D. 2018. Worst Wildfire Year since When? More California Acres Have Burned in 2018 Than the Past Decade. *Sacramento Bee*, November 16, 2018. Available at: https://www.sacbee.com/latest-news/article221788220.html (Accessed 7 March 2019).
7. Cox, L. 2019. Warning for Extreme Heatwave for Southern Australia after New Year Reprieve. *The Guardian*, January 2, 2019. Available at: https://www.theguardian.com/australia-news/2019/jan/02/warning-for-extreme-heatwave-for-southern-australia-after-new-year-reprieve (Accessed 7 March 2019).
8. For an overview of Kiribati's response to climate change threats and infrastructure, see (a) Office of Te Beretitenti. 2013. *National Framework for Climate Change and Climate Change Adaptation*, especially pp. 38–39. Available at: http://www.president.gov.ki/wp-content/uploads/2014/08/National-Framework-for-Climate-Change-Climate-Change-Adaptation.pdf (Accessed 31 March 2019); (b) Government of Kiribati. 2014. *Kiribati Joint Implementation Plan for Climate Change and Disaster Risk Management*. Available at: http://extwprlegs1.fao.org/docs/pdf/kir154056.pdf (Accessed 31 March 2019).
9. (a) Office of the President, Republic of Kiribati. Kiribati Climate Change. 2010. Seawalls to Protect Kiribati Shorelines. August 10, 2010. Available at: http://www.climate.gov.ki/2010/08/10/new-guidelines-to-protect-kiribatis-shoreline/ (Accessed 7 March 2019); (b) Hardwick, L. 2010. *South Tarawa Coastal Condition Assessment*. Prepared by Beca Infrastructure Ltd. for Office of Te Beretitenti, Government of Kiribati. February 16, 2010. Available at: http://www.climate.gov.ki/wp-content/uploads/2013/05/South-Tarawa-Coastal-Condition-Assessment.pdf (Accessed 31 March 2019).
10. Stone, G. 2015. Cyclone Pam Signals Slow-Motion Disaster in Kiribati. Humanature: Conservation International Blog. March 15, 2015. Available at: https://blog.conservation.org/2015/03/cyclone-pam-signals-slow-motion-disaster-in-kiribati/ (Accessed 7 March 2019).
11. Floodlist News. January 8, 2019. Pacific Islands – Thousands Affected by Floods after Heavy Rain and Storm Surge. Available at: http://floodlist.com/australia/pacific-islands-penny-mona-fiji-solomon-january-2019 (Accessed 7 March 2019).
12. See for example: Intergovernmental Panel on Climate Change. 2018. *Global Warming of 1.5 Degrees Centigrade: Summary for Policymakers*. Available at: https://www.ipcc.ch/site/assets/uploads/sites/2/2018/07/SR15_SPM_version_stand_alone_LR.pdf (Accessed 31 March 2019).
13. (a) Kench, P. S., Ford, M. R. and Owen, S. D. 2018. Patterns of Island Change and Persistence Offer Alternate Adaptation Pathways for Atoll Nations. *Nature Communications*. Vol. 9, Article number 605. Published February 9, 2018; (b) RadioNZ. April 4, 2018. Climate Change in the Pacific – What's Really Going On? Available at: https://www.radionz.co.nz/national/programmes/saturday/audio/2018640643/climate-change-in-the-pacific-what-s-really-going-on (Accessed 7 March 2019).
14. Storlazzi, C., Gingerich, S. B., van Dongeren, Ap. et al. 25 April 2018. Most Atolls Will Be Uninhabitable by the mid-21st Century Because of Sea-Level Rise Exacerbating Wave-Driven Flooding. *Science Advances*. Vol. 4, No. 4. DOI: 10.1126/sciadv.aap9741. Available at: http://advances.sciencemag.org/content/4/4/eaap9741 (Accessed 31 March 2019).
15. Office of Te Beretitenti. 2013. *National Framework for Climate Change and Climate Change Adaptation*. Available at http://www.president.gov.ki/wp-content/uploads/2014/08/National-Framework-for-Climate-Change-Climate-Change-Adaptation.pdf (Accessed 31 March 2019).
16. Oakes, R., Milan, A., and Campbell, J. 2016. *Kiribati: Climate Change and Migration – (Relationships between Household Vulnerability, Human Mobility and Climate Change.)* Report No. 20. Bonn: United Nations University Institute for Environment and Human Security (UNU-EHS). Available at: https://www.unescap.org/sites/default/files/Online_No_20_Kiribati_Report_161207.pdf (Accessed 31 March 2019).
17. Wyett, K. 2013. Escaping a Rising Tide: Sea Level Rise and Migration in Kiribati. *Asia & The Pacific Policy Studies*. October 11, 2013. DOI: 10.1002/app5.7
18. For an overview of Van Oord's climate initiatives and coastal defence projects, see: https://www.vanoord.com/sustainability/climate. For a roster of the company's engineering projects, see: https://www.vanoord.com/projects (Accessed 31 March 2019).

19. For examples of Pacific nations unifying on climate change vulnerability and actions, see: (a) the United Nations Economic and Social Commission for Asia and the Pacific's Sub-Regional Office for the Pacific. Available at: https://www.unescap.org/subregional-office/pacific/about (Accessed 31 March 2019); (b) UN Environment. 2017. Informing Action: Pacific Nations Unite on the Environment. 14 December 2017. Available at: https://www.unenvironment.org/news-and-stories/story/informing-action-pacific-nations-unite-environment (Accessed 31 March 2019).
20. For Ocean Thermal Energy Conversion, see: (a) https://www.eia.gov/energyexplained/index.php?page=hydropower_ocean_thermal_energy_conversion and (b) Liu, C. C. K. November 2018. Ocean Thermal Energy Conversion and Open Ocean Mariculture: The Prospect of Mainland-Taiwan Collaborative Research and Development. *Sustainable Environment Research*. Vol. 28, No. 6, pp. 267–273. DOI: 10.1016/j.serj.2018.06.002. For the Korean project, see OTC Foundation. 2013. Completion of 20 kw OTEC Pilot in South Korea. October 26, 2013. Available at: http://www.otecnews.org/2013/10/completion-20kw-otec-pilot-south-korea/ (Accessed 18 March 2019).
21. Bianchi, C. 2018. A Floating Pacific Island Is in the Works with its own Government, Cryptocurrency and 300 Houses. May 28, 2018. CNBC Asia-Pacific News. Available at: https://www.cnbc.com/2018/05/18/floating-island-is-planned-with-government-cryptocurrency-and-houses.html (Accessed 31 March 2019).
22. (a) Tautai Foundation. 2015. Borrow Pits Rehabilitation in Tuvalu. April 26, 2015. Available at: http://tautai.com/borrow-pits-rehabilitation-in-tuvalu/ (Accessed 18 March 2019); (b) Radio New Zealand. 2016. Tuvalu's Borrow Pits Finally Filled in. March 22, 2016. Available at: https://www.radionz.co.nz/international/pacific-news/299471/tuvalu%27s-borrow-pits-finally-filled-in (Accessed 18 March 2019).
23. See the Green Climate Fund website for detailed information: https://www.greenclimate.fund/home (Accessed 18 March 2019).
24. (a) World Ocean Review. 2014. Mineral Resources. Manganese Nodule Treasures. Available at: https://worldoceanreview.com/en/wor-3/mineral-resources/manganese-nodules/ (Accessed 18 March 2019); (b) Miller, K. A., Thompson, K. F., Johnston, P. and Santillo, D. 2018. An Overview of Seabed Mining Including the Current State of Development, Environmental Impacts, and Knowledge Gaps. *Frontiers in Marine Science*. January 10, 2018. DOI: 10.3389/fmars.2017.00418 (Accessed 18 March 2019).
25. (a) Marawa Research and Exploration Ltd. Undated. Exploration. Available at: http://marawaresearch.com/exploration.html (Accessed 18 March 2019); (b) World Ocean Review. 2014. Mineral Resources. Manganese Nodule Treasures. Available at: https://worldoceanreview.com/en/wor-3/mineral-resources/manganese-nodules/ (Accessed 18 March 2019); (c) Clark, A. L. 2018. A 'Golden Era' for Mining in the Pacific Ocean? Perhaps Not Just Yet. East-West Center. East-West Wire, April 6, 2018. Available at: https://www.eastwestcenter.org/system/tdf/private/ewwire010clark.pdf?file=1&type=node&id=36597 (Accessed 19 March 2019).
26. Standing, A. 2018. Blue Bond ... Saving Your Fish or Bankrupting the Oceans? CFFA-Cape, April 14, 2018. Available at: https://cape-cffa.squarespace.com/new-blog/2018/4/14/blue-bondsaving-your-fish-or-bankrupting-the-oceans (Accessed 18 March 2019).
27. (a) Oakes, R., Milan, A., and Campbell, J. 2016. *Kiribati: Climate Change and Migration – (Relationships between Household Vulnerability, Human Mobility and Climate Change.)* Report No. 20. Bonn: United Nations University Institute for Environment and Human Security (UNU-EHS). Available at: https://www.unescap.org/sites/default/files/Online_No_20_Kiribati_Report_161207.pdf (Accessed 31 March 2019). See also: (b) New Zealand Immigration, New Zealand Now. Undated. Moving from the Pacific Islands. Available at: https://www.newzealandnow.govt.nz/choose-new-zealand/compare-new-zealand/pacific-islands (Accessed 18 March 2019).

# 6

## 'When God Put Daylight on Earth We Had One Voice' Kwakwaka'wakw Perspectives on Sustainability and the Rights of Nature

**Douglas Deur, Kim Recalma-Clutesi and Clan Chief Kwaxsistalla Adam Dick**

**CONTENTS**

Introduction ..... 89
Speaking with One Voice ..... 92
An Introduction to Kwakwaka'wakw Relationships with the Natural World ..... 92
The Wolves and the Mountains ..... 96
Special Obligations to Places and Beings ..... 96
Salmon, Eulachon, Clams and Plants ..... 98
Relationships and Obligations to Game Species ..... 98
Orca and Cedar ..... 101
Relationships and Obligations to Other Species ..... 101
Having One Voice ..... 104
A Conclusion ..... 104
Acknowledgements ..... 107
Notes ..... 107
Appendix: Common and Scientific Names of Plants and Animals ..... 111

### Introduction

This book, and the intellectual and legal movement summarised within its pages, charts a bold alternative course for humanity. That there are certain 'rights of Nature' intrinsic to landscapes and life-forms around the world is a revolutionary assertion, yet an assertion with abundant and venerable precedents. By the logic of this movement, nonhuman beings have intrinsic existential rights and, by extension, should possess certain rights protecting their survival and interests within the evolving legal practises of modern nations. Concepts akin to human rights are thus extended to populations of wild nonhuman species, and to landforms such as mountains or rivers, on which many other lives depend. These entities might then possess rights to representation in legal arenas akin to personhood – so that certain keystone landforms or living beings cannot be destroyed for the profit of human individuals without overwhelmingly compelling reasons, nor damaged without efforts to directly compensate nonhuman 'claimants' for damages.

The Rights of Nature movement has proven compelling as a critique. One can now see the tentative but transformative effects of this foment internationally as legal challenges, applying concepts of personhood to nonhuman entities, upend a variety of destructive

DOI: 10.1201/9780429505959-6

land and resource regimes. Litigation asserting the intrinsic rights of mountains and rivers, forests, birds and orcas has extended meaningful protections to these landforms and life-forms. At the time of this writing, additional litigation is pending.[1]

As these instances demonstrate, extending legal rights to these entities beyond their role as commodities and disposable human 'properties' has broad conservation outcomes that are frequently prosocial and sustaining of human life. The present volume posits that by formally extending such rights to certain categories of landforms and life-forms around the world, nations will support the linked goals of environmental sustainability and biosocial resiliency – ultimately supporting some of humanity's most urgent shared interests and needs. In time, by acknowledging the intrinsic values of key landscapes and landforms, this realignment of legal tradition might benefit humanity, such as by protecting 'ecosystem services' that benefit all of humanity and extend well beyond whatever benefits are accrued by one individual with the unfettered right to exploit for private gain.

This literature, and the foment underlying it, draws significant inspiration from a few key precedents. Among these, Western legal history provides examples of revolutionary changes effectively extending rights of personhood to categories of individuals formerly treated as 'property' and denied such rights. To name a few cornerstone examples, changes in the legal status of women, slaves and indigenous peoples over the last two centuries provide compelling reminders of how concepts of 'personhood' have evolved and expanded in ways leading to overwhelmingly positive social outcomes.[2]

Beyond this, the Rights of Nature movement draws foundational inspiration, sometimes explicitly and sometimes implicitly, from the perspectives of indigenous peoples. At certain times, perhaps all human societies have extended concepts of personhood to nonhuman beings and viewed nonhuman beings as being on parallel, even coequal, life trajectories. This is ostensibly an ancient part of human experience and worldview, much eclipsed in the industrialised world. Yet, modern indigenous societies still uphold such values. On this basis, some modern writers suggest indigenous peoples have a clear and edifying perspective of the Rights of Nature by virtue of their animistic and holistic worldviews, and especially by virtue of direct connections to the land and life-forms with which they coexist. Generally such writings are quite empathetic with Native peoples and their worldviews, and sometimes draw genuine insights from Native precedents. Yet too often, these depictions of aboriginal concepts of the 'rights of nature' remain shallow and unexamined – a kind of obligatory preface to broader philosophical arguments, a simplified caricature of primordial virtue to be used as both inspiration and as a counterpoint to the crude materialism of the industrialised capitalist world. Corrective steps, including a more nuanced and careful examination of genuine aboriginal perspectives, seems in order. The present chapter is but one step in that direction.

Accordingly, in this chapter we ask: How are the 'Rights of Nature' truly manifested in an indigenous context? We contend that one especially illuminating example can be found in the teachings of the Kwakwaka'wakw (Kwakiutl) people of coastal British Columbia. The Kwakwaka'wakw are among the most studied indigenous people in the Americas, and are thus a key reference point, providing a rich tradition that is widely known and accessible through the accounts of past anthropologists and a handful of living experts. Though so much remains unclear, or was misconstrued in early accounts, Kwakwaka'wakw culture provides a universally known example, worthy of attention as Native communities, researchers and policy-makers seek to advance an aboriginal perspective on the Rights of Nature movement.[3]

The authors of this article all speak from a deep grounding in Kwakwaka'wakw tradition. No person in our time was, however, as knowledgeable on these traditions as co-author *Kwaxsistalla wa-thla* – the Clan Chief, Adam Dick, who held the chiefly name *Kwaxsistalla.*

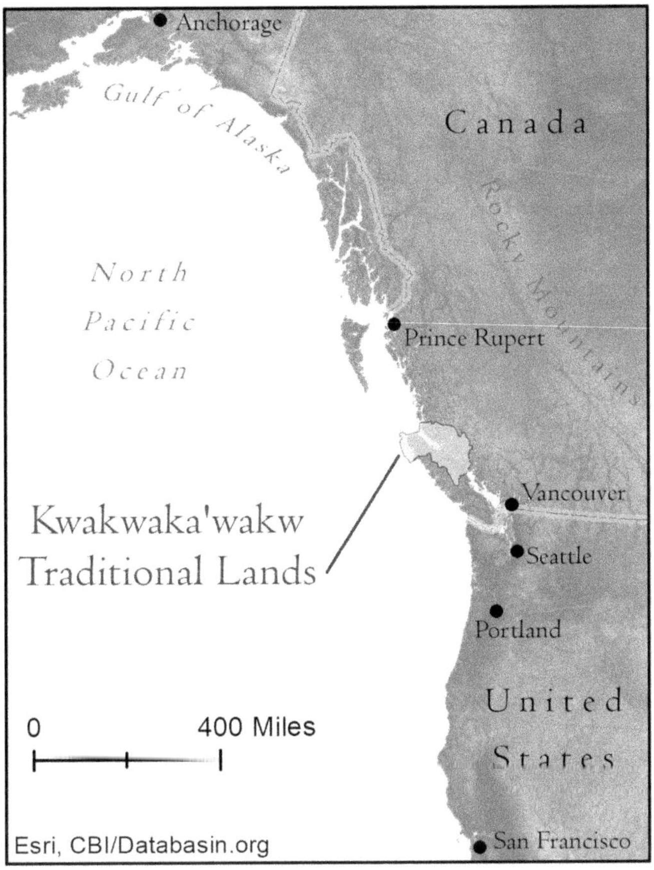


Map of Kwakwaka'wakw Territory. (Map produced by Eric Owen.)

He long served as chief of the *Qawadiliqalla* [wolf] clan of the *Dzawada'enuxw* [Tsawataineuk] Kwakwaka'wakw – hailing from Kingcome Village, on the mainland coast of British Columbia, Canada. Chief Adam passed away as the present manuscript was being prepared, but was able to provide his co-authors with sufficient guidance to make his message clear in the pages that follow. We assert that many important clarifications can be found in the teachings carried by Chief Adam into our time.

Our co-author and teacher, Kwaxsistalla[4] Clan Chief Adam Dick, was the last chief of the Kwakwaka'wakw world to be fully trained in the traditional way. As the peoples of the British Columbia coast were increasingly persecuted and even arrested for their practises and religious beliefs in the early twentieth century, their children taken by force to residential schools for colonial re-education, many Kwakwaka'wakw actively resisted colonial control. Ceremonies had to be held 'underground', often with lookouts posted to scan the horizon for approaching police boats; regalia and even children were sometimes hidden to avoid confiscation.[5] Within this context, a prophesy arose – that a child would be born who would serve as a special bearer of cultural and spiritual knowledge through the very difficult times ahead, to carry these things forward for the benefit of future generations. Young Adam was chosen, and by age 4 he began his focused training to serve this keystone role within the sweep of Kwakwaka'wakw history. Intentionally isolated from the non-Native world for most of his youth, clan chiefs (*oqwa'mey*) and other specialised knowledge-holders systematically educated young Adam in every domain of traditional chiefly knowledge, from the most sacred to the most mundane. Overseeing his education was an association of four clan chiefs born in the nineteenth century. Secluding and training this young boy, they entrusted their central cultural teachings to Adam, urgently hoping that he might carry forth this information through his life and beyond – like a time capsule in human form. The faithful retelling of these cultural lessons was key to Chief Adam's mission; he often asserted that he was not the originator of the ideas transmitted to him, but the person appointed with a sacred duty to convey ancient teachings to the modern world. As he often remarked, he was the designated living conduit of this knowledge, though credit for this knowledge was to be attributed to the ancestors: 'when you honor me, you honor my teachers'.[6]

The wealth of knowledge conveyed by Chief Adam – especially knowledge pertaining to traditional environmental practise and values – provided Native and non-Native researchers with a wealth of detail omitted from past writings on Northwest Coast Native cultures. In recent decades, his teachings on topics from fishing ethics to the cultivation of native plants have prompted a revolutionary reinterpretation of traditional human-environment relations along the entire coast.[7] It is from that authority, rooted in the teachings of the deep past, that we offer comment and clarification on how 'Rights of Nature' have been engaged in this indigenous context since time immemorial. It is from that authority, rooted in the teachings of the ancestors, that we explore what lessons Kwakwaka'wakw teachings might hold for modern legal concepts and frameworks across the modern industrialised world.

## Speaking with One Voice

### An Introduction to Kwakwaka'wakw Relationships with the Natural World

Around the globe, an understanding is mounting regarding how indigenous cultural values shape Native societies' understanding, engagement and modification of natural landscapes and life-forms within their homelands.[8] This nexus between core cultural values and traditional resource practises certainly defines much of Kwakwakwa'wakw social, ceremonial and economic life, as is true for indigenous societies around the world. But although generations of anthropologists have produced a vast literature on Kwakwaka'wakw or 'Kwakiutl' cultural life, lucid writings on the topic of human–environmental relationships remain scant. This reflects idiosyncrasies within anthropological praxis over the last century and a half. Researchers of Northwest Coast societies commonly focused on the most striking ceremonial traditions and visibly exotic aspects of the culture, bringing their own theoretical proclivities and agendas to bear.

When natural resource topics were addressed in this corpus, they were commonly dissociated from the larger system of cultural values and practises of which they are a part.[9] Even discussions of traditional religious concepts, woven intricately into our natural resource practises and values, were commonly presented with attention to the most exotic aspects of traditional ceremonialism, but in curious isolation from relationships with the natural world.[10] The teachings carried by Chief Adam have provided an important correction to this oversight. Indeed, Kwakwaka'wakw relationships with the natural world are central within the growing body of literature rooted in his teachings, which point toward a number of fundamental concepts – 'underlying principles' providing critical context for the discussion to follow.

According to the Kwakwaka'wakw worldview, animals, plants and landforms – indeed, all living things – certainly possess distinctive identities, as well as spiritual lives and power, and this significantly influences how our people treat these beings. All beings are said to possess a fundamental similarity to humankind, in marked contrast to traditionally dualistic Western views of nature that set humankind apart. All beings are said to 'breathe the same air', to share the same basic fears and motivations, to have the same genetic processes and imperatives, to depend on their communities and to experience the same basic arc of life and death. Living beings have a spirit and sentience, and many things the Western world sees as inert are 'living' in the Kwakwaka'wakw view. Like humans, living beings are permeated and animated in part by *nawalux*, the spiritual energy latent within the universe and flowing from the Creator and through all of creation – a power traditionally engaged and cultivated by highly trained specialists including certain healers and clan chiefs. In all these ways, we are one. To underscore this point, in the oldest Kwakwaka'wakw oral traditions, all beings and even certain landscapes are said to have been capable of open intercommunication, of fundamentally 'speaking the same language' in ancient times. As Chief Adam reminds us, 'When God put daylight on this Earth, we all had one voice'.

In this respect, our values may stand apart from certain Western philosophical and religious traditions, which have asserted varying degrees of human separateness from the natural world – especially from the time of the Enlightenment and Western industrialisation. In the Kwakwaka'wakw world, our traditional values resonate with other, eclipsed domains of the Western tradition that situate humanity closer to the natural world, as well as relatively recent developments in Western science, cosmology and values that place humanity back in its biophysical context, back in its sprawling web of deep kinship, back into the flows of nature. Indeed, we are pleased that Western science is beginning to catch up with Kwakwaka'wakw teachings on this point and hope the Western mind continues to evolve in this regard for our common good.

This appreciation of the importance and sanctity of nonhuman life is embodied within all aspects of resource management. Clan chiefs and their associates, who control land and resource decisions in our world, all traditionally start their work from a set of core values and assumptions asserting the essential value of nonhuman life. Our patterns of property ownership and exchange; our mechanisms for resource monitoring, management and harvest and adjudication that occurs within and between communities are all permeated with this understanding. A clan chief effectively controls the resources within his territory, but also upholds profound obligations: to other human communities, but also to nonhuman communities and the biotic systems on which their lives depend. This reverence, along with an intimate and detailed understanding of local environments, contributes to sustainable patterns of use over deep time in indigenous societies, in Kwakwaka'wakw country and beyond – a point receiving more attention in the pages that follow.[11] But this observation, so simple yet fundamental, is only part of the story.

These values are also manifested in every Kwakwakw'wakw social, ceremonial and economic institution. Prominent among these is what anthropologists have often called the 'potlatch' – an inexact term that encompasses many ceremonies, but especially the system called the *pa'sa* in the Kwak'wala language.[12] The term 'potlatch' is deeply woven into academic discourse and is commonly depicted as a ceremonial tradition involving lavish displays of hoarded wealth through reciprocal gift-giving between communities, as well as the destruction of property, purportedly to allow chiefs to advertise their wealth and thus elevate their relative status in society. Though quite famous, this representation is also shockingly misrepresentative of the cultural realities of Kwakwaka'wakw. Indeed, the gluttony and wanton destruction of property described in some old anthropological writings might suggest that our ancestors were not wise resource stewards but megalomaniacal buffoons, obsessed with conspicuous resource consumption and little else.[13] These writings also suggested our people live in a land of resource superabundance, where resource stewardship was simply unnecessary and wasteful consumption would have negligible effects on our relationships with the natural world.[14] Nothing could be further from the truth. The roots of this misrepresentation are deep, reflecting persisting biases and transmission errors. Though this topic is beyond the scope of this chapter, the origins of the bias have been addressed in prior publications by the authors, and in other venues.[15] Tragically, this misrepresentation has, until very recently, eclipsed the potential gift of Kwakwaka'wakw environmental wisdom to the wider world.

A generation or two ago, environmental anthropologists began to realise that these misrepresentations could not be entirely correct. Cultural ecologists began asking whether potlatch traditions might actually have ecological influences or positive environmental consequences. Some noted that higher-status Kwakwaka'wakw clans and chiefs were those who possessed the most abundant natural resources, and resource abundance, they suggested, contributes to the status of clans. Such wealth allowed chiefs to enhance their standing relative to other chiefs not just through displays of wealth in the potlatch. Careful management and redistribution of the resources was also key.[16] Authors such as Piddocke concluded that the potlatch tradition 'had a very real pro-survival or subsistence function' that worked 'to counter the effects of varying resource productivity by promoting exchanges of food from those groups enjoying a temporary surplus to those groups suffering a temporary deficit'.[17] This was closer to the truth, but still not quite correct.

We counter that the idea of reciprocal exchanges, as made famous in the potlatch literature, permeate almost all other traditional Kwakwaka'wakw institutions, and help organise every aspect of social, ceremonial and economic life. These reciprocal exchanges are guided by a system of ethics and belief asserting the importance of maintaining 'balance' in all relationships. The chiefs, especially the Clan chiefs, serve as mediators and managers of these relationships, aided by an entourage of specialists – 'potlatch recorders' who carefully monitor specific exchanges, along with resource specialists, public speakers, spiritual practitioners and others. Clan chiefs traditionally work in diverse arenas to maintain balance between communities through reciprocal exchanges – on the potlatch floor and beyond. With the goal of achieving 'balance', and the assistance of a coterie of specialists, clan chiefs actively monitor and correct imbalances between communities through the giving of material gifts, the repayment of specific debts, the offering of ceremonial honours and praise and other mechanisms. When neighbours experience hardship, they are given support through these exchanges; when one's home village experiences hardship, our community expects their support in return – making all communities stronger, our wealth greater than the sum of its parts. The long-term equity and stability in the social, economic and ceremonial relationships affecting our home communities are the goal of the potlatch, not flamboyant displays of wealth and status. All of the interventions by clan chiefs, interventions both material and intangible, are means to achieve these greater ends, including the avoidance of 'imbalance' within relationships of mutual benefit and dependence.

The 'gifts' that are exchanged in the Kwakwaka'wakw world are understood to come in many forms – not only as gifts of property exchanged between communities in the potlatch. The food one harvests is a gift; even a single fish is understood as a gift both from the Creator and the fish that gave its life, a gift to the many beings and living systems dependent on fish for survival. The weight of this gift is even greater, recognising that any living being that a person might consume is traditionally understood to be sentient, possessing a spiritual identity all its own.[18] Killing is a weighty act, even as it must be an everyday act. Life-sustaining relationships may be upended if humans take life casually, without acknowledging the weight of that gift. Embedded in Kwakwaka'wakw values is the understanding that if one shows disrespect toward other species, if a person unbalances our relationships with those on which our communities depend, those species are likely to reciprocate in kind. Receiving a gift, such as that of a life given for food, requires repayment and deep demonstrations of respect – as it would with a gift received from a human neighbour. Relationships with our neighbours, human or otherwise, are 'systematically monitored and rebalanced' over time to protect mutually beneficial relationships in the long term. The essential kinship between species is assumed. Our ancestors have recognised that many mutual obligations link communities – not only human, but also nonhuman communities, obligations that must be monitored and maintained over the long term.[19]

Western concepts of the individual, of individualism, are alien and potentially dysfunctional in the Kwakwaka'wakw view. Traditional knowledge-holders recognise that humans serve in no small part as conduits (of genetic material, of culture and knowledge, of values, of water and matter). To eat something is to commune with it – to bind your life and its life together, even at a molecular level.

Connections to nonhuman species were not arbitrary, then, but represent systematic and strategically negotiated relationships carried out repeatedly over generations. Poor care of living things by humans could profoundly affect our relationships with nonhuman communities. Over time, this could create instabilities likely to undermine our own wealth and standing, our relationships with other human communities and our relationships with the Creator and the *nawalux*. Wise resource stewardship requires seeking 'balances on every ledger', with all of our human and nonhuman neighbours, resulting in a healthier local environment and a richer human community, as well as peace and resilience. The cultural ecologists were correct on this count: when Kwakwaka'wakw communities practise wise 'resource stewardship', we thrive and become wealthy in myriad ways.

Importantly, this is done with a time perception quite different from that of the Western world. Kwakwaka'wakw tradition asserts that – in all endeavours – we are operating within long-term relationships that extend into the very distant past and far into the distant future. Our clans and communities are bounded, existing in place over deep time. If the ancestors overexploited local resources, they did not traditionally have the option of picking up and moving to another undamaged place. Inevitably, one's children, one's children's children and beyond become direct beneficiaries or victims of the resource decisions being made today. The great-great grandparents of the salmon in the stream beside our village fed our great-great grandparents; if we honour all obligations and show due respect, the great-great-grandchildren of today's salmon will make themselves available to feed our own great-great-grandchildren.

Yet, the obligations across generations are even more pressing than this suggests. The clan chiefs, indeed all Kwakwaka'wakw nobility, hold names and identities that first appear at the time of creation. When one ceremonially receives those names and titles, one is not simply being ennobled by a chiefly moniker; a person takes on what is arguably a 'symbiotic relationship' with an eternal identity, an identity that came before our present time and will live on in perpetuity. One receives the name and works to 'keep our name good' for future generations in life. When one dies, our spirits endure and the noble identities one held are passed on in good condition to the next generation. Even now, in the wake of Chief Adam's passing, the name and chiefly identity of Kwaxsistalla is being transmitted to his successor.

In this light, human obligations to future generations seem especially urgent. To some extent our identities are eternal. And even as these identities persist long after human bodies fail and perish, overexploitation can have effects haunting individuals and communities for generations. Conversely, wise resource stewardship can have benefits that persist and benefit our entire community, and communities beyond, all for generations.

## The Wolves and the Mountains

### Special Obligations to Places and Beings

In the Kwakwaka'wakw world, as in many Native societies, people hold singular obligations to certain animals by virtue of enduring connections encoded in our most ancient ceremonies and oral histories. In Kwakwaka'wakw tradition, humans are declared to be bound to these living beings from the beginning of remembered time. This is especially true as an outcome of one special branch of our oral tradition – the *gilgalis*, which is the cornerstone of our 'creation story cycle' describing how humans took shape on the land.

Each clan's *gilgalis* story cycle is transmitted and owned as chiefly property. Each describes a first ancestor of the clan's chiefly lineage arriving in the world in the form of a living being. The being takes human or humanlike form and in that form becomes ancestral to the lineage of clan chiefs who follow across the generations. These beings appear in the dramatic crests and regalia of Kwakwaka'wakw people and others along the coast; images of these ancestral beings are featured in the regionally iconic 'totem poles' and other totemic art, which are to be displayed only by the chiefly lineages possessing rights to use those crests.

The ancestral being of Chief Adam's clan, the *Qawadillikala*, is the wolf. Based on this cornerstone of the clan's origin story, the human bond with wolves is one of important, enduring connections across the generations – for members of the clan in general, but for the clan chief in particular. Within each generation, the clan chief is understood to be a lineal descendent of the wolf, possessing unique rights to use wolf crests and other clan images within carvings and regalia. Wolves are treated as near-kin by the larger community, while the living line of chiefs holds a unique sense of kinship with the wolf. As Chief Adam asserted, 'I am the wolf man ... that's where I came from'.

In the spirit of kinship, traditions prohibit the hunting of this animal. As Chief Adam often observed of his teachers, 'they say don't ever hurt the wolf, or you hurt yourself'. Members of the clan are said to have learned key lessons by watching wolves. The enduring practise of using Sitka spruce (*Picea sitchensis*) pitch as a salve and sealant for wounds, for example, is said to have been taught to the Qawadillikala by watching wolves rub their own wounds against pitchy trees. The wolf educates, and clan members reciprocate. Elders of Chief Adam's youth attest that his grandfather, the former holder of the chiefly title Kwaxsistalla, once aided a wolf that ran into the longhouse seeking assistance, with a bone stuck in its teeth. When the chief removed the bone, the wolf darted out the door, pausing briefly in front of the ceremonial building to howl. Oral tradition teaches that even in ancient times these mutualistic connections existed between the wolf and its human kin.

So too, the *gilgalis* of each clan mentions key landmarks throughout their homeland – mountain peaks, rivers, glaciers, rock outcrops and more – that are shaped by the events of creation described within the story cycle. These places of origin are invoked in songs, stories, teachings and traditional rites relating to the clan, all owned by the clan and managed by the chief as clan property. These places are sometimes represented artistically in clan crests and regalia, and are even sometimes depicted with modern artistic styles, produced

Kwaxsistalla Clan Chief Adam Dick, in chiefly regalia, standing beside a carved 'totem pole' at his home. Carved on poles, painted on houses or worn as regalia – wolf crests are chiefly property, reflecting a sense of deep kinship with wolves and enduring connections between species. (Photograph by Bert Crowfoot.)

by contemporary clan members for more secular purposes. These landmarks are treated with special reverence and respect, as places of origin and as the handwork of the Creator and ancestral beings – by the members of a clan, but also by others who comprehend their importance. These landmarks are sources of validation of chiefly prerogatives and events within oral tradition. They also serve as genuine sources of strength and power to traditionally trained clan chiefs who might tap into those powers to support efforts to heal and support the larger clan.

As in their relationships to wolves and other clan ancestors, Kwakwaka'wakw people are duty-bound to respect and protect these places and to keep their significance ever present through invocation in stories and ceremonies linking key oral traditions to key values and social relationships. These obligations to other species and landmarks are recorded in our most ancient origin narratives. Being at the root associated with the very moment of creation, these interspecific relationships are woven into the fabric of our culture and society.

Other places are treated with similar reverence, even as they are less directly related to the *gilgalis* story cycles. Some, for example, are linked to oral traditions of a great flood that swept across the coast, effectively ushering in a new era in human time. There are mountains such as *gwa'gwayems*, 'the whales', that resemble a pair of humpback whales near Kingcome Village, where it is said a pair of whales were trapped as the waters receded. And there are others of even greater significance. Chief Adam especially invoked the apex of one mountain looming over Kingcome Village, a mountain clearly topped with a rectangular rock outcropping. In Kwakwaka'wakw oral tradition, this is the giant box in which all chiefly possessions – including rights, regalia and chiefly knowledge – were held and protected from harm during the deluge. Through his life, when publicly performing chiefly ceremonies or presenting chiefly crests far from home, Chief Adam referred to the acts as 'opening the box'. The meaning of his reference was understood by his clan and broader cultural circle: he was bringing forth true chiefly property while simultaneously invoking the landmark looming high above Kingcome Village. Like the wolf, this mountain has been inextricably linked to *Qawadillikala* clan chiefs. It is foundational to their identity and has a power, life and identity intrinsically worthy of reverence and protection.

Other powerful places are recognised for their capacity to teach and empower all people, for their significance encoded in oral tradition. This includes places created by the transformer, *Hethla'tusla*, whose name means 'the one who makes things right'. Kwakwaka'wakw oral tradition describes how he travelled across the land in ancient times. As he travelled, he shaped landmarks into their present forms, teaching humanity lessons by his actions and pronouncements at certain points along his route. Specific landmarks shaped by his handiwork hold moral, social and environmental lessons still instructing human observers today. They also possess an enduring power brought by *Hethla'tusla*, who continues to uplift Kwakwaka'wakw people with those powers and teachings if one approaches the locations with knowledge and reverence.

Many other places hold special identities and powers, too – other categories of what might generally be called 'sacred places'. There are special healing places, prayer places and training places. There are certain mountains visited by young chiefs, and sometimes shamans or others for prayer and meditation. There are specific waterfalls where young Chief Adam was taken as part of his training, to help him expand his abilities and sharpen his focus as a chief-in-training. There are certain rivers and streams considered to be unique sources of 'holy water', *kwelth'esta*, bringing forth strength, cleansing and success to people during especially intense healing or ceremonial work. These places dispel darkness, enrich the soul and allow trained people a portal into the *nawalux* to enliven and to heal.

These landmarks are all respected and revered, and are sometimes invoked in the songs, stories and even regalia of Kwakwaka'wakw clans. Each place, each waterway, is understood to be its own unique animated thing, with its own character and identity. They have unique potentials to enliven, heal, empower, inspire and enlighten. At the most powerful landmarks, one only approaches with preparation and reverence; everyday visits to these sites for mundane purposes are prohibited. To approach the landmarks without due respect is to invite danger, as accidents happen when people travel casually and disrespectfully through these places. As with other peoples or species of power, harming the landmarks almost inevitably brings harm in return. To destroy or deface them is to disrupt fundamental powers and balances in our world. Indeed, such acts would be unthinkable, in the way destroying a centuries-old church would be to a devotee of a Christian faith. This pattern of special reverence and specialised ceremonial use of distinctive landmarks is consistent with what is known of other Native societies along the Northwest Coast and, indeed, around the world.[20]

## Salmon, Eulachon, Clams and Plants

### Relationships and Obligations to Game Species

We return to the question of how Kwakwaka'wakw people relate to the species on which our subsistence depends. Our relationships with these species are shaped by an appreciation of their spiritual identities and integrities, to be sure. Yet, they are also immediate and direct – between a specific community of people and, say, a specific community of salmon that returns annually to maternal streams immediately beside our home villages. Human communities were not bound directly to all salmon, but were bound with particular directness to the community of salmon who returned to the clan territories – human and fishy fates linked in part by our shared geography, but also by the choreography of countless generations, repeated rounds of mutual giving and taking that shapes both communities' fates.

If treated disrespectfully, the salmon simply do not return. This is an inexorable fact of life, a fundamental law of the universe. Our oral traditions hint at how greed and overexploitation inevitably result in 'equal and opposite reactions' at once biological, social and spiritual. With the guidance of clan chiefs and sometimes their court of shamans and other resource specialists, Kwakwaka'wakw traditionally halt fishing when our catch is sufficient, in order not to 'offend' the fish and our Creator. Our people have transported smolts to blighted streams and sometimes removed obstacles to fish passage such as logjams. Even today, our people hold ceremonies to honour their sacrifice. In all these actions, Kwakwaka'wakw respectfully seek the enduring consent of the Creator and the fish, so that the fish might still participate in the ancient relationship that links our two communities. In the aggregate, over time, this may measurably sustain or even enhance the population of fish.[21] So too with the seals, the berry bushes, the deer, the ducks and all other living communities that Kwakwaka'wakw rely on for survival. Kwakwaka'wakw people traditionally negotiate our own well-being from within a web of interdependency, linked to myriad species around us. Humans cannot unilaterally dictate terms to this vast network of natural sovereignties. In the natural world, just as within the potlatch ceremony, human stewards must therefore seek 'balance on every ledger'.[22]

These values come into play in almost every aspect of our traditional resource management, especially through our intentional cultivation of natural resources across our clan territories and through the seasons. These cultivation practises were encapsulated by Chief Adam in a single term, "*qwak'qwala'owkw*", or literally 'keeping it living' – a concept that implies many things. The term implies mechanical efforts undertaken by the ancestors to sustain our most important native food species, based on their nuanced understandings of environmental cause and effect within our homeland. Yet, the term also implies the respect extended to these species, the efforts to help them thrive – in part as reciprocation for their many sacrifices on our behalf.

To demonstrate the practical and philosophical implications of *qwak'qwala'owkw*, Chief Adam often spoke of the traditional care of cultural 'keystone' species.[23] For example, he explained how his grandfather, in his role as clan chief, long ago served as a *de facto* 'fish warden' – monitoring not only salmon but also the eulachon smelt, an oily anadromous fish eaten whole or rendered into an oil that has long been a staple food along the Northwest Coast. Along Kingcome River, people awaited the arrival of the fish with respect bordering on reverence, even being careful to only speak respectfully about the fish as they ascended the river from the sea. Special precautions are taken to not interfere with the species' spawning, for reasons both ecological and spiritual: 'you don't even touch them until they start spawning', to be sure that they have the opportunity to reproduce before being caught. With fishing underway, Adam's grandfather monitored the fishing stations along Kingcome River, consulting with his entourage of specialists. At once, he would declare that all fishermen must remove their fishing gear from the water – determining that enough fish had been caught, and more fishing might overexploit and alienate the fish. When other rivers along the coast experienced cataclysmic damage to eulachon runs, Chief Adam recalls that families sometimes gathered eulachon from Kingcome River in wooden boxes and paddled them by canoe to these other rivers to transplant.[24] Only if people observed these precautions and exhibited this kind of respect did 'the fish come back' as well as, or better than, before.

Such patterns are found in all manner of resources within Kwakwaka'wakw tradition. The same values and practises are expressed within the traditional cultivation and harvesting of clams. In suitable tidelands, our ancestors rolled rocks out of natural clam beds and into the low intertidal zone – at once improving clam habitat while also entrapping sediment that expanded the clam beds seaward. These specially managed places are sometimes termed 'clam gardens' – *luxiwey* in Kwak'wala. In these places, harvesters look after the clams: leaving young clams in place, intentionally aerating the soil and ensuring the clams are well. If done correctly, harvesters traditionally understand that the clams appreciate the changes and reciprocate by making themselves more available for harvest. Indeed, recent research confirms that these cultivated clam beds materially improved the quality and quantity of clams, beyond the conditions in naturally occurring clam beds.[25] When cultivated correctly, the *luxiwey* became both a source of everyday food and a risk-reducing resource to use when our ancestors experienced temporary downturns in productivity of salmon or other species. If the Kwakwaka'wakw kept our side of the bargain, the clams were there to keep theirs.

Throughout the Kwakwaka'wakw world, traditional resource management was carried out with similar objectives and outcomes – guided by understandings of ecological process and interspecific reciprocity calibrated over countless generations on the land. These values allowed our most important food species to thrive. This was seen in the management and care of maritime plant species as well. The production of estuarine 'root gardens', *tekilakw*, containing plants such as Pacific silverweed and Springbank clover, as taught to Chief Adam by his grandparents, followed similar protocols. By using selective harvest, soil amendments, soil aeration and other techniques that demonstrated respect for the plants and the people who depended on them, cultivators of these gardens verifiably enhanced production of these important root vegetables.[26] So too, the traditional harvesting of submerged beds of eelgrass, carried out selectively with long poles from canoes, also helped maintain and even enhance the productivity of these beds above and beyond the output of natural plots.[27] Berry patches and crabapple groves, burned and cared for in myriad ways by the ancestors, also follow this pattern.[28] In all these cases, Western science seems to confirm that *qwak'qwala'owkw* – 'keeping it living' – as a suite of biomechanical practises guided by consistent philosophical principles simply works. By longstanding attention to our obligations within and between communities, by applying concepts of reciprocity and even sovereignty within interspecific relationships, our people and the species on which we depend have been able to thrive on this coast since time immemorial.

In our intentional engagements with these beings, Kwakwaka'wakw stewards focus especially on our relationships with, and care for, communities of living beings with whom we have direct and enduring ties. The focus is significantly on what some have termed 'cultural keystone' species,[29] reflecting concepts and terminology used by Western land managers. In these modern Western contexts, keystone species and their habitats have often been used as a proxy for environmental health writ large. Yet, traditional Kwakwaka'wakw resource stewards understand that each of our 'keystone species' are themselves interwoven into bonds of interdependency and reciprocity with entire constellations of species beyond their linkages to humanity. To focus on keystone species in the traditional sense is not to forget the integrity of the whole, and the innumerable webs of life linking back to us. No community lives in isolation. To honour the salmon, one must also honour the insects, the plankton, the squid, shrimp and small fishes that they consume; one must honour the cleanliness of waters instream and offshore, the temperature of the waters, the wellbeing of riparian trees and brackish bayshore meadows. Tracing our extrapolated connections out into the world, we find that the webs of interdependence call for respects to innumerable species and environmental systems that expand beyond our distant horizons, spreading out into the larger world.

## Orca and Cedar

### Relationships and Obligations to Other Species

Beyond those outlined above, many other respects are shown between species, rooted in deeply multigenerational reciprocities that link human communities to communities of other living beings. These are not only relationships contingent on specific clan obligations, or on the mutualism between predator and prey, but also on other reciprocal relationships rooted in deeply shared interests and respects.

To illustrate this point, Chief Adam often spoke of our close relationship with orcas, the 'killer whales' that ply the Northwest Coast. Kwakwaka'wakw people have long had a sense of kinship with these whales, and even have oral traditions suggesting certain whales are reincarnations of human hunters from long ago. As manifestations of these connections, enduring patterns of cooperation exist between orcas and human hunters. Among their foremost mutual prey species are harbour seals, which often come ashore to sun themselves on the salt marsh tide flats at the mouth of Kingcome River, or to congregate in the adjacent shallow waters. In Kingcome village, when orcas were seen near the flats, human hunters were summoned. In these intertidal areas, orcas and human hunters upheld mutual obligations: orcas flushed the seals shoreward and onto land, while hunters hiding behind drift logs flushed seals back into the water for waiting orcas, each side killing a few in turn. Both humans and orcas ate better because of this arrangement. When families harvested the seal meat from these hunts, they did so with deep respect – not only for the orcas who helped in the hunt, but also for the lives of the seals. Seal meat was divided ceremonially between the four clans of Kingcome village, each clan receiving a designated portion of the animal.

Orcas honoured their relationship with humans in other ways as well. Kwakwaka'wakw oral tradition abounds with accounts of orcas helping humans who were in distress. These whales might, for example, help people lost in the fog when travelling by canoe. As a boy, Chief Adam witnessed just such an event when lost in the fog while canoeing the open water with his grandfather. Sighting a pod of orcas, his grandfather asked Adam to be still in the canoe, and then stood and 'spoke to the orca at the top of his voice … "Look after us friend!" he said … he gave a halibut to that orca … and asked it to take us back'. The orcas came alongside the canoe on either side, parallel to the craft, and began swimming slowly. His grandfather paddled along at the orca's pace. The moment they could see land clearly through the fog, the orcas dropped into the water and disappeared. The ancestors explained that the orcas – highly intelligent beings of spiritual significance – expect respect from humanity, and reciprocate what they receive. Receiving praise and food, the orcas recognise this respect as genuine and are compelled to assist. As a corollary, the ancestors shared accounts suggesting that people who disrespected orcas, such as in recent times by shooting at them, have been 'corrected' by orcas – even having their canoes sunk by the whales. The reciprocity human communities maintain with the species works both ways.

In addition to conferring success in the hunt, or helping Kwakwaka'wakw people in other tangible ways, some animals are said to grant humans special knowledge or power if we show proper respect, or have the potential to show respect. Chief Adam possessed the rights to one of the most sacred masked dance cycles of our people, the Atlikimma. It retells an account of a powerful vision dream that a grouse spirit brings to a young man who had killed grouse wantonly. The Atlikimma recalls the sequence of forest spirits called forward by this grouse to appear to the young man in turn, teaching him ethical and spiritual lessons that positively transformed his outlook and behaviour, allowing him to share these transformative lessons with others along the coast.

Mountain goats encountered in the rugged peaks along the coast are also said to confer powers. Chief Adam's grandfather reported receiving such powers from a mountain goat encountered high in the mountains above Kingcome Village – the animal teaching him skills, as well as a song he used in potlatches. As Chief Adam admonished, the ancestors taught that this aspect of human-animal encounters was to be honoured: it is 'not to be played with', but is a sacred power that has helped sustain humankind so long as the recipients are knowledgeable, prepared and respectful of the message and nonhuman messenger. Such potent communications between species are often mentioned as pivotal moments in the lives of individuals and communities. Owls, if treated respectfully, may carry information – such as when they appear and make a call sounding like a person's name, informing that person's friends and family of their death. Ravens are sometimes said to convey various messages to careful human observers as well. Similar examples are too numerous to list in a single chapter.

Chief Adam trains a young man who will dance the Atlikkima cycle in the grouse mask – a ceremonial event that reminds humankind of our shared obligations to the sentient and life-giving creatures of the forest. (Photograph by Kim Recalma-Clutesi.)

Yet, there are less dramatic examples of our mutual relationships with other species. Kwakwaka'wakw traditionally do many things to show respects and to maintain balance – even with dangerous animals who might not share our interests. Bears, for example, are a persistent part of Kwakwaka'wakw life and often congregate at the very places we might go for traditional resource harvests. Unlike many animals, bears have a diet like our own. Thus, at salmon fishing stations, berry patches and many other places throughout our traditional lands, bears are a persistent and potentially dangerous presence. This is often true at crabapple (*Malus fusca*) groves, where humans and bears both find a favourite plant food. To address the dangers inherent in our mutual love of crabapples, pickers traditionally approach crabapple groves with caution and respect. Often paddling to these groves by canoe, young people have been taught to sing a special crabapple picking song – sung loudly enough that it temporarily disperses the bears from the grove while broadcasting a promise: 'we will leave a little for the bears'. When human harvesters pick at these groves, they honour the bargain, leaving a few crabapples for the bears so they might not suffer from the harvest.

Even plants are traditionally afforded these types of respect. For those outside the Kwakwaka'wakw world, this tradition may be known by practises relating to Western red cedar. When Kwakwaka'wakw harvesters take cedar bark for use in clothing, ceremonial regalia or other purposes, they only harvest a portion, in part to avoid killing the tree and thus to keep up the human side of the relationship, ensuring materials made from bark and wood are blessed by their living source. Bark peelers also offer a blessing, a statement of deep thanks, to the tree and the Creator as they prepare to remove the bark from a living tree. As quoted by Boas, and retranslated for this chapter by Kwakwaka'wakw linguist and cultural knowledge-holder, Daisy-Sewid Smith, the bark peeler speaks to the tree:

> Go ahead and look at me, friend, for I have come to beg for your protection robe, for this is the reason you were created so that you may help us, you can be used for so many things, this is the reason you came to this world, we use your protective robe for everything whenever you are willing to give it to us. The reason I have come to beg for your protective robe long life maker is because I am going to make a basket for lily roots out of you.
>
> Now, I ask for mercy from you my friend so you will not be uneasy for what I am about to do to you.
>
> Now, I am now begging you my friend, to tell your friends that I will continue to beg for their protection robes.
>
> Now, my friend, be careful, you will protect me so that I will not catch any of the sickness and to be in pain.
>
> Now it is finished my friend.
>
> Now, this is the praise spoken by those who peel cedar bark from young cedar trees and old cedar trees.[30]

While past authors such as Franz Boas referred to these blessings as a 'prayer' to cedar, Kwakwaka'wakw people understand this to be a statement of praise and mutual respect, demonstrating gratitude and ensuring balance with a species that will suffer a little for the well-being of human harvesters.[31]

Such statements are offered at the harvest of almost any species of profound cultural significance, and there are many. Our ancestors have even held prayers of thanks for environmental phenomena; such statements of blessings and thanks are offered to falling snow, to thank it and the Creator for blanketing the land and the plants, 'letting them get the rest they need' before the next season. In all things, traditional Kwakwaka'wakw recognise the blessings bestowed on us, and seek to express thanks and to reciprocate in kind.

## Having One Voice

### A Conclusion

The Kwakwaka'wakw experience confirms the spirit and central thesis of this volume. Traditional concepts and values asserting intrinsic 'rights of nature' – rights extending well beyond those employed in current Western legal and philosophical traditions – have been essential to the long-term integrity of our homeland environments and the long-term resiliency of our society and culture. We contend that the following is verifiably true, even by the methods and standards of Western science: applying certain 'rights', akin to personhood, to nonhuman landscapes and life-forms has contributed to their reverential treatment in the Kwakwaka'wakw case, and this reverence is linked to their sustainable management and care. As a common practise among aboriginal peoples worldwide, this reverential treatment of nonhuman landscapes and life-forms reflects a deeper, even universal human perspective that has been lost through various historical developments in the formation of modern industrialised societies.

The Kwakwaka'wakw understanding of nonhuman life-forms and landscapes as having a fundamental sentience, spirit and consciousness is key. This perspective creates significant barriers to overexploitation or other forms of 'disrespectful' engagement. Treatment of living things is instead rooted in notions of mutualism, our relationships with them negotiated to some degree as one might negotiate with human counterparts. Yet, human decision-makers do serve as advocates and stewards for living things, such as salmon runs or orcas. As with most Rights of Nature writing, the Kwakwaka'wakw system does not presume an absence of human stewardship. Still, nonhuman beings are not considered inert or convertible 'commodities' at the disposal of human owners for unlimited exploitation. Instead, ancient bargains define these relational ties. Human communities are bound to other species – by a sense of ancient kinship, by mutual interdependence over deep time and by firm obligations to future generations of humans and nonhumans alike. Our ancestors and trained nobility admonish: if we do not observe these connections and respect them, we must live with the effects. The 'negative externalities' of disrespectful behaviour cannot be sidestepped but will be experienced quite directly by ourselves, our children and our children's children yet to come. So each generation shows respect to the orca, to the wolf, to the salmon and eulachon, to the clams and seals and cedar – indeed, to all the fish and plants and living things to sustain our people – and in doing so also shows the depth of our respect for our own ancestors, our descendants yet to come and the living human communities of today.

Of course, it is important to reassert that in the Kwakwaka'wakw world, many landforms and lifeforms are understood to be 'property' of a sort – lands attributed to the clan in the *gilgalis*, for example, or places such as estuarine root gardens and berry grounds that were significantly the product of clan labour and investment. Yet this concept of property ownership differs markedly from property as it is understood in the context of Western industrial capitalism. Property is 'owned' by a clan chief but managed and stewarded on behalf of the larger clan; the clan chief inherits these things but must 'keep his name good' in part by ensuring that the standing of the clan, the chieftainship and the lands and resources in their control are passed on to the next generation in good condition. These relationships shift the objectives and the timeframes significantly, promoting sustainable harvests and long-term planning.

In all actions, the Kwakwaka'wakw employ a concept of time that peers into the dimly lit past and also into the dimly lit future. In our traditional laws, our understandings of environmental process and our views of the universe, it is obligatory to consider the effects of actions across vast spans of time. Chiefly titles and identities are rejuvenated, and our identities and concerns are spread across deep time. Obligations to future generations are thus much less abstract and contingent on goodwill. Just as one is obligated to be equitable with people of distant places, one is also compelled to be equitable with people of distant future times.

So too, it is important to reaffirm that in the Kwakwaka'wakw world, humans remain consumers and predators—no matter one's place or diet, and that killing is surely a part of life. With nonhuman species being on a cosmological plane akin to humans, Kwakwaka'wakw find ourselves in a context where 'to kill' is not necessarily a sin; however, to kill recklessly, without intention or ability to repay the great debts so incurred, is among the gravest of offences. Kwakwaka'wakw people are reminded of this by the grouse in the Atlikimma dance cycle, and in all manner of other stories, songs and sacred rites spanning across remembered time. Our people recognise the sacrifice with transcendent gratitude, respect and a sense of direct obligation. In myriad ways, the Kwakwaka'wakw honour those beings that give their lives for our sake: with songs, with ceremonies and with material actions to ensure their well-being. And when we do this well, these beings reciprocate.

If our people take too much from salmon, the salmon will not return in the times to come. If our people overharvest clams, they will disappear as well. So it is with all the other living things on which our lives depend. Their responses are arbitrated independent of human judgment, by inexorable laws humans cannot meaningfully control. These beings come and go in response to our actions, and in this way are unavoidably their own sovereigns.

To ignore these facts is to be either naïve or arrogant. Industrial societies tend to forget these facts by virtue of the sheer mobility of capital and the ability to move on to 'greener pastures' after overexploiting a resource, or by disconnecting the point of consumption from the place where the damage occurs. The natural world always reciprocates, however, always responds to our actions, and operates with will and autonomy no matter what a society might wish or prescribe by law. To have a legal system that embraces this fact and adapts to it – as is true of Kwakwaka'wakw traditional law – is one way to ensure a society thrives and does not ultimately collide with some of the most fundamental laws of the universe.

This way of living has served the Kwakwaka'wakw well for millennia and is worthy of consideration as humanity seeks new models for rebalancing human-environment relations worldwide. In seeking 'balances on all ledgers', any act of taking must be balanced by an act of giving. Damage requires remediation – that includes damage inflicted on other communities, whether or not they are human. With clan chiefs such as Kwaxsistalla Clan Chief Adam Dick as mediators, our people have vigilantly monitored for signs of imbalance, and pre-emptively sought to identify and redress imbalances where they occur. These 'rights' are encoded in chiefly rules and modes of conduct.

In these ways, the ancestors have continuously 'minimised and mitigated' anthropogenic environmental damage, to use contemporary terms from the Western industrial world. Modern Western resource management commonly seeks to mitigate damage to species and habitats through 'wetlands mitigation', 'mitigation banks', and similar mechanisms; sometimes these strategies work, but very often they do not. In contrast, Kwakwaka'wakw tradition focuses on ultimate outcomes across generations: mitigation is only acceptable if it is truly and fully restorative over deep time. If it is done poorly, or as an excuse for bad behaviour, the fish will know, the clams will know, the plants will know. The Creator will know. Ultimately one's descendants will know – and suffer. Poor mitigation is like theft, and is punished as theft by a jury whose domain spans the land and waters, and the cosmos beyond.

Extending concepts of reciprocity and empathy to nonhuman species and landscapes produces reciprocal obligations that are interspecific and mutually sustaining. Even the Western industrial nations would seem likely to appreciate the protection of one's own interests in these concepts. If they find the cosmological foundations unfamiliar, Western thinkers might still recognise the general wisdom of the approach, embracing a kind of 'rational anthropomorphism' in legal and resource planning arenas. Such a concept demonstrates our acceptance of humans' position in the webs of causality and mutual obligations among species; it reverses centuries of missteps in Western legal tradition, each predicated on a false ontology that set humans fully apart from their position in the natural order.

What forms might Kwakwaka'wakw-influenced environmental law take? In the Kwakwaka'wakw world, there have traditionally been many legal mechanisms designed to review and arbitrate human effects on the natural world: clan chiefs, and their chiefly counsellors consisting of such people as shamans and resource specialists, conferred on how specific actions might affect the delicate balance with other species. When there was a need to seek opinions on these matters that were of great importance, the ancestors might debate these effects in specially organised ceremonial and social contexts. They might, for example, bring in those called the *Kw'kwikw* – literally the 'eagles' or 'eagle sentinels' – a sort of Supreme Court of specially trained nobility who assembled to review the facts of a case, making pronouncements that helped ensure balanced relationships between communities, human and otherwise, for the common good.

Yet, the strength of traditional values, the environmental and social systems that immediately pushed back against bad behaviour: these things limited the potential for extreme transgressions, and with it the need for organised arbitration. Here too, we might seek inspiration for the modern industrial world. Fostering an ethic of interspecific reciprocity and 'balance on every ledger' seems as urgent today as ever. Rooted in such an ethic, policy and legal mechanisms might facilitate the active monitoring of human relationships with those nonhuman entities on which we depend, those many sovereigns with which we are bound in a never-ending cycle of giving and taking. Communities of keystone species, the habitats that support them and even our 'ecosystem services and infrastructure', such as clean air and water – Kwakwaka'wakw experience suggests that our relationships with these things must be 'systematically monitored and rebalanced' continuously over time in order to protect mutually beneficial relationships in the long term. If this systematic evaluation reveals that we are indebted, that we owe more than we are owed, that this debt causes or may someday cause imbalance, similarly systematic mechanisms are prescribed to meaningfully repay the debt. These repayments cannot be mere 'window dressing' to satisfy short-term needs; the repayments must verifiably facilitate balance and the wellbeing of future generations, human and nonhuman, into the distant reaches of imaginable future time.

In all things, the Kwakwaka'wakw clan chiefs traditionally seek to balance and rebalance myriad relationships guided by an intricate understanding of cause and effect within the full web of life that supports us. Yet, modern technology now allows environmental causes and effects to play out at global scales. The Earth now begins to push back in response to the greed and disrespect of the industrialised world, on its own terms and by rules humans do not control. Resource stewards, seeking balance, must begin to think on unprecedented scales and with an understanding of vast webs of cause and effect that exist on a global scale. In this way, the Earth is its own sovereign. Will humanity soon grant this shared home of ours rights befitting its importance to our shared survival? Will peoples around the world seek 'balance on every ledger', as the Kwakwaka'wakw do within our own homelands? We offer Kwakwaka'wakw cultural values, outlined here, as one source of inspiration as people worldwide endeavour to answer these urgent questions. In light of the scale and urgency of the task, humanity may need to learn to 'speak with one voice'. Humanity may need to relearn ancient teachings that place concepts of respect and reciprocity at the centre of all relationships – including those with our nonhuman kin.

## Acknowledgements

The authors wish to thank Daisy-Sewid Smith (*Mayanilth*), Kwakwaka'wakw cultural specialist and teacher, for her generous assistance with Kwak'wala translation and spellings, as well as her input on the structure of the traditions sometimes called the 'potlach'—the *pa'sa* and the *yaqw'qwa*. Klaqwagila Clan Chief Mark Recalma and William White (*Kusulid/ Xelimulh*) also provided extensive and valuable editorial review on matters of cultural values and practise. Tricia Gates Brown also provided helpful editorial assistance with the original manuscript.

## Notes

1. See, for example, overviews of past litigation in (a) Boyd, D.R. 2017. *The Rights of Nature: A Legal Revolution That Could Save the World*. ECW Press, Toronto; (b) Pecharroman, L.C. 2018. 'Rights of Nature: Rivers That Can Stand in Court'. *Resources* 7(1); (c) Babcock, Hope M. 2016. 'A Brook with Legal Rights: The Rights of Nature in Court'. *Ecology Law Quarterly* 43. (https://scholarship.law.georgetown.edu/facpub/1906); and (d) Stone, C.D. 2010. *Should Trees Have Standing?: Law, morality, and the Environment*. Oxford Univ. Press. At the time of this writing, some of the most compelling litigation centres on efforts to extend rights to the Colorado River in the American Southwest. A U.S. district court dismissed a 2017 lawsuit brought against the State of Colorado for environmental damages, based on the precept that the River holds 'legal personhood'. While this case was dismissed, NGOs are presently coordinating on potential litigation to protect the integrity of the Colorado River utilizing a refined 'rights of nature' case.
2. For an overview of the philosophical foundations of the Rights of Nature movement, and its linkages to expanded notions of 'personhood' and indigenous precedents, see for example (a) Nash, R. 1989. *The Rights of Nature: A History of Environmental Ethics*. Univ. of Wisconsin Press, Madison, WI; (b) Biggs, S. 2011. *The Rights of Nature: The Case for a Universal Declaration on the Rights of Mother Earth*. The Council of Canadians, Fundación Pachamama and Global Exchange,

Ottawa; (c) Boyd, D.R. 2017. *The Rights of Nature: A Legal Revolution That Could Save the World.* ECW Press, Toronto, ON; and (d) La Follette, C. and Maser, C. (eds.). 2017. *Sustainability and the Rights of Nature: An Introduction.* CRC Press, Boca Raton, FL.

3. For an overview of key Kwakwaka'wakw values and concepts addressed in this manuscript, see also Deur, D., Recalma-Clutesi, K. and Dick, A. 2019. 'Balance on Every Ledger: Kwakwaka'wakw Resource Values and Traditional Ecological Management' in *Handbook of Indigenous Environmental Knowledge: Global Themes and Practice.* T. Thornton and S. Bhagwat. (eds.), Routledge, London.
4. *Kwaxsistalla* is an ancient name that appears in oral traditions relating to the origin of his clan; the name denotes Adam Dick's chiefly status – akin to a royal title within a European context – and was bestowed upon him in adulthood. The full name and title '*Kwaxsistalla* Clan Chief Adam Dick' is the formal and proper form address. We recognize that the use of that full name throughout this chapter would be cumbersome, so use the name 'Chief Adam' to refer to him less formally; when speaking of his childhood, we simply call him 'Adam'.
5. On this period and its implications, see for example (a) Cole, D. and Chaikin. I. 1990. *An Iron Hand upon the People: The Law against the Potlatch on the Northwest Coast.* Douglas & McIntyre, Vancouver, B.C. and (b) Sewid-Smith, D. 1979. *Prosecution or Persecution.* Nu-Yum-Baleess Society, Cape Mudge, B.C.
6. This, and related philosophical pronouncements, can be found in Deur, D., Recalma-Clutesi, K. and White, W. 2019. 'A Benediction: The Teachings of Kwaxsistalla Clan Chief Adam Dick' in *Plants, People and Places: The Roles of Ethnobotany and Ethnoecology in Indigenous Peoples' Land Rights in Canada and beyond,* N.J. Turner (ed.), McGill-Queens University Press, Montreal, Canada.
7. A full summary of Chief Adam's contributions to the academic literature on traditional ecological knowledge would require a chapter-length review. Summaries and keystones may be found in Mathewes, D. and Turner, N.J. 2017. 'Ocean Cultures: Northwest Coast Ecosystems and Indigenous Management Systems' in *Conservation for the Anthropocene Ocean: Interdisciplinary Science in Support of Nature and People,* P.S. Levin and M.R. Poe (eds.), 169–199. Academic Press, Cambridge, MA; and Deur, D. and Turner, N.J. (eds.) 2005. *'Keeping It Living': Traditions of Plant Use and Cultivation on the Northwest Coast of North America.* University of Washington Press, Seattle and University of British Columbia Press, Vancouver, B.C. A wide range of resource-specific studies have also been developed with his significant guidance. See, for example (a) Deur, D., Dick, A. Recalma-Clutesi, K. and Turner, N.J. 2015. 'Kwakwaka'wakw "Clam Gardens": Motive and Agency in Traditional Northwest Coast Mariculture'. *Human Ecology* 43(1): 201–212; (b) Cullis-Suzuki, S., Wyllie-Echeverria, S., Dick, A. and Turner, N.J. 2015. 'Tending the Meadows of the Sea: A Disturbance Experiment Based on Traditional Indigenous Harvesting of *Zostera marina* L. (*Zosteraceae*) in the Southern Region of Canada's West Coast'. *Aquatic Botany* 127: 26–34; (c) Wyllie-Echeverria, S.R. 2013. *Moolks (Pacific Crabapple, Malus fusca) on the North Coast of British Columbia: Knowledge and Meaning in Gitga'at Culture.* Unpub. MSc. thesis. University of Victoria (BC) School of the Environment; (d) Lloyd, T.A. 2011. *Cultivating the Tekkillakw, the Ethnoecology of Tleksem, Pacific Silverweed or Cinquefoil (Argentina egedii (Wormsk.) Rydb; Rosaceae): Lessons from Kwaxsistalla, Clan Chief Adam Dick, of the Qawadiliqella Clan of the Dzawadaenuxw of Kingcome Inlet (Kwakwaka'wakw).* Unpub. M.Sc. thesis. School of Environmental Studies, University of Victoria, Victoria B.C.; and (e) Sewid-Smith, D., Dick, A. and Turner, N.J. 1998. 'The Sacred Cedar Tree of the Kwakwaka'wakw People' in *Stars above, Earth below: Native Americans and Nature,* M. Bol (ed.), 189–209. The Carnegie Museum of Natural History, Pittsburgh, PA.
8. See, for example (a) Berkes, F. 2008. *Sacred Ecology: Traditional Ecological Knowledge and Resource Management,* 2nd ed. Routledge, New York; (b) Turner, N.J. 2005. *The Earth's Blanket. Traditional Teachings for Sustainable Living.* Douglas & McIntyre, Vancouver and Seattle, and University of Washington Press, Seattle; and (c) Turner, N.J. 2014. *Ancient Pathways, Ancestral Knowledge: Ethnobotany and Ecological Wisdom of Indigenous Peoples of Northwestern North America,* 2 vols. McGill-Queens University Press, Montreal.

9. This tendency was significantly due to the influence of Franz Boas, the father of American anthropology, who published on Kwakwaka'wakw topics throughout his career. Efforts to address environmental topics were mixed. See Boas, F. 1921. *Ethnology of the Kwakiutl*. 35th Ann. Rept. of the Bureau of American Ethnology, Parts 1 and 2. U.S. Govt. Printing Office, Washington, D.C.; and Boas, F. 1966. *Kwakiutl Ethnology*. H. Codere. (ed.). University of Chicago Press; cf. Deur, Recalma-Clutesi and Dick, *op cit*.
10. For example: (a) Boas, F. 1930. *The Religion of the Kwakiutl Indians*. Columbia University Press, New York; (b) Goldman, I. 1975. *The Mouth of Heaven: An Introduction to Kwakiutl Religious Thought*. John Wiley and Sons, New York; and (c) Walens, S.D. 1981. *Feasting with Cannibals: An Essay on Kwakiutl Cosmology*. Princeton University Press, Princeton, NJ.
11. Turner 2014, 2005 *op cit*; see also Turner, N.J. and Wilson, B.J. (Kii'iljuus) 2008. 'The Culture of Forests: Haida Traditional Knowledge and Forestry in the 21st Century' in *Wild Foresting: Practicing Nature's Wisdom*, A. Drengson and D.M. Taylor (eds.), 130–137. Island Press, Washington, D.C.
12. Boas often blurred the distinction between the pa'sa and the yaqw'qwa – conflating both under the term 'potlatch'. Pa'sa was an intricate system, in which contributed resources accrue value or dividends through their exchange. The yaqw'qwa involved a series of ceremonial payments to witnesses in order to validate and accurately recall ceremonies, namings, and transactions. Both organized Kwakwaka'wakw social, economic, and ceremonial life for countless generations until the 'potlatch ban' that forbade both traditions from the nineteenth through the mid-twentieth century (Daisy Sewid-Smith, pers. comm. 2005, 2018).
13. There is a significant literature that depicts the potlatch in this manner, some portion of it rooted in the work of Boas' students. See, for example, (a) Benedict, R. 1934. *Patterns of Culture*. Houghton Mifflin, Boston; and (b) Codere, H. 1950. 'Fighting with Property' *Monographs of the American Ethnological Society, 18*, M.W. Smith (ed.), J.J. Augustin, New York.
14. Among the more sophisticated examples were the influential works of Suttles. See (a) Suttles, W. 1960. 'Affinal Ties, Subsistence, and Prestige among the Coast Salish'. *American Anthropologist* 62: 296–305; (b) Suttles, W. 1968. 'Coping with Abundance: Subsistence on the Northwest Coast' in *Man the Hunter*, R.B. Lee and I. DeVore (eds.), 56–68. Aldine, Chicago; and (c) Suttles, W. 1974. 'Variation in Habitat and Culture on the Northwest Coast' in *Man in Adaptation: The Cultural Present*, Y. Cohen (ed.), 93–106. Aldine, Chicago.
15. Deur, Recalma-Clutesi and Dick, *op cit*.; Deur and Turner 2005, *op cit*; see also Deur, D. 2000. *A Domesticated Landscape: Native American Plant Cultivation on the Northwest Coast of North America*. Unpub. Ph.D. diss. Louisiana State University Department of Geography and Anthropology.
16. See (a) Vayda, A.P. 1961. 'A Re-Examination of Northwest Coast Economic Systems'. *Transactions of the New York Academy of Sciences*, 23: 618–624; (b) Piddocke, S. 1965. 'The Potlatch System of the Southern Kwakiutl: A New Perspective'. *Southwestern Journal of Anthropology* 21(3): 244–264; and (c) Suttles 1960, *op cit*.
17. Piddocke, *op cit*: 244.
18. Walens, *op cit*; Goldman, *op cit*.
19. Deur, Recalma-Clutesi, and Dick, *op cit*.
20. For other Northwest Coast examples, see (a) Thornton, T.F. 2008. *Being and Place among the Tlingit*. University of Washington Press Seattle; (b) 'Scientific Panel for Sustainable Forest Practices in Clayoquot Sound'. 1995. *First Nations' Perspectives on Forest Practices in Clayoquot Sound*. Report 3. Victoria, B.C.
21. Thornton, T.F., Deur, D. and Kitka, Sr., H. 2015. 'Cultivation of Salmon and Other Marine Resources on the Northwest Coast of North America'. *Human Ecology* 43(2): 189–199.
22. Deur, Recalma-Clutesi, and Dick, *op cit*.
23. Garibaldi, Ann and Turner, Nancy J. 2004. 'Cultural Keystone Species: Implications for Ecological Conservation and Restoration'. *Ecology and Society* 9(3).
24. Thornton, Deur, and Kitka, *op cit*.

25. (a) Groesbeck A.S., Rowell, K., Lepofsky, D. and Salomon. A.K. 2014. 'Ancient Clam Gardens Increased Shellfish Production: Adaptive Strategies from the Past Can Inform Food Security Today'. *PLOS ONE*. 9(3): e91235; (b) Deur, D., Recalma-Clutesi and Turner, *op cit*.
26. (a) Deur, D. 2005. 'Tending the Garden, Making the Soil: Northwest Coast Estuarine Gardens as Engineered Environments'. in *Keeping It Living: Traditions of Plant Use and Cultivation on the Northwest Coast of North America,* D.E. Deur and N.J. Turner (eds.), 296–330. University of Washington Press, Seattle, WA and University of British Columbia Press, Vancouver, B.C.; (c) Deur 2000, *op cit;* Boas 1921, *op cit.;* Lloyd, *op cit.*
27. Cullis-Suzuki et al., *op cit.*
28. (a) Lepofsky, D., Hallett, D., Lertzman, K., Mathewes, R. McHalsie, A. and Washbrook. K. 2005. 'Documenting Precontact Plant Management on the Northwest Coast, an Example of Prescribed Burning in the Central and Upper Fraser Valley, British Columbia'. in *Keeping It Living: Traditions of Plant Use and Cultivation on the Northwest Coast of North America,* D.E. Deur and N.J. Turner (eds.), University of Washington Press, Seattle, WA and University of British Columbia Press, Vancouver, B.C.; (b) Wyllie de Echeverria, *op cit.*
29. Garibaldi and Turner, *op cit.*
30. Retranslated by Daisy Sewid-Smith, February 2019, from the original Kwak'wala language version of the blessing printed in Boas 1921, *op cit*: 619.
31. *Ibid.* For comparison, see Sewid-Smith, *op cit.*

## Appendix: Common and Scientific Names of Plants and Animals

| | |
|---|---|
| **GRASSES AND GRASSLIKE PLANTS** | |
| Eelgrass | *Zostera marina* |
| **FORBES** | |
| Lily | *Lilium* spp. |
| Pacific silverweed | *Argentina pacifica* |
| Springbank clover | *Trifolium wormskioldii* |
| **TREES AND SHRUBS** | |
| Crabapple | *Malus fusca* |
| Sitka spruce | *Picea sitchensis* |
| Western red cedar | *Thuja plicata* |
| **INVERTEBRATES** | |
| **MOLLUSKS** | |
| Clams | Mollusca |
| Squids | Cephalopoda |
| **CRUSTACEANS** | |
| Shrimp | Pleocyemata |
| **VERTEBRATES** | |
| **FISH** | |
| Eulachon smelt | *Thaleichthys pacificus* |
| Salmon | *Salmo* spp. |
| **BIRDS** | |
| Eagles | Accipitridae |
| Grouse | Tetraoninae |
| Owls | Strigiformes |
| Raven | *Corvus corax* |
| **MAMMALS** | |
| Bears | Ursidae |
| Harbour seal | *Phoca vitulina* |
| Mountain goat | *Oreamnos americanus* |
| Orca | *Orcinus orca* |
| Wolf | *Canis lupus* |

# 7

# *Environmental Sustainability: The Case of Bhutan*

**Dechen Lham**

**CONTENTS**

Country Background ........ 113
Bhutan and Its Biodiversity ........ 114
Conservation History: Protected Area System ........ 115
Legacy of the Monarchs in Sustainable Environmental Conservation ........ 116
  The Philosophy of Gross National Happiness ........ 118
Stewardship and the Rights of Nature: Conservation Policies and Laws ........ 118
  Bhutan's International Commitments and National Policies ........ 120
  National Planning and Institutional Frameworks ........ 121
Conservation Milestones in Bhutan ........ 123
  Bhutan for Life ........ 124
  Buddhism and Its Role in the Rights of Nature ........ 125
  Conservation Awards Received in Bhutan ........ 125
Coping with Challenges to Sustainable Conservation ........ 126
  Human-Wildlife Conflict ........ 126
  The Problem of the Medical Fungus ........ 128
  Waste Management and the Sustainable Landscape ........ 130
Desuung: 'Guardians of Peace and Harmony' ........ 131
Other Conservation Challenges ........ 131
  The Shingkhar-Gorgan Road ........ 131
  The Green Bench: The Environment in Court ........ 133
Conclusion ........ 133
Notes ........ 134

## Country Background

Bhutan, best known for its developmental philosophy of Gross National Happiness, is a small, landlocked country of 38,394 square kilometres with a small population of 779,666 people.[1] Within its extensive forests, currently covering about 72% of the land base, Bhutan harbours some of the best remaining representatives of the Himalayan wildlife and habitat. More than half (51.44%) of the country's total geographic area is designated as part of the protected areas network.[2] As of 2016, Bhutan had a forest cover of 71%, based on the recent National Forest Inventory conducted between 2012 and 2015.[3] Sixty-nine percent of the population depends on agriculture for livelihood.[4] Bhutan has less than 8% total arable land, and 70% of this is dry land farming.[5] The forestry, livestock and agriculture sector contributes 16.52% of the gross domestic product as of 2015–2016.[6] Sustainable hydropower is considered Bhutan's major income source, and tourism is also becoming a

DOI: 10.1201/9780429505959-7

notable contributor. Buddhism is the state religion, and the majority of the people follow it.[7] Religion has a lot of influence in the daily lives of the Bhutanese. Before moving to how conservation has progressed over time, let me share briefly about Bhutan's biodiversity, and the rights of people and Nature.

## Bhutan and Its Biodiversity

Bhutan is known as one of the top 10 biodiversity hotspots globally.[8] This rich biodiversity results from the country's tremendous altitudinal ranging from 150 metres above sea level to more than 7,000 metres above sea level.[9] The forest ecosystems are equally diverse, with species from subtropical, temperate and alpine levels.[10]

Jomolhari mountain range from Chelala, the highest road point in Bhutan, 2016. (Photograph by Dechen Lham.)

Forests provide everything for the Bhutanese: food, timber, fibre and medicines and a huge range of ecosystem services.[11] To the rest of the world, Bhutan's forests are carbon sinks, absorbing up to an estimated 6.3 million tonnes of carbon.[12] Bhutan is home to a wide array of wild and domestic species. More than 200 species of mammals (such as elephants, rhinoceros, wild dogs, leopards, macaques, gorals, takins, serow and blue sheep) are said to occur in Bhutan, including 27 globally threatened species.[13] This includes 11 cat species, the two best known of which are tigers and snow leopards. In Bhutan, tiger habitat overlaps with snow leopard range, which has never been recorded elsewhere. Bhutan is also home to 780 different species of birds, 61 species of amphibians, 124 species of reptiles and more than 91 species of freshwater fish.[14]

There are more than 5,600 vascular plants, of which 94% are considered to be native to Bhutan, and 105 are endemic to Bhutan alone.[15] The country also has 411 species of ferns and their allies. Based on records, Bhutan has an estimated 282 species of nonvascular plants like mosses and liverworts. There are also more than 350 species of fungi, of which 53 are edible.[16] The most important insect-fungi association is the fungus called in Bhutan 'yartsa goenbub' (*Ophiocordyceps sinensis*), a widely sought-after medicinal fungus that grows in the highlands of Bhutan.[17] In Bhutan it is known as the 'gold of the mountains', but in the West it is popularly referred to as 'Himalayan Viagra'.[18] Later I will discuss this species further, and its connection to the Rights of Nature. Although only 286 species of lichens have been recorded so far, experts estimate Bhutan has more than 1,100 kinds of lichens.[19]

## Conservation History: Protected Area System

Bhutan's first Five Year Plan was initiated in the 1960s,[20] which coincided with the beginning of formal conservation programmes via establishment of Northern and Southern Wildlife Circles.[21] These consisted of the Forest Department divisions mandated to oversee wildlife conservation and protected areas management.[22] This was followed by the establishment of the first protected area of Bhutan, popularly known as Manas Wildlife Sanctuary, in 1966.[23] The Forest Act of Bhutan, enacted in 1969, was the first modern law enacted by the government.[24] It mandated that a minimum of 60% of Bhutan's forests be protected in perpetuity.[25] This mandate is also reflected in the Constitution of the Kingdom. Conservation continues to be a central focus in Bhutan. The Paro Resolution of May 1990 highlighted the need to balance nature conservation and development, by ensuring conservation is part of every 5-year plan and all government policies.[26] In 1992, the two Wildlife Circles were merged to form a nature conservation division of the Forestry Department, with a mandate for the overall management of nature conservation and protected areas management.[27]

The government in the early 1990s evaluated the Northern and Southern protected area system and realised that these regions were poorly represented among areas protecting the country's ecological diversity.[28] In 1993, Bhutan expanded its protected areas network, establishing nine areas with a total area of 10,513 sq.km (26% of the country) based on IUCN criteria.[29] In 1999, the government created a network of biological corridors that connects the protected areas, covering an additional 8.6% of the country. This combined network of protected areas and biological corridors is called the Bhutan Biological Conservation Complex.[30] Thus, the protected areas increased to 35% of Bhutan's land base overall, consisting of four national parks, four wildlife sanctuaries, one strict nature reserve and 12 corridors.[31] Three of the biological corridors were later subsumed into Wangchuck Centennial National Park after its formation in 2008, the country's largest protected area. The remaining eight biological corridors provide linkages and habitat connectivity for wildlife. They were created to ensure sustained ecological and environmental integrity, prevent inbreeding of species, secure protected migratory habitat, and provide additional wildlife habitat.

The Rights of Nature, or perhaps more accurately, the stewardship of Nature to ensure its flourishing, have always been part of government policy and intervention.[32] Not only conservation policies but all public policies have to be screened to ensure they are in alignment with the principles of the Gross National Happiness philosophy. During the tenure of the second elected government (2013–2018), the Minister of Agriculture and Forests formed a task force to review the protected areas network – 51.4% of Bhutan's land area – and look into the possibilities of increasing the protected areas to 60%. The Department of Forests and Parks Services manages the protected areas. All are governed by 5-year conservation management plans. However, due mainly to budget constraints, many of the programmes could not be satisfactorily completed; by 2017 only one biological corridor had a conservation plan, which expired before it was even implemented. This was also one of the main findings from the evaluation of Bhutan's protected areas network conducted in 2015–2016.[33]

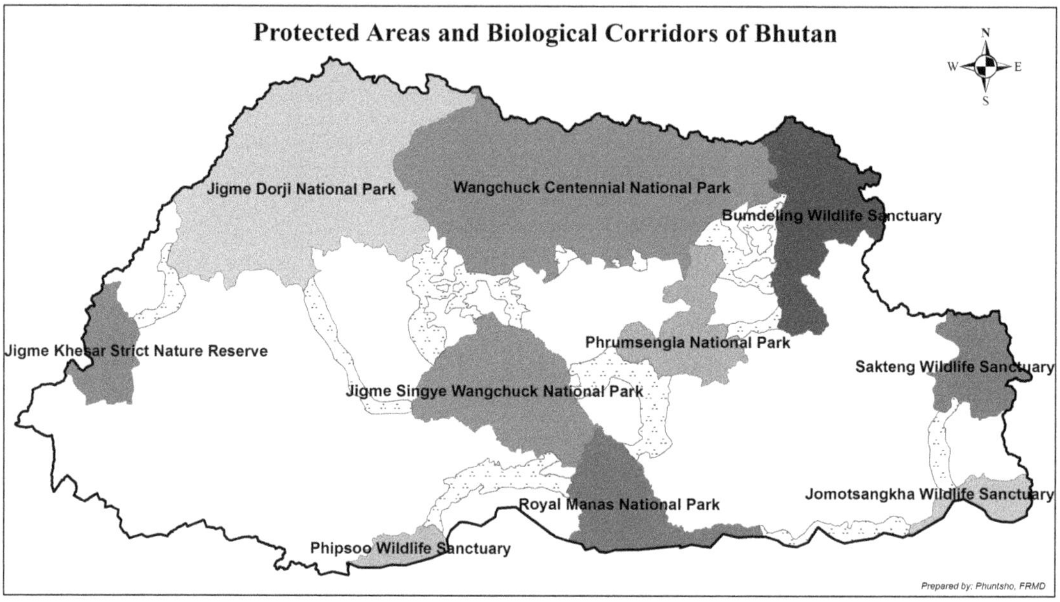


Protected Areas of Bhutan. (Map courtesy of Department of Forests and Parks Services, Royal Government of Bhutan.)

This first-ever assessment of Bhutan's protected areas was an eye-opener for national officials, who came to understand the current situation better, including the emerging threats and challenges.[34] This information provided the government an opportunity to address threats as early as possible. This first assessment indicated that much still remains to be done for biodiversity and environmental management in Bhutan, and also set a performance baseline of the country's protected areas. Now progress can be tracked over time to ensure sustainable conservation is achieved.

## Legacy of the Monarchs in Sustainable Environmental Conservation

I will begin with background on the conservation history of Bhutan, how it has formed over the years to what it is now, and discuss the challenges the government faced in addressing these issues linked to the Rights of Nature.

With the reunification of the country under the dynamic leadership of Jigme Namgyel (father of the First King), the first Monarch of the Kingdom of Bhutan was crowned on 17 December 1907 as His Majesty King Ugyen Wangchuck, the First King of the Kingdom of Bhutan.[35] Through several generations the Wangchuck dynasty has selflessly contributed to the development of Bhutan and its people. The Second King, His Majesty Jigme Wangchuck, ruled between 1926–1952, and the late Third King, His Majesty King Ijgme Dorji Wangchuck, ruled from 1952 until 1972. The first and second kings' acquaintance with the British botanists George Sheriff and Frank Ludlow, who were at that time stationed in India, could have potentially helped instil in their Majesties an understanding of the importance of the environment.[36] The late King was known for his passionate outdoor activities and expeditions, for which he also travelled to Africa. The late King passed away young due to illness, and the Fourth King of Bhutan was crowned on 2 June 1974, when he was only 18. The Fourth King initiated and institutionalised many conservation initiatives during his reign, and His Majesty is known for his conservation legacy in Bhutan.

In 2005, His Majesty the Fourth King of Bhutan King Jigme Singye Wangchuck announced his abdication in favour of his son, the present His Majesty King Jigme Khesar Namgyel Wangchuck. The present King ascended to the throne on 9 December 2006, while His Majesty's public coronation was celebrated on 6 November 2008 coinciding with the 100 years of Monarchy in Bhutan. Further, His Majesty the Fourth King selflessly gifted democracy to the people of Bhutan in 2006, and the first elected government formed in 2008. Most importantly, His Majesty the Fourth King was the person behind Bhutan's Gross National Happiness concept, which empowers people for their own betterment and also led to the 'gift of democracy'.[37]

The importance of conservation and stewardship of Nature is reflected in the Constitution's language. Article 1, Number 12, states:

> The rights over mineral resources, rivers, lakes and forests shall vest in the State and are the properties of the State, which shall be regulated by law.[38]

Most importantly, the Constitution has an entire article focusing on the environment. Under Article 5, each and every citizen of Bhutan is responsible for taking care of the country's environment, making every Bhutanese an environmental trustee. The phrase entrusting this responsibility states:

> Every Bhutanese is a trustee of the Kingdom's natural resources and environment for the benefit of the present and future generations and it is the fundamental duty of every citizen to contribute to the protection of the natural environment, conservation of the rich biodiversity of Bhutan and prevention of all forms of ecological degradation including noise, visual and physical pollution through the adoption and support of environment friendly practices and policies.[39]

Another extremely important Constitutional provision, Article 5.3, requires the country to maintain 60% of forest cover for all time to come. As of now I do not know of any other country that has such a declaration in its constitution to conserve the rights of the environment to exist. Articles 7 and 8, listing the fundamental duties and rights, further require the citizens of Bhutan to 'preserve, protect and respect the environment, culture, and heritage of the nation'. These two articles clearly reflect the rights of people, be it in terms of speech or rights with respect to natural resources and the need to abide by rules and regulations.

Article 13 regulates the process of legislation. This creates checks and balances in the government, as the King must assent to any bill passed by the Parliament. It is very unlikely that any elected government will propose anything that will be detrimental to the country's pristine environment and the citizens' rights and responsibilities over resources. Nonetheless, given future uncertainties, Article 13 eliminates bills that could negatively impact the country's environment. The hereditary Monarchs have always recognised the importance of our environment and its conservation. The strong language in the Constitution, and long before it, is clear evidence of this concern. Nature has always been linked to the lives of the Royal family and Bhutan's people. This is even evident from the selection of the national bird and plant. The raven is believed to safeguard the country from adversity and thus adorns the crown of the King. The medicinal cypress (*Cupressus corneyana*), often burned as incense, is the national tree of Bhutan.

### The Philosophy of Gross National Happiness

Living in harmony with Nature is one of the concepts ingrained in Buddhist philosophy.[40] Buddhism, the dominant religion in Bhutan, has a big influence in how the citizens think about, behave towards and respect Nature.[41] The focus on honouring Nature was strengthened with establishment of the country's development philosophy of the Gross National Happiness. This standard is composed of four main pillars: (a) sustainable environmental conservation, (b) promotion of equitable and sustainable socio-economic development, (c) preservation and promotion of cultural values and (d) good governance.[42] The Gross National Happiness index measures societal well-being through nine domains: living standard, good governance, education, health, ecological diversity and resilience, community vitality, time use, cultural diversity and resilience and psychological well-being.[43] All four pillars are linked, and achieving one is integral to, and contingent on, achieving the others. Such a holistic strategy has especially shaped the Eleventh Five Year Plan, which linked all programmes to the Gross National Happiness standard, as well as measured them against environment, climate change and poverty indicators for the first time.

The idea is to ensure that environmental conservation is never compromised at the cost of development activities. Bhutan's planning agency is called the Gross National Happiness Commission, mandated to develop and monitor the country's policies. According to its mission, the Commission 'evaluates, monitors, sets goals and raises national consciousness about what conditions are conducive to the happiness and wellbeing of the people and country as a whole'. The agency's screening process is applied in two phases: at the project level and at the policy level. For example, it evaluates good governance for all ministries and sectors, such as education, health and employment. Any policy initiative starts with submission to the Commission for endorsement, and later goes through the Commission's screening process. If approved, it is shared with the appropriate sector, and finally recommended for final approval at the Cabinet level.

## Stewardship and the Rights of Nature: Conservation Policies and Laws

Bhutan's conservation is supported by powerful set of government policies and commitments. Due credit goes to the visionary Monarchs of this country, particularly His Majesty King Jigme Singye Wangchuck, the Fourth King, and the continued leadership and immense dedication by His Majesty the Fifth King. Below is a timeline of the forest and/or conservation policies the Bhutanese government has approved:

| Sl.No | Law | Year |
|---|---|---|
| 1 | Thrimshung Chenmo (Country's plan) | 1958 |
| 2 | Bhutan Forest Act | 1969 |
| 3 | National Forest Policy | 1974 |
| 4 | Notification for Wildlife Sanctuaries | 1974 |
| 5 | Nationalization of logging | 1979 |
| 6 | Shifting of Forest Department from Ministry of Trade and Industry to Ministry of Agriculture | 1983 |
| 7 | Master plan for forest development | 1990 |
| 8 | Forest and Nature Conservation Act | 1995 |
| 9 | Biodiversity Action Plan | 1997 |
| 10 | Free market ban on export of logs | 2000 |
| 11 | Forest and Nature Conservation Rules | 2003 |
| 12 | Revision of Forest and Nature Conservation Rules | 2006 |
| 13 | Revision of Forest Policy | 2011 |
| 14 | Forest and Nature Conservation Rules | 2017 |

Although Bhutan does not want to compromise its environment at the cost of development, it is a challenge to have any development without negatively impacting the environment. Furthermore, it is an even more difficult task when the country's socio-economic growth is dependent on the environment (especially hydropower) and more than 69% of the population depends on agriculture and natural resources.[44] Bhutan's conservation policy is unique in that it embraces people as part of the larger landscape conservation and allows local communities to reside inside the protected areas (backed by the Nature Conservation Act of 1995); but pressure on nature is inevitable unless people's needs are taken care of.

In view of these issues, the concept of 'poverty and environment mainstreaming' was introduced through a 2007 United Nations Development Programme project in Bhutan.[45] This philosophy ensures the integration of environment, climate and poverty in the country's policies, plans, programmes and budget to ensure a green, more inclusive sustainable development path. Bhutan developed a framework to incorporate these concerns into the long-term planning process, and chart how each ministry would incorporate these elements into their respective programmes.[46]

Poverty is an essential element of focus, since the poor and the vulnerable are the first to be affected when environmental and development challenges occur. This is particularly important with climate change and its negative impacts, especially as the majority of Bhutanese depend directly on natural resources for their livelihoods. Highlighting these elements in the country's plan, at both national and local levels, will ensure that the well-being of the most vulnerable people is considered, and tools to tackle conservation challenges are in place. This focus on poverty and the environment started as a joint project with funding from the United Nations programmes in development and the environment, and the governments of Denmark, Switzerland and the European Union.[47]

The present phase, building on the earlier focus of mainstreaming at a national level, is called the Local Governance Sustainable Development Program, and its goal is to promote inclusive and equitable socio-economic development, sustainable use of natural resources and strengthening local governance. This programme formed an 'environment climate change and poverty reference group', which consists of members from all nine government ministries of Bhutan. The team conducted a series of awareness workshops with each ministry to acquaint officials with the process and identify opportunities for collaborating among ministries.

Let me share an example: forest fire is a key issue under the 'environment, climate change and poverty' policy. First, officials analysed the trends and impacts of forest fire, which leads to forest degradation and biodiversity loss, making land vulnerable to soil and water erosion. Second, officials identified options and opportunities to mitigate forest fires, such as advocacy and campaign, rehabilitation of degraded forests, improving fire monitoring systems and research to improve forest fire management. Finally, officials analysed the impacts of the identified opportunities, including how they would be monitored and evaluated.

Thus, impacts of these forest fire interventions will contribute to reducing forest and biodiversity loss, and improving soil and water conservation. These can be measured through the percentage reduction of fire incidents and the research-based management schemes in place. Finally, these are linked to national and sectoral key results areas such as carbon neutrality and sustainable forests.

The national plans had identified 16 National Key Results Areas with an objective to 'self-reliance and inclusive green socio-economic development', and for these the government created baselines, targets and indicators to monitor performance. Through all these, Bhutan aims to strive for an environmentally sustainable economic progress that fosters low-carbon and socially inclusive development. The Twelfth Five Year Plan (2018–2023) aims to take decentralisation to the grassroots level, while giving more emphasis to local government and its funding. Bhutan's Twelfth plan objective is 'Just, Harmonious and Sustainable Society through enhanced decentralisation'.[48] In an effort to encourage decentralisation, this plan has local government agencies determining their own key result areas and performance indicators. Such an approach is expected to build local ownership and synergize between results and the programmes. This plan builds on flexibility to accommodate financial emergence and unexpected events.

## Bhutan's International Commitments and National Policies

Bhutan's commitment to global conservation is evident from its membership in more than 15 regional and international environment agreements and treaties covering topics such as biological diversity, wetlands, wildlife enforcement, agricultural genetics, desertification and the law of the sea, many sponsored by the United Nations.

At the 2009 United Nations Framework Convention on Climate Change, Bhutan issued a declaration, 'The Land of Gross National Happiness to Save our Planet', and committed to be carbon neutral at all times.[49] Accordingly, the country's 11th FYP prioritised environmental enforcement to maintain carbon neutrality.

There are about 15 national policies and frameworks that exemplify Bhutan's commitment to the international community while fulfilling the national focus on Nature's Rights. Let me give an example, from the National Forest Policy of 2011: 'Bhutan's forest resources and biodiversity are managed sustainably and equitably to produce a wide range of social, economic and environmental goods and services for the optimal benefit of all citizens while maintaining 60% of the land under forest, thereby contributing to Gross National Happiness'.

The policy guides forest use in a sustainable manner, moving away from the former protection paradigm, which encouraged unsustainable utilisation in an attempt to alleviate poverty. Policy strategies include such objectives as community forestry, which empower rural communities to manage their own forests to provide high-quality drinking water, irrigation water and timber. The local sale of forest products aids in reducing rural poverty, a long-standing problem in Bhutan.

Bhutan, in 2015, adopted the 2030 Agenda for Sustainable Development Goals. Due to its strong conservation and reform history, the country was able to align these Goals into its own 5-year plans without difficulty; the majority of the Goals were already part of existing Bhutanese plans. For example, 'poverty reduction' continues to be a goal of Bhutan's plans, and the country has already reduced national poverty from 12% to 8.2% from 2012 to 2017 using strategies such as land reform, expanded land tenure and associated rights.[50] Up to 71% of Bhutanese households now hold land tenure, and of this 94.1% is owned by poor households.[51] Nevertheless, unforeseen challenges, such as the impacts of climate change, affect the farming communities where many of these reforms took effect.

Despite such setbacks, Bhutan is in the midst of graduating from a Least Developed Country to a Middle Income Country, per the United Nations System.[52] This highly important shift means that Bhutan will have less budgetary assistance from overseas development agencies to finance its development programmes, and thus needs to plan accordingly. Some of the greatest challenges in this transition arise in the environmental sector. Until now, environmental protection has been mostly a donor-financed programme. It is a major challenge for Bhutan to finance all its Constitutional mandates to maintain the sustainable use of natural resources and continue to steward Nature's rights to flourish. However, Bhutan's transition did not come as a surprise, and the country had done its homework in planning and preparation. Briefly, Bhutan has created a transitional fund for conservation. Today it is known as 'Bhutan for Life', and earlier, as the Bhutan Trust Fund for Environmental Conservation, which was the first of its kind globally.[53]

## National Planning and Institutional Frameworks

National policy planning is an integral part of any development task, and is well linked from the national to the lowest level of local institutions. Bhutan's national policies are aligned with the national needs and goals of the country, and also to its international commitments. Agencies are in place to oversee specific mandates. For example, the National Environment Commission (NEC) is a separate, high-level coordinating and decision-making body that does not belong to any of the nine ministries of the government. Its mission is 'to conserve and protect environment through the NEC as an independent apex body on all matters relating to environment to regulate environmental impacts and promote sustainable environment'. One of its mandates is to develop, review and revise environmental policy, plans and programmes, such as the 2011 Water Act of Bhutan and the National Environment Protection Act of 2000. The Commission oversees Bhutan's environmental policies and guidelines, (such as those prepared for industrial projects), sets standards and publishes rules for drinking water, waste management and environmental discharges.

Another example is the National Biodiversity Centre, an agency established for the implementation in Bhutan of the International Convention of Biological Diversity. Through the Centre's initiative, species previously unknown for their medicinal or other properties came to have commercial value, such as the edible orchid *Cymbidium erythraeum*. In collaboration with a Swiss pharmaceutical company, the cellulose from the cells of this orchid are used in the development of an anti-aging cream called 'Redeem'. This created a demand for huge quantities of the orchid. Since it had never been cultivated, locals often gathered the flowers from the wild and sold them on the illegal market, because they would be fined by government agencies if they were found selling them.

Through the Centre's initiative, Wangdi and Trongsa community groups were formed to domesticate and propagate the orchids. The Centre provided a seed fund to start the community farm, and technical staff trained the farmers how to grow orchids and manage them in the greenhouses. The Centre developed a scheme to buy the produce and sell to the Swiss company while farmers are allowed to sell their excess. The funds from the sales went to the community development and a percentage to the Government's exchequer.

Realising conflicts would increase with more species being studied for medicinal purposes, the Centre in 2014 drafted an access and benefit-sharing mechanism, creating national regulations to oversee such practises. This further aligns with Bhutan's commitment to Nagoya Protocol on Access and Benefit Sharing.

The needs of the people are an important guiding principle in Bhutan's planning. For example, Druk Holding and Investment Corporation, formed in 2007, is the commercial arm of the Royal Government to manage the government's investments and safeguard national wealth management for the long-term benefit of the Bhutanese people. Its Natural Resources wing, first created in 1979, focuses on sustainable management of natural resources. It has been reorganised several times, having started as a timber and forest-oriented agency. Finally in 2007, when His Majesty requested creation of the best possible system to ensure fair and equitable distribution of Bhutan's scarce resources – especially sand, timber and stone – the agency took its present form with expanded responsibilities to be the nation's premier supplier of these essential construction materials at an affordable and sustainable manner.

The Corporation and the Department of Forest and Parks Services collaborate closely, as the Department oversees all the natural resources of Bhutan. Established in 1952, it was one of the first government departments in Bhutan, and was initially staffed by Indians.[54] The Department carries out the Constitutional mandates relating to natural resources, including the best-known mandate to maintain at least 60% of the land under forest cover for all time to come.

There are conflicts, as the Natural Resources Development Corporation supplies the hands-on management, while the government agency oversees protections and use, although both are involved in developing forest management unit plans. For example, one big challenge is the national requirement of timber for the renovation of fortresses. There are many fortresses, the so-called dzongs, in Bhutan that used to be the heads of local government. Maintaining the culture and traditional architecture of these dzongs leads to a huge requirement for special class hardwood timber, such as sal (*Shorea robusta*), teak (*Tectona grandis*), walnut (*Juglan regia*) and champ (*Michelia champaca*).

Does Bhutan have that much timber? Of course, dzong renovation needs throughout the country will not all arise at the same time, but the high demand for particular species of timber presents a huge challenge to the resource-based agencies. What if Bhutan does not have enough timber or standing trees of the required species for a dzong renovation? Is Bhutan going to harvest all that it has? Will it buy the timber from outside, or substitute with second-best timber species? One partial solution has been the establishment of a glue-laminated timber plant. The pilot project is presently located at Pambisa, Paro, to supply laminated woods to a special project of the Academic Centre of Excellence for now, and will later be shifted to Ramtokto, Thimphu to cater to larger clients across the country.

Punaka dzong. 2016. (Photograph by Dechen Lham.)

(Wangchuk, S. 2018. Chief Executive Officer, Natural Resources Development Cooperation Limited, DHI [personal communication]). This glue-laminated timber fulfils the needs of different shapes, sizes and lengths required for dzong renovation or other construction.

With booming construction of buildings and roads, the requirement for sand and stone has increased tremendously over the past decade. So far, Bhutan has been able to peacefully fulfil its natural resource needs through agency oversight of private stone and sand quarries. The agencies' task is to regulate the pricing of natural resources, and ensure they are available and affordable for the general public as mandated. The Bhutanese believe that natural resources are state-owned and should be sustainably utilised to fulfil the needs of the people.

All agencies, institutions and departments directly or indirectly involved in the conservation and development sectors are linked. However, that does not mean the situation is picture-perfect. Challenges frequently arise, as development is always at the cost of Nature's ability to flourish unimpeded, no matter how minimal the impact may be. Before sharing the challenges in stewardship of Nature's right to flourish in Bhutan, let me briefly present the conservation milestones and role of Buddhism in protecting Bhutan's environment.

## Conservation Milestones in Bhutan

Setting up more than 51% of the entire country as a protected area gave Bhutan both a tremendous opportunity and many challenges. Managing half the country under a conservation regime, with locals residing within these protected areas, requires both fulfilling people's needs and taking care of Nature. The Forest and Nature Conservation Act of 1995 clearly calls for the management of biological resources within the protected areas system. From the separation of the Forest Division in the 1970's from the Trade Ministry to the establishment of its own agency, the Department of Forest and Parks Services has come a long way. With further new divisions focused on nature conservation, watershed, forest resource management and social forestry, every aspect of overseeing the use of our natural resources is well structured. Initially, the Department did not have enough trained forest officials; today it has more than 1,200 foresters spread across the country to conserve Bhutan's biodiversity. The department also has a research wing, called the Ugyen Wangchuck Institute for Conservation of the Environment, that trains foresters in forest resource and wildlife management.

The Department has had many notable achievements. First is the national forest inventory of 2016, conducted by Bhutanese nationals, which surveyed the entire country.[55] Other conservation milestones include the first-ever national tiger survey (2014–2015), snow leopard survey (2014–2016) and elephant survey (2016–2018).[56] Bhutan also recently developed a national tiger action plan,[57] and established national animal rescue and rehabilitation centres in north and southern Bhutan to cater to the needs of injured wildlife. Conservation research in Bhutan has matured from not having enough trained Bhutanese to now having well-trained professionals to undertake both field surveys and other environmental studies.

Besides the baseline research, many ongoing studies respond to the urgent needs that arise during successful implementation of sustainable conservation. For example, Bhutan recently opened a regional tiger conservation and research centre, which serves as the regional hub for tiger research. This is a great achievement, and shows the knowledge and commitment of Bhutanese biologists.

## Bhutan for Life

One of the biggest conservation milestones for ensuring sustainable environmental conservation is the project Bhutan For Life (BFL). Pressure to secure internal financial support for the environmental sector has been very strong during Bhutan's transition to a mid–income generating country. World Wildlife Fund's brilliant idea of sustainable financing conservation came to the rescue. Today the Bhutan For Life project is one of Bhutan's strategic and long-term innovative solutions to secure environmental sustainability. This 14-year project aims to provide a sustained flow of funds to maintain Bhutan's protected areas system in perpetuity (after 14 years, the Royal Government will take over the full financial responsibility).

Such a permanent-finance model is the first in Asia, and its uniqueness is that it is a single deal-closing project. Groups of donors commit funds towards Bhutan for Life, and the funds are held until the entire sum has been raised. The fund then begins to make annual payments, high in the beginning and ending at zero, while the Royal Government increases its share of investment annually, until it fully takes over the financial management of the half of Bhutan that is forever conserved. Government contributions come from sources such as hydropower, ecotourism and a green tax levied on import of vehicles as a sort of a plough-back mechanism.

Bhutan for Life is composed of a comprehensive conservation plan and a financial plan. The two plans have been developed over 3 years (2013–2017) through rigorous consultation with all relevant stakeholders, particularly the foresters working in the protected areas. In order to ensure sustainable conservation that secures people's well-being and increases climate resilience, the BFL conservation plan has five themes: (a) sanctuary for the diversity and persistence of life; (b) purveyor of sustainable, resilient ecosystem goods and services; (c) reservoir for carbon and adaptation to climate change; (d) centre of economic opportunity and community well-being and (e) centre of effective management and efficient services. Bhutan was able to raise US$43 million, and begin implementation of the first-year plan in 2018. For the next 14 years, Bhutan for Life will contribute tremendously to strengthening the enforcement and safeguarding of protected areas, ecotourism, improving livelihoods and protecting and monitoring wildlife.[58]

## Buddhism and Its Role in the Rights of Nature

Buddhism has an immense influence on the citizens of Bhutan.[59] The basic principle that every living being has a life is the core of the concept of Nature's rights. Most Bhutanese believe that whether you are a human, an animal or a tree, all have life; all are born, live and grow and ultimately die. Just because they cannot speak for themselves, it does not mean they have no rights. Like each of us have rights, all living beings have rights. We who receive or enjoy the benefits of Nature should ensure Nature's rights are always protected. The interconnectedness of all living beings, popularly known as the 'Thuenpa Phuen Zhi', is what most Bhutanese believe in.

Thuenpa Phuen Zhi is the story of four living beings: a tree, an elephant, a bird (usually a partridge or hornbill) and a monkey. It is known as the 'living in harmony story' or as the 'four harmonious animals'. Thuenpa literally means cooperation, and Phuen Zhi means four relatives. The story is about living together harmoniously. Though there are several versions, the ultimate moral is the same: the importance of living in cooperation and respecting each other. The concept of karma and the cycle of rebirth are widely believed in Bhutan. That is to say, if you harm any living being in this life, you will be harmed that same way in your next life. These beliefs greatly influence Bhutan's focus on environmental safeguarding and stewardship, so that Nature can fully flourish.

## Conservation Awards Received in Bhutan

In 2005, the United Nations Environment Programme presented its prestigious Champions of the Earth Award to His Majesty the Fourth King in recognition of his placing environmental conservation at the centre of the country's Constitution and development plans. World Wildlife Fund subsequently awarded His Majesty the J. Paul Getty Conservation Leadership Award for his 30 years of continued support for environmental conservation in Bhutan.[60] The award especially singled out His Majesty's leadership in placing in the Constitution the requirement that Bhutan maintain 60% of forest cover for all time, and also His Majesty's leadership in developing many conservation policies and laws. His Majesty received the award on behalf of the people of Bhutan, underscoring that all citizens had worked together to ensure sustainable environmental conservation in the country.

In March 2018, Bhutan received the Earth Award after being selected as the first amongst 100 top destinations globally in recognition of its 'sustainable tourism and distinctive appeal'.[61] Bhutan was in the top list in 2016 and 2017 as well. The then-Prime Minister Tshering Tobgay was awarded the German Sustainability Award in 2016 for his outstanding contributions in promoting ecological and social sustainability. Currently, Her Majesty the Gyaltsuen plays a crucial role as the United National Environment Programme Ozone Ambassador and the Royal Patron of the environment in the country.

## Coping with Challenges to Sustainable Conservation

Let me begin by quoting,

> We Bhutanese are good at writing plans, speaking well and expounding ideas. But implementation fall short of commitments. There is gap between commitments made and output delivered. We are not able to deliver results of expected quality in a timely manner.[62]

Of course, nothing is perfect. With a changing climate, Nature is taking its toll. Bhutan has set itself a high benchmark, and strives for the best stewardship and caretaking possible. Phenomena beyond control are difficult to prepare for; Nature's next unexpected step can bring down an unanticipated disaster. Despite conservation successes, and amid its transition to a middle-income country, Bhutan is affected by many environmental challenges. These make stewardship of Nature's right to flourish difficult to balance with development.

### Human-Wildlife Conflict

To quote Dr. Pema Gyamtsho (Minister of Agriculture and Forests of the first elected government), 'When we harm nature, we harm ourselves'. Human-wildlife conflict has been the biggest challenge in conservation and in respecting Nature's rights in Bhutan.[63] In the past three decades, cases of human conflict with wildlife such as the tiger, Himalayan black bear, Assamese macaque and wild boar have been on the rise.[64] Population growth and development activities have led to encroaching on wildlife habitat. The degradation and fragmentation of habitats lead to humans and wildlife competing for the same limited space and resources. This is despite the stringent environmental safeguards placed on development activities in protected areas, including a ban on shifting cultivation, limited extraction of timber and nontimber produce from the forest and limits on grazing areas.

In the past, when Bhutan had little mass media coverage, outsiders knew little about these conflicts. Today the country is well connected, and one can often read about human-wildlife conflicts. Recognising an urgent need to address this issue, the government reviewed the on-the-ground situation by bringing all the stakeholders together, and then drafted the 'Bhutan National Human Wildlife Conflicts Management Strategy 2008'.[65] The objective of the plan was to ensure reduction of human wildlife conflicts, enhance the livelihood of farmers and offset farmers' losses from wildlife damages. As the Strategy points out, 'had we taken better solutions and taken care of nature's need appropriately then, we would not have such a big issue now'. Human-wildlife conflict is often perceived as one of the major causes of poverty in rural Bhutan.[66] Livestock predation by wild carnivores is a major problem across the country, as agriculture and animal husbandry are the mainstays of the economy, the principal livelihood of 79% of the population. Many of the poorest farmers live in protected areas, where predator concentration is higher, thus exacerbating the problem.[67]

For example, in the 1980s, dhole (Asian wild dog) predation decimated livestock, and the farmers decided to eradicate this predator through mass poisoning of livestock carcasses.[68] This led to drastic increases in wild pig numbers, a favoured prey species of dholes. Subsequently, wild pigs became notorious for causing crop damage in Bhutan. The government then had to reintroduce dholes in the early 1990s to increase their populations to tackle this problem.[69]

Overall, locals suffered huge financial losses due to crop damage by wild pigs and macaques,[70] and livestock loss to predators like tigers, Himalayan black bears and leopards.[71] But even in such situations, religion still played a crucial role. For example, farmers performed rituals to protect their crops, as a study conducted in Jigme Singye Wangchuck National Park (JSWNP) found. Of the 274 households surveyed, 21.2% reported a loss of 2.3% of domestic livestock to wild predators in the year 2000, and this loss was equivalent to 17% of their per capita annual income.[72] But since farmers were left to guard their livestock independently, life became harder each time livestock was lost to predators.

The study found that livestock predation increased after the park was established. About half of the farmers surveyed blamed the stringent Forest and Nature Conservation Act, which banned wildlife killing and limited grazing areas for livestock.[73] However, it was also reported that a lax herding system by farmers contributed equally to livestock loss to wild carnivores. The study found 67.5% of the farmers were willing to exterminate problem animals, while 76.3% valued the integrated conservation development programmes initiated by park for enhancing livelihood opportunities.[74]

In the 1990s, 23% of the farmers in Zhemgang Dzongkhag stopped growing rice due to enormous crop damage, and 39% of farmers abandoned dry land agriculture altogether.[75]Several cases of elephant crop-raiding were recorded in Southern Bhutan in 2018, and this is nothing new, as elephant conflicts have been studied for quite some time now.[76] Crops were damaged, properties destroyed and often lives were lost. What was the government doing? Following the Strategy, government agencies designed interventions to handle conflicts with the different wildlife species, along with indicators to measure success over specific time frames.

Ura Valley in Bumthang. (Photograph by Dechen Lham.)

Since the Strategy is a living document meant to be adapted as situations change, it continues to play a crucial role in conflicts. For example, tiger and leopard depredation on livestock caused 82% of farmer's losses in Jigme Singye National Park in 2000, which was a huge loss to poor households.[77] Farmers who never complained before started to demand action and compensation from the government.[78] In order to ensure that locals remained pro-conservation and would not retaliate against wildlife, the government initiated the Human Wildlife Conflict Management Endowment Fund in 2010.[79] Before this fund was created, the government compensated all livestock losses from tiger, leopard, bear and snow leopard predation through donations. This initiative, known as the Tiger Compensation Fund, was developed in collaboration with WWF Bhutan in 2003.[80] On review this was found to be unsustainable, as the government could not continue providing compensation once donations ended.

Learning from the experiences of the Tiger Fund, the new fund was created as a more sustainable approach to mitigate human-wildlife conflict at a local level, through a community-based insurance scheme. A local community would then be more prepared to handle conflict situations, and at the same time be compensated for their losses. This local compensation fund is called a 'gewog environmental conservation committee fund' (GECC). Although government policy has proposed that more than 200 such funds be established across Bhutan, the question is whether it would be more effective to focus them just on the conflict areas. Furthermore, in 2016, the National Plant Protection Centre, together with WWF Bhutan, initiated the Human Wildlife Conflict SAFE Strategy in 2016.[81] This is a holistic integrated-management approach aiming for five outcomes: safe person, safe wildlife, safe assets, safe habitat and effective monitoring.[82] The pilot project took place in nine gewogs of Bhutan.[83]

The government has also initiated mass electric fencing, particularly for elephants, ungulates and primates. Bhutan so far has installed 2,773 kilometres of electric fencing across the country.[84] But is electric fencing the solution? What about trans-boundary elephants that move between India and Bhutan? Can Bhutan afford, and is it practical, to fence the entire southern border? Elephants flourish in Bhutan, but there are conflicts.[85] Based on the results of a recent national elephant survey, Bhutan has 678 migratory elephants.[86] To better understand the conflicts and potential solutions, four elephants were collared in order to study their movement patterns and behaviour.[87] Many Bhutanese revere elephants as gods, but when farmers incur severe economic losses and at times endure life-threatening confrontations, it is difficult to garner their support for elephant protection. In situations like this, conflict between people's livelihood and Nature's right to flourish require creative solutions to enormously difficult challenges.

## The Problem of the Medical Fungus

*Ophiocordyceps sinensis* (referred to as Cordyceps) is a nonwood forest plant, initially listed as a totally protected species under the 1995 Forest and Nature Conservation Act. It is a highly valued medical fungus that grows abundantly in northern Bhutan. Well known for its medical properties, it fetches US$25,400 per kilogram in the market.[88] His Majesty the Fourth King only legalised Cordyceps harvest as of 2004. Before then, people were not allowed to collect the plants, except for one local group of villages, the *gewog* of Lunana. The people residing in these high mountains have always had difficulty finding a reasonable livelihood, but since Cordyceps gathering was legalised, their income increased and their livelihoods improved.

When local livelihoods improved, this greatly reduced the pressure on natural resources, and allowed replacement of fuel wood by cooking stoves and solar lighting. For instance, from 2004–2009, some 394 collectors from five *gewogs* earned Nu. 57 million, which is about Nu. 0.14 million per household on average.[89] Farmers used this income to purchase household needs and renovate or construct new homes. People pay a small royalty for a permit each year, as Cordyceps is an annual species, and collection is variable each year depending on growth and availability. The peak period is usually between April and August. The government limits the collection period to just 1 month (mid-May to mid-June) to ensure against overharvesting.[90] The permit initially allowed only one person per household to gather the plants, but was later expanded to allow three per household.

However, challenges remain. One major problem is illegal collectors from across the country's border, despite the strict monitoring by foresters and army officials. Another ongoing issue is waste management in the mountains due to Cordyceps collectors.[91] Most bring packaged or canned food items, and dispose the trash in the vicinity without any sense of responsibility. At the forefront of livelihood improvement, we cannot afford to let the alpine ecosystem perish due to insensitive trash disposal.

Not only this, but the local people notice that each year the amount of Cordyceps harvested changes; at times there is plenty and at other times there is very little. The Parks and Forestry Department developed a sustainable harvesting management plan, and trains all collectors in proper harvesting methods to ensure future sustainability. But in the face of greed – who collects first gets the most – and the fear that illegal collectors will harvest all the plants, leaving an area unharvested for next year is a hard decision. During the Cordyceps collection period, forest officials are in the field monitoring whether people carry valid permits, collect only within their jurisdictions and follow the regulations. However, there are never enough forest officials to monitor every eligible collector every day. At one time Forestry officers recruited Desunng officials ('Guardians of Peace' volunteers) to help with the monitoring, but this ended up being expensive and was discontinued.

Often those with permits collect from ineligible areas or encroach into others' collection territories, and there are too few foresters to solve such conflicts. Researchers at the Ugyen Wangchuck Institute for Conservation and Environment (UWICE) conducted a study on the impact of the Cordyceps collection on people's livelihood and alpine ecosystems from 2004–2009.[92] Apart from the income gain and livelihood improvement, the study found that the overuse of rhododendron and juniper as fuel wood could result in their vanishing from the ecosystem. The study recommended reducing the number of collectors, instituting proper garbage management, requiring use of stoves and conducting further research on Cordyceps biology to ensure this species continues to thrive in alpine ecosystems.

On the positive side, local residents are aware of the negative impacts current practises are having on the environment, and are willing to cooperate with government officials. The government creates awareness of the need for species conservation, fragile Cordyceps habitat and how best to avoid negative environmental impacts. Residents understand the need for use of gas stoves, managing waste properly and harvesting Cordyceps sustainably.[93] Based on past harvesting trends, many fear that one day Cordyceps might disappear and their livelihoods will be affected, so they support the government's interventions. As an incentive, the government initiated a pilot project providing free gas stoves for cooking and heating, and tents for temporary housing, during collection periods to reduce pressure on the alpine ecosystem.

In addition, due to the uncertainty of Cordyceps' future, the government initiated alternative livelihoods, such as ecotourism-related activities. In this way policies remain adaptive and farsighted enough to make the local communities resilient.

### Waste Management and the Sustainable Landscape

In general, waste management is a big challenge for Bhutan.[94] Landfills are being filled sooner than planned, and improper waste segregation and management have negatively impacted the environment by polluting the air and water. Waste degrading the quality of downstream rivers affects aquatic biodiversity. To cope with this mounting problem, the government passed the Waste Prevention and Management Act of 2009, which implements rules at the local dzongkhang level. The government is supporting these local government bodies through state and donor funding to prepare for initial plan implementation.

Bhutan's cities have increasing nondegradable trash.[95] Cities like Thimphu and Phuntsholing in the south face major problems due to their high populations.[96] Nondegradable waste was originally an issue mainly in urban areas, but now is of concern everywhere, particularly with the increasing consumption of nonbiodegradable wastes. A study conducted in 2007 reported that 47% of 43,700 tonnes of municipal solid wastes from urban Bhutan was domestic or household waste, followed by nonhousehold waste from commercial sources at 23%, and nonhousehold office wastes at 12%. Schools, other institutions and vegetable markets created 10% or less of the overall waste.[97] Household waste was 58% organic, presenting untapped opportunities for greater composting.[98] In fact, organic waste composting has been initiated in Thimphu.[99]

Waste management is struggling with increasing population. In addition to the national and local waste management strategies and plans, 'Clean Bhutan' was established in February 2014 under the royal patronage of Her Majesty, the Gyaltshuen Jetsun Pema Wangchuck, the Queen of Bhutan.[100] Clean Bhutan has a vision of zero waste Bhutan by 2030 and a mission 'to change the mindset of every Bhutanese to be responsible citizens and practice sustainable consumption lifestyle by using available resources most efficiently'. Their strategy is to 'reduce greenhouse gas emission from landfill and prevent river pollution from waste by advocating behavioural change on sustainable consumption lifestyle'. For the short term, they work closely with local governments, municipal authorities and communities to management waste more efficiently.[101]

The United Nations Development Programme supported a study in Thimphu to analyse the solid waste management interventions.[102] The study reported that with proper waste segregation and good management of equipment, the present system of waste collection and disposal could be improved without cost implications. The study recommended the employment of a private-sector partnership for waste management so that the city could increase segregated waste collection efficiency.[103] Further solutions to waste management at the local level included the encouragement of informal waste collectors and scrap dealers, and independent private sector initiatives such as Greener Ways and ReCiTi. Greener Ways has been a success story of waste collection, segregation and disposal of wastes following the principles of the 4Rs (refuse, reduce, reuse and recycle), while ReCiTi has been a failure, mainly due to management issues and lack of business wisdom.[104]

A nationwide cleanup campaign was conducted on 9 December 2016, a national holiday coinciding with His Majesty's 10 years of enthronement.[105] The ninth day of every month is now a cleanup day for the entire country.[106] To cite one recent new form of waste monitoring, Sarpang dzongkhag is using closed circuit television, as other means of encouraging residents to dispose of waste properly have not succeeded well.[107]

## Desuung: 'Guardians of Peace and Harmony'

> The ultimate objective of nation-building is peace and harmony for Bhutan and her people.
>
> **His Majesty the Fifth King of Bhutan.**[108]

Desuung is an initiative of His Majesty the Fifth King, and clearly shows His Majesty's concern for his people and country's well-being. He instituted the Desuung integrated training programme to encourage all people to be active in nation-building. The underlying principles of Desuung include the spirit of volunteerism, ethics and values of community service, integrity and civic responsibility. Desuung volunteers play a big role, especially during disasters and many other activities. Desuung volunteers help tackle the challenges of environmental sustainability, particularly disaster risk management. Bhutan is often faced with natural disasters such as earthquakes, the danger of glacial lake outburst floods, landslides, flash-foods and forest fires.[109] Desuung are in place mainly to fill the gap of lack of skilled manpower able to tackle rescue operations, relief operations, rehabilitation and reconstruction works.[110]

Desuung volunteers have been involved in assisting during forest fires, cleanup campaigns and crowd management of government events.[111] Bhutan sent a team of 63 Desuung volunteers to Nepal in 2015 when a major earthquake hit the country.[112]

## Other Conservation Challenges

Other conservation challenges include population increase and its pressure on natural resources; increasing rural-urban migration, leading to more pressure on available resources; forest fire; the illegal wildlife trade, which often uses traditional trade routes; overgrazing; unsustainable harvest of nonwood forest products and infrastructure developments such as roads and telecommunication towers that are not well planned.[113] All these contribute to soil erosion, habitat fragmentation and degradation, increased wildlife conflicts and socio-economic burdens on the most vulnerable: the poor.

### The Shingkhar-Gorgan Road

The 56-kilometer Shingkhar-Gorgan road, which has so far cost Nu.779 million, is the most controversial road in Bhutan.[114] This proposed road passes through pristine tiger habitat in Phruemsengla (Thrumshingla) National Park and if constructed would be Bhutan's highest-elevation road. It was planned in 2010, but even 8 years later, the government has not approved the construction. The main reason for the delay is controversy over a section of the proposed route, which would pass through the core area of the Park. Phruemsengla is known for its pristine natural habitat and is home to many thriving wildlife species, including the Bengal tiger. The first image of a tiger from Bhutan was captured on a camera trap from this park. Environmentalists have consistently maintained that this road would violate forest and nature conservation rules, which do not permit any construction in the core of any protected area. Concerned Bhutanese noted that if the case reached court, this issue could be easily solved, but at that time there was no environmental court.

Narphung Valley under Samdrupjongkar dzonkha, on the road to Mongar, 2017. (Photograph by Dechen Lham.)

Despite many public debates, stakeholder consultations and discussions, the road was not approved. The National Environment Commission argued that the 36-kilometre stretch that would cut through the Park's core requires the government to assess the road's legality. The Commission also pointed out that the information in the Environmental Impact Assessment were all from secondary sources drawn from outside the project area. The previous government argued that this road would shorten the distance to the capital by 100 kilometres for people living in very remote areas of the country, thus connecting them to the rest of the nation more effectively.[115]

When former Prime Minister Tshering Tobgay and former Minister Lyonpo Dorji Choden visited Lhuntse in 2017, the issue of this road came up. Tobgay responded that he would look into it, and construction would start immediately if the road were not illegal.[116] But Bhutanese government officials know the road is illegal and environmentally disastrous, and the need for it very questionable. The former Forests and Agriculture Minister, Dr. Pema Gyamtso from the first democratic government, said at the time that he would resign from the Cabinet should this road be approved. Today, the neighbouring roads are constructed and only the stretch that goes through the core area awaits approval.[117] Originally the road construction was awarded to the Border Roads Organization, popularly called DANTAK, and former Minister Dorji Choden confirmed that the road would be completed by 2018. However, the National Environment Commission decided the project has to be re-evaluated with a new environmental assessment before the agency could review it again.

The former government directed the Ugyen Wangchuck Institute for Conservation and Environment to undertake an independent study. When the Institute conducted a survey in 2016 and found that the road fell within the multiple zones of the National Park, but not the core zone, the Forest Department – which had earlier refused approval – suddenly granted permission to proceed with the road construction.[118] Twenty kilometres of the road outside the Park are complete, but despite having the money and equipment available, the government has not approved the critical 36-kilometre stretch through the Park's core area, which contains the endangered Bengal tiger habitat.[119] A new interim government is in place (late 2018), but the road still does not have the required approvals. We hope this road will never get approved and never be constructed.

### The Green Bench: The Environment in Court

In view of the challenges facing the environmental sector, and to ensure that the rights of Nature are heard and given justice, the Supreme Court in 2015 established the so-called 'Green Bench'.[120] This court specifically allows public-interest litigation for environmental issues. The Green Bench was established in honour of His Majesty the Fourth King, and is chaired by the chief justice and high court justices.[121] Thus far (2018), not a single environmental case has been registered. However, through this court, any person, even someone not directly affected by an environmental problem, can file a case. This right, embedded in the Bhutan Constitution (Article 21, section 18), empowers every citizen to fight for Nature's rights.[122]

There are environmental conflicts – but none have yet risen to the level of a court case. In 2014 alone, the National Environment Commission imposed fines of Nu.855, 666 for noncompliance with provisions of the National Environment Protection Act.[123] These penalties were levied for activities that have not complied with requirements, such as beginning development without environmental clearances, late renewal of clearances and operation of mines and quarries outside of the approved plans.

The Supreme Court (2018) is in the midst of drafting the Green Bench book to describe the court's rationale, and sign a memorandum of understanding with the National Environment Commission, which is the sole authority on all matters related to protection and conservation of the natural environment. The Green Bench will seek the Commission's opinion before hearing a case.[124] In July 2018, the judiciary of Bhutan, in collaboration with the Asia Development Bank, United Nations Environment Programme and the U.S. Environmental Protection Agency, conducted a workshop on environmental adjudication for the judiciary of Bhutan.[125] The workshop brought together lawyers and practitioners from India, Pakistan, New Zealand and the United States to equip the judges sitting on the Green Bench with skills needed to tackle potential environment and climate change cases. Their input greatly helped Bhutan finalise its draft rules of procedure for environmental cases and bench book for the Green Bench judges.

## Conclusion

Being a small country with a manageable population, Bhutan has brought sustainable environmental conservation to the forefront of the country's development. The government has pledged co-financing to manage the protected areas system, which covers more than half the country, for the next 14 years (until 2032). The present government has committed to investing the same funding level required to maintain the country's protected areas at the most optimal level.

Bhutan stands out from the rest of the world's nations by its exemplary sustainable environmental management, which provides for the right of Nature to flourish without compromising the welfare and rights of its citizens. With Bhutan's sound conservation policies, in combination with dynamic leadership from the highest level of the government, and with full support from the visionary monarchs, the country is far ahead in respecting the rights of Nature. If all the other countries followed in Bhutan's footsteps, we could all contribute to a better state of our Mother Earth, and future generations could enjoy the same Nature we enjoy today.

Let me end with His Majesty the Fifth King's quote at his address at Keio University, Japan, in 2011:

> The problems facing the world today – they challenge all of us equally, And the solutions to these challenges must come from a real sense of concern and care for others, for all sentient beings and, for future generations. We must care about what happens to this earth.

## Notes

1. National Statistics Bureau, 2017. Bhutan at a Glance 2017. http://www.nsb.gov.bt/publication/files/pub9wt9959wh.pdf (Accessed 6 August 2018).
2. (a) Wangchuk, S., Lham, D., Dudley, N., & Stolton, S., June 2017. 'Half Bhutan: The Evolution and Effectiveness of Protected Areas in a Country Recognizing Nature Needs Half. International Perspectives'. *International Journal of Wilderness*, 23(1); (b) Ministry of Agriculture and Forests (MoAF, 2016). Bhutan State of Parks 2016. Department of Forest and Parks Services, Ministry of Agriculture and Forests, Royal Government of Bhutan; (c) National Forest Inventory Report: Stocktaking Country's Natural Resources. NFI, Volume I. 2016. Department of Forest and Parks Services, Ministry of Agriculture and Forests, Royal Government of Bhutan.
3. DoFPS, 2016. National Forest Inventory Report: Stocktaking Country's Natural Resources. NFI, Volume I. 2016. Department of Forest and Parks Services, Ministry of Agriculture and Forests, Royal Government of Bhutan.
4. DoFPS, 2016. *National Snow Leopard Survey of Bhutan 2014–2016* (Phase II): Camera Trap Survey for Population Estimation. Department of Forests and Parks Services, Ministry of Agriculture and Forests, Thimphu, Bhutan.
5. Gross National Happiness Commission, 2013. Tenth Five-Year Plan (10 FYP) (2008–2013), 2013. Volume I. Royal Government of Bhutan.
6. National Statistics Bureau, 2017. Bhutan at a Glance 2017. http://www.nsb.gov.bt/publication/files/pub9wt9959wh.pdf (Accessed 6 August 2018).
7. (a) Wangchuk, S., Lham, D., Dudley, N., & Stolton, S., 2017. 'Half Bhutan: The Evolution and Effectiveness of Protected Areas in a Country Recognizing Nature Needs Half. International Perspectives'. *International Journal of Wilderness*, 23(1), (June 2017); (b) Planning Commission, 1999. *Bhutan 2020: A Vision for Peace, Prosperity and Happiness*. Government of Bhutan.
8. (a) National Biodiversity Centre, 2014. *National Biodiversity Strategies and Action Plan of Bhutan* Ministry of Agriculture and Forests, Royal Government of Bhutan; (b) National Environment Commission, 1994. *Conservation in Bhutan. Royal Government of Bhutan.*
9. (a) Lham, D., Wangchuk, S., Stolton, S., & Dudley, N., 2018. 'Assessing the Effectiveness of a Protected Area Network: A Case Study of Bhutan'. In *Oryx*, March 2018; (b) DoFPS, 2016. National Snow Leopard Survey of Bhutan 2014–2016(Phase II): Camera Trap Survey for Population Estimation. Department of Forests and Parks Services, Ministry of Agriculture and Forests, Thimphu, Bhutan.
10. (a) National Forest Inventory Report: Stocktaking Country's Natural Resources. NFI, Volume I. 2016. Department of Forest and Parks Services, Ministry of Agriculture and Forests, Royal Government of Bhutan; (b) *National Biodiversity Strategies and Action Plan of Bhutan (NBSAP, 2014)*. National Biodiversity Centre, Ministry of Agriculture and Forests, Royal Government of Bhutan.;

11. Ministry of Agriculture and Forests, 2016. *Bhutan State of Parks 2016*. Department of Forest and Parks Services, Ministry of Agriculture and Forests, Royal Government of Bhutan.
12. Bhutan INDC, 2015. Communication of INDC of the Kingdom of Bhutan. https://www4.unfccc.int/sites/ndcstaging/PublishedDocuments/Bhutan%20First/Bhutan-INDC-20150930.pdf
13. (a) Lham, D., Wangchuk, S., Stolton, S., & Dudley, N., 2018. 'Assessing the Effectiveness of a Protected Area Network: A Case Study of Bhutan'. In *Oryx*, March 2018; (b) *National Biodiversity Strategies and Action Plan of Bhutan (NBSAP, 2014)*. National Biodiversity Centre, Ministry of Agriculture and Forests, Royal Government of Bhutan.
14. National Biodiversity Centre, 2014. *National Biodiversity Strategies and Action Plan of Bhutan*. Ministry of Agriculture and Forests, Royal Government of Bhutan.
15. Ibid.
16. Ibid.
17. Wangchuk, S., Norbu, N., & Sherub, 2012. *Impacts of Cordyceps Collection on Livelihoods and Alpine Ecosystem in Bhutan as Ascertained from Questionnaire Survey of Cordyceps Collectors. Royal Government of Bhutan*. UWICE Press, Bumthang. (Accessed 19 August 2018)
18. Ibid.
19. National Biodiversity Centre, 2014. *National Biodiversity Strategies and Action Plan of Bhutan*. Ministry of Agriculture and Forests, Royal Government of Bhutan.
20. Planning Commission, 1999. *Bhutan 2020: A Vision for Peace, Prosperity and Happiness*. Royal Government of Bhutan.
21. (a) Nature Conservation Division, 2005. *Tiger Action Plan for the Kingdom of Bhutan* 2006–2015. Department of Forests, Ministry of Agriculture, Royal Government of Bhutan; (b) National Environment Commission (NEC), 1994. *Conservation in Bhutan*. Royal Government of Bhutan.
22. Nature Conservation Division, 2005. *Tiger Action Plan for the Kingdom of Bhutan* 2006–2015. Department of Forests, Ministry of Agriculture, Royal Government of Bhutan.
23. (a) Nature Conservation Division, 2005. *Tiger Action Plan for the Kingdom of Bhutan* 2006–2015. Department of Forests, Ministry of Agriculture, Royal Government of Bhutan; (b) National Environment Commission, 1994. *Conservation in Bhutan*. Royal Government of Bhutan.
24. (a) National Environment Commission, 1994. *Conservation in Bhutan*. Royal Government of Bhutan; (b) Planning Commission, 1999. *Bhutan 2020: A Vision for Peace, Prosperity and Happiness*. Royal Government of Bhutan.
25. (a) Wangchuk, S., Lham, D., Dudley, N., & Stolton, S., 2017. 'Half Bhutan: The Evolution and Effectiveness of Protected Areas in a Country Recognizing Nature Needs Half. International Perspectives'. *International Journal of Wilderness*, 23(1), (June 2017); (b) Ministry of Agriculture and Forests, 2016. *Bhutan State of Parks 2016*. Department of Forest and Parks Services, Ministry of Agriculture and Forests, Royal Government of Bhutan.
26. National Environment Commission, 1994. *Conservation in Bhutan*. Royal Government of Bhutan.
27. Nature Conservation Division, 2005. *Tiger Action Plan for the Kingdom of Bhutan 2006–2015*. Department of Forests, Ministry of Agriculture, Royal Government of Bhutan
28. National Environment Commission, 1994. *Conservation in Bhutan*. Royal Government of Bhutan.
29. Ibid.
30. Wangchuk, S., 2007. 'Maintaining Ecological Resilience by Linking Protected Areas through Biological Corridors in Bhutan'. International Society for Tropical Ecology. *Tropical Ecology*, 48(2), 176–187.
31. Ibid.
32. National Environment Commission, 1994. *Conservation in Bhutan*. Royal Government of Bhutan.
33. Ministry of Agriculture and Forests, 2016. *Bhutan State of Parks 2016*. Department of Forest and Parks Services, Ministry of Agriculture and Forests, Royal Government of Bhutan.
34. Ibid.
35. Zurick, D., 2006. 'Gross National Happiness and Environmental Status in Bhutan'. *American Geographical Society: Geographical Review*, 96(4), 657–681.
36. Major Sheriff, G. (1898–1697). Gazetteer for Scotland. http://www.scottish-places.info/people/famousfirst1916.html (Accessed on 18 August 2018).

37. Sinpeng, A., 2007. 'Democracy from Above: Regime Transition in the Kingdom of Bhutan'. *Journal of Bhutan Studies*, 17(2), 21–48. http://www.dspace.cam.ac.uk/handle/1810/226946
38. The Constitution of the Kingdom of Bhutan, 2008. Royal Government of Bhutan.
39. Ibid.
40. (a) Wangchuk, S., 2007. 'Maintaining Ecological Resilience by Linking Protected Areas through Biological Corridors in Bhutan'. International Society for Tropical Ecology. *Tropical Ecology*, 48(2), 176–187; (b) Zurick, D., 2006. 'Gross National Happiness and Environmental Status in Bhutan'. *American Geographical Society: Geographical Review*, 96(4), 657–681.
41. (a) Zurick, D., 2006. 'Gross National Happiness and Environmental Status in Bhutan'. *American Geographical Society: Geographical Review*, 96(4), 657–681; (b) National Environment Commission, 1994. *Conservation in Bhutan*. Royal Government of Bhutan.
42. Zurick, D., 2006. 'Gross National Happiness and Environmental Status in Bhutan'. *American Geographical Society: Geographical Review*, 96(4), 657–681.
43. Ura, K., Alkire, S., Zangmo, T., & Wangdi, K., 2012. *A Short Guide to Gross National Happiness Index*. The Center for Bhutan Studies, Thimphu, Bhutan.
44. Sangay, T., & Vernes, K. 2014. 'The Economic Cost of Wild Mammalian Carnivores to Farmers in the Himalayan Kingdom of Bhutan'. *Proc Bhutan Ecol Soc*, 1(January), 98–111.
45. Gross National Happiness Commission, 2013. *Framework to Mainstream Environment, Climate Change and Poverty (ECP) Concerns in the Eleventh Five Year Plan*. Royal Government of Bhutan.
46. Ibid.
47. http://www.unpei.org/sites/default/files/dmdocuments/Bhutan%20Brochure.pdf (Accessed 31 December 2018).
48. Gross National Happiness Commission, 2018. *Twelfth Five-Year Plan (2018–2023)*, Royal Government of Bhutan.
49. Gross National Happiness Commission, 2013. *Tenth Five-Year Plan (2008–2013)*, Volume I. Royal Government of Bhutan.
50. The World Bank and the National Statistical Bureau, 2017. *Bhutan Poverty Analysis Report*. Royal Government of Bhutan.
51. Ibid.
52. Marshall, R., 2013. 'Graduation from the Group of Least Developed Countries: Prospects and Challenges for Bhutan'. 12th Round Table Meeting, Thimphu, Bhutan
53. Bhutan For Life Prospectus. http://www.bfl.org.bt/resources/BFL_Prospectus.pdf (Accessed 19 August 2018).
54. National Environment Commission, 1994. *Conservation in Bhutan*. Royal Government of Bhutan.
55. DoFPS, 2016. *National Forest Inventory Report: Stocktaking Country's Natural Resources*. NFI, Volume I. 2016. Department of Forest and Parks Services, Ministry of Agriculture and Forests, Royal Government of Bhutan.
56. (a) DoFPS, 2016. *National Snow Leopard Survey of Bhutan 2014–2016 (Phase II): Camera Trap Survey for Population Estimation*. Department of Forests and Parks Services, Ministry of Agriculture and Forests, Thimphu, Bhutan; (b) NCD, 2018. *National Elephant Survey Report*. Nature Conservation Division, Department of Forests and Parks Services, Ministry of Agriculture and Forests, Thimphu, Bhutan.
57. DoFPS, 2005. *Tiger Action Plan for the Kingdom of Bhutan 2006–2015*. Nature Conservation Division, Department of Forests, Ministry of Agriculture, Royal Government of Bhutan.
58. Bhutan For Life Prospectus. http://www.bfl.org.bt/resources/BFL_Prospectus.pdf (Accessed 19 August 2018).
59. Zurick, D., 2006. 'Gross National Happiness and Environmental Status in Bhutan'. *American Geographical Society: Geographical Review*, 96(4), 657–681.
60. WWF Communications, 2006. 'King of Bhutan Receives Prestigious Getty Conservation Award', WWF Press Release. https://www.worldwildlife.org/press-releases/king-of-bhutan-receives-prestigious-getty-conservation-award (Accessed 31 December 2018).
61. Tshomo, D., 2018. 'Bhutan wins Earth Award'. Kuensel online, 3/9/18 http://www.kuenselonline.com/bhutan-wins-earth-award/ (Accessed 31 December 2018).

62. His Majesty the Fifth King of Bhutan. National Day Address, 2013.
63. Sangay, T., & Vernes, K. 2014. 'The Economic Cost of Wild Mammalian Carnivores to Farmers in the Himalayan Kingdom of Bhutan' *Proc Bhutan Ecol Soc*, 1(January), 98–111.
64. (a) Wang, S. W., & Macdonald, D. W., 2006. 'Livestock Predation by Carnivores in Jigme Singye Wangchuck National Park, Bhutan'. *Biological Conservation*, 129(2006), 558–565; (b) Sangay, T., & Vernes, K. 2014. 'The Economic Cost of Wild Mammalian Carnivores to Farmers in the Himalayan Kingdom of Bhutan'. *Proc Bhutan Ecol Soc*, 1(January), 98–111; (c) Tshering, K., & Thinley, P. 2017. 'Assessing Livestock Herding Practices of Agro-Pastoralists in Western Bhutan: Livestock Vulnerability to Predation and Implications for Livestock Management Policy'. *Pastoralism*. https://doi.org/10.1186/s13570-017-0077-1.
65. DoFPS, 2008. *Bhutan Human Wildlife Conflict Management Strategy*. Nature Conservation Division. Department of Forest and Park Services. Ministry of Agriculture and Forests. Royal Government of Bhutan.
66. (a) Penjore, D. n.d.. 'Is National Environment Conservation Success a Rural FailureThe Other Side of Bhutan's Conservation Story' pp. 66–87. Centre for Bhutan Studies and Gross National Happiness Research. (Accessed 31 December 2018) from: http://citeseerx.ist.psu.edu/viewdoc/download?doi=10.1.1.729.8759&rep=rep1&type=pdf
(b) World Wildlife Fund Bhutan. 'A Holistic Answer to Human-Wildlife Conflict'. 4/25/16, http://www.wwfbhutan.org.bt/?266210/A-holistic-answer-to-human-wildlife-conflict (Accessed 31 December 2018).
67. (a) Sangay, T., & Vernes, K., 2008. 'Human-Wildlife Conflict in the Kingdom of Bhutan: Patterns of Livestock Predation by Large Mammalian Carnivores'. *Biological Conservation*, 141, 1272–1282; (b) Sangay, T., & Vernes, K. 2014. 'The Economic Cost of Wild Mammalian Carnivores to Farmers in the Himalayan Kingdom of Bhutan'. *Proc Bhutan Ecol Soc*, 1(January), 98–111.
68. Wang, S. W., & Macdonald, D. W., 2006. 'Livestock Predation by Carnivores in Jigme Singye Wangchuck National Park, Bhutan'. *Biological Conservation*, 129, 558–565.
69. Ibid.
70. Tshering, K., & Thinley, P. 2017. 'Assessing Livestock Herding Practices of Agro-Pastoralists in Western Bhutan: Livestock Vulnerability to Predation and Implications for Livestock Management Policy'. *Pastoralism*. https://doi.org/10.1186/s13570-017-0077-1
71. Wang, S. W., & Macdonald, D. W., 2006. 'Livestock Predation by Carnivores in Jigme Singye Wangchuck National Park, Bhutan'. *Biological Conservation*, 129, 558–565.
72. Ibid.
73. (a) Wang, S. W., Lassoie, J. P., & Curtis, P. D. 2006. 'Farmer Attitudes towards Conservation in Jigme Singye Wangchuck National Park, Bhutan'. *Environmental Conservation*, 33(2), 148–156. https://doi.org/10.1017/S0376892906002931; (b) Wang, S. W., & Macdonald, D. W., 2006. 'Livestock Predation by Carnivores in Jigme Singye Wangchuck National Park, Bhutan'. *Biological Conservation*, 129, 558–565.
74. Wang, S. W., Lassoie, J. P., & Curtis, P. D. 2006. 'Farmer Attitudes towards Conservation in Jigme Singye Wangchuck National Park, Bhutan'. *Environmental Conservation*, 33(2), 148–156.
75. Pannozzo, L., Dorji, C., Choden, U., Zangmo, D., Hayward, K., Om, T., Raftis, L., Morales, K., Colman, D., Aikens., Doukas, A., Oddy, M., & Colman, R., 2012. 'Profile of Samdrupjongkhar Initiative', Appendices (2012). http://gpiatlantic.org/bhutan/docs/samdrup-jongkhar-profile-appendices-may.pdf (Accessed 10 January 2019).
76. (a) Lhamo, N., 2008. 'Extent of Human Elephant Conflicts and the Threat of Elephant Populations in Southern Bhutan'. Masters thesis. University of Natural Resources and Applied Life Sciences, Vienna, Austria. (b) Kuensel online articles (31 Accessed 2018):
3/9/17, http://www.kuenselonline.com/elephants-compel-farmers-to-leave-their-fields-fallow-in-sipsu/;
5/24/18, http://www.kuenselonline.com/victim-of-elephant-attack-discharged-from-hospital/;
9/12/18, http://www.kuenselonline.com/elephants-rampage-crops-in-gelephu/;
9/3/18, http://www.kuenselonline.com/frustrations-build-up-as-elephants-cause-menace-in-gelephu/;

7/9/18, http://www.kuenselonline.com/elephants-rampage-nursery-farm-and-maize-field-in-bhur/

77. Wang, S. W., Lassoie, J. P., & Curtis, P. D. 2006. 'Farmer Attitudes towards Conservation in Jigme Singye Wangchuck National Park, Bhutan'. *Environmental Conservation*, 33(2), 148–156.
78. Sangay, T., & Vernes, K. 2014. 'The Economic Cost of Wild Mammalian Carnivores to Farmers in the Himalayan Kingdom of Bhutan'. *Proc Bhutan Ecol Soc*, 1(January), 98–111.
79. Jigme, K., & Williams, A. C. 2011. 'Current Status of Asian Elephants in Bhutan'. *Gajah*, 35, 25–28.
80. Sangay, T., & Vernes, K. 2014. 'The Economic Cost of Wild Mammalian Carnivores to Farmers in the Himalayan Kingdom of Bhutan'. *Proc Bhutan Ecol Soc*, 1(January), 98–111.
81. NPPC and WWF-Bhutan, 2016. *Human Wildlife Conflict Strategy: Nine Gewogs of Bhutan*. National Plant Protection Centre (NPPC), Thimphu, and World Wildlife Fund Bhutan, Thimphu.
82. Ibid.
83. Ibid.
84. Kuensel online, 5/9/18, http://www.kuenselonline.com/2773km-of-electric-fencing-installed-nationwide/ (Accessed 31 December 2018).
85. Jigme, K., & Williams, A. C. 2011. 'Current Status Asian Elephants in Bhutan'. *Gajah*, 35, 25–28.
86. Nature Conservation Division, 2018. *National Elephant Survey Report*. Nature Conservation Division, Department of Forests and Parks Services, Ministry of Agriculture and Forests, Thimphu, Bhutan.
87. (a) Ministry of Agriculture and Forests. 3/28/14. http://www.moaf.gov.bt/gps-collaring-of-asian-elephants-in-southern-bhutan/ (Accessed 31 December 2018);
(b) Ministry of Agriculture and Forests. 12/4/18. http://www.moaf.gov.bt/gps-collaring-operation-of-elephants/ (Accessed 31 December 2018);
(c) Kuensel online, 1/6/15, http://www.kuenselonline.com/collaring-elephants-to-conserve-them/ (Accessed 31 December 2018);
(d) Nature Conservation Division, 2018. *National Elephant Survey Report*. Nature Conservation Division, Department of Forests and Parks Services, Ministry of Agriculture and Forests, Thimphu, Bhutan.
88. Wangchuk, S., Norbu, N., & Sherub, 2012. *Impacts of Cordyceps Collection on Livelihoods and Alpine Ecosystem in Bhutan as Ascertained from Questionnaire Survey of Cordyceps Collectors*. Royal Government of Bhutan. UWICE Press, Bumthang.
89. Ibid.
90. Ibid.
91. Ibid.
92. Ibid.
93. Ibid.
94. (a) National Biodiversity Centre, 2014. *National Biodiversity Strategies and Action Plan of Bhutan*. Ministry of Agriculture and Forests, Royal Government of Bhutan; (b) Ministry of Agriculture and Forests, 2016. *Bhutan State of Parks 2016*. Department of Forest and Parks Services, Ministry of Agriculture and Forests, Royal Government of Bhutan; (c) Bhutan State of the Environment Report, 2016. National Environment Commission, Royal Government of Bhutan.
95. Thethirdpole.net. 6/4/18, https://www.thethirdpole.net/en/2018/04/06/as-bhutans-economy-grows-so-does-its-waste-problem/
96. Phuntsho, S., Dulal, I., Yangden, D., Tenzin, U. M., Herat, S., Shon, H., & Vigneswaran, S. 2010. 'Studying Municipal Solid Waste Generation and Composition in the Urban Areas of Bhutan'. *Waste Management and Research*, 28(6), 545–551. https://doi.org/10.1177/0734242X09343118.
97. Ibid.
98. Ibid.
99. Kuensel online, 6/12/18, http://www.kuenselonline.com/thimphu-thromde-working-to-reduce-organic-waste-at-source/ (Accessed 31 December 2018).
100. Clean Bhutan. http://cleanbhutan.org/?page_id=57 (Accessed 31 December 2018).

101. Ibid.
102. United Nations Development Programme, 2012. 'Piloting Public Private Partnership on Solid Waste Management'. https://open.undp.org/projects/00058476 (Accessed 31 December 2018).
103. United Nations Development Programme, 2012. 'Knowledge, Innovation, Capacity: Private Sector Collaboration in Sustainable Development: Experiences from Asia and the Pacific'. http://www.asia-pacific.undp.org/content/dam/rbap/docs/Research%20&%20Publications/KIC/APRC_KIC_PS%20collab%20in%20SD.pdf (Accessed 10 January 2019).
104. Kuensel online, 6/9/18, http://www.kuenselonline.com/thromde-begins-efficient-waste-collection-in-thimphu/ (Accessed 31 December 2018).
105. (a) The Bhutanese, 12/3/16 https://thebhutanese.bt/national-cleaning-campaign-to-mark-10th-anniversary-of-his-majestys-golden-reign/; (b) BBS, 12/10/16 'Country's Mass Cleaning Campaign' http://www.bbs.bt/news/?p=64317 (Both Accessed 31 December 2018).
106. (a) Kuensel online, 7/30/18, http://www.kuenselonline.com/saving-the-environment-6/; (b) Ministry of Agriculture and Forests. 10/8/17. 'Sanam Lyonpo Joins Mass Cleaning Campaign in Bumthang'. http://www.moaf.gov.bt/sanam-lyonpo-joins-mass-cleaning-campaign-in-bumthang/ (Accessed 31 December 2018).
107. Kuensel online, 2/19/18, http://www.kuenselonline.com/cctv-to-monitor-waste-in-damphu-town/
108. De-Suung, Guardians of Peace. https://desuung.org.bt/ (Accessed 2 January 2019).
109. (a) National Biodiversity Centre, 2014. *National Biodiversity Strategies and Action Plan of Bhutan.* Ministry of Agriculture and Forests, Royal Government of Bhutan; (b) National Environment Commission, 2016. *Bhutan State of the Environment Report, 2016.* Royal Government of Bhutan.
110. De-Suung, Guardians of Peace. https://desuung.org.bt/ (Accessed 2 January 2019).
111. Kuensel online: (a) 4/4/18, http://www.kuenselonline.com/fire-above-bajo-town-razes-350-hectares-of-forest/; (b) 2/17/16, http://www.kuenselonline.com/forest-fires-in-chuzom-and-wangdue-yet-to-be-contained/; (c) 3/27/18, http://www.kuenselonline.com/forest-fire-in-semtokha/ (All Accessed 2 January 2019).
112. Kuensel online, 4/28/15, http://www.kuenselonline.com/bhutan-contributes-usd-1m-to-nepal/ (Accessed 2 January 2019).
113. (a) National Biodiversity Centre, 2014. *National Biodiversity Strategies and Action Plan of Bhutan.* Ministry of Agriculture and Forests, Royal Government of Bhutan; (b) National Environment Commission, 2016. *Bhutan State of the Environment Report, 2016.* Royal Government of Bhutan; (c) Ministry of Agriculture and Forests, 2016. *Bhutan State of Parks.* Department of Forest and Parks Services, Ministry of Agriculture and Forests, Royal Government of Bhutan.
114. 'Shingkhar-Gorgan Road Waits Environmental Clearance'. Bhutan Times, June 18, 2017. https://www.pressreader.com/bhutan/bhutan-times/20170618/281496456278703 (Accessed 2 January 2019).
115. (a) 'Shingkar-Gorgan Road: Economic Benefits vs. Conservation Laws'. BBS, August 1, 2014. http://www.bbs.bt/news/?p=42389 (Accessed 2 January 2019); (b) 'Bhutan's Shingkhar-Gorgan Highway: Development versus Environment?' The thirdpole.net, December 14, 2017, https://www.thethirdpole.net/en/2017/12/14/bhutans-shingkhar-gorgan-highway-development-versus-environment/ (Accessed 2 January 2019).
116. 'Issue on Shingkhar-Gorgan Bypass Dominates Lhuentse DT Meeting'. BBS, February 4, 2017, http://www.bbs.bt/news/?p=90144 (Accessed 2 January 2019).
117. Kuensel online, 8/2/18, http://www.kuenselonline.com/governments-tenure-a-big-failure-opposition/ (Accessed 2 January 2019).
118. Kuensel online: (a) 6/5/16, http://www.kuenselonline.com/shingkhar-gorgan-road-construction-starts/; (b) 6/27/18, http://www.kuenselonline.com/shingkhar-gorgan-road-awaits-environment-clearance/
119. (a) 'Shingkar-Gorgan Road: Economic Benefits vs. Conservation Laws'. BBS, August 1, 2014. http://www.bbs.bt/news/?p=42389 (Accessed 2 January 2019); (b) Kuensel online, 7/28/18, http://www.kuenselonline.com/roads-sector-a-work-in-progres/ (Accessed 2 January 2019).

120. Annual Report, 2015. The Judiciary of the Kingdom of Bhutan. http://www.judiciary.gov.bt/publication/ar2015.pdf (Accessed 2 January 2019).
121. Ibid.
122. The Constitution of the Kingdom of Bhutan, 2008. Royal Government of Bhutan.
123. Kuensel online, 3/28/15, http://www.kuenselonline.com/the-greening-of-the-judiciary/ (Accessed 2 January 2019).
124. Ibid.
125. Asian Judges Network on Environment. July 20, 2018. 'Workshop on Environmental Adjudication for the Judiciary of Bhutan'. https://www.ajne.org/event/workshop-environmental-adjudication-judiciary-bhutan (Accessed 2 January 2019).

# 8

# *The Restoration of the Caledonian Forest and the Rights of Nature*

**Alan Watson Featherstone**

**CONTENTS**

Summary ........ 141
Introduction: Caledonian Forest ........ 141
Historical Background ........ 142
Initial Steps in Forest Restoration ........ 144
The Founding of Trees for Life ........ 144
Reweaving the Web of Life ........ 147
Deepening the Restoration Process ........ 151
Reconnecting People with Nature ........ 154
Looking Ahead: If the Rights of Nature Become Paramount ........ 156
Notes ........ 161
Appendix: Common and Scientific Names of Plants and Animals ........ 164

## Summary

This chapter describes a case study in the Highlands of Scotland concerning the native Caledonian Forest, and the work to help restore it from its degraded state, initiated in the late twentieth century. It begins with an introduction to the forest, including its ecology and historical background, and then describes its depletion by humans. This is followed by details of the restoration work that has been carried out, particularly by the conservation charity Trees for Life, which aims to restore the forest as a fully functional ecosystem, complete with all its constituent species. The case study concludes with a vision for the future, in which the Rights of Nature would be paramount, extrapolating how the Caledonian Forest restoration work could achieve its full potential, accompanied by a paradigm change in wider Scottish human culture and society.

## Introduction: Caledonian Forest

The Caledonian Forest is the natural ecosystem that covered most of the Highlands of Scotland from soon after the end of the last Ice Age until about 4,000 years ago, when it began to decline significantly. It forms the westernmost extent of the boreal forest in Europe, and is characterised by the Scots pine, which is the largest and longest-lived tree in the ecosystem.

DOI: 10.1201/9780429505959-8

Although Scots pine has one of the widest natural distributions of any tree on the planet, extending from eastern Siberia to the west coast of Scotland and from well north of the Arctic Circle in Scandinavia to the Mediterranean, the pinewoods of Scotland are distinctive and different from all others within the species' geographic range. Elsewhere, it occurs with other conifers, such as Norway spruce in Scandinavia and with Siberian fir and larch further east.

In Scotland, however, it is the only coniferous tree in the Caledonian Forest, which otherwise consists of broadleaved species such as downy birch, silver birch, European aspen, rowan, alder, hazel, holly, several species of willows, bird cherry, wild cherry and, where the soils are better, sessile oak, ash and wych elm. Understory trees and shrubs include juniper, hawthorn, blackthorn or sloe and, at the treeline, dwarf birch. This unique mixture of tree species, plus a range of associated special plants and wildlife, including the capercaillie, the Scottish crossbill (the only bird endemic to Scotland), red squirrel, twinflower, one-flowered wintergreen and narrow-headed wood ant, as well as various lichens and mosses, has led to the Caledonian Forest's recognition as a priority habitat in the European Union's Habitats and Birds Directive, which requires member states to take practical action to ensure its conservation and restoration.

## Historical Background

At its maximum extent, the Caledonian Forest covered an estimated 1.5 million hectares and was home to such large mammal species as the European brown bear, grey wolf, Eurasian lynx, European beaver and wild boar. About 4,000 years ago the forest went into a period of decline, coinciding with both a change in climate, in which conditions became cooler and wetter, and the spread of humans into more of the woodland areas, as evidenced by charcoal deposits. It is likely that both climate and humans played a role in the reduction of the forest, with each factor compounding the other.

Over a period of centuries, the forest was gradually reduced in extent, becoming confined to more remote and less accessible sites. Peat began to spread, even on sloping ground, and the resulting wet soil conditions inhibited the return of the forest once it had been lost. The more ecologically sensitive woodlands with slower-growing and height-restricted species, such as bog woodland and montane scrub, disappeared almost completely, as did the temperate rainforests that flourished in wetter areas near the west coast and in sheltered river gorges. Periodic instances of more deliberate deforestation occurred in the mediaeval era, where records show that forests were set alight to drive wolves out into the open so they could be killed.[1]

Then in the eighteenth and early nineteenth centuries, significant areas of trees were felled for industrial exploitation[2] because the forests further south in Britain were all depleted. All the large mammals were hunted to extinction and disappeared with the loss of their forest habitat, whilst the populations of smaller species that survived, such as the pine marten and red squirrel, were reduced to tiny fractions of their previous numbers.

As a result, by the late 1700s, the native pinewoods of the Caledonian Forest had been reduced to scattered and fragmented remnants distributed across their former range, covering about 17,000 hectares – just over 1% of their maximum extent. The extermination of the wolf (reputedly in 1743),[3] followed by the introduction of large numbers of domestic sheep and the increase of the red deer population that was encouraged by landowners for 'sporting' trophy purposes, created a huge ecological imbalance in the remaining forest patches, causing the recruitment of new young trees to come to an abrupt halt.[4] As a consequence, by the late twentieth century, the last new trees to be established, about 200 years previously, were nearing the end of their lives, leaving only 'geriatric' woodlands as forest remnants. These contained only old and dying trees, which were geographically isolated from each other by large areas of deforested ground and missing a significant proportion of their constituent species of flora and fauna.

Many key ecological processes such as nutrient cycling, predator-prey dynamics and natural disturbance regimes ceased to function. The vast majority of the Highlands had been reduced to treeless expanses of heather and grass, intensively overgrazed by red deer and sheep. They were devoid of most of their original wildlife and in many cases scarred by peat hags – areas of land where the underlying peat, and the stumps of old Scots pine trees preserved within them, had become exposed by the elements and erosion. Gone from most of the Highland glens as well were the former inhabitants. Their involuntary departure in the second half of the eighteenth and early nineteenth centuries at the hands of the mostly absentee landowners who evicted them, and brought in sheep instead, have become notorious as the Highland Clearances. Those traumatic events created deep psychological scars in the human culture of the Highlands that mirrored both the physical scars of peat hags on the land and the forced removal of indigenous peoples from their land in other parts of the world, such as North America and Australia.

In many former areas of the Caledonian Forest, such as this one in Glen Strathfarrar, only dead or dying Scots pines remained by the beginning of the twenty-first century. Overgrazing by red deer prevent the growth of any seedlings, resulting in entirely treeless landscapes like this. (Photograph by Alan Featherstone.)

Scotland has never experienced any significant process of land reform, and large areas of land in the Highlands (typically from 10,000 to over 50,000 acres each) are still owned by absentee landlords who only visit for a few weeks of the year, treating them effectively as playgrounds for the rich.[5] They are used for trophy hunting of deer or shooting red grouse, and the land is either neglected for the rest of the year or deliberately managed to benefit those activities in ways that continue to damage and deplete the ecosystems. One example is the regular burning of heather moorland as a management technique to increase the number of grouse. The burning kills any young tree seedlings that have started to grow and thereby prevents the natural process of ecological succession from moorland to pioneer woodland. By the late twentieth century, the last Scots pines to germinate in the Caledonian Forest remnants were nearing the end of their lives, and it was clear that action was required immediately, or this unique ecosystem would be lost as the old trees died without being replaced.

## Initial Steps in Forest Restoration

The plight of the Caledonian Forest was brought to public attention in 1959 with the publication of the landmark book, *The Native Pinewoods of Scotland*,[6] written by Henry Steven, a professor of forestry at Aberdeen University, and his student Jock Carlisle. That stimulated some pioneering work, and from the early 1960s onwards several projects were initiated, for example by the Nature Conservancy Council at Beinn Eighe[7] and the Forestry Commission at Glen Affric,[8] on a relatively small scale to help the regeneration and restoration of the Caledonian Forest. In some of the best pinewood remnants, deer and sheep were excluded by deer-fenced exclosures so that young trees could grow successfully without being eaten. In the 1980s, in other areas such as Creag Meagaidh[9] and Abernethy,[10] the deer numbers were drastically reduced through intensive culling operations, so that regeneration of the trees could take place naturally, without the need for fences. Both of these methods resulted in the recruitment of a new generation of young trees – the first ones to grow successfully in the Caledonian Forest in over 200 years.[11]

At about the same time, some initial wildlife reintroduction projects were implemented by the Nature Conservancy Council for the white-tailed eagle (or sea eagle),[12] and by the Joint Nature Conservation Committee and the Royal Society for the Protection of Birds (RSPB) for the red kite,[13] which resulted in the return of some of Scotland's missing bird species. Concerted conservation action also ensured that the osprey, which recolonised Scotland naturally in 1954 after having been hunted to extinction in 1916, became successfully re-established.

These initial steps in helping the native forest to recover prepared the way for a substantial development and expansion of ecological restoration work in Scotland in the last two decades of the twentieth century. Increased public attention to environmental issues in general, and deforestation in particular (especially in the tropical rainforests), formed the backdrop for a more widespread recognition that the Highlands of Scotland were environmentally depleted and in urgent need of assistance to return to greater ecological health. Sir Frank Fraser Darling, a pioneering British ecologist in the 1950s, had coined the term 'wet desert'[14] to describe the degraded condition of the northwest Highlands in particular, and now this understanding began to expand outwards from specialist scientific circles into the more general public consciousness.

## The Founding of Trees for Life

I first became aware of the Caledonian Forest in 1979, when I visited Glen Affric, the site of one of the largest remaining remnants of the native pinewoods. Like most Scots, I had grown up thinking that the treeless glens and heather-covered hills were the natural state of the land in Scotland. For me, and many other people, I suspect, trees in the Highlands were more familiar from the regimented ranks of non-native conifers that had been widely planted in uniform blocks in the post-war era, forming 'cellulose crops' that more closely resembled industrial farming than a natural forest. It therefore came as something of a revelation to discover a large area of wild forest in Glen Affric, replete with trees of all ages and varied shapes, many swathed in lichens. The forest was growing in a landscape amongst old mountains, beautiful lochs and a river with cascading waterfalls, reminding me of forests I had visited in both northwest Canada and the Patagonia region in South America for the wildness and natural aesthetics.

An area on the south side of Glen Affric had been fenced off to exclude red deer by a visionary forester, Finlay McRae, in the early 1960s; consequently, a new generation of young trees was flourishing amongst the old pines. Outside the fence it was a different story, and I could see that the old trees were dying off without being replaced. I wondered why nothing was being done to extend the protection to replicate what had been achieved on the south side.

The previous year, in 1978, I had joined the Findhorn Community,[15] situated about 65 miles from Glen Affric, which had become known internationally since the 1960s for its pioneering work of cooperation with the spirits of Nature. The members of the Findhorn community recognised that all plants and animals have an innate intelligence, consciousness and purpose, and during their meditations some of them received clear messages from the spirits, or 'devas', of individual plant species.[16] These messages emphasised the need for humanity to develop a different relationship with Nature, one that was based on cooperation, harmony and co-creation, instead of domination and exploitation. This had resonated deeply with me, and after joining the community I began working in one of the gardens, seeking to develop and extend my own personal inner connection with Nature.

In subsequent visits to Glen Affric, I began to get a strong sense that the old trees, and the land itself, were calling out for help. It felt as though the trees were silently saying, 'do something to help us recover', and I kept thinking that someone should indeed take action to reverse the decline of the forest, before it was too late and it disappeared with the death of all the old trees.

However, nobody was doing anything, and after 2 or 3 years, it began to dawn on me that perhaps the 'someone' needed was me. In 1986, I spent the year working as the main organiser of a week-long major international conference held at Findhorn, entitled 'One Earth: A Call to Action', which focussed on the need for urgent, practical action to address the world's environmental problems. In the final session, we asked anyone who was inspired to stand up in front of the 300 delegates and make a personal commitment to do something positive and practical for the planet. I stood up and committed myself to launching a project to help restore the Caledonian Forest. It was the defining moment for me, when I fully embraced those feelings that had being growing inside me for several years and said, 'Yes, I am going to respond to this inner calling'.

At the time, I had no experience or training in ecology or about forests, no access to any of the areas of land where the Caledonian Forest remnants still clung on and no funding or resources to help me achieve a very ambitious commitment. On the outer level I had nothing, but with hindsight I realise that I had the most important things I needed – a deep passion for the forest, an unshakeable dedication to follow my heart's calling and a strong inner trust that, by sharing my inspiration, I would attract all the support necessary to succeed.

Thus was born Trees for Life (the only organisation specifically dedicated to the restoration of Scotland's Caledonian Forest), although it was almost 3 years before any practical action took place, and another 4 years before the registered charity known as Trees for Life would be officially established. In the intervening period I educated myself about the forest and its ecology, made contact with landowners, raised some initial funds for practical work and began recruiting some volunteers to help. In 1989 the first practical projects began when some volunteers and I protected naturally occurring Scots pine seedlings with individual tree guards (plastic tubes), to keep them safe from being grazed by deer. This was followed in 1990 by the erection of a deer-proof fence encompassing 50 hectares (125 acres) on the edge of an old, native pinewood area in Glen Affric, in partnership with Forestry Commission Scotland (the government's forest agency in Scotland, which owns most of the land in Glen Affric).

Naturally regenerating Scots pines growing around the snag of an old dead pine in Glen Affric, 24 years after the area was fenced by Trees for Life (in partnership with Forestry Commission Scotland) to prevent overgrazing by red deer. (Photograph by Alan Featherstone.)

A research project we organised by a forestry student[17] had shown the area had an estimated 100,000 pine seedlings, self-sown from the remaining old trees, with an average age of 10 years, but an average height of just 8.5 cm (3.3 inches) because deer had browsed 95% of them. By excluding the deer, the fence would enable those pine seedlings to grow – the first new generation of trees to get established in the area for about two centuries. This was the spirit of co-creation with Nature in practical action, letting the land do what it wanted to do itself (i.e. be restored to natural forest). For me it was a major accomplishment, and a tangible demonstration of giving the right of self-expression back to Nature, instead of it being suppressed by human 'management'. It also gained substantial attention in the media, with radio interviews, news items on both Scottish television channels and a major article in *The Guardian* newspaper all covering the story.

That first significant achievement became the launch-pad for the growth and expansion of the work of Trees for Life, and for an ongoing partnership with Forestry Commission Scotland that still continues to this day. In the following years, a number of other areas in Glen Affric were fenced to exclude deer,[18] both to facilitate the natural regeneration of existing seedlings and protect young planted Scots pines where there was no seed source remaining. The larger aim was to establish a series of areas, or 'islands', of new young native forest strategically sited in the glen to enable an extension of the existing pinewoods and to form future seed sources from which the trees could expand naturally by themselves. This was predicated on there being a reduction in deer numbers while the young trees were growing, so browsing pressure would be lowered to a level at which their self-seeded progeny would in the future be able to grow successfully without being eaten, and without the need for further fencing.

Over the following years, the work of the charity developed to include partnerships with other nearby landowning organisations, including the National Trust for Scotland, the Royal Society for the Protection of Birds and a number of private landowners. Trees for Life also expanded geographically into the surrounding landscape. A core area of about 1,000 square miles, situated to the west of Inverness and Loch Ness,[19] was the focus of the work, as it was a largely roadless and remote region, with no human settlements and very little economic activity. Containing mountains, lochs, rivers and some of the best remnants of the Caledonian Forest, it provides what is possibly the best opportunity in the whole United Kingdom for the restoration of a large-scale wild and natural forest, and is of a size that would be adequate to support populations of the missing mammals whose reintroduction will be necessary to return the ecosystem to full health and functional self-sustainability. It is of course a long-term project to restore a healthy ecosystem to such a large area of degraded land, and Trees for Life has always operated with a 250-year vision, as it will take that long for healthy mature forest to once again flourish on areas that are currently treeless.

Although the work initially focussed on Scots pine (the largest and longest-lived tree in the Caledonian Forest), which is a keystone species in the ecosystem, it quickly developed to include all of the native tree species. From the mid-1990s onwards, Trees for Life has operated its own tree nursery,[20] concentrating on scarcer tree species and those that are harder to grow, as they are difficult or impossible to source from regular commercial nurseries. A major step forward came in 2008, when the charity purchased the 10,000-acre Dundreggan Estate[21] in Glenmoriston, the next valley south of Glen Affric, thanks to a significant legacy from a supporter, supplemented by a major fundraising initiative. This enabled the development and implementation of a long-term management plan for forest restoration, free of the constraints of differing objectives from partner organisations. For the first time it also gave the charity a role in the management of red deer, as that lies exclusively with landowners. While planting and protecting young trees behind fences gets the process of ecological recovery underway, it is not the ideal solution. It is much better to reduce the number of large herbivores, both deer and sheep, to a level that is compatible with regeneration of the vegetation, including trees, without the need for fencing, and that process is now underway at Dundreggan.

## Reweaving the Web of Life

The planting of a tree,[22] or the protection of a naturally occurring seedling so that it can grow, may seem like a simple and basic act, but it has profound and long-lasting consequences. It is a key step in stimulating the re-establishment of natural habitats, which many other species benefit from, and in reinstating crucial ecological processes, which have not been functioning in Scotland for at least the past two centuries. Those processes include: nutrient cycling and soil enrichment (e.g. through nitrogen fixation), ecological succession, predator-prey dynamics and natural disturbance regimes.

In the areas protected by fences, or where deer densities had been reduced to very low numbers, vegetation begins to grow again to its natural potential, instead of being held at the height of a few centimetres by herbivore pressure or burning. This is the key step in restoring the forest and other natural habitats such as bogs and mires, because it is the vegetation that absorbs the sun's energy through photosynthesis, converting it into organic compounds that both enable the plants and trees to grow, and also provides the food source for the other, higher trophic levels within the ecosystems. Whereas in the past the Highlands have been described as being a sheep-shaved landscape, or a large-scale heavily contoured lawn (because only grasses that are adapted to high levels of grazing were able to survive amongst the high densities of large mammalian herbivores), they are now experiencing a new growth of lush plants and trees. This re-establishment of varied vegetation communities contrasts dramatically with the former heather- and grass-dominated monotony.

Scots pines planted 24 years previously, with naturally regenerating heather and bog myrtle, inside a fence to exclude red deer in Glen Affric. Outside the fence, only grass is growing, and peat hags are visible in the overgrazed and highly degraded landscape. (Photograph by Alan Featherstone.)

The growth of vegetation is quite varied at first, with the best results being seen in areas of better soils having good protection from the prevailing winds, and where there are nearby seed sources for trees and other native plants to spread from. The process of natural ecological succession begins to get re-established. In the most depleted areas, where only grasses had been able to survive the relentless grazing pressure of the previous centuries, heather begins flourish, and in it young pioneer trees, such as rowan, silver and downy birch and eared willow, appear. As they grow and spread, improving the soil with their fallen leaves each autumn, they prepare the way for the slower growing later successional species, including Scots pine and various broadleaved trees such as oak, hazel, wych elm, ash, goat willow and bird cherry.

In wetter areas, alder trees and bog myrtle, a woody deciduous shrub, begin to flourish, drying out the ground as they grow, and also improving the soil through the nitrogen-fixing ability of the symbiotic bacteria that grow on their roots. Over time, this enables other, less wet-tolerant species to spread there as well. In areas where bog myrtle has been able to grow unchecked by heavy grazing, it can reach a height of 2 metres, and large patches can be seen with young birches growing in their midst, because of the improved soil fertility and drier soil conditions that the bog myrtle has created.

As the vegetation and trees grow, they provide the habitat for increased numbers and a much greater diversity of invertebrate species. Insects such as moths, beetles and flies take advantage of the new food sources, spreading rapidly through their power of flight, and attracting in turn their suite of predators and parasites. Even invertebrates with very limited dispersal abilities, such as spiders and mites, reappear in unexpected places. Although both groups are flightless, spiders are able to drift on the wind, attached to their lines of silk, for long distances. Some species of mite are phoretic, meaning that they hitchhike rides on winged insects, such as beetles and ants during their flight period each year to reach new sites.

The burgeoning invertebrate populations attract insectivorous birds, some of which also carry seeds in their guts that germinate in their droppings and accelerate the further recovery of the vegetation. Birds such as black grouse, which thrive on the edge of young woodlands of Scots pines, are doing particularly well, and their previously fragmented and isolated populations are now beginning to connect up and interbreed, as woodland corridors begin to get re-established across whole landscapes and watersheds. As young trees get established and begin to grow successfully, their symbiotic fungal partners also reappear and flourish.

Through the interconnection of tree roots and fungal hyphae in the soil, the trees provide sugars and carbohydrates to the fungi (which have no chlorophyll and therefore cannot photosynthesise those nutrients themselves), while the fungi provide minerals that they take up from the soil, which the trees cannot access directly on their own. This mutually beneficial mycorrhizal partnership is a fundamental basis of most natural forests (and many other terrestrial ecosystems as well), and in Glen Affric we have observed species such as *Russula emetica* growing at the base of Scots pines we have planted as part of the forest restoration work there – the mycorrhizal relationship between those two is well documented.[23]

Even the peat hags, which had formerly scarred the land as open wounds in many of the Highland glens, are beginning to revegetate and return to a condition of ecological health[24] in areas where the grazing pressure of deer has been greatly reduced or eliminated through fencing. This is a slow process, because of the nutrient-poor, acidic conditions of the hags, but over time plants such as sphagnum mosses, sundews, butterworts and bog cotton grow over the bare peat. They are all well adapted to those conditions and are followed by species such as cross-leaved heath and bog myrtle that are able to flourish in the drier conditions, and where some more nutrients have accumulated (e.g. through the nitrogen-fixing ability of the bacteria growing on the bog myrtle's roots, and from the nutrients deposited on the soil surface each autumn by the shed leaves of that deciduous shrub species). The insectivorous sundews and butterworts also aid the soil enrichment process, by catching insects and recycling the nutrients from them through their own annual leaves and then into the ground.

This process of vegetation recovery and functional ecosystem re-establishment is taking place at varying rates, spreading outwards from different locations where it has been able to get started quickly. Some of those places are around the periphery of the surviving fragments of the original Caledonian Forest, especially those where some human-assisted regeneration had been initiated in the late twentieth century. However, for other sites, particularly in the northwest Highlands, that are more exposed to the prevailing westerly winds and have poorer soils and greater rainfall, it is much slower because of the harsher environmental conditions. In many of those cases, the work of Trees for Life and other organisations involves actively assisting the recovery process through the targeted planting of pioneer tree species to form nuclei, or 'islands', of new young forest at strategically selected sites in the bleak landscape, from which further natural regeneration can spread in due course.[25]

By 2018, Trees for Life had planted over 1.5 million native trees[26] and had facilitated the natural regeneration of a larger number of trees, either through the use of deer-proof fencing or substantial reductions in the numbers of deer. Fencing is not an ideal solution, but is instead more of an emergency action that buys time while deer numbers are brought down. It causes problems of its own, for example by causing the mortality of woodland birds such as the capercaillie and black grouse, which have relatively poor eyesight and fly low through the woodlands, colliding with the fences. Wherever possible, the charity removes redundant fencing as soon as possible, returning the land to a more natural condition, free of the intrusive barriers created by 2-metre-high deer fences.

A key part of the success of Trees for Life and other organisations working to restore the Caledonian Forest and Scotland's other native ecosystems has been the ability to draw on supportive pieces of legislation in support of their efforts. In particular, the inclusion of the Caledonian Forest as a priority habitat for conservation and restoration in the European Union's Habitats and Birds Directive[27] has added considerable weight to project proposal funding applications, both to national bodies and in some cases to the European Union's LIFE Fund, which is the EU's financial instrument supporting environmental, nature conservation and climate action projects throughout the European Union. This has led to funds from the European Union being made available by the Scottish government for native woodland expansion through the Forestry Grant Schemes administered by Forestry Commission Scotland, which have funded many restoration projects.

The internationally recognised importance of the Caledonian Forest has also supported the availability of funds from the country's National Lottery, via the Heritage Lottery Fund, for restoration of the forest and many of its species. In the years running up to new millennium on 1 January 2000, a consortium of Scottish conservation groups succeeded in obtaining significant funding from the Lottery for the 'Millennium Forest for Scotland' project,[28] which involved many tree planting and forest restoration schemes all across the country. This was predicated on the basis that the best way Scotland could celebrate the arrival of the new millennium was by helping to repair some of the ecological damage that been wrought during the previous centuries.

Another key element in supporting the movement to restore the forest was the publication of the UK Biodiversity Action Plan in 1994,[29] as the UK government's response to the Convention on Biological Diversity (CBD), which the country signed in 1992 in Rio de Janeiro. This designated priority habitats and species for conservation action, with specific action plans being drawn up for many of them, as a guide for practical work to follow. Native pine woodlands and many of the species that depend on the Caledonian Forest were included as priorities in the Biodiversity Action Plan, and this again helped to garner funding and support for restoration projects.

## Deepening the Restoration Process

With a new generation of young pines and birches beginning to get established as a result of our initial projects, we began turning our attention to other components of the ecosystem, particularly those which were unable to recover by themselves. In 1991 Trees for Life launched a pioneering project[30] to aid the recovery of the European aspen, which had been lost from most of its former range in Scotland, threatening the survival of the unique assemblage of rare mosses, lichens and invertebrates specifically associated with it.[31] For reasons that are still not fully understood, aspen rarely flowers in Scotland, and because it is a dioecious species, with male and female flowers being borne on different trees, which are often located many miles apart from each other, seed production has become a very rare occurrence. The effects of this characteristic have been compounded by the fact that aspen is one of the most palatable of all species to deer, resulting in no young seedlings being recruited, except where they could grow as suckers or ramets off the root system of a parent tree and inside a fenced area, where deer were excluded. In many cases the surviving small stands of aspen consisted of just one or a very small number of clones, with each clone being composed of a number of separate trunks, all connected through their roots and therefore all the same (single-sexed) organism.

To address this issue, vegetative propagation of aspen was developed from root cuttings, taking advantage of the species' ability to grow clonally, with new young trees being nurtured from severed root sections of a mature tree. This is an intrusive and labour-intensive method of reproduction, but it provides young trees for planting out, again inside fenced areas to protect them from deer, so that aspen can return to areas where it is missing. By planting out mixed groups of young aspens, sourced from a number of different parents, the aim is to achieve new stands of aspens containing both male and female trees in close proximity to each other, so that pollination and seed production will become possible again in future.

Tens of thousands of young aspens have now been grown by this method, and Trees for Life staff and volunteers have planted them on a wide range of sites, both to enhance the genetic (and sexual) diversity of existing aspen stands and to create new stands where the species is absent. In addition to restoring the tree for its own sake, these sites have been targeted with two larger purposes in mind. One of those is to provide future habitat for rare aspen-dependent biota,[32] such as the aspen hoverfly,[33] which is currently confined to a very few sites in Scotland, and which requires an area of at least 4.5 hectares of aspen woodland in which to live. The other purpose is to provide a future habitat for the European beaver, which relies on aspen bark as one of its key winter foods, and whose reintroduction Trees for Life (and many other organisations) have long advocated.[34]

In addition to the propagation of aspens from root cuttings, Trees for Life has also engaged in experimental work to stimulate the aspen to flower more often than it normally does, by stressing some aspens in controlled conditions. These efforts, and other projects to graft branches from aspens that had a greater proclivity to flower on to stock aspens in the charity's nursery, holds considerable promise for a gradual increase in the availability of aspen seed for new planting projects.[35] This is important because aspens grown from seed inherently have a greater genetic diversity compared to aspens grown from root cuttings, which are each genetically identical to the parent tree. Over time, it is hoped this work will lead to the growth of aspen trees that are more frequent in their flowering and will result eventually in a significant increase in the natural production of seeds by trees in the wild.

Trees for Life has also developed similar assisted regeneration techniques for other species that are severely constrained in their ability to disperse naturally. These include some of the rare montane willow shrubs, such as dwarf birch, downy willow and woolly willow, the latter of which is confined to about a dozen small isolated populations at high elevations in remote mountain sites. Again, this is due to the palatability of the species and intensive grazing pressure from deer. Like aspen, the willows are dioecious, so very little seed is produced because of the wide geographic separation between male and female plants.

Trees for Life extended the assisted regeneration[36] beyond trees to include flowering plants that have a tendency to spread vegetatively rather than by seeding. One such case is twinflower,[37] a species that is circumboreal in its distribution. Although it is very common and abundant in forests in both Scandinavia and western Canada, it has become extremely rare in Scotland, surviving at a few sites where it seldom, if ever, produces any seeds. Trees for Life has collected propagation material from some of the few small patches of twinflower that have survived in Glen Affric, and has grown new plants in the charity's nursery that will be planted out both to enrich the existing remnant patches and to create new populations in suitable areas of restoring native woodland.

However, it is not just trees and plants that struggle to expand from the isolated pockets of suitable habitat, where they have managed to survive the widespread deforestation that has occurred in the Highlands. The red squirrel also suffered from the loss of its native forest home, and has been greatly reduced both in numbers and its geographic range. In addition, by the early twenty-first century, it had been displaced from much of the United Kingdom by the introduced non-native grey squirrel,[38] which out-competes its native relative for food and also carries a virus that is lethal to the reds, but doesn't affect the greys. The Caledonian Forest is the stronghold for the remaining population of red squirrels,[39] but many of the isolated remnants, particularly in the northwest Highlands, have lost their squirrels completely due to past exploitation.

With the red squirrels unable to spread again across large tracts of treeless ground, these remote woodland patches had been devoid of their arboreal bushy-tailed rodents for 50 years or more. Squirrels fulfil an important ecological role in forests, as dispersers of tree seeds. Because they do not recover all of the seeds they cache for the winter, some of the forgotten ones germinate and grow, thereby providing a new generation of young trees. In the absence of squirrels, this function was missing from these isolated forest remnants, compounding the existing problem of lack of tree regeneration due to overgrazing by deer and sheep. In partnership with another conservation charity, the Roy Dennis Foundation for Wildlife,[40] Trees for Life initiated translocation projects to bring red squirrels back to some of these squirrel-less forests in the Northwest Highlands from areas where they were still abundant elsewhere in Scotland. This is a good example of how restoring a species (the red squirrel) can bring additional benefits to the recovering forest ecosystem by re-establishing a key ecological function – seed dispersal. This is known in scientific terms as a positive trophic cascade, whereby action on one tier or trophic level of an ecosystem produces knock-on benefits to another level.

The other issue affecting red squirrels – displacement by the introduced non-native grey squirrels – has been the focus of many labour-intensive and costly conservation initiatives for several decades, which were in essence rear-guard actions that merely slowed down and delayed the extent of the problem. However, recent research has shown that another (and unexpected) positive trophic cascade is helping the red squirrel. This is due to the increase in both numbers and range of the pine marten, a native mustelid species that had been virtually exterminated in Scotland in the early twentieth century because it was viewed as 'vermin' that preyed on game birds and other species deemed to be of economic value to humans.[41] However, following protection granted to the species in the last quarter of the twentieth century, pine martens began to increase in numbers and range again.[42] Scientific research that was initially carried out in Ireland and has subsequently been replicated in Scotland has shown that as pine martens regained some of their former territory, grey squirrels disappeared and reds recovered.

Although the pine marten preys upon red squirrels, the two species have co-evolved over a long period of time so that, although martens take some reds, the two species live together in ecological balance. By contrast, grey squirrels had no experience of a mustelid predator like the pine marten in their native habitat in the eastern United States. Being larger and slightly less agile than the reds, and preferring to spend more time on the ground, they are more easily caught by pine martens, and therefore tend to avoid areas where martens are present. Thus, as pine martens continue their spread throughout Scotland, they have now been documented as providing natural control on grey squirrels, which should enable the red squirrel to recolonise more of its former range.[43]

Red deer have been positively affected by the restoration of native forests in the Highlands. Although the species is generally considered an animal of the open hillside in Scotland, it is by nature forest-dwelling for much of the year. Because of deforestation in Scotland, the red deer has been forced to adapt to a largely treeless landscape, but one effect of this has been a significant reduction in its body size. A study examined[44] the difference in size between mature male red deer (stags) in the Cairngorms National Park in Scotland and those of a comparable region in southwest Norway, with similar soils, temperatures and rainfall. The Norwegian deer have a natural forest habitat in which they spend much of their time, and they were, on average, 40% larger and heavier than the Scottish stags, because of the healthier woodland conditions in which they live. Now, in the Highlands, some estates that have begun to carry out native forest restoration projects, and a parallel reduction in deer numbers, are finding that the average weight of their red deer is increasing.[45]

In sites such as Glen Affric, where forest restoration work has been underway since the 1960s and concerted action has been taken (at least on the land managed by Forestry Commission Scotland) to reduce the deer numbers to a level compatible with the recovery of the vegetation, the deepening of the restoration process is clearly apparent. At the eastern end of the glen, where the remnant forest was in better ecological condition, the shift from pioneer woodland to later successional forest species is well underway. A few individuals of a number of hardwood tree species – sessile oak, ash and wych elm – survived in inaccessible locations in the Affric River gorge, and their seedlings are now growing up through the pioneer birch woodland, along with understory trees such as hazel, hawthorn, bird cherry and grey willow.

Accompanying them are a wide range of flowering plants, which had previously been suppressed by the deer, including species such as hogweed and angelica, whose large umbels provide a vital nectar source for a whole suite of pollinating insects, from sawflies and beetles to hoverflies and bees. Over time, the large-leaved dense-canopied species, such as oak, hazel and wych elm, should also begin, through shading, to reduce the prevalence of bracken fern. That is a rhizomatous fern which has taken advantage of a lack of competition in the deforested landscape. Because its fronds are unpalatable to most herbivores, it has taken over large areas of ground in an impenetrable monoculture of 2-metre-tall fronds in the summer, suppressing the growth of flowering herbaceous plants and indeed tree seedlings themselves.

In areas such as this in Glen Affric, the ecological balance has been righted, shifted away from the excessive overgrazing from deer and sheep in the past, and the numerous strands of the natural forest ecosystem are being reconnected, and the associated ecological processes are beginning to get re-established.

## Reconnecting People with Nature

Key to this success has been the active engagement of people, particularly volunteers, in the process. I founded Trees for Life as a volunteer myself, working on the project in my spare time in the evenings and on weekends to get it established. Inspired by the vision, and opportunity of engaging in practical and positive action to help the return of the Caledonian Forest, others joined me, and this became formalised in 1991, with the establishment of a programme of volunteer Work Weeks (now called Conservation Weeks). Based initially in a remote bothy – an old, stone-walled uninhabited cottage – with no running water, flush toilet or electricity, these volunteer weeks offer the opportunity of spending 7 days out in a wild landscape, away from all the distractions and luxuries of modern life, working with like-minded people to help Nature recover from past damage by planting trees as part of the restoration of the Caledonian Forest. It is an elemental experience, with the volunteers working willingly in all weathers, often with long walks over difficult terrain to reach the planting sites, and then talking together in the evenings about the value of what they're doing, as they prepare shared meals throughout the week.

A volunteer on a Trees for Life Conservation Week planting a downy birch seedling as part of the forest restoration work on the charity's Dundreggan Conservation Estate in Glenmoriston. (Photograph by Alan Featherstone.)

Thousands of people have taken part in these weeks since 1991, with large numbers coming back repeatedly over the years because they find them so meaningful. In many cases participation has been deeply transformative, with some volunteers changing their careers as a result. Others have been trained up to lead the weeks, giving them an opportunity to engage more fully with the work. Some volunteers have been inspired to found their own forest restoration projects elsewhere, such as the charity Moor Trees in 1999,[46] which works to restore native forest on Dartmoor in the southwest of England. Volunteers on these 'work weeks' have planted almost all the over 1.5 million native trees planted by Trees for Life to date.[47]

The key factor behind their success is that the weeks provide an opportunity for the volunteers to engage directly in a hands-on way with both Nature and the nurturing of new life. Most people today grow up deprived of one of the essential birthrights of all humanity – daily contact with wild Nature – and that separation is one of the main causes of many of the problems we now face in the world. The volunteers who take part in these weeks learn about the ecology and special species of the Caledonian Forest, and see for themselves that the old remnants are dying out because of the human-created imbalance of overgrazing in the landscape. They understand in a very direct and personal way that the trees they plant are part of the recovery of the forest and will grow to provide the habitat for all its species. Many people return year after year, so that they can see how 'their' trees have grown, and can observe the insects, plants, birds and fungi that are living in the young forest that has resulted from their labours.

In addition to the Conservation Weeks, Trees for Life also runs Conservation Days for more locally based people, and frequently hosts site visits and tours of the forest areas where it works. Participants in the latter include groups of students, academic staff, other conservation practitioners, researchers and the general public. All of these activities and programmes serve to connect people with the Caledonian Forest (which until the 1980s was little known, even in Scotland), and provide them with a tangible experience of how it is possible to help Nature heal, and assist a highly degraded ecosystem to regain its vitality, diversity and abundance.

By the end of the first decade of the twenty-first century, numerous forest restoration projects were underway in most of the main remnants of the old Caledonian Forest. These were initiated and run by a wide range of bodies, ranging from government agencies such as Forestry Commission Scotland[48] and Scottish Natural Heritage,[49] to conservation groups such as the National Trust for Scotland (NTS), the Royal Society for the Protection of Birds (RSPB),[50] Trees for Life and the Woodland Trust,[51] local community groups and private landowners. Although the country's inequitable and archaic system of land ownership is a barrier to achieving restoration in many instances, some landowners, especially newer ones who have different values, have been amongst the most effective initiators of large scale forest recovery, because of their control over significant tracts of land.

It is not surprising, in my view, that during the period when Trees for Life and other organisations have been working to reverse the long process of ecological decline that has taken place in Scotland, there has been a renewed drive for the country to regain more control over its own affairs in a political context. The loss of Nature in the Highlands was mirrored by the lack of local decision-making, with absentee landlords and the distant government in London holding all the power over what happened on the land. Just as restoring the almost-lost forest necessitates reconnecting with Nature, so does it also require regaining responsibility and authority for the decisions which enable that to occur. Thus, after a gap of almost 300 years, Scotland regained its own parliament in 1999, with substantial powers being devolved from the UK government in London, and land reform is now being given serious consideration. There is a strong interest amongst many people in Scotland for independence from the rest of the United Kingdom, and although the 2014 referendum led to a 'no to independence' result,[52] it is likely another vote will be held in the coming years. One very practical consequence of the re-establishment of the Scottish government was that it authorised a 5-year trial reintroduction of European beavers,[53] beginning in 2009. Following its successful conclusion, the government then accepted the beaver as an officially reinstated native species in 2016.[54] This is the first time any part of the United Kingdom has officially carried out a successful reintroduction of an extirpated large mammal species, and it set a precedent that is now being followed by England, where a beaver trial reintroduction is underway in Devon,[55] and potentially also by Wales.

## Looking Ahead: If the Rights of Nature Become Paramount

While much has been achieved in terms of laying the groundwork for the restoration of the Caledonian Forest and other natural ecosystems (e.g. raised peat bogs) in Scotland, there is still a long way to go before the land, and indeed the coastal areas and adjacent marine areas, will be returned to full ecological health and diversity. Although it has a relatively progressive government and well-educated population that demonstrably cares for the environment in many ways, the country is still run primarily on a model of society that is dominated by monetary values and the outdated, patently unsustainable goal of unlimited and endless economic growth. Everything else is subservient to that financially based dogma, despite the fact that it flies in the face of ecological realities and represents a collective form of delusional madness that many countries in the world still subscribe to.

The shift that is required now from that old destructive philosophy and mentality is a profound one, and human survival, in Scotland and indeed through the world, relies on it taking place. We have reached a turning point where our future, both here in Scotland and globally, depends on us stopping to look at Nature and other species as merely objects to be exploited, used and enjoyed for human gain. Instead, we need to recognise that natural ecosystems and all their constituent species have both a right to exist in and of themselves, and a role to play in the ongoing evolution of all life on Earth. This transformation in values is the key to the full recovery of the Caledonian Forest. It will lead to a whole set of decisions and actions that reinforce and strengthen one other, and at the same time will form a human cultural equivalent of a positive trophic cascade in an ecosystem. This of course will have to be part of a much larger, global movement, in which humans collectively begin to adapt and model our culture on natural systems and processes again, in contrast to the separation from, and domination of, Nature that has prevailed until now.

The seeds of such a transformation already exist, and many people now recognise that major changes are required in human endeavours. This is being driven to a considerable extent by the concerns about anthropogenic climate change, caused by our industrially based culture, and is exemplified in such varying movements as the spread of veganism, the rapid deployment of renewable energy in many countries, the establishment of eco-villages on every continent, the widespread utilisation of Permaculture principles and techniques, the use of alternative systems of decision-making such as sociocracy (a system of dynamic governance that operates with consent at all levels of decision-making)[56] and the significant traction gained by rewilding projects in recent years (rewilding is a relatively new popular term that is largely synonymous with ecological restoration).[57] Projecting those trends forward in time enables us to envision a positive future for both humanity and the planet, and the remainder of this chapter imagines what could happen when they are applied to Scotland and the Caledonian Forest. Just imagine …

*****

It was only when a widespread shift in public attitudes and values occurred, which placed the Rights of Nature as paramount in all human endeavours and activities, that the Caledonian Forest truly began to recover and return to its former extent, diversity and ecological health. When it took place, the change was dramatic. For example, whilst there had been government grants on offer since the 1980s for practical work to help restore the forest remnants and replant areas where the trees had already vanished, many landowners never took up the grants, because their priorities lay elsewhere. Now, a concerted, coordinated and fully funded programme was launched to facilitate the regeneration and expansion of all the surviving remnants of the original Caledonian Forest, combined with a targeted scheme of re-establishing the forest through tree planting in areas that had been completely stripped of their trees. Planting was done in patterns that mimicked the natural regeneration of woodland, with irregular and varied spacing between the trees, which were selected to match the soil conditions that were suitable for them. The goal was for any areas of new planted woodland to be indistinguishable in their appearance and composition from naturally regenerated forests. These two strands of action – natural regeneration and tree planting – were integrated together in an overall plan, designed on natural ecological principles, that would enable the natural recovery of the forest as quickly as possible.

Viewed from above, as snapshots over time, the process could be seen as green waves of new vegetation and trees spreading outwards from numerous source points, linking up with each other and covering the land with a rich and abundant tapestry of diverse natural habitats and species. As such, it replicated in many ways what had occurred in the Highlands about 10,000 years previously, at the end of the last Ice Age, when plants, trees, birds, animals and insects all spread to create healthy natural ecosystems on the bare land as it was exposed by the retreating ice sheets. For at least the previous two millennia, that natural recovery process had been actively prevented by the way people lived on and used the land. However, as soon as humanity turned away from our destructive (and self-destructive) course, the planet's inherent natural ability to recover from wounds or massive disturbance and create the conditions for maximising the abundance and diversity of life, re-asserted itself. Now, the returning young forests also provided an expanded habitat for many mammals, including wood mice, voles, badgers, red squirrels and pine martens.

It was widely recognised that red deer in the Highlands had suffered greatly from deforestation, and were in fact the 'runts of the glen' rather than the 'monarchs of the glen' of their previous public image, due to reduction in body size because of loss of their natural forest habitat. A coordinated programme of a reduction in deer numbers was launched. As this took effect, more natural regeneration of the forest occurred spontaneously. The deer benefitted too; the remaining animals were able to grow to a larger size, as there was more food for the smaller population in the increased area of native woodlands. Their individual weights began to approach those of deer in southwest Norway, which had been shown to be 40% heavier than those in deforested areas of Scotland.[58]

The sheep that had roamed freely over most of the Highlands, due almost entirely to the 'perverse' subsidies[59] from the European Union's Common Agriculture Policy (CAP), were largely removed, once the subsidies were redirected towards ecological restoration of the land, rather than its ongoing depletion. The sheep took up nutrients from the already impoverished soil through their grazing, which were then lost permanently to the ecosystem when the animals were sent for slaughter further south in the United Kingdom. The greatly reduced numbers of sheep that remained were concentrated in lower-lying areas, where they were more productive, and where it was simpler to ensure that the reintroduced terrestrial carnivores did not prey upon them.

Other ecologically destructive activities, such as the practise of muirburn for the purpose of promoting the growth of young heather plants as food for red grouse, and the persecution of raptors such as the hen harrier, red kite and golden eagle for their perceived threat to grouse-shooting interests, came to an end surprisingly quickly. As the populations of these birds of prey began to recover, their presence in the landscape produced knock-on (cumulative) ecological benefits to other species, as a result of positive trophic cascades getting re-established again. This also helped to make the case for the reintroduction of the larger, extirpated, terrestrial predators that followed.

After an absence of many centuries, first the Eurasian lynx and then the wolf were successfully reintroduced to the Highlands, restoring the missing role of apex predator to the forests. They played only a minor role in the substantial reduction of deer numbers, particularly red deer, as the majority of that took place at the hands of people before the carnivores were returned. However, once the lynx and wolf were reinstated, their other ecological functions, apart from simply killing prey, soon became apparent. Their presence improved the fitness of the remaining deer by keeping the animals on the move, which in turn facilitated the further regeneration of the vegetation, particularly highly palatable species such as rowan and aspen. The overall health of the deer also improved, as the predators removed the sick, weak, old and some very young animals from the population.

Crucially, the presence of the apex predators also enabled the reinstatement of a closed-loop nutrient cycle in the Highlands, instead of the previous ongoing depletion of natural biological wealth represented by the export of hundreds of thousands of animal carcasses (both deer and sheep) every year. This in turn provided ecological benefits to the host of species that function as scavengers and decomposers in the ecosystem. In an unexpected parallel with the pine marten/red squirrel experience, the return of the lynx resulted in a similar effect. Its presence in the ecosystem shifted the balance between the native roe deer and the introduced non-native sika deer, with the former benefitting and the sika being progressively edged out from more and more of the Highlands.

As the forest and other natural ecosystems recovered, ecological relationships that had been sundered for centuries or longer became spontaneously re-established. Thus, robins began following wild boar around, once the latter had become widespread again, finding worms and grubs in the disturbed soil that the boar left behind from their rooting activity. Similarly, once the European brown bear had been reintroduced, it began catching some of the increased numbers of Atlantic salmon that were thriving and spawning in Scottish rivers again, because of the greatly enhanced riparian woodland cover that provided the leaf litter for the aquatic invertebrates that the fish feed on. Significantly, the feeding on the fish by bears resulted in a net transfer of nutrients from the water (both freshwater and seas) to the land, offsetting to some extent the tremendous loss of nutrients to the seas that had taken place during the centuries when the land was deforested and ecologically impoverished.

Some of the changes that occurred were quite surprising. For example, the bluethroat, a small passerine bird common in Scandinavia, had only been known in Scotland as an occasionally passing migrant in coastal areas. However, once its preferred habitat, montane scrub,[60] became re-established, it began breeding successfully in the Highlands, as it must have done in the past. Other species that occur in similar pinewood areas in Scandinavia, and are circumpolar in distribution, such as the calypso orchid, which had previously been unrecorded in Scotland, began to grow spontaneously in the Highlands. Even the notorious Highland biting midge, which seriously impacted outdoor activities on windless damp days because of its huge numbers and disproportionately bad bite, started to decrease somewhat in numbers. This was as a result of the drying out of much of the land in the Highlands when trees and other vegetation grew again, which reduced the availability of waterlogged soils for it to breed in, and also through the substantial increase in numbers of the insectivorous birds and bats that eat midges.

Another interesting consequence of the widespread restoration of the Caledonian Forest that occurred after the Rights of Nature became paramount was the confirmation of a theory that had sought to explain why aspen in Scotland at the turn of the twenty-first century flowered so rarely in comparison with aspen trees in similar ecological conditions just 120 miles away across the North Sea, in southwest Norway. The theory proposed that within any given population of European aspen, there was a natural variation in reproductive strategy, with some trees flowering readily, whilst others had a greater tendency to reproduce via root suckering. Because of the near total deforestation that had occurred in Scotland, and subsequent intensive pressure of overgrazing by large herbivores for several centuries, it was suggested that the aspens that reproduced by flowering were removed from the population.

The theory was that any trees that invested their future in seedlings, as against suckers, saw their offspring quickly eaten by deer, and over time the parent trees died out without leaving any surviving progeny. By contrast, an aspen that invested its reproductive future in suckering would just produce more suckers to replace those eaten by deer, and eventually environmental conditions, such as an occasional very cold winter, would reduce the deer numbers so that some suckers survived and grew to maturity. It was thought that the intense pressure of overgrazing for a period of several centuries had skewed the genetic spread of aspen in Scotland, away from trees that preferred to reproduce by seeding, leaving only those that flourished by suckering. This theory was eventually proved correct when a new generation of aspens grown from seed began flowering. Once the browsing pressure from herbivores was reduced to a level compatible with natural regeneration, aspen began reproducing by seed again in the Highlands.

In tandem with the recovery of the land, concerted action by people took place to drastically reduce the ongoing human impacts on nature in the Highlands. For example, projects were launched to eradicate invasive non-native species, including rhododendron, giant hogweed, Japanese knotweed, American mink and sika deer amongst others. This was a major effort, coordinated across a wide spectrum of society, which included conservation groups, scout and girl guide groups, armed forces personnel, local community organisations, hill-walkers and countless informal volunteers. It also led to the establishment of a National Voluntary Environmental Service programme that all young people were encouraged to sign up for as part of a gap year between the end of their formal education and beginning regular employment. Older people were also welcomed to take part in the Service, which ran a wide range of projects that facilitated and aided the ecological recovery of Highland ecosystems and also enabled the participants to establish and deepen a meaningful personal connection with Nature.

As people began to fully grasp the scale and scope of the changes to human culture that were necessary for the restoration of the Caledonian Forest and other ecosystems, major shifts took place. People made personal decisions to simplify their lifestyles, to ask less of the Earth, so for example the consumption of meat and animal products fell dramatically as they opted for plant-based and vegan diets, using organic and much more locally grown food than previously. The land that was freed up by these changes enabled the Caledonian Forest to expand significantly again.

Scotland also began a massive programme of retrofitting of existing buildings, including insulating them to high standards and installing solar thermal and photovoltaic panels to reduce their needs for external energy supplies. That in turn facilitated the removal of significant energy-generating infrastructure that had been installed in the Highland glens from the middle of the twentieth century onwards, first dams and similar hydro-electric schemes, and more recently for large-scale windfarms. Those installations had noteworthy adverse environmental effects, both directly through their impact on local hydrology in the case of dams and by the widespread mortality of birds and bats caused by wind turbines, and indirectly through the associated network of roads and tracks that scarred many mountain landscapes and provided ideal conditions for the spread of invasive species. The Scottish people developed a whole new set of skills and practical techniques to enable removal of this infrastructure from key areas identified as providing the best restoration sites for returning landscapes to a natural condition.

An overall land use zoning system was developed for the whole of Scotland, prioritising the most suitable areas for large-scale ecological restoration of the Caledonian Forest and other ecosystems. It also focussed land uses such as agriculture, commercial forestry and renewable energy schemes in areas closer to where the majority of the people lived and the wild qualities of the landscape were already the most heavily compromised.

Wider aspects of society in Scotland changed, too, with a particular case being that of landownership. The country had never experienced any meaningful land reform, and in the early twenty-first century most of the Highlands were owned by a relatively small number of mostly absentee landlords, many of whom lived abroad and only visited for a few weeks of the year to shoot deer and grouse, and to fish in the then still-depleted rivers.[61] However, after the country gained its independence and took responsibility for restoring its degraded ecosystems, Scotland became one of the first countries in the world to discard the whole concept of landownership entirely, replacing it with a stewardship principle instead. Under this, people were recognised as stewards of land when they took active measures to restore and maintain its ecological health and biological diversity, receiving grant aid and tax incentives to support them in the work. As a result of this, people began to feel they belonged to the land, rather than the land belonging to people.

The country's education system was also transformed, putting the connection with Nature at the heart of the curriculum in schools and universities, and incorporating significant practical elements involved with hands-on ecological restoration, biological recording and environmental site monitoring, amongst other activities. Public attitudes to Nature changed dramatically as well. For example, people embraced a newfound respect and admiration for the wolf, replacing the former irrational and misplaced fear of this apex predator. By these means, the Scottish people became truly indigenous again, connected to the land and its life in a way that had not been possible for centuries. The recovering Caledonian Forest became a source of considerable national pride and an inspiration to people of Scotland.

Crucially, Scotland was also able to make a significant contribution to the movement for 'Nature needs half'[62] that began in the early part of the twenty-first century, and which posited that humans should give half of the planet back to natural processes and as a habitat for all the other species we share the world with. While that 50% target was a challenge for many of the densely populated parts of Europe to achieve, it was a different situation in Scotland, especially in the Highlands. At the time of gaining independence, the country set a target of meeting the 50% for Nature goal, but in fact surpassed that slightly overall.

However, in the Highlands, where the human population had always been low and the renewed Caledonian Forest now covered much of its former range, a remarkable 70% of the land had its natural processes prevailing, and healthy populations of all the native species thriving. As a result, by the latter part of the twenty-first century, the Caledonian Forest was recognised as a good example of ecosystem recovery and of the new human relationship with the Earth that is based on the Rights of Nature being paramount. With Scotland having been one of the countries to have pioneered relatively large-scale ecological restoration and rewilding, the expertise of its practitioners was sought for advice on similar projects round the world, as the movement to heal the ecological wounds on the planet became a concerted global effort as the twenty-first century progressed.

## Notes

1. See for instance, Short, J. 'Wolf's Tale – The history of the wolf in Scotland'. Available at: http://www.wolvesandhumans.org/wolves/history_of_wolves_in_scotland.htm (accessed 20 January 2019).
2. Smout, C. 2014. 'The History and Myth of Scots Pine'. *Scottish Forestry*, 68:1.
3. Crumley, J. 2010. *The Last Wolf.* Birlinn, Edinburgh.
4. Steven H. M. and Carlisle A. 1959. *The Native Pinewoods of Scotland.* Oliver & Boyd, Edinburgh.
5. Wightman, A. and Hunter, J. 1996. *Who Owns Scotland?* Canongate Books Ltd.
6. Steven H. M. and Carlisle A. 1959. *The Native Pinewoods of Scotland.* Oliver & Boyd, Edinburgh.
7. Clifford, T. and Forster, A. 1997. 'Beinn Eighe National Nature Reserve: Woodland Management Policy and Practice 1944–1994' in *Scottish Woodland History*, Smout, T. C. (ed). Scottish Cultural Press, Edinburgh.
8. MacRae, F. 1980. 'The Native Pinewoods of Glen Affric', *Arboricultural Journal*, 4:1, 1–10.
9. Ramsay, P. 1996. *Revival of the Land: Creag Meagaidh National Nature Reserve.* Scottish Natural Heritage.
10. Summers, R. W., Proctor, R., Raistrick, P. and Taylor, S. 1997. 'The Structure of Abernethy Forest, Strathspey, Scotland' *Botanical Journal of Scotland*, 49:1, 39–55.
11. Newton, A. C., Stirling, M. and Crowell, M. 2001. 'Current Approaches to Native Woodland Restoration in Scotland' in *Botanical Journal of Scotland*, 53:2, 169–195.
12. (a) Love, J. A. 1983. *The Return of the Sea Eagle.* Cambridge University Press, Cambridge; (b) Love, J. A. 1988. *The Reintroduction of the White-Tailed Sea Eagle to Scotland: 1975–1987* Nature Conservancy Council.
13. Evans, I. M., Dennis, R. H., Orr-Ewing, D. C., Kjellen, N., Andersson, P.-O., Sylven, M., Senosiain, A. and Carbo, F. C. 1997. 'The Re-Establishment of Red Kite Breeding Populations in Scotland and England' *British Birds* 90:123–138.
14. Fraser Darling, F. 1955. *West Highland Survey: An Essay in Human Ecology.* Oxford University Press.
15. Findhorn Community, The. 1976. *The Findhorn Garden Story.* 1st Edition. Turnstone Books and Wildwood House Ltd.
16. Findhorn Community, The. 2008. *Findhorn Garden Story: A Brand New Colour Edition of the Black & White Classic.* 4th Edition. Findhorn Press. pp. 53–99.
17. Blanchflower, P. May 1990. 'A Survey of Tree Regeneration within a Native Pinewood, Coille Ruigh na Cuileige, with Particular Reference to Ground Vegetation'. Student dissertation, Edinburgh University.
18. See, for instance: 'Iconic Highlands Bothy Reborn as Eco-Friendly Rewilding base' 2016, available at: https://treesforlife.org.uk/news/article/iconic-highlands-bothy-reborn-as-eco-friendly-rewilding-base/ (accessed 20 January 2019).
19. Watson Featherstone, A. 1997. 'The Wild Heart of the Highlands' *ECOS*, 18:48–61.
20. See: https://treesforlife.org.uk/work/tree-nursery/ (accessed 21 January 2019)
21. See: https://treesforlife.org.uk/work/dundreggan/ (accessed 21 January 2019).
22. Lipkis, L. 1990. *The Simple Act of Planting a Tree: A Citizen Forester's Guide to Healing Your Neighborhood, Your City, and Your World*, J. P. Tarcher.
23. Termorshuizen, A. J. 1991. 'Succession of Mycorrhizal Fungi in Stands of *Pinus sylvestris* in the Netherlands' *Journal of Vegetation Science*, 2:4, 555–564.

24. Rosenburgh, A. E. 2015. Restoration and Recovery of Sphagnum on Degraded Blanket Bog. PhD thesis, Manchester Metropolitan University (available at: https://e-space.mmu.ac.uk/615952/)
25. Watson Featherstone, A. 2004. 'Rewilding in the North-Central Highlands – An Update' *ECOS*, 25:4–10; Taylor, P. 2005. *Beyond Conservation: A Wildland Strategy*. Routledge, London.
26. See: https://treesforlife.org.uk/news/article/trees-for-life-25th-anniversary/
27. European Union. 1992. *Council Directive 92/43/EEC of 21 May 1992 on the Conservation of Natural Habitats and of Wild Fauna and Flora*. European Union, Brussels.
28. Roe, A. 1999. *Restoration of Scotland's Forests: The Millennium Forest for Scotland Initiative*. University of Minnesota, Department of Horticultural Science. University of Minnesota Digital Conservancy, (accessed 30 November 2018).
29. Joint Nature Conservation Committee 1994. *UK Biodiversity Action Plan (UK BAP)*. UK Government, London.
30. (a) Watson Featherstone, A. 2002. 'The Trees for Life Aspen Project' in Cosgrove, P. and Amphlett, A. (eds.) 2002. *The Biodiversity and Management of Aspen Woodlands: Proceedings of a one-day conference held in Kingussie, Scotland, 25th May 2001* The Cairngorms Biodiversity Action Plan, Grantown-on-Spey, UK. (b) https://treesforlife.org.uk/work/aspen-project/ (accessed 21 January 2019).
31. Cosgrove, P. and Amphlett, A. (eds.) 2002. *The Biodiversity and Management of Aspen Woodlands: Proceedings of a one-day conference held in Kingussie, Scotland, 25th May 2001* The Cairngorms Biodiversity Action Plan, Grantown-on-Spey, UK; Cosgrove, P., Amphlett, A., Elliot, A., Ellis, C., Emmett, E., Prescott, T. and Watson Featherstone, A. 2005. Aspen: Britain's Missing Link with the Boreal Forest. *British Wildlife* 17:2, 107–115; Parrott, J. & MacKenzie, N. (eds) 2008. *Aspen in Scotland: Biodiversity and Management: Proceedings of a conference held in Boat of Garten, Scotland* 3 – 4 October 2008.
32. MacGowan, I. 1993. 'The Entomological Value of Aspen in the Scottish Highlands' *Malloch Society Research Report No. 1*, pp. 1–43.
33. Rotheray, E. L., MacGowan, I., Rotheray, G. E. et al. 2009. 'The Conservation Requirements of an Endangered Hoverfly, *Hammerschmidtia ferruginea* (Diptera, Syrphidae) in the British Isles', *Journal of Insect Conservation* 13:569.
34. Watson Featherstone, A. 1997). 'The Wild Heart of the Highlands' *ECOS*, 18:48–61.
35. https://treesforlife.org.uk/news/article/conservation-breakthrough-for-scotlandrsquos-rare-aspen-tree/ (accessed 9 January 2019).
36. https://treesforlife.org.uk/work/woodland-ground-flora-project/ (accessed 9 January 2019).
37. Scobie, A. R. and Wilcock, C. C. 2009. 'Limited Mate Availability Decreases Reproductive Success of Fragmented Populations of *Linnaea borealis*, a Rare, Clonal Self-Incompatible Plant' *Annals of Botany*, 103:6, 835–846.
38. Gurnell J. and Pepper, H. 1993. 'A Critical Look at Conserving the British Red Squirrel *Sciurus vulgaris*' *Mammal Review*, 23:3–4, 127–137.
39. Saving Scotland's Red Squirrels. https://scottishsquirrels.org.uk/scotlands-red-squirrels/
40. Roy Dennis Foundation for Wildlife. http://www.roydennis.org
41. Pine Marten Recovery Project. https://pine-marten-recovery-project.org.uk/our-work/conservation-status
42. See: https://www.nature.scot/professional-advice/safeguarding-protected-areas-and-species/protected-species/protected-species-z-guide/protected-species-pine-martens
43. Sheehy E., Sutherland C., O'Reilly C. and Lambin X. 2018. 'The Enemy of My Enemy Is My Friend: Native Pine Marten Recovery Reverses the Decline of the Red Squirrel by Suppressing Grey Squirrel Populations' *Proceedings of the. Royal Society B: Biological Sciences* 285:doi:10.1098/rspb.2017.2603.
44. Halley, D. 2016. Personal communication to author.
45. Wright, A. 2016. Corrour: Estate History, Progress and Plans (available at: https://www.corrour.co.uk/wp-content/uploads/2017/03/2016-corrour-environmental-sheet.pdf) (accessed 23 January 2019).
46. Moor Trees, https://www.moortrees.org/about-us/
47. Trees for Life, https://treesforlife.org.uk/volunteer/

48. Forestry Commission Scotland. 2008. 'Action for Scotland's Native Woodlands Edinburgh' (available at: https://scotland.forestry.gov.uk/images/corporate/pdf/fcfc116websmaller.pdf)
49. Scottish Natural Heritage. 2015. *The Management Plan for Beinn Eighe and Loch Maree Islands NNR 2015-2025* (available at: https://www.nature.scot/sites/default/files/2018-03/Management%20Plan%20for%20Beinn%20Eighe%20NNR%202015-2025_0.pdf)
50. Royal Society for the Protection of Birds. 1993. *Time for Pine: A Future for Caledonian Pinewoods.* RSPB, Sandy.
51. The Woodland Trust. https://www.woodlandtrust.org.uk/visiting-woods/wood-information/loch-arkaig-pine-forest/
52. See, for example: https://www.open.edu/openlearn/people-politics-law/the-2014-scottish-independence-referendum-why-was-there-no-vote
53. (a) Jones, S. and Campbell-Palmer, R. 2014. *The Scottish Beaver Trial: The Story of Britain's First Licensed Release into the Wild. Final Report.* Scottish Wildlife Trust and Royal Zoological Society of Scotland. (Available for download at: https://www.scottishbeavers.org.uk/beaver-facts/publications/
54. Scottish Government. 2016. 'Beavers to Remain in Scotland' (available at: https://news.gov.scot/news/beavers-to-remain-in-scotland)
55. Devon Wildlife Trust 2018. *River Otter Beaver Trial Third Annual Report* (available at: https://www.devonwildlifetrust.org/sites/default/files/2018-06/ROBT%20Annual%20Report%20April%202018%20-%20Final.pdf)
56. See: https://www.sociocracy.info/about-sociocracy/what-is-sociocracy/
57. See: https://rewildingeurope.com/what-is-rewilding/
58. Halley, D. Personal communication to author.
59. Myers, N., and J. Kent. 2001. *Perverse Subsidies: How Tax Dollars Can Undercut the Environment and the Economy.* Island Press, Washington, DC.
60. Gilbert, D. (ed.) 2002. *Montane Scrub: The Challenge above the Treeline.* Highland Birchwoods, Munlochy.
61. Wightman, A. and Hunter, J. 1996. *Who Owns Scotland*? Canongate Books Ltd.
62. Nature Needs Half. https://natureneedshalf.org (accessed 24 January 2019).

## Appendix: Common and Scientific Names of Plants and Animals

| | |
|---|---|
| **FUNGI** | |
| Sickener mushroom | *Russula emetica* |
| **LICHENS** | |
| Lichens | Mycophycophyta |
| **MOSSES** | |
| Mosses | Bryophyta |
| Sphagnum moss | *Sphagnum* spp. |
| **FERNS** | |
| Bracken fern | *Pteridium aquilinum* |
| **GRASSES AND GRASSLIKE PLANTS** | |
| Bog cotton | *Eriophorum* spp. |
| Grasses | Gramineae |
| **FORBES** | |
| Angelica | *Angelica sylvestris* |
| Butterwort | *Pinguicula vulgaris* |
| Calypso orchid | *Calypso bulbosa* |
| Giant hogweed | *Heracleum mantegazzianum* |
| Hogweed | *Heracleum sphondylium* |
| Japanese knotweed | *Fallopia japonica* |
| One-flowered wintergreen | *Moneses uniflora* |
| Sundews | *Drosera* spp. |
| Twinflower | *Linnea borealis* |
| **TREES AND SHRUBS** | |
| Alder | *Alnus glutinosa* |
| Ash | *Fraxinus excelsior* |
| Bird cherry | *Prunus padus* |
| Blackthorn (a.k.a. sloe) | *Prunus spinosa* |
| Bog myrtle | *Myrica gale* |
| Cross-leaved heath | *Erica tetralix* |
| Downy birch | *Betula pubescens* |
| Downy willow | *Salix lapponum* |
| Dwarf birch | *Betula nana* |
| Eared willow | *Salix aurita* |
| European aspen | *Populus tremula* |
| Goat willow | *Salix caprea* |
| Grey willow | *Salix cinerea oleifolia* |
| Hawthorn | *Crataegus monogyna* |

| | |
|---|---|
| Hazel | *Corylus avellana* |
| Holly | *Ilex aquifolium* |
| Juniper | *Juniperus communis* |
| Larch | *Larix sibirica* |
| Norway spruce | *Picea abies* |
| Rhododendron | *Rhododendron ponticum* |
| Rowan | *Sorbus aucuparia* |
| Scots pine | *Pinus sylvestris* |
| Sessile oak | *Quercus petraea* |
| Siberian fir | *Abies sibirica* |
| Silver birch | *Betula pendula* |
| Wild cherry | *Prunus avium* |
| Willow | *Salix* spp. |
| Woolly willow | *Salix lanata* |
| Wych elm | *Ulmus glabra* |
| **INVERTEBRATES** | |
| **BACTERIA** | |
| Bacteria | *Frankia* spp. |
| **SPIDERS AND ALLIES** | |
| Mites | Acari |
| Spiders | Arachnida |
| **INSECTS** | |
| Ants | Formicidae |
| Aspen hoverfly | *Hammerschmidtia ferruginea* |
| Sawflies | Hymenoptera |
| Bees | Hymenoptera |
| Beetles | Coleoptera |
| Highland biting midge | *Culicoides impunctatus* |
| Hoverflies | Syrphidae |
| Narrow-headed wood ant | *Formica exsecta* |
| **VERTEBRATES** | |
| **FISH** | |
| Atlantic salmon | *Salmo salar* |
| **BIRDS** | |
| Black grouse | *Tetrao tetrix* |
| Bluethroat | *Luscinia svecica* |
| Capercaillie | *Tetrao urogallus* |
| Golden eagle | *Aquila chrysaetos* |
| Hen harrier | *Circus cyaneus* |
| Osprey | *Pandion haliaetus* |

| | |
|---|---|
| Red grouse | *Lagopus lagopus scotica* |
| Red kite | *Milvus milvus* |
| Robin (a.k.a. European robin) | *Erithacus rubecula* |
| Scottish crossbill | *Loxia scotica* |
| White-tailed eagle (a.k.a. sea eagle) | *Haliaeetus albicilla* |
| **MAMMALS** | |
| American mink | *Neovison vison* |
| Bats | Chiroptera |
| Domestic sheep | *Ovis aries* |
| European badger | *Meles meles* |
| European beaver | *Castor fiber* |
| European brown bear | *Ursus arctos arctos* |
| Eurasian lynx | *Lynx lynx* |
| Eurasian wolf | *Canis lupus lupus* |
| Non-native grey squirrel | *Sciurus carolinensis* |
| Pine marten | *Martes martes* |
| Red deer | *Cervus elaphus* |
| Red squirrel | *Sciurus vulgaris* |
| Roe deer | *Capreolus capreolus* |
| Sika deer | *Cervus nippon* |
| Voles | Cricetidae |
| Wild boar | *Sus scrofa* |
| Wood mouse | *Apodemus sylvaticus* |

# 9

# *The Significance of the Stewardship Ethic of the Indigenous Peoples of Nigeria's Niger Delta Region on Biodiversity Conservation*

**Ngozi F. Unuigbe**

**CONTENTS**


Background ........ 167
The Niger Delta Ecosystem ........ 168
Wastewater ........ 168
Gas Flaring ........ 169
Seismic Surveys and the Construction of Roads and Pipelines ........ 169
Dredging ........ 169
Inadequate Clean-Up ........ 169
The Niger Delta Indigenous Peoples and Their Sacred Sites ........ 171
Adibe and Esiribi Lakes ........ 171
Boupere Lake ........ 172
Okpagha and Ogriki Trees ........ 172
Ovughere Shrine ........ 173
Obi Pond ........ 174
Usede Pond ........ 174
Ode Evil Forest ........ 174
Umuaja Shrine ........ 175
The Impact of Environmental Degradation on Indigenous Sites ........ 176
Indigenous Conservation Culture and the Rights of Nature ........ 178
Indigenous Conservation Laws versus Statutory Conservation Laws in the Niger Delta Region ........ 180
Preserving the Stewardship Ethic of Indigenous Peoples: Win-Win for Culture and Nature ........ 182
Conclusion ........ 184
Notes ........ 185


## Background

The Niger Delta region of Nigeria is inhabited by a variety of small ethnic groups that have a deep bond with Nature. At the heart of this bond between indigenous peoples and Nature is the perception that all living and nonliving things, and natural and social worlds, are intrinsically interrelated. Indigenous peoples draw their spirituality from their environment; thus, many aspects of the natural world are sacred to them.[1] Humans are seen as one form of life participating in a wider community of living beings.[2] Vine Deloria,

DOI: 10.1201/9780429505959-9

Jr., has put this relationship between indigenous peoples and Nature into an apt equation: Power + Place = Personality. According to Deloria, power is defined as the 'living energy that inhabits and/or composes the universe', while place refers to the 'relationship of things to each other'.[3]

This ideology of indigenous peoples combines a holistic approach to nature with a sense of deep respect for it. Sadly, this indigenous culture is being subjected to constant threat from several factors. One of these is the strictly anthropocentric nature of Nigeria's biodiversity conservation regulations; another is the massive pollution resulting from fossil fuel extraction in the Niger Delta. The objective of this paper, therefore, is to examine the significance of the ethos of respect, within the context of the existing biodiversity conservation laws in Nigeria, between the Niger Delta indigenous peoples and Nature. The paper will then explore how the indigenous ethos of respect could impact biodiversity conservation in Nigeria if adopted. The paper will begin with an overview of the ecosystem of the Niger Delta Region, which will be followed by a discussion on specific sites important to the indigenous peoples, and the impacts of environmental degradation on these sites. In part five, the indigenous conservation laws vis-a-vis statutory conservation laws in the Niger Delta region will be analysed. Part six will make a case for the preservation of the indigenous ethos of respect.

## The Niger Delta Ecosystem

The Niger Delta region of Nigeria, in West Africa, extends over more than 70,000 square kilometres of the southeastern part of Nigeria, making up about 7.5% of the country's total land mass. The Niger Delta is one of the world's largest deltas, comparable to the Mekong, the Amazon, and the Ganges, and is home to Africa's most extensive mangrove swamp forest. The region has a population of more than 12 million people, with more than 20 distinct indigenous ethnic groups, such as the Efik, Ibibio, Ogba, Itsekiri, Urhobo, Isoko, Anang, Ijaw and Ogoni, who spread across 9 of the 36 states that constitute Nigeria.[4] The Niger Delta region, the location of one of Africa's most significant oil and gas deposits, is predominantly populated by indigenous ethnic groups.[5]

Nigeria is the largest oil and gas producer in Africa, with estimated oil reserves of 37.20 billion barrels and an estimated 5. 11 billion cubic metres of natural gas reserves, making it one of the top 10 natural gas deposits in the world. Constituent states of the Niger Delta region produce more than 75% of Nigeria's total oil and gas production output, providing more than 80% of the national government's annual revenue.

Oil spills in the Niger Delta region destroy crops and damage the quality and productivity of soil that communities use for farming. Oil in water damages fisheries and contaminates water that people use for drinking and other domestic purposes.[6]

The increasing degradation of the Niger Delta environment can be broken down into five major activities – waste disposal, gas flaring, seismic surveys and the construction of roads and pipelines, dredging and inadequate clean up.

### Wastewater

Wastewater is one of the major sources of waste material.[7] When oil is pumped out of the ground, a mixture of oil, gas and water emerges. Following treatment – and in some cases without any treatment – much of this wastewater (known as 'produced water' or 'formation water') is discharged into rivers and the sea. Experts have queried the quality of the treatment in some cases.[8] Only some of the oil can be removed from the water before it is discharged, and, along with oil, produced water may also contain heavy metals and other potentially dangerous substances. Hundreds of tonnes of oil together with other potentially toxic substances are released into the Niger Delta in wastewater.[9]

## Gas Flaring

When oil is pumped out of the ground, the gas produced is separated and, in Nigeria, most of it is burnt as waste in massive flares. This practise has been going on for more than five decades. The burning of this 'associated gas' has long been acknowledged as extremely wasteful and environmentally damaging.[10] Nigeria has become the world's biggest gas flarer, both proportionally and absolutely, with around 2 billion standard cubic feet (scf), perhaps 2.5 billion scf, a day being flared.[11] Communities and nongovernmental organisations (NGOs) have raised concerns about the impact of gas flaring on human health.

## Seismic Surveys and the Construction of Roads and Pipelines

The construction of access roads has resulted in deforestation in the Niger Delta, cutting through the region's mangrove forests. In some cases, access roads have been constructed in such a way as to block the natural flow of water.[12] When this happens, one side of the road may become flooded or waterlogged, while plant life on the opposite side is starved of water.[13] On the flooded side, forests die of asphyxiation, while on the other side, vegetation dies of desiccation. Experts believe that the entire hydrology of the Delta ecosystem has been significantly altered by oil development.[14] The construction of access channels through waterways and swamps is both damaging in itself, and in some cases has caused salt water to flow into freshwater systems, destroying freshwater ecosystems. The incursion of salt water into freshwater is highly damaging to fisheries, and once saltwater enters freshwater, it is no longer usable for drinking or other domestic purposes.[15]

## Dredging

Oil companies also dredge rivers to facilitate navigation and obtain sand for construction. Dredging causes serious environmental damage, with direct repercussions for human rights, since it harms fisheries and can significantly degrade water quality.[16] During dredging, sediment, soil, creek banks and vegetation along the way are removed and deposited as dredge spoils.[17] Sediment introduced into the water system as a result of dredging and other related activities can destroy fish habitats.[18]

Toxic substances attached to sediment particles can enter aquatic food chains, cause fish toxicity problems and make the water unfit for drinking.[19] The waste material from dredging has often been dumped on the riverbanks, which disrupts the environment. Moreover, the waste is often acidic, and if it leaches into the water, is a further source of contamination.[20]

## Inadequate Clean-Up

Clean-up of oil pollution in the Niger Delta is frequently both slow and inadequate, leaving people to cope with the ongoing impacts of the pollution on their livelihoods and health. Failure to swiftly contain, clean up and remediate oil spills can increase the danger of fires breaking out and causing damage to life and property.[21]

One of the worst incidents on record is the Jesse explosion of 1998, when more than 1,000 people reportedly lost their lives. On 18 October 1998, a pipeline explosion occurred in the community of Jesse, southeast of Lagos, Nigeria. The cause of the blast has been debated. The Nigerian government stated the explosion took place after scavengers intentionally ruptured the pipeline with their tools and ignited the blaze; however, others have stated the pipeline ruptured due to a lack of maintenance and neglect, with a cigarette igniting the fire. The bottom line is that a ruptured pipeline was ignited, with a total of 1,082 deaths attributed to the blast.[22]

The Niger Delta environment can be broken down into four ecological zones: coastal barrier islands, mangrove swamp forests, freshwater swamps and lowland rainforests.[23] This incredibly rich and productive ecosystem contains one of the highest concentrations of biodiversity on the planet, supporting abundant flora and fauna, as well as arable terrain that can sustain a wide variety of crops, trees used for lumber or agricultural products and more species of freshwater fish than any ecosystem in West Africa.[24] The Niger Delta is characterised by a great variety of plant species found in a complex vertical structure in dense forest canopies. Some economically important rainforest trees include mahoganies, Mansonia (African black walnut) and a number of others.[25] Many other nontimber forest products extracted from these forests have significant value as food items and medicines, as well as for domestic uses by local residents. The Delta forests also contain a number of rare or endangered rainforest animal species, including finfish, primates, forest antelopes, rodents and birds.[26]

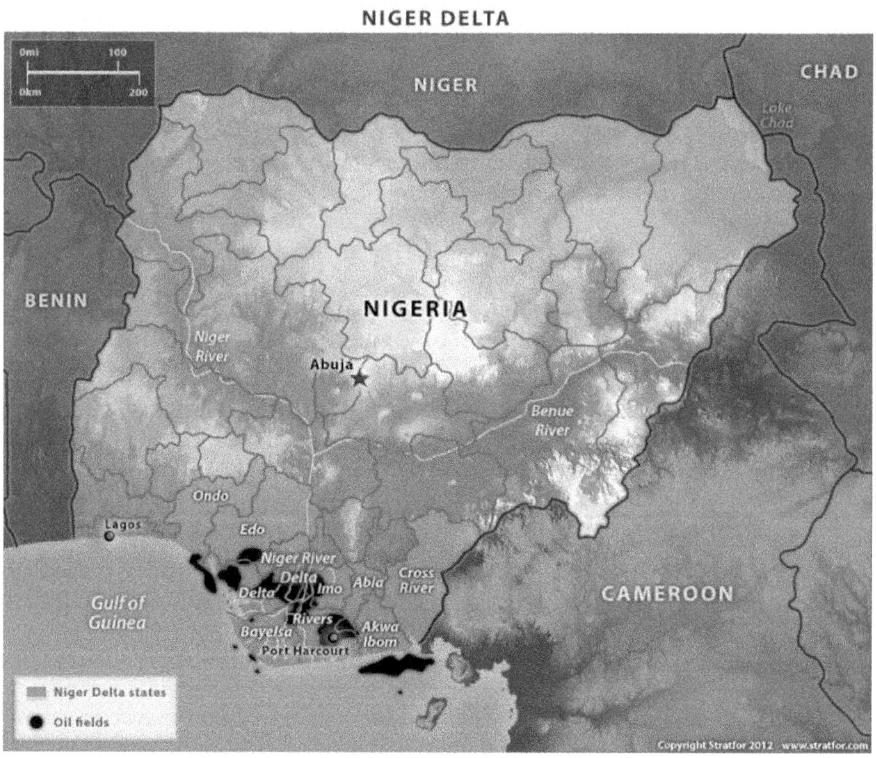


The graphic 'Niger Delta States and Oil Fields' (2012) is republished with permission of Stratfor, a leading global geopolitical intelligence and advisory firm. https://www.researchgate.net/figure/Map-of-Nigeria-Showing-the-Exact-Position-of-the-Niger-Delta-States-Source-Stratfor_fig1_319205214.

## The Niger Delta Indigenous Peoples and Their Sacred Sites

Sacred forests are important sites for biodiversity conservation in many parts of the world. Sacred sites are part of a broader set of cultural values that social groups' traditions, beliefs or value systems attach to places, which fulfil humanity's need to understand, and connect in meaningful ways to the environment of its origins and to Nature.[27]

In Nigeria over the past three or four decades there has been a tremendous decline in biodiversity, and the sacred sites have not been spared the devastation. These sacred sites have been threatened with deterioration due to decline in knowledge about and respect for traditional cultural values, increased demand for forest products, population growth, road construction, local agricultural activities, the advent of Christianity and the exploitation by multinational oil companies in the Niger Delta.[28]

The tribes within the Niger Delta region share similar belief systems, culture and occupations.[29] Fishing, as an occupation, is a significant aspect of their lifestyle; communities have a close affinity with and knowledge of the habits of fish and move seasonally in search of better fishing grounds.[30] These communities are deeply spiritual and their belief systems and relationships with the environment are intimately connected, as in many other African communities.[31] Hence, indigenous practises can be viewed as socio-ecological entities,[32] where the landscape or seascape is protected by human behaviour influenced by spiritual values.[33]

Over the last 400 years, the ecologically rich Niger Delta has played an important role in the global economy through palm oil plantations, the slave trade and fossil fuel extraction. Economically powerful stakeholders still pose threats to the Delta's natural ecosystems, thereby endangering the cultural traditions and territories of the local indigenous peoples. Some of the major sacred sites in the Niger Delta region are Adibe and Esiri lakes, Boupere Lake, Okpagha and Ogriki trees, Ovughere shrine; Obi pond, Usede pond, Ode evil forest, Umuaja shrine.

### Adibe and Esiribi Lakes

The Biseni indigenes occupy the upper regions of the Delta, while the Osiama people inhabit the lower marsh forest zone. Both groups perceive their lands, and particularly their lakes, as sacred. Two important lakes are Adibe (inhabited by the Osiama people) and Esiribi (inhabited by the Biseni people). Both are the homes of their sacred brothers – the Crocodiles.[34]

In the cosmology of both communities, two realms of reality exist: the visible world (kiri) and the invisible world, or the land of spirits (Teme). The visible world is perceived with the physical senses and contains humans, plants and animals.[35] The invisible world is composed of spirits, which are not perceived by the physical senses. The custodian of the sacred lakes also ensure that protocols are observed regarding the lakes which are forbidden to access (Aweye), and lakes which are accessible (aweaya).[36] As a result of these traditional regulations, specific animals, such as crocodiles and lizards, are not harmed and have long been protected in the region. These animals are seen as brothers, and local people do not want to hurt them. They are anthropomorphised, and seen as brothers. If a human being accidentally or purposefully kills a crocodile, it receives full funeral rites as would a human, and must be replaced with a living specimen.[37]

Ovwian community river, which links several communities like Okorobi and Ovu. (Photograph by Ngozi Unuigbe.)

## Boupere Lake

This Lake is situated deep in the rich mangrove ecosystem, and is of immense spiritual value to the Ijaws, especially those in the Central Niger Delta, who trace their origin to Oporoma. They believe the Lake to be the home of the Oporoma, the god of war who defends them against their enemies. It is also reckoned as a place to seek favours, blessings and spiritual powers. The beginning of the fishing year is also celebrated at this site.[38]

## Okpagha and Ogriki Trees

The *Okpagha* and *Ogriki* trees are very significant to the indigenes of the Ethiope East local government area.[39] The *Okpagha* tree is linked with the *aziza* spirit, said to have received enlightenment under the tree. *Aziza* is essentially considered by the people a deity of the woods, whose province is to guard the fields, crops and herds of the peasantry and to drive away their enemies. The indigenous people regard the location of these trees as a sacred place, where trees and plants were allowed to grow undisturbed and where reptiles, birds and animals live without fear of poaching or interference by humans.[40] Groves of this tree are sacred and hence no axe may be laid to any tree, no branch broken, no firewood gathered, no grass burnt; wild animals that have taken refuge there may not be molested. Underneath the trees, cocks, sheep and goats are sacrificed, and the people offer prayers for rain or fine weather, or on behalf of sick children. In addition, a number of the shrine priests have their powers associated with the *Aziza* spirit.[41]

Although the species is strictly protected in the community, religious use is allowed. Wood from the sacred tree is believed to keep its magical powers when fashioned into other objects and was used for making a variety of objects, such as statues of gods, staffs and sceptres. Its bark is reported to be very medicinal, and efficacious in the treatment of unknown illnesses.[42]

## Ovughere Shrine

At Ovu Inland in the Ethiope East Local Government Area is the *Ovughere* (village deity), who is highly regarded as the god of war. The *Ovughere* shrine is located within a very thick forest of Iroko (a large hardwood tree, sometimes known as African teak), mahogany and *Ogriki*. According to oral literature, these trees are well over 70 years old. The forest is the property of the god of the village in which they are situated, and the trees ought not to be cut. Location of residential settlements close to the shrine is not permitted, thereby checking deforestation and farming and protecting the forests. This promotes the conservation of biodiversity in sacred groves through selective limits or prohibitions on the use of forest species. Secondly, information about ritual, medicinal or commercial value of forest species is kept secret from outsiders. This finding is consistent with that reported in traditional African societies, in which many people believed that trees and forests were the manifestation of the power of the Supreme Being.[43] While the area occupied by the forest may be relatively small, the level of biodiversity is high, and thus is of some environmental significance, as well as of ritual importance, to the community.

Shrine near Obi Pond in Ethiope East Local Government Area. (Photo by Ngozi Unuigbe.)

## Obi Pond

At Okorobi village in the Ethiope East Local Government Area is the Obi pond, popularly called Obi Lake. It is the main source of water for drinking and domestic purposes in the community. It is believed that the Obi spirit inhabits the water body. Harvesting of fish there is strictly prohibited, but is allowed when fish leave the lake for other waterways. Felling of trees or fuel-wood collection within a 30-metre radius of the pond is strictly prohibited.[44] These religious principles also preserve the watershed and the surrounding vegetation, and consequently check the amount of evapotranspiration. This in turn maintains tolerable water temperature for both micro and aquatic organisms to survive and continue to enrich the soil, maintain the water supply and aid in continuing forest health.[45] The vegetation cover also helps to keep the water cool and fresh for drinking. Bathing and washing of clothes around, near or in the pond where drinking water is fetched is not allowed; fishing or harvesting any aquatic animals in the pond is also prohibited.

Reasons abound for these restrictions, such as showing respect for the *Obi* god who protects the pond and its organisms and purifying the pond and keeping it alive, as well as controlling the spread of diseases.[46] Silence is observed while fetching water from the pond, because it is believed that an infected person might spill or splash saliva while speaking. For instance, an infected person with tuberculosis or whooping cough spills infected saliva containing bacteria in the water.[47]

In addition, these laws ensured the gods were not provoked to anger. Their anger could result in streams or rivers drying up. Other taboos, such as prohibiting menstruating women from collecting water from the pond, prevent the defilement of the pond deity. Anthropology has extensively explored the concept in traditional belief systems that menstrual blood is a source of potent force.[48]

## Usede Pond

In Ase village, in the Ndokwa East Local Government Area, the *Usede* pond is a mysterious pond harvested for fish by the entire Ase kingdom, but only once every 10 years. The only species of fish caught, collected and shared by everybody present is the mudfish. Any other fish caught on this day are owned by whoever catches them during the harvesting period.[49] Nobody goes to *Usede* pond to fish alone within this 10-year period; punishment for a violation begins with a bloated stomach followed by death, no matter the sacrifices the trespasser makes. However, even within the permitted 10 years, entry to the pond is contingent upon the performance of certain rituals.

Community fishing at the pond is usually carried out in the dry season, and two peculiar events happen in the course of the harvesting.[50] First, heavy rainfall precedes the fish harvesting from *Usede* pond, showing that the spirit of the gods and the activities of the living are in tandem. Second,, the appearance of a big fish with a string of cowry signifies the end of the fishing festival. The big fish with the cowry string will come to where the people are fishing. Once anybody sights the mother fish with the cowries, the harvesting must stop and everybody in the pond must leave. Although the species diversity has not been documented; it is believed that the pond is rich with different species of fish.[51]

## Ode Evil Forest

In Ode Itsekiri, in the Warri – South Local Government Area, there is an evil forest in the mangrove swamp forest where 'dead bad people' are thrown.[52] The dead bad people may include those who have confessed to acts of witchcraft (men or women), and those who died by suicide, cancerous wound, falling from a tree or palm tree and pregnant dead women.[53] In addition, when a person dies as a result of mysterious sickness, and did not confess to any act of evildoing, the oracles are consulted to inquire into the cause of the death. If the deceased was not a 'good citizen', the corpse is often thrown into this forest.[54] This is done to prevent the reincarnation of the person's spirit. The number of wild beasts that live in this area is awesome. No human activity of any kind is carried out in the evil forest, as it is believed that the spirits will not take it kindly to anybody disturbing the peace of the dead or the spirits themselves. So the area remains a very thick forest, in which a large array of animals lives unmolested by humans.[55]

Juju priests at River Ethiope Source Umuaja. (Photograph by Irikefe V. Dafe.)

## Umuaja Shrine

At Umuaja, in the Ukwani Local Government Area, is the origin of River Ethiope, where water is seen to come out from underground through the roots of a very big tree.[56] The tree is regarded as sacred and an abode of gods. The chief priest, who is in charge of the god's abode and also is the messenger of the god in human form, controls the spiritual activities at the sacred grove. The source of the river is also revered and protected, because it is regarded as the source of life and fertility; barren women go to bathe in these waters in the hope of becoming able to bear a child.[57] The river and the immediate surroundings, especially forest, are protected because the spirit of the river resides in the area. The responsibility for protecting the grove is vested in the entire community, but a selected group of people normally takes the duty to enforce the rules.[58]

The conservation strategy, which is one of preservation, is enshrined in taboos, totems, sacrifices, other numerous cultural and religious rites, and is maintained through reverence for the gods and ancestral spirits. The traditional guards regularly patrol the periphery of the grove and arrest intruders, who are reported to the chief priest and must make reparation via customary sanctions. The sanctions, which are required for pacifying and purifying the gods and spirits, vary depending on the gravity of the offence. However, they usually consist of a cash fine, bottles of hot drinks, goats, sheep, chicken, kola nuts and alligator pepper as sacrifice to the gods.[59]

Actions that attract the wrath of the god include: (1) Felling of trees or fuel wood collection within a 20-metre radius from the sacred area. These ritual taboos also preserve the watershed and the surrounding vegetation, which help keep the water cool and fresh for drinking. In other words, this system protects the watershed from stress and degradation. (2) Bathing and washing of clothes around, near or inside the source of the river where drinking water is fetched. (3) Fishing or harvesting of any aquatic animals within the source of the river.[60]

This sacred grove has survived all these years purely because of the strong religious/cultural beliefs held by the local people and their spiritual, religious and cultural attachments to the grove. The major virtue of this strong culture-based practise is that it encourages community participation in natural resource conservation and sustains positive awareness of Nature and the linkages between people and Nature.[61] This tradition has been used to protect the headwaters of several river bodies, especially those that served as potable water sources for a community or group of communities.[62] The River Ethiope watershed may be small in size, but it is of biological significance, with a high potential for biodiversity conservation.

## The Impact of Environmental Degradation on Indigenous Sites

There is increasing evidence that degradation of the environment in areas populated with indigenous peoples[63] results not only in human rights violations, but also the loss of cultures.

Several factors threaten the biodiversity, as well as the cultures of the indigenous peoples of the Niger Delta region. For example, community members increasingly engage in modern livelihood opportunities, such as dynamite fishing, which require changes to their way of life that cause them to compromise their traditional beliefs. The weakening of traditional leadership provides opportunities for power and influence to social or political groups that hold other values. Migrants, who enter the region in search of job opportunities at oil companies, pose higher demands on local fisheries for food supply. Changes in religious orientation decrease the traditional and iconic values of sacred crocodiles and respect for the local flora and fauna.

The Niger Delta, a lush region of mangrove swamps, rainforests and marshes, is also the site of rich oil and natural gas reserves in Nigeria, where a great deal of extraction is taking place, endangering many of the natural and cultural ecosystems. Despite being the richest geopolitical region in its natural resources, the Niger Delta's potential for sustainable development not only remains unfulfilled but is now also increasingly threatened by environmental devastation and worsening economic conditions. Particularly threatened are the mangrove forests of Nigeria, the largest in Africa, 60% of which flourish in the Niger Delta. Also facing extinction are the freshwater swamp forests of the Delta, which cover 11,700 square kilometres – the most extensive in West and Central Africa. Importantly, the local people, as well as many wildlife species, depend on them for sustenance.

Environmental quality and sustainability are fundamental to the overall well-being of the Niger Delta indigenes and the development of the region. They find aspects of the ecosystem useful for their cultural, emotional and spiritual satisfaction.

The economy of the region is traditionally dependent on fishing and farming. The network of creeks and rivers of the delta, and the coastal seas, provide the basis for a peasant fishing industry. Although there is fishing for subsistence in all parts of the delta, the scale of operation in the outer delta (by the coast), where fishing is the main source of livelihood, is along commercial lines.[64] The estuaries and the coastline were renowned as the suppliers of the fish consumed in the urban centres before the inception of the modern fishery of the country. Whereas the mangrove swamps region of the outer delta depends on fishing, the inner delta area of rain forest is a farming region.[65] Except for the rubber and oil palm plantations, the agricultural landscape is one of the small staple food-crop farms. The oil palm and rubber economies of the area were important foreign exchange earners before the country became a major exporter of crude oil.[66]

The oil industry was superimposed on this predominantly peasant economy. Although the urban centres have grown rapidly because of the oil industry, it is in the rural areas, the actual production centres of petroleum, that the environmental crisis consequent on the exploitation of the resource is really evident.[67] A disturbance of the environment occurs not only during the search for oil but also in the process of production, storage and transportation. Prospecting activities necessitate the provision of various routes, pits, stream diversions and embarkation facilities, and the use of explosives in rivers and seas among others. These, in varying degrees, deprive people of their livelihood.[68]

The ultimate effect of environmental impacts of oil production activities is therefore a reduction in the standard of living of the people in the area of primary activities. The overall economic effects are extensive and include the dislocation of traditional economic activities and associated livelihood pursuits as well as danger to human health.[69] All the economic effects translate to pecuniary effects, which can be measured in terms of reduced income and the loss of alternative uses of resources consumed by oil communities.[70]

The oil exploration activities in the Niger Delta region also have cultural implications for cultures of the indigenous peoples of the region. For example, Boupere Lake is a majestic site located in the hinterlands of Bayelsa State, in the Niger Delta region of Nigeria. Situated deep

An abandoned shrine in Egiri following new religious orientation. (Photograph by Ngozi Unuigbe.)

in the rich mangrove ecosystem, the lake is of immense spiritual value to the Ijaws, especially those in the Central Niger Delta who trace their origin to Oporoma. Boupere Lake is said to be the abode of the war god of the Oporomas. It is from here that the popular Egbesu god of the Ijaws originated. The people believe the god provides them the power of remaining bulletproof in defence against enemies. Boupere is a place of refuge – people who have their abode here in war times are guaranteed safety and protection. Boupere is also home to the spirits of their ancestors, a place to seek favours and blessings, a place to seek supernatural powers, a place to perform rituals to signal the commencement of a fishing year and rituals to bring peace and security to Oporoma and her people.[71] The constant pollution of this lake has deprived the indigenes of their culture, leaving them with a sense of vulnerability.

## Indigenous Conservation Culture and the Rights of Nature

It is commonly believed that the recognition of legal rights of Nature is a codification of indigenous culture into law (as will be discussed below), thus reaching back into thousands of years of human history.[72] More recently, though still more than a century ago, environmentalist John Muir wrote that we must respect 'the rights of all the rest of creation'. In 2015, Pope Francis called for a new era of environmental protection, stating in a speech before the United Nations, 'A true "right of the environment" does exist ...'[73]

Much of the more modern-day discussion heralds back to Professor Christopher Stone's 1972 law review article titled 'Should Trees Have Standing – Toward Legal Rights for Natural Objects', in which he considered why we might want, and what it might mean, to recognise legal rights of Nature.[74] Stone described how under the existing structure of law, Nature is considered 'right-less', having no legally recognised rights to defend and enforce. Thus, Nature – much like slaves once were – is treated by the law as a thing, as property, existing for the use of its owner.

Moving from the idea of rights of Nature to the codification of those rights in law first occurred in Tamaqua Borough in Schuylkill County, Pennsylvania, in 2006.[75] Now communities in a number of states in the United States have such laws in place, the largest being Pittsburgh, where the City Council unanimously passed the Rights of Nature law in 2010. These laws recognise that ecosystems and natural communities have the legal right to exist and flourish, and that residents and their government have the authority to enforce and defend those rights.[76]

The question of constitutional recognition and protection for the natural world has moved to the forefront in several European countries and at least one Latin American country. However, only Ecuador provides constitutional rights for the entirety of Nature. Some countries in the European Union have made commitments at different levels to prioritise environmental protection over economic development. The European Commission (EC) has also been called upon by a group of lawyers, environmentalists and academics (the European Citizens' Initiative) to adopt the right of Nature to exist.[77] Spain has granted rights to animals, but the rest of the European states merely protect animals, or both plants and animals, without giving them any legal standing.

On 25 June 2008, the Spanish Parliament's Environmental Committee passed a resolution to grant basic rights to all primates, specifically the 'right to life, freedom, and the absence of torture'. The resolution would end all harmful experiments on apes and ban their use in circuses, television commercials, and films.[78] In Switzerland – a country that has included animal protection in its constitution for over 100 years – Article 80 of the 1999 Swiss Federal Constitution grants the government power to enforce animal protection. In

2002, an amendment was added to Article 120, covering gene technology. The new language specifically named 'plants, animals, and other organisms' to recognise 'the dignity of the creature in the security of man, animal, and the environment'.[79]

Although this amendment seems somewhat narrow in context and scope, subsequent interpretation strongly indicates that animals and plants have been granted actual legal standing under Swiss law. In 2008 the Federal Ethics Committee on Non-Human Biotechnology (ECNA) – a federally established group of scientists brought together specifically to interpret the 'wurde der keatur' (the dignity of living beings) – in Swiss constitutional law and to advise the federal government as to its implementation, released a remarkable 22-page report entitled *The Dignity of Living Beings with Regard to Plants: Moral Consideration of Plants for Their Own Sake.* In this report the Committee, after consulting with numerous outside experts from a broad diversity of disciplines – from law to science to theology – determined that arbitrary harm caused to plants is morally impermissible. It went on to say that simply collecting plants requires moral justification, that plants were excluded from absolute ownership by humans, and that any action with or towards plants that serves the self-preservation of humans was morally justified only as long as it is appropriate and follows the principle of precaution. The report concluded by stating that 'plants may not be arbitrarily destroyed … the majority (of the committee) considers this to be morally impermissible because something bad is being done to the plant itself without rational reason and thus without justification'.[80]

In the summer of 2002, Germany's parliament voted 545 to 19 to amend its Constitution to specifically protect animals. The amendment – supported by 80% of the German population – added the words 'and animals' to Article 20a of the German Basic Law to read, 'The state takes responsibility for protecting the natural foundations of life and animals in the interest of future generations'. Previously the courts had interpreted 'life' to mean only human life.[81] In 2004, Austria passed similar legislation, writing animal welfare protection into its Federal Constitution.[82] Also in August 2010, Kenya adopted a new Constitution in which article 69 declares that the 'state shall be responsible for maintaining tree cover at least 10% of the nation's land …'[83]

Thus far, the most notable innovations specifically on the Rights of Nature have taken place in South America. In September 2008, the Constitutional Assembly of Ecuador became the first political body to recognize – by a vote of 92–12 – constitutional rights for the natural world. This has made the Ecuadoran Constitution the first in the world to expressly protect Nature's rights. Here are the words from the relevant sections of the new Ecuadoran Constitution, adopted by Ecuadorans in the fall of 2008:

Chapter 7, Articles 71-72, provides as follows:

> Article 71: Nature or Pachamama, where life is reproduced and exists, has the right to exist, persist, maintain and regenerate its vital cycles, structure, functions and its processes in evolution. Every person, people community or nationality, will be able to demand the recognitions of rights for nature before public bodies. The application and interpretation of these rights will follow the related principles established in the Constitution.
>
> Article 72: Nature has the right to an integral restoration. This integral restoration is independent of the obligation on the natural and juridical persons or the State to indemnify the people and the collectives that depend on the natural systems. In the cases of severe or permanent environmental impact, including the ones caused by the exploitation of non-renewable natural resources, the state will establish the most effective mechanisms for the restoration, and will adopt the adequate measures to eliminate or mitigate the harmful environmental consequences.

In March 2011, there was a successful case against the Government of Ecuador on behalf of Nature. Richard Frederick Wheeler and Eleanor Geer Huddle demanded the observance of the Rights of Nature, based on Article 71 of the Ecuadoran Constitution. They obtained a Constitutional injunction from the Provincial Court of Justice of Loja in favour of Nature, specifically the Vilcabamba River, against the Provincial Government of Loja.[84]

The Provincial Court of Loja ruled in favour of Nature, particularly the Vilcabamba River, granting a Constitutional Injunction[85] that established (among other things) that the argument of the Provincial Government that the population needs roads did not apply, because the case did not question the widening of the Vilcabamba-Quinara road, but rather the respect for the constitutional Rights of Nature. The court further held that the Constitutional injunction was the only suitable and effective way to remedy the environmental damage, focusing on the undeniable, elemental and essential importance of Nature, and taking into account the evident progress of degradation.[86] The Court further ordered that the Provincial Government of Loja must present within 30 days a remediation and rehabilitation plan for the Vilcabamba River and the populations affected by the dumping and accumulation of rubbish from the project. The court also required the government to implement corrective actions, such as construction of security fences to prevent oil spills around the fuel storage tanks and machinery, cleaning soils contaminated by fuel spills, implementation of an adequate road sign system and creation of a location to store the rubbish from the construction.[87]

Similarly, in a 2015 decision related to illegal shrimp farming, the Constitutional Court of Ecuador explained that enshrining the Rights of Nature in the country's constitution establishes a 'biocentric vision in which nature is prioritized, as opposed to the classical anthropocentric conception in which the human being is the center and measure of all things, whereas nature was considered a mere provider of means'.

The Court ruled that the lower court, which had earlier ruled in the case, had failed to consider the Rights of Nature in its decision. In not considering these rights, the Court explained, the lower court failed to recognise that the Rights of Nature are 'transversal', such that 'all the actions of the State, as well as of individuals, must be in observance … to the rights of nature'.

The Court explained that under Ecuador's Constitution, Nature is now a holder of rights, and that government and the people have a responsibility to uphold and protect those rights. This comes with the recognition that until and unless we establish a harmonious relationship with Nature – much different than humankind's relationship with the natural world today – we will continue to see the decline of ecosystems and species, the very fabric of life.

Following Ecuador's legal template, on 22 April 2011, Bolivia passed 'The Law of Mother Earth' (*Ley de Derechos de la Madre Tierra*). The Bolivian law formalised the belief system of the indigenous people of the Andes, who pay homage to Pachamama, the female spirit of Nature.[88] One article in the new law prohibits commercialisation: 'Neither living systems nor processes that sustain them may be commercialized, nor serve anyone's private property'.[89]

## Indigenous Conservation Laws versus Statutory Conservation Laws in the Niger Delta Region

In addition to national and international laws, the indigenous peoples in Nigeria have customary laws that regulate the protection of forests in many ways, and which have been enforced by the people many centuries before colonialism and the emergence of statutory laws. For example, there are local laws that provide for communal declaration of certain forests and groves as sacred; for instance, there is a Yoruba customary law on the sanctity of *Igbo Oro* (forest of shrines).

There are further examples of the delineation of forests as burial grounds for good and evil people, the recognition given to, and observed in, boundary forests between neighbouring communities, family heritage forests, forests of common use and the essential habitat forests. Furthermore, there are customary conservation laws on fishing, hunting, water and animals.[90]

In many cultures, some species are prohibited for human use. This may reduce hunting pressure. But these customary prohibitions may not be regarded as a conservation based by those who practise them, and are not honoured in anticipation of a future benefit to the species. Such a community may, therefore, show little concern if human or natural catastrophe destroys the species. Community members consider the prohibited species to be unclean, so, presumably, there would be no reason to mourn their demise. Examples of such animals are alligators, chameleons, bats, owls and snails. In the same way, other customary laws protect species like crocodile, tortoise, monkeys and African giant rat, as they are regarded as either 'brothers' (e.g. the Egbesu indigenes of the Niger Delta), or 'gods' (like the African giant rat to the Era and Ase indigenes of the Niger Delta).

Another example is provided by sacred groves and forests. These play a critical role in biodiversity conservation. However, most of them are not purposely designated for conservation, but rather for religious and cultural reasons, such as burial sites and holy places for ceremonies and initiation rites such as the *igbo oro* forests mentioned above.

In spite of their relevance to conservation, customary laws will not be applicable where they are incompatible with statutory law, and where they are 'repugnant to natural justice, equity and good conscience'.[91] This is evidenced in the incompatibility of customary laws on biodiversity conservation with the statutory (and constitutional) vesting of ownership of oil in the Federal Government of Nigeria, which includes rights to regulate issues affecting or which may affect oil exploration.[92] Yet one of the strategies for achieving national environmental policy on forestry, wildlife and protected natural areas under the revised Nigerian National Policy on the Environment of 1999 is 'combining desirable features of traditional approach with modern scientific methods of conservation'.[93] Thus the customary laws cannot be enforced unless the provisions of the statute are brought into conformity with the present National Policy on the Environment.

Whether traditional practises and values are conservation oriented or not, the importance of recognising the unique ways they affect biodiversity must not be overlooked. Current conservation efforts spearheaded by the government might be flawed due to neglect of these values in planning. This negligence is historical, following the restriction of interactions between local people and their environment by existing laws. The indigenous peoples of the Niger Delta region have a long custom of ancestry that involves handing down of lands from one generation to the other in a family and community configuration. However, the Land Use Act of 1978[94] has eroded the inheritance system by reducing family and community perpetuity over lands to term of years 'in-absolute'. In essence, the children of a deceased right-holder may inherit the residue of the interest in the land, but there is no guarantee that the government will renew the same statutory right of occupancy upon expiration of the residual interest.[95] Thus, the right to freely alienate land by the communities and families is now being severely restricted by the Act. This is because it is now mandatory to obtain the consent of the governor for the transaction to be fully valid.[96] This is contrary to the indigenes' cultural and social heritage of land administration, where they could transfer land (with the cultures attached) from generation to generation. Arguably therefore, greater recognition of indigenous people's rights over land, water natural resources, based on traditional occupancy or use, may be a key path towards greater pollution prevention and sustainable development.[97]

Nigeria has far-reaching legislations on biodiversity conservation. They include: the National Parks Act,[98] Endangered Species (Control of International Trade and Traffic) Act,[99] the Sea Fisheries Act[100] and the various state Forestry Laws.[101] Some of these laws[102] were enacted as a fulfilment of Nigeria's obligations under biodiversity conservation treaties, such as the African Convention on the Conservation of Nature and Natural Resources 1968,[103] the Agreement on the Joint Regulation of Fauna and Flora on the Lake Chad Basin,[104] Convention on the International Trade on Endangered Species of Wild Fauna and Flora (CITES)[105] and the Convention on Biological Diversity.[106]

Nigeria's domestic biodiversity conservation laws adopt different methods to protect biodiversity. The Sea Fisheries Act regulates the system of fishing in order to protect certain fish stocks, and also to prevent harmful fishing and overfishing. The Forestry Laws protect the forests (thereby protecting the flora and fauna in them) by the designating certain areas as forest reserves and protected forests.[107] In such areas, certain activities are prohibited, such as burning, unlawful taking of forest products and uprooting or stripping the bark or leaves from a tree.[108] The National Parks Act concerns similar offences in section 27(1). In addition, it provides that any person who introduces any chemical or otherwise causes any form of pollution; drives, stampedes or in any way disturbs unnecessarily any animal; carries out any undertaking connected with forestry, agriculture, grazing, mining, excavation or prospecting; does any drilling, levelling the ground or construction or any tending to alter the configuration of the soil or the character of the vegetation; or does any act likely to harm or disturb the fauna, flora or animals, in the National Park, is guilty of an offence under the Act. As profound as this provision may seem, its violation exacts an inconsequential fine not exceeding 1,000 naira[109]

Basically, the Forestry and National Park laws protect certain areas and species, but leave an area like the Niger Delta region unprotected. This is because the region has no forest reserve or national park, notwithstanding that it contains abundant species of flora and fauna. The Endangered Species (Control of International Trade and Traffic) Act would have been helpful, as some of the species listed for protection are found in the Niger Delta region. However, the problem with this law is its silence on habitat destruction[110] by human activities such as oil extraction. As succinctly put by Kaniye Ebeku, 'by human activity, important habitats are destroyed both directly through forest clearance or wetland drainage and indirectly through pollution'.[111]

The consequence of this narrow and specific focus in the conservation laws has been a rift preventing use of traditional knowledge to further conservation efforts. The relevance of such knowledge to environmental protection has therefore been relegated to the background.

## Preserving the Stewardship Ethic of Indigenous Peoples: Win-Win for Culture and Nature

Awareness of indigenous socio-religious rituals and belief systems is imperative in making decisions that are respectful of their taboos and unique values. Understanding of indigenous knowledge, values and practises may provide an opportunity for using them to complement the current strategies seeking to address such conservation problems as resource overexploitation, conflicts and limited budget for law enforcement.

Indigenous peoples have an abundance of information about their environment, which could be used as baseline data to conduct comparisons of pre- and postdevelopment environmental conditions. These types of comparisons assist in identifying and mitigating environmental damage before it reaches irrevocable proportions. This clearly has implications for sustainable development, because it tries to reduce environmental damage so future generations can inherit a clean and abundant Earth.[112] Thus, so long as indigenous peoples are kept outside the realm of environmental decision-making, natural resources will constantly be overexploited and mismanaged.

Some potential roles for traditional rules and values include:

- Regulate overexploitation of resources and habitat loss
- Complement economic incentive strategies for conservation
- Minimise the costs of law enforcement
- Complement scientific knowledge

Some practical constraints to the use of traditional practises include:

- Complicated method of acquiring indigenous knowledge
- Population growth and change of demands on the environment by indigenous communities
- Poverty
- Lack of appreciation of traditional institutions among conservation planners

Clearly, certain traditional practises may be quite extreme, and so qualify as being 'contrary to natural justice, equity and good conscience'. Examples of such practises are the 'oro' and 'aremo' festivals, during which the movement of women is restricted; this has been deemed contrary to the fundamental right to freedom of movement under section 41 of the Constitution of the Federal Republic of Nigeria. In the same way, some factors that precipitate constraints on use of traditional practises (e.g. changes in religion, or adoption of a more mundane lifestyle) may also be difficult to modify. Both extremes are unfavourable to a conservation ethic. It is against this backdrop that a middle ground is proposed to take into consideration modern and traditional practises in creating environmental conservation policies.[113]

The 1992 Convention on Biological Diversity recognises the importance of traditional knowledge, innovations and practises of indigenous and local communities for the conservation and sustainable use of biodiversity and aims to respect, preserve, and promote such traditional knowledge.[114] The Convention also recognises the interdependence of indigenous and local communities and biodiversity. Environmental conservation planning should take into account both the rights and traditional knowledge of indigenous and local communities. The main method for achieving this is through the effective participation of indigenous peoples and local stakeholders in decision-making and governance processes, on the basis of free, prior and informed consent to any projects, plans or changes that affect their communities, traditional lifestyles and environment.

Not all biodiversity can be maintained. Setting realistic, measurable and locally relevant biodiversity objectives will provide a sound basis for the negotiation of trade-offs. For instance, it is important to recognise that the indigenous peoples of Niger Delta may have their own priorities for biodiversity, which differ from those of outside conservation groups. Building on these may provide a sound basis for securing local buy-in.[115] It is also important to recognise that sustainable use of an ecosystem may be a more attractive option for indigenous peoples than total protection. If people can benefit from using a species, they are more likely to conserve it.

Recent initiatives for reducing emissions resulting from deforestation and forest degradation have the potential to provide benefits to local and indigenous communities. One such initiative is the United Nations Development Programme's Territorial Approach to Climate Change in Delta State (2012–2014), which targeted restoration of mangroves as one of its project strategies. However, a number of conditions would need to be met for these co-benefits to be achieved. Indigenous peoples are unlikely to benefit from restrictions on deforestation where they have no secure land tenure; if there is no principle of free, prior and informed consent concerning the use of their lands and resources and if their identities are not recognised or they have no space to participate in policy-making processes and/ or lack the capacity to engage on an equal footing. There is a need for different types of capacity-building on indigenous issues and rights, both on the side of governments and for indigenous peoples and local communities. This should include education and awareness-raising, indigenous-to-indigenous transfer of knowledge, and capacity-building.

Sustainable management of freshwater ecosystems in Nigeria's Niger Delta region, especially areas with large water bodies like the Aside, Esiribi Boupere and similar areas, will be greatly improved by adapting management to the traditional stewardship model of the local communities. Strategies should centre on maintaining the traditional system of rotational fishing, as the time lapse between fishing events encourages the rejuvenation of fishery and aquatic resources. This in turn leads to higher productivity and contributes to the high biological diversity of the lakes.

Also, more organisation of local events and recognition of traditional leadership would likely encourage more successful collaboration. Some scientific studies, assessing indigenous knowledge of the region, show nature and culture are closely related and evince the value of traditional governance and management. They recommend enforcement of the international Ramsar Convention's resolutions regarding cultural values and community participation.[116]

Local government policy and legal instruments accord very little protection to the rights of indigenous people in areas traditionally conserved by them. New laws providing integration of indigenous traditions and customary laws are vital for sustainable development in the Niger Delta area, which is now so endangered by fossil fuel extraction. Nigeria must create the legal framework that provides the tools to implement Resolution 61/295 of the United Nations Declaration on the Rights of Indigenous Peoples.[117] Also important is the amending the nation's Land Use Act to favour indigenous custodians and create the methods to integrate their stewardship with conservation laws.

## Conclusion

Indigenous peoples have spiritual, social and sustenance-related interests in a healthy environment. All three levels are associated with the concept of intergenerational equity and sustainable development. Spiritually, indigenous people believe that their ancestors have given them this land, and the onus is on them to protect it and hand it over to their descendants. Thus, reckless hunters who hunt animals to extinction without regard for the security of unborn generations may provoke the wrath of the ancestors, for such hunters would be viewed as abandoning their sacred responsibilities. In this respect, indigenous people are motivated to manage ecological resources responsibly and sustainably.[118]

Policy reforms and legal instruments that recognise and legitimise indigenous conserved areas and customary laws are important to reaching goals of sustainability.[119] It is therefore imperative for Nigeria to give priority to fulfilling its obligations under international biodiversity conventions. Also important is the review of the nation's Land Use Act, recently proposed to the National Assembly. However, to what extent the review will resolve tenure rights in the Delta is difficult to determine.

In conclusion, evidence from research in the Niger Delta region is indicative of the people's holistic view of the ecosystem. Their interaction with natural resources allows for constant observation and feedback to meet the challenges of changes in the natural system. Hence, involving the Niger Delta indigenous groups in monitoring and governance of land and water resources in the region is germane, particularly in the face of climate change, global biodiversity loss and (un)sustainable development generally.

Biodiversity conservation in diverse ecosystems like the Niger Delta region holds the key to mitigate vicious environmental crises that are gradually leading to environmental catastrophe. Some of the traditional ways evolved by devout indigenous peoples stand as a lesson for us to learn. We are forgetting some tried and tested practises that have helped us over the generations. Today, as the world faces a depleted environment, it is time to look back at some of these practises, analyse them and, if need be, reclaim them.

## Notes

1. Appiah-Opoku, S. 2005. *The Need for Indigenous Knowledge in Environmental Impact Assessment: The Case of Ghana*, Edwin Mellon Press Ltd., p. 70.
2. (a) Chapin, M. 2004. 'A Challenge to Conservationists' *World Watch* (November/December 2004); (b) Persoon, G., Minter, T., Slee, B. and Van Der Hammen, C.M. *The Position of Indigenous Peoples in the Management of Tropical Forests* Tropenbos Series 23 (Wageningen: Tropenbos International 2004); (c) Aranbiza, E. and Painter, M. 2006. 'Biodiversity Conservation and the Quality of Life of Indigenous Peoples in the Bolivian Chaco' 65(1) *Human Organisation*, p. 20.
3. Deloria, V. and Wildcat, D. 2001. *Power and Place: Indian Education in America* (Fulcrum Publishing 2001) 22; see also Rothschild, D. (ed), *Protecting What Is Ours: Indigenous Peoples and Biodiversity* (SAIIC 1997) ch 1; Watson, A., Alessa, L. and Glaspell, B. 2003. 'The Relationship between Traditional Ecological Knowledge, Evolving Cultures and Wilderness Protection in the Circumpolar North' 8(1) *Conservation Ecology* 2.
4. United Nations Development Programme (UNDP). 2006. *Niger Delta Human Development Report* (UNDP 2006), p. 15.
5. Ibid, 41; see also Eregha, P. and Irughe, I. 2009. 'Oil Induced Environmental Degradation in Nigeria's Niger Delta Region: The Multiplier Effects' 11(4) *JSDA* 160, 163.
6. For a detailed discussion on the effect of oil spills on land, see (a) Mackay, D. and Matsuju, R. 1973. 'Evaporation Rates of Liquid Hydrocarbon Spills on Land and Water' 51(4) *CJChE* 434, 439; (b) Amadi, A., Abbey, S. and Nma, A. 1994. 'Chronic Effects of Oil Spill on Soil Properties and Microflora of a Rain Forest Ecosystem in Nigeria' 86(1–4) *Water, Air & Soil Pollution* 1; (c) Anoliefo, G. and Vwioko, D. 1995. 'Effect of Spent Lubricating Oil on the Growth of *Capsicum annum L.* and *Lycopersicon Esculentum Miller*' 88(3) *Environmental Pollution*, 361–364; (d) Osuji, L., Adesiyan, S. and Obute, G. 2010. 'Post-Impact Assessment of Oil Pollution in Agbada West Plain of Niger Delta, Nigeria: Field Reconnaissance and Total Extractable Hydrocarbon Content' 1(10) *Chemistry & Biodiversity* p. 1569.
7. Ayotamuno, M., Akor, A. and Igho, T. 2002. 'Effluent Quality and Wastes from Petroleum Drilling Operations in the Niger Delta, Nigeria' 13(2) *Environmental Management & Health* 207, 208–213.

8. Ibid., 210; see also (a) Bello, O., Sonibare, J., Macaulay, S. Okelana, A. and Durojaiye, A. 2004. 'Incineration of Hazardous Wastes from the Petroleum Industry in Nigeria' 22(6) *IJEP* 710; (b) Osuji, L. and Uwakwe, A. 2006. 'Petroleum Industry Effluents and Other Oxygen-Demanding Wastes in Niger Delta, Nigeria' 39(7) *Chemistry & Biodiversity.* 705, 708–710.
9. See Ayotamuno, M. et al. (n 8) pp. 208–209; Osuji, L. and Uwakwe A. (n 77) p. 706.
10. Ayotamuno, M. et al. (n 8) p. 210.
11. Platform London, 'Gas Flaring in the Niger Delta'. http://www.platformlondon.org/carbonweb/showitem.asp (accessed 19 November 2018); see also Ikari, B. 2008. 'The Contribution of Gas Flaring in Nigeria to Global Warming' (46) *Synthesis/Regeneration.* http://www.greens.org/s-r/46/46-06.html (accessed 19 November 2018).
12. See Ecumenical Council for Corporate Responsibility, *Shell in the Niger Delta: A Framework for Change: Five Case Studies from Civil Society* (ECCR 2010); Federal Ministry of Environment, *Niger Delta Natural Resource Damage Assessment and Restoration Project* (FME 2006).
13. Ibid.
14. Ibid.
15. (a) Ibid.; (b) Manny, B. 1985. *Ecological Effects of Rubble-Mound Breakwater Construction and Channel Dredging at West Harbor, Ohio (Western Lake Erie)*(US Army Corps of Engineers, Environmental Laboratory) 6–10; (c) Nwilo, P. and Onuoha, A. 1993. 'Environmental Impacts on Human Activities on the Coastal Areas of Nigeria' in Awosika, L., Ibe, A. and Shroader, P. (eds), *Coastlines of West Africa* (American Society of Civil Engineers 1993), pp. 220–234; 223–226.
16. ECCR (n 13).
17. Fagbami, A., Udo, E. and Odu, G. 1988. 'Vegetation Damage in an Oilfield in the Niger Delta of Nigeria'. 4 *Journal of Tropical Ecology,* 65, 65–67.
18. Ohimain, E., Imoobe, T. and Benka-Coker, M. 2002. 'Impacts of Dredging on Zooplankton Communities of Warri River, Niger Delta'. 1 *AJEPH* 37, 40–42.
19. Nwilo, P. and Onuoha, A. (n 16) 223; Moffat, D. and Linden, O. 1995. 'Perception and Reality: Assessing Sustainable Development in the Niger River Delta' 24 *Ambio* 527, 530.
20. Nwilo, P. and Onuoha, A. (n 16), p. 224.
21. (a) Jibrin, B. and Ugwoke, N. 2009. 'Niger Delta: Human Rights Tragedy' *The Market* (Nigeria, 6 July 2009) 10; (b) Irin News, 'Nigeria: Poor Oil Spill Clean-Up Methods Affect Niger Delta Communities. http://www.irinnews.org/Report.aspx?ReportId=76635 (accessed 19 November 2018).
22. See 'A Chronology of Pipeline Fires in Nigeria' *The Nation* (Nigeria, 16 May 2008) http://www.thenationonlineng.com/dynamicpage.asp?id=51129 (accessed 19 November 2018).
23. Eregha, P. and Irughe, I. (n 6), p. 164.
24. Ibid.
25. USAID, *Nigeria Biodiversity and Tropical Forestry Assessment* (Report on Maximizing Agricultural Revenue in Key Enterprises for Targeted Sites (Markets) USAID (2008) 6–10.
26. Ibid.
27. Putney, A. 'Building Cultural Support for Protected Areas through Sacred Natural Sites', in McNeely, J. *Friends for Life,* (IUCN, Gland, 2005), p. 132.
28. Omokhua, G.E., and Asimiea, A.O. 2015. 'Biodiversity Conservation and the Sacred Forests of Emohua, Rivers State, Niger Delta Region – A Review' 4(1) *International Journal of Science and Technology,* 37–44.
29. Osunade, M.A. 1988. 'Nomenclature and Classification of Traditional Land Use Types in Southwestern Nigeria', (1988) 9 *Savanna,* 50–63.
30. NEST, *Nigeria's Threatened Environment: A National Profile* (NEST Publications 1991) 22–26.
31. See generally, (a) Schoffeleers, J.M. 1978. *Guardians of the Land: Essays on Central African Territorial Cults* (Mambo Press, Harare); (b) Omari, C.K. 'Traditional African Land Ethics' in Engel, J.R. and Gibb, E.J. (eds) *Ethics of Environment and Development: Global Challenge, International Response* (University of Arizona Press 1990).
32. Berkes, F. Colding, J. and Folke, C. 2003. (eds) *Navigating Social-Ecological Systems: Building Resilience for Complexity and Change.* Cambridge University Press.

33. Bisht, S. and Ghildiya, J.C. 2007. 'Sacred Groves for Biodiversity Conservation in Uttarakhand Himalaya' 92 *Current Science*, 711–712.
34. Anwana, E.D., Cheke, R.A., Martin, A.M., Obireke, L., Asei, M., Otufu, P. and Otobotekere, D. 2010. 'The Crocodile Is Our Brother: Sacred Lakes of the Niger Delta, Implication for Conservation Management' in Vershuren, B., Wild, R., McNeely, J. and Oviedo, G. *Sacred Natural Sites: Conserving Nature and Culture* (Earthscan), p. 126.
35. Ibid.
36. Ibid.
37. Ibid.
38. Adekola, O. 2011. 'Oil Smears the Gods of the Niger Delta' (*Earth Collective Magazine*, 20 March 2011) http://www.earthcollective.net/category/art/ (accessed 16 August 2018); see also Alagoa. E. 1999. (ed) *The Land and People of Bayelsa State: Central Niger Delta* (Onyoma Research Publications).
39. Rim-Rukeh, A., Irerhievwie, G. and Agbozu, I.E. 2013. 'Traditional Beliefs & Conservation of Natural Resources: Evidence from Selected Communities in Delta State, Nigeria'. 5(7) *International Journal of Biology and Conservation*, 426, 429.
40. Ibid.
41. Ibid.
42. Ibid.
43. Eneji, C.V., Gubo, Q., Xiaoying, J., Oden, S.N. and Okpiliya, F.I. 2009. 'A Review of the Dynamics of Forest Resources Valuation and Community Livelihood: Issues, Arguments and Concerns', 2(2) *Journal of Agriculture, Biotechnology and Ecology* 210–231; see also, Wei, F., Wang, S., Fu, B., Zhang, L., Fu, C. and Manga, E. 2018. 'Balancing Community Livelihoods and Biodiversity Conservation of Protected Areas in East Africa'. 33 *Current Opinions in Environmental Sustainability* 26–33.
44. Rim-Rukeh, A. n. 23, p. 430.
45. Ibid.
46. Ibid.
47. Ibid.
48. McLeod, M.D. *The Asante*, (British Museum Publication, 1981).
49. Rim-Rukeh, A. n. 23, p. 431.
50. Ibid.
51. Ibid.
52. Ibid.
53. Ibid.
54. Ibid.
55. Ibid.
56. Ibid.
57. Ibid.
58. Ibid.
59. Ibid.
60. Ibid.
61. (a) Tunon, H. and Bruhn, J.G. 1994. 'Drugs in Ancient Texts'. *Nature*, 369:702; (b) Tupper, M. 2002. 'Marine Reserves and Fisheries Management'. 295 *Science*, 1233.
62. Falconer, J. 1992. 'People's Uses and Trade in Non-Timber Forest Products in Southern Ghana: A pilot study'. ODA Report.
63. The U.N working group on Indigenous Populations defines 'indigenous people' as 'the distinguished descendants of those peoples that inhabited a territory prior to the formation of a state'. The term 'indigenous' may be defined as a characteristic relating to the identity of a particular people to a particular area and distinguishing them culturally from other people or peoples. As the notion of 'indigenous peoples' has gained currency, many marginalised social groups, struggling to assert their rights to land and to identity, and to regain control over their lives, have redefined themselves as indigenous. The ILO Convention of 1989 (art 2) provides that self-identification as indigenous or tribal shall be regarded as a fundamental criterion for determining the groups to which the

provisions of the convention will apply. Also, art 8 of the Draft Declaration on The Rights of Indigenous Peoples provides that: 'indigenous peoples have the collective and individual right to maintain and develop their distinct identities and characteristics, including the right to identify themselves as indigenous and to be recognized as such.'
64. Ikporukpo, C. 1983. 'Petroleum Exploitation and the Socio-Economic Environment in Nigeria'. 21 *International Journal of Environmental Studies*, 193.
65. Ibid.
66. Ibid., p. 194.
67. Ibid.
68. Ibid
69. Babatunde, A. 2010. 'The Impact of Oil Exploitation on the Socio-Economic Life of the Ilaje-Ugbo People of Ondo State, Nigeria'. 12 (5) *JSDA* 61, 71.
70. Ibid.
71. Adekola, O. 2011. 'Oil Smears the Gods of the Niger Delta' (*Earth Collective Magazine*, 20 March 2011) http://www.earthcollective.net/category/art/ (accessed 19 November 2018).
72. Weiner, James F. and Glaskin, Katie. 2006. 'Introduction: The (Re-)Invention of Indigenous Laws and Customs', *The Asia Pacific Journal of Anthropology*, 7:1, 1–13.
73. 'Pope Francis's Speech to the UN in Full'. https://www.theguardian.com/environment/2015/sep/25/pope-franciss-speech-to-the-un-in-full (accessed 14 September 2018).
74. (a) Stone, Christopher. 1972. 'Should Trees Have Standing? – Toward Legal Rights for Natural Objects' 45 *Southern California Law Review* 450, 456; (b) Tribe, L. 1974. 'Ways Not to Think about Plastic Trees: New Foundations for Environmental Law'. 83 *Yale Law Journal*, 1315, 1343; (c) Stone, Christopher. 1985. 'Should Trees Have Standing? Revisited: How Far Will Law And Morals Reach?' 59 *Southern California Law Review*.
75. 'On the Rights of Nature' February 24, 2017, https://www.earthlawcenter.org/newsfeed/2017/2/on-the-rights-of-nature (accessed 14 September 2018).
76. Ibid.
77. See Municipal Chamber of Sao Paolo. Projeto de Emenda a Lei Organica 04-00005/2015 do Vereador Eduardo Tuma (PSDB). Available online: http://cmspbdoc.inf.br/iah/fulltext/projeto/PLO0005-2015.pdf
78. Roberts, M. 2008. 'Spanish Parliament to Extend Rights to Apes'. www.reuters.com/article/idUSL2 565863 20080625?irpc=932 (accessed 23 August 2018). This has, however, still not been incorporated into the Constitution.
79. Paragraph 2.
80. Willemsen, A. (ed). 2008. 'The Dignity of Living Beings with Regard to Plants: Moral Consideration of Plants for Their Own Sake' (Paper presented to the Federal Ethics Committee on Non-Human Biotechnology, Berne, Switzerland April 2008).
81. (a) Nattrass, K. 2004. 'Und Die Tiere: Constitutional Protection for Germany's Animals' 10 *Animal Law* 283; (b) Haupt, C. 2010. 'The Nature and Effects of Constitutional State Objectives: Assessing the German Basic Law's Animal Protection Clause' 16 *Animal Law* 213.
82. 'Tough Animal Rights Laws Enacted in Austria – Regulations Protect a Range of Creatures from Hens to Dogs'. www.msn.com/id/5077350/ns/health-pet_health/t/tough-animal-rights-laws-enacted-austria/#T3BO1flTajc (accessed 26 August 2018)
83. Article 69, Chapter 5 Part 2 of the Constitution of Kenya, 2010. http://www.kenyalaw.org/lex/actview.xql?actid=Const2010 (accessed 19 November 2018).
84. Greene, N. 'The First Successful Case of the Rights of Nature Implementation in Ecuador'. www.therightsofnature.org (accessed 19 November 2018).
85. Ibid.
86. Ibid.
87. Ibid.
88. See the Preamble of the Bolivian Constitution of 2009.
89. Art 2(6) of the Law of Mother Earth. 2010. http://f.cl.ly/items/212y0r1R0W2k2F1M021G/Mother_Earth_Law.pdf (accessed 22 August 2018).

90. (a) Adewale, O. 'Customary Environmental Law' in M.A. Ajomo, and O. Adewale (eds), *Environmental Law and Sustainable Development in Nigeria* (NIALS, Lagos 1994) 157–180, noted in Ebeku, K. 2004. 'Biodiversity Conservation in Nigeria: An Appraisal of the Legal Regime in Relation to the Niger Delta Area of the Country.' 16 *JEL*, 361, 364; (b) Kankara, A.I. 2013. 'Examining Environmental Policies and Laws in Nigeria' 4(3) *International Journal of Environmental Engineering and Management*, 165–170.
91. As provided by *section 20(1)* of the Eastern Nigeria High Court Law which is applicable to the Niger Delta states of Rivers, Bayelsa and Delta.
92. The combined effect of *section* 1of the Petroleum Act, *section 2* of Territorial Waters Act *section 2 of* Exclusive Economic zone Act, *section 14* of the Land Use Act and *Section 44(3)* of the 1999 Constitution of Nigeria vests ownership and right of exploitation of minerals and natural resources in the territorial waters and exclusive economic zone of Nigeria in the federal Government of Nigeria.
93. Paragraph 4.9.i.
94. Cap. L5 Laws of the Federation of Nigeria, 2004.
95. Kingston, K. and Oke-Chinda, M. 2016. 'The Land Use Act: A Curse or Blessing to the Anglican Church and the Ikwerre Ethnic People of Rivers State'. 6(1) *African Journal of Law and Criminology*, 147, 154.
96. Sections 21 and 22 of the Land Use Act 1978.
97. Emejuru, C. and Izzi, M. 2015. 'Environmental Justice and Sustainable Development in Nigeria.' 1(1) *Donnish Journal of Biodiversity and Conservation*, 1.
98. Cap N65 Laws of the Federation of Nigeria 2004.
99. Cap E9 Laws of the Federation of Nigeria 2004.
100. Cap S4 Laws of the Federation of Nigeria 2004.
101. All the State Forestry Laws contain similar provisions.
102. For example, the preamble of the Endangered Species (Control of International Trade and Traffic) Act 1985 states that it is 'an Act to provide for the conservation and management of Nigeria's wildlife and the protection of some of her endangered species in danger of extinction as a result of over-exploitation, as required under certain international treaties to which Nigeria is a signatory'.
103. Signed by Nigeria in 15 September 1968 and ratified thereafter.
104. Signed by Nigeria on 3 December 1977 and ratified thereafter.
105. Signed by Nigeria on 11 February 1974 and ratified thereafter.
106. Signed by Nigeria on 13 June 1992 and ratified thereafter.
107. For example, *section 4(1)* of the Forestry Laws of Edo State (Cap. F4 Laws of Edo State) gives the governor power to constitute forest reserves and protected forests.
108. These offences are provided for in *section 30 (a)-(k)* of the Edo state Forestry Law. The *section* provides that anyone who in any way commits the offences in *sub-section (a)*, that is, takes any forest produce, shall be liable on summary conviction to a fine of two hundred thousand Naira and, in the case of the offences covered in *subsections (b)-(K)*, fifty thousand Naira and imprisonment for two years respectively.This amounts to approximately $2.50 from a defendant oil company whose average daily income (e.g. Exxon Mobil) is over $400 billion, as against the average income of less than 2 dollars a day of a typical Niger Delta peasant.
109. 1,000 naira is approximately US$2.80.
110. *Article 1 (1)(g)* of the Bonn Convention on the Conservation of Migratory Species of Wild Animals defines 'habitat' as 'any area in the range of a migratory species which contains suitable living conditions for that species'. It has also been defined as 'an ecological or environmental area that is inhabited by a particular animal or plant species'. See M Abercombie, CJ Hickman and ML Johnson, *A Dictionary of Biology* (London: Penguin Reference Books, 1966).
111. Ebeku, K. (n. 92), p. 366.
112. Dietz, T. 2003. 'What Is a Good Decision? Criterion for Environmental Decision Making', *Human Ecology Review* 10(1), 30–39.
113. See Doremus, H. 1991. 'Protecting the Ark: Improving Legal Protection of Biological Diversity' 18 *Ecology Law Quarterly*, 265–333.

114. See also articles 8(1), 11(1), 12, 25 and 33 of the United Nations Declaration on the rights of Indigenous Peoples, 2007; the preamble of the Declaration also instructively provides that '*Recognizing* that respect for indigenous knowledge, cultures and traditional practices contributes to sustainable and equitable development and proper management of the environment'.
115. See Sheil, D., Puri, R., Wan, M., Basuki, I., Heist, V., Liswanti, M., Rukmiyati, N., Rachmatika, I. and Samsoedin, I. 2006. 'Recognizing Local People's Priorities for Tropical Forest Biodiversity'. 35 (1) *Ambio*, 17.
116. *Culture and Wetlands: A Ramsar Guidance Document*. http://www.ramsar.org/sites/default/files/documents/library/cop10_culture_group_e.pdf (accessed 07 November 2016); Nigerian Conservation Foundation, *Critical Sites for Biodiversity Conservation in Nigeria* (NCF, 2002).
117. http://www.un.org/esa/socdev/unpfii/documents/DRIPS_en.pdf (accessed 07 September 2018).
118. (a) Jones, G'Nece. 'The Importance of Indigenous Knowledge and Good Governance to Ensuring Effective Public Participation in Environmental Impact Assessments', 2012. *ISTF News, 2012*; (b) McDaniel, L. 2002. 'Spirituality and Sustainability' 16 *Conservation Biology*, 1461–1464.
119. (a) Oviedo, G., Jeanrenaud, S. and Otegui, M. 2005. 'Protecting Sacred Natural Sites of Indigenous and Traditional Peoples: An IUCN Perspective' (Gland, Switzerland 2005); (b) Mgumia, F.H. and Oba, G. 2003. 'Potential Role of Sacred Groves in Biodiversity Conservation in Tanzania'. 30 *Environmental Conservation*, 259–265; (c) Heinämäki, L. and Herrmann, T.M. 'From Knowledge to Action: How to Protect Sacred Sites of Indigenous Peoples in the North?' In: L Heinämäki, T Herrmann (eds) *Experiencing and Protecting Sacred Natural Sites of Sámi and Other Indigenous Peoples*. (2017). Springer Polar Sciences. Springer, Cham.

# 10

# *German* Energiewende*: A Way to Sustainable Societies?*

**Michael W. Schröter and Dani Fössl**

**CONTENTS**

Summary ............ 191
Introduction ............ 191
A Brief History of the German *Energiewende* ............ 192
Opportunities and Problems ............ 196
Possible Contribution to Sustainable Societies ............ 199
Conclusion ............ 201
Notes ............ 202

## Summary

The German energy turnaround (*Energiewende*) was not a spontaneous event but the result of a long, ongoing debate. This debate is rooted in a societal dispute regarding the peaceful use of nuclear energy and, with growing impetus, climate protection. Thus, from the start the debate focused on finding ways of energy generation and supply harmless to the people and Nature. From this perspective, the turnaround is the political confession to start the technical and societal challenge to find such ways. This thesis begins with a brief history of the German energy debate, and then discusses some aspects of the current state of the energy turnaround, highlighting opportunities and problems. The paper concludes with a discussion of the possible contribution of the energy turnaround to sustainable society and the Rights of Nature.

## Introduction

Striving for sustainable societies is an endeavour every state that participated in the 1992 United National Conference on Environment and Development, held in Rio de Janeiro, Brazil (popularly known as the Rio Summit), process has to tackle. Since then, many ideas have been discussed, further global conferences have taken place and more international treaties have been signed, but the vision of a sustainable society remains out of reach. One reason may be the Triple Bottom Line, which was embraced in national and international policy and which considers the three dimensions of sustainability (environmental – economical – social) equal. Frequently, the discussions about and visions of sustainable societies only aim for shallow sustainability.[1] According to an ecological understanding,

DOI: 10.1201/9780429505959-10 

'sustainability must mean creating and living in a society in which the integrity, resilience, and productivity of Nature's Laws of Reciprocity come first.'[2] Such an understanding must include a concept of Rights of Nature.

The German *Energiewende* is not a direct result of such discussions. But it might have the potential to enhance international discussions around sustainability towards a more ecological understanding, since it appears to be the first serious attempt of an industrial nation to overcome its dependence on fossil and nuclear energy by use of renewable energies. This attempt may provoke further discussions and eventually bring to the table the whole way of living that human societies have taken for granted in the industrial era. We would like to examine whether the German *Energiewende* has this potential.

## A Brief History of the German *Energiewende*

The German Parliament (*Deutscher Bundestag*) opted to end the use of nuclear energy on 30 June 2011 with 513 to 79 votes. This vote was not the start of the energy turnaround, but it was the political acknowledgment of public opinion.[3] This opinion had grown over decades, starting in the 1970s with its movements, such as the protests against nuclear power plants. However, it would be wrong to consider the opinion change a consequence of one movement but, rather, of heterogenic causes:[4] security of energy supply, social compatibility and acceptance and impacts on Nature were some other motives for societal discourse in the 1970s. Therefore, speaking of a linear development from the movements in the 1970s up to the political decision in 2011 towards the energy turnaround would be far too simplistic.[5] But there are some milestones we would like to discuss briefly regarding our focus on the German *Energiewende* and its possible contribution towards sustainable societies.

As already mentioned, a first milestone was the antinuclear movement of the 1970s. This movement was not limited to Germany, but the way its effects acted on German society was unique. For example, it provoked one of the biggest demonstrations, with over 100,000 people, that Germany had seen so far.[6]

Antinuclear demonstration in the Hofgarten, Bonn, Germany on 14 October 1979. https://commons.wikimedia.org/wiki/File:ANTIAKW2.jpg

It also triggered a new realm of sociological research in the so-called risk society. A 'risk society' is characterised as an advanced industrial society which not only produces wealth but more and more risks, which generate a complete new realm of distribution conflicts.[7] The characteristic trait of the antinuclear movement was not to protect an actual place, but to try to make human society aware of threats against Nature that might or might not take place. Threats that cannot be easily grasped by the senses can be brought to light and understood by scientific investigation.[8] This is a feature nuclear power shares with other new environmental threats, such as genetic engineering, climate warming and modern pesticides, for example.

Environmental politics and environmental law-making in Germany reacted to those risks and the obviously deteriorating environmental conditions by implementing the Precautionary Principle. The idea of this principle is historically closely related to the principle of sustainability which was known in Germany for at least two hundred years; Carl von Carlowitz was considered the creator of the sustainability principle in its modern sense when he published his *Sylvicultura oeconomica* in 1713. Von Carlowitz aimed at the most important resource of this time – wood – and suggested, briefly, not to cut more trees than can regrow. One main motivation for him was caring for the coming generations, or, in more modern terms: precaution.[9] The German Federal Government, a coalition of the Social-Democratic Party (SPD) and the Free Democratic Party (FDP) under Chancellor Willy Brandt introduced the Precautionary Principle in its first environmental programme in 1971, and further explained it in its environmental report in 1976.[10]

As a political maxim it served as a focal point in the law-giving processes of the environmental laws, like the Federal Nature Protection Act, the Atomic Energy Act and similar statutes. In those laws the Precautionary Principle provided individual risk configurations for the different realms of environmental law, while extending to the whole branch a character of risk prevention. Thus, the Precautionary Principle shaped the further development of environmental law in Germany and greatly influenced the public discussion around nuclear energy.

When the German Federal Constitutional Court (*Bundesverfassungsgericht*) had to decide on 8 October 1978 whether the peaceful use of nuclear energy was compatible with the German constitution (*Grundgesetz*), it stated that the requirements of the Precautionary Principle were measurable, as long as the technical and scientific knowledge to safely construct nuclear power plants was recognised in the administrative process of approval. But the remaining risk (*Restrisiko*), which is not measurable or easily understandable, had to be considered a 'social-adequate burden' (*sozialadäquate Lasten*).[11] Legislators, the court said, cannot be obliged to exclude every possible threat of technical innovation with absolute certainty, because that is beyond the limit of human cognitive faculty, and would hinder official use of the Precautionary Principle. For the maintenance of the social order, estimates of practical reason have to be sufficient. Uncertainties, the court went on, beyond this limit have to be considered inescapable and as such to be born as a social burden by all citizens.[12] These remaining social-adequate risks, not graspable by practical reason, became a focal point in further discussions around the Precautionary Principle and nuclear energy, genetic engineering and similar subjects.

Therefore, the 1970s and the following years were shaped by an increasing consciousness of risks which are neither easily measurable nor understandable, and which cannot completely be controlled by human technology and the normative dimension of the Precautionary Principle. This increasing risk consciousness was especially fertile in Germany since, after the Second World War, following partition and creation of the Iron Curtain, the public had a high awareness of threats.

One other consequence of the protest movements of the 1970s in Germany was the founding of a new political party, the Greens (*Die Grünen*). This new party soon gained political power and had to be taken as seriously as other political parties. Thus, the protest was brought into the institutions of the German state. The Green Party started in the end of the 1970s, with party founding at the local and regional level. Its first political success was the entry into the parliament of the German State of Bremen in 1979. The federal party was constituted at a federal convention at in January 1980 in Karlsruhe.

The first federal executive speakers, as they were called at those times, were August Haußleiter, Petra Kelly and Norbert Mann. Petra Kelly played an essential role in the first years. Founding members were mostly people from grass-roots initiatives, and especially from antinuclear movements. Although the party was constituted on conservative, liberal and left-wing pillars, a branch of the party focusing on more realistic goals soon took shape. While the more realistic wing considered the predominant political structure capable of reforming from inside the system, which included taking over governmental responsibility, the more idealistic wing aimed for system change as an oppositional power. One potential goal was to deindustrialise Germany. This difference in Green Party philosophy continues today, although the current chairpersons, Annalena Baerbock and Robert Habeck, try to overcome it.

The first entry of the Greens into the German parliament began in 1983. After the German reunification 1990, the Green Party and the East-German movement Alliance 90 merged to form Alliance 90/The Greens (*Bündnis 90/Die Grünen*). From 1998 to 2005, Alliance 90/The Greens formed a coalition with the Socialistic-Democratic Party under Chancellor Gerhard Schröder (SPD); Joschka Fischer became Vice-Chancellor and Minister for Foreign Affairs, while Jürgen Trittin led the Ministry for Environment, Protection of Nature, and Safety of Reactors. Today (2018) Alliance 90/The Greens participates in the governmental responsibilities in 9 of 16 German States; in Baden-Wuerttemberg, for the first time in their history, the State Premier, Winfried Kretschmann, is a member of this party.

The next step towards the *Energiewende*, following this very brief survey of the political and environmental landscape in Germany in the 1970s, happened at the end of 1990 and the beginning of 1991, when in the German Parliament the conservative parties (Christian Democratic Party [CDU] and Christian Social Party [CSU]) and Alliance 90/The Greens agreed to a new law: the *Stromeinspeisungsgesetz* (Act of the Input of Power).[13] Comprising only five articles, the Input of Power Act obliged the big power companies – Germany had four at that time – to accept all power generated by renewable sources into the power grid and to have the person or company feeding this energy into the grid reimbursed at fixed prices. Thus, the aim of the *Stromeinspeisungsgesetz* was to encourage private people to invest in renewable energy technologies, such as solar panels. We might characterise this act as a law for the people, a characteristic that was further enhanced by the later Renewable Energy Act (*Erneuerbare-Energien-Gesetz*; see below). The Input of Power Act was mainly justified on grounds of resource conservation and climate protection.[14]

The big power companies reacted in two ways: on the one hand they endeavoured to disparage the possible effects of the law. On the other hand, they also took legal action against it. But the German Federal Constitutional Court and the European Court of Justice decided the act was in accordance with the German constitution and the European treaties.[15] The effect of the *Stromeinspeisungsgesetz* surpassed even the most optimistic expectations; the share of renewable power surged from around 3% to 4% (mostly from hydropower plants) to nearly 7% by the year 2000.[16] This increase was mainly the effect of renewable installations set up by private persons. At that time, most of the renewable installations were produced by small companies, some of them already established in the 1970s. This development encouraged government policies to go further down the path of renewable energy.

In 1998 the coalition of the Social Democratic Party and Alliance 90/The Greens accelerated the development of the German *Energiewende*. For the first time in German history, the responsibility for energy policy shifted from the Economics Ministry to the Environment Ministry.[17] This so called 'red-green coalition' agreed on a nuclear phase-out by 2022 – the process was implemented in 2002 – and a dynamic expansion of renewable energy, encouraged by the Renewable Energy Act (*Erneuerbare-Energien-Gesetz;* EEG) that the Bundestag passed in 2000. Remarkably, this act was not a law of the government, but of the Parliament. Although the governing coalition of SPD and the Greens had a general consensus about the law, the Minister of Economic Affairs opposed some planned regulations. Thus the SPD and Green members of the German Parliament essentially brought the Renewable Energy Act into being. This is a further sign that the Renewable Power Act is an act for the people, as described above.

This law, among others, set distinct expansion targets for renewable energy sources, and guaranteed not only the acceptance but also the payment for renewable power production through reimbursement for private persons or companies by fixed prices depending on the source (wind, solar etc.). Up to this time, renewable energy in Germany was mainly a complement to the existing fossil-nuclear based energy system. From then on it was regarded as an alternative to the status quo. The primary goal of the *Energiewende,* therefore, was the replacement of the current energy system in the first half of the twenty first century.[18]

Consequently, the government not only set targets for the expansion of renewable energy, but also targets for a more efficient and economical use of energy overall. High energy consumption for decades indicated economic growth and hence the prosperity of society. Then, suddenly, the decision to decouple energy consumption and economic growth became an important pillar of German energy policy.[19]

Different political coalitions have since amended the Renewable Energy Act several times, adjusting the framework of further expansion for renewable energies in Germany. There were notable amendments in 2004, 2009, 2012, 2014 and 2017.[20] But beyond different points of view about the right pace and targets of further expansion, there has been a common-sense agreement concerning the necessity of the *Energiewende* across all democratic political parties ever since. There has been no fundamental modification of the federal energy policy or the *Energiewende's* objectives. The coalition of the Union parties and the liberal Free Democratic Party, which took over the reins of government in 2009, reversed the first nuclear phase-out in 2010 by granting a lifetime extension for Germany's nuclear reactors. But just half a year later, in the beginning of 2011, this decision was withdrawn as consequence of the Fukushima nuclear disaster in Japan.[21]

The share of renewable energy in total energy production has been increasing ever since. In 2018 renewable energy sources reached a share of 36% of total energy production, up from 7% in 2000. In comparison, coal provides 32% of total energy sources, down from 51% in 2000. Nuclear energy now comprises only 11% of the total energy picture, down from 29% in 2000.[22] In addition, the fact that today 42% of renewable energy plants are owned by private individuals and farmers shows that the original idea of the Act of the Input of Power as well as the EEG has been – at least partially – successful. Thus, one goal of the *Energiewende* was and is a democratisation of the electricity system. Consumers are turning into producers. At the same time the market share of large corporations is falling.[23]

Wind and solar park in Germany. https://pixabay.com/de/solarpark-windpark-1288842/

## Opportunities and Problems

The effectiveness of the German *Energiewende* as an instrument for climate and environmental protection is undisputed. The increasing use of renewable energy sources leads to a displacement of fossil fuels and thus to an increasing avoidance of climate-damaging greenhouse gases. Due to renewables in electricity production, an equivalent of 138 million tonnes of greenhouse gases have been avoided just in 2017.[24] Yet the sustainable transformation of the German energy system has turned out to be a tough process – the overall success has fallen behind the objective. The total output of greenhouse gases declined comparatively slowly. Current surveys show that Germany will miss its greenhouse gas reduction targets for 2020 by a significant amount, which the German Government recently officially confirmed.[25] This is due to several causes, which we will briefly describe below.

The German *Energiewende* is a complex project. It includes the phase-out of nuclear power and significantly lowering the share of fossil energy, substituting both with renewable energy sources. More efficient and economical use of energy are important pillars of this new policy framework; the overall objective is a significant reduction of greenhouse gas emissions across all sectors (energy supply, buildings, transport, industry and business, agriculture and forestry).[26]

So far, political and public discourse and actions have mainly concentrated on energy supply. Other sectors, like the buildings and the transport sectors, lag behind. Although the transport sector, for instance, has to provide relevant reduction contributions in order to achieve the political targets, they have achieved hardly any noticeable reductions in the last decades. In 2017 the total emissions of the transport sector, as well as industry and business sectors, even rose slightly in comparison with the previous year.[27] Monitoring the efforts towards a more efficient and economical use of energy shows similar results. Officially the German government wants to cut primary energy consumption by half by the middle of the twenty first century. As a half-time goal the government seeks to reduce primary energy consumption 20% by 2020, compared to 2008. By 2017, however, a reduction of only 5% to 6% had been achieved.[28] Consequently, the current partial focus on the production of electricity within Germany's energy policy is obviously not sufficient to reach the overall targets. To some extent it seems as if Germany's current energy policy is changing piecemeal rather than pursuing a long-term cross-sectoral strategy. Instead of developing a detailed, long-term and cross-sectoral strategy, the government often seems to focus solely on individual subareas – relying mostly on technological springboard innovations to solve upcoming challenges.

We have to be aware that the legal affirmation of the energy turnaround in 2011 did not actually follow a political plan, but occurred in consequence of the Fukushima catastrophe. The support of renewable energies was a political aim before, as shown above, and an energy system relying mainly on renewable sources for the long term was discussed, but reaching this goal was thought to need a lot more time. That was the reason the German government, at the end of 2010, prolonged the approvals for the nuclear energy plants. The idea was to take the money earned through the prolongation and put it into renewable energy research. After the disaster of Fukushima this ceased to matter. After Chernobyl and Fukushima, both large-scale nuclear disasters, there was no longer a public consensus in Germany favouring the use of nuclear energy. In polls Alliance 90/The Greens reached nearly 30%, since it was the only political party that consistently opposed nuclear power and opted for substituting fossil energies with renewable ones. An important legal debate also came up, as to whether a technique (nuclear power) which was supposed to be safe, and now appeared to be not so controllable, would still be consistent with the German constitution and the way the Federal Constitutional Court interpreted it (see above). But politics reacted quickly, and just half a year after prolonging the nuclear plants' lifespans, the decision-makers changed direction. Thus the political decision to tackle the challenges of an energy turnaround in 2011 came without a long-elaborated plan. Rather, it came as an adventure. But the majority of Germans were ready and willing to begin this adventure, even if energy might cost them more money in consequence.

The transformation of the energy-supply sector did not and does not take place without resistance, and hence complications. It is obvious that shifting a centralised system, limited to a few players, to a decentralised and broad-based energy market will provoke resistance. The major German energy companies saw and still see themselves and their business models as being endangered by the government's policies. As explained above, opening the electricity market for citizens and other market actors is a core element of the German *Energiewende*. Hence, the competitive pressure on the large energy companies is increasing. At the same time, the substitution of renewable for nuclear and fossil energy sources leads to a devaluation of the companies' capital, which is bound up in nuclear and/or fossil based power plants. This explains the existence of a strong lobby created by the energy companies that is influencing politics to prevent a dynamic expansion of renewables.

Due in part to political pressure from the major energy corporations, the expansion of renewable energy today is slower than technically feasible. At the same time fossil-based power capacity is not being reduced to the extent that the growing capacity of renewable energy sources would allow. This generates inefficiencies in the energy system, which in recent years has been expressed by constantly increasing energy exports.[29] To some extent two energy systems are currently running in parallel, causing inefficiencies and additional costs. That also implies that high levels of greenhouse gas emissions from the energy sector are only declining slowly, despite the continued expansion of renewable energies.

Germany, however, has a special feature due to the nature of its economy. The sustainable transformation of the energy system, combined with the simultaneous phasing-out of nuclear energy, is a major challenge for a highly industrialised country. Whereas many countries seize nuclear energy as an important contribution to climate protection (since nuclear power plants, unlike fossil fuel power plants, do not emit carbon dioxide, at least not in the energy generation process), Germany opted for the nuclear phase-out due to the risk assessments, and with increasing concern regarding the costs of nuclear power. At the same time, current surveys show that compensation for the loss of capacity can be compensated for by the expansion of renewable energy without any fear of a collapse in supply, as long as investments and developments into grid infrastructure and energy storage systems continue.

The big energy companies, despite massive resistance, did not succeed in reversing the *Energiewende* itself. One explanation for this is the broad support of the *Energiewende* from large sectors of the economy, the medium-sized economy particularly, which is very strong in Germany. Medium-sized companies especially recognised and used the advantages of the energy transition at an early stage, by investing in renewable energy plants and in the production of climate-friendly technologies and accompanying services to lower their production costs. Another reason is certainly the growing share of citizens and cooperative energy companies investing in and benefiting from renewable energy plants (see above). Overall, it is apparent that the *Energiewende* is widely supported across large parts of society.

The energy turnaround has radically changed traditional thinking about energy generation, supply and consumption. Within the population the acceptance of the energy transition, as well as the understanding of the necessity, is quite high. Current representative surveys show that 88% of Germans basically support the transformation of the energy landscape towards renewable energy – independent of education, income and age. In addition, two-thirds of the Germans believe that the decision to get out of nuclear energy is right.[30] In addition, political and public debates have contributed to sensitising people to the needs for climate and environmental protection, as well as to the issue of sustainability as a whole.

However, renewable energy installations still have impacts on Nature. The debate over wind turbines killing birds and bats is especially sensitive. And renewable energy projects frequently require rare minerals, which are often mined in countries that show little concern for human health or the environmental pollution that usually accompanies the mining process. On the other hand, these critical debates show that the life of birds and bats and human health questions in foreign countries matter now, which they did not at this magnitude before the turnaround. This is a first but important step in the public debate around how best to arrive at sustainable solutions.

It appears that Nature conservation is being exploited to some extent by those who, for various reasons, oppose the construction of renewable energy plants. This is predominantly a local phenomenon, also known as the NIMBY ('not in my backyard') phenomenon. Local resistance to individual renewable energy plants often forms because of aesthetic dislike of these projects, but also because of fear of negative health effects, such as the sound emissions of wind turbines. Thus individuals founding initiatives especially for this purpose, focusing on Nature and Nature conservation, take action against the construction of renewable energy plants. In these instances, the stated reasons for opposition may not wholly reflect the original motivation for resistance.[31]

On the other hand, there is legitimate criticism and resistance to individual projects. One cannot deny that renewable energy plants have negative impacts on Nature and the environment. Examples include land-use conflicts between renewable energy plants and agriculture, for example with regard to the construction of open-space photovoltaic systems. The effects of wind turbines on bird populations and forests, which have to be partially cleared for their installation, should also be mentioned in this context. The massive increase in maize cultivation necessary for bioenergy plants and the consequences provide another example.[32]

At the same time, the further development of the framework conditions for the expansion of renewable energy continues to address these problems. Conflicts are anticipated by using existing planning instruments, as Nature and species protection are an integral part of the respective planning and approval procedures. Distance rules and exclusion areas serve to minimise the negative effects in sensitive areas. In addition, the urgent Nature and species protection questions were and are examined in various studies. The findings serve as a basis for the further development of planning and approval procedures, for example by redefining safety margins. The proportion of maize allowed to be cultivated for bioenergy plants has also been regulated in order to respond to existing environmental concerns.[33]

Energy generation always has negative effects on Nature and the environment. Hence it is important that these effects be publicly and critically discussed. Roughly speaking, the increased awareness of these negative effects on Nature and the environment and the vital discussion about it are also a success of the *Energiewende.*

## Possible Contribution to Sustainable Societies

While transforming society and its economic systems, especially industrial ones, towards a sustainable state forms one of the biggest obstacles to the energy turnaround, it can also generate an effort to maintain, or even enhance, the standard of living. The prevailing economic covenant promises constant growth, even though it cannot always keep this promise. Although the ecological insight that we all share one Earth, on which we have to flourish together with all other inhabitants,[34] implies a strong appeal in a moral and practical sense,[35] people's worry of losing their jobs and experiencing a social descent could thwart every attempt at urgent action – it could even undermine the acceptance of the needed changes. The social transformation process requires political reconfiguration. As we have tried to show above, the German *Energiewende* is such an attempt at reconfiguring the governing structure. But is it comprehensive and strong enough to weather the challenges against building a sustainable society?

As a highly industrialised country, Germany tackles the challenges by focusing on one important pillar of the needed transformation process: a sustainable energy supply. If this venture is successful, it will encourage other nations to follow the path; if it fails, it will be devastating in the struggle to take further steps towards sustainable societies.[36] Thus, the energy turnaround has a very important symbolic significance. With that in mind, the above-described problems the energy turnaround actually faces could be interpreted as fundamental obstacles on the path to sustainability. They could be seen as the logical consequences Germany had to experience, since it began the adventure of the turnaround while lacking a sophisticated plan, but retained a somewhat naïve belief in scientists and engineers to solve all problems. Doing so, Germany risks its wealth and its high-ranking scientific reputation. From this perspective, it could be argued the risk towards sustainability on the path of the German turnaround is too high; there is just too much at stake.

On the other side, this interpretation regarding the actual problems of the German turnaround appears much too pessimistic. It seems to be no coincidence that many of those critics also deny human-caused climate warming. Germany still has one of the strongest economies in the world, and the turnaround did not hinder the prosperity; rather, it created an important new branch of the economy. The acceptance of the turnaround by the public remains on a high level, as shown above, although energy prices have increased. The diversity of critically important discussions about the energy turnaround can, therefore, also be understood as the engagement of a broader public that is highly interested in the success of the *Energiewende*. As evidence supporting this thesis, we note that most of the criticism has been constructive. It does not try to argue against the turnaround, but focuses on particular aspects and tries to show alternatives that might cause fewer problems – for humans and for Nature. This is common in democratic discourse, and shows that high engagement underlies the energy turnaround, and shows its nature as a project of the people. And it gives us a hint for answering the question of whether the *Energiewende* contributes to a sustainable society.

The public discourse in Germany rests on a special consciousness of risk awareness that started to evolve since the 1970s, if not before. Consequently, the German public greeted all technologies having known and unknown risks, like nuclear power and genetic engineering, with a critical stance. Environmental politics and law translated this risk awareness into the Precautionary Principle and implemented it at an early stage. But even after this implementation, the societal debate remained, and remains, largely sceptical regarding so-called risk technologies.

This sceptical stance does not express an irrational unease against all technical improvement, but reasonably argues for technologies that will not harm people and Nature. It represents the difference between practical (and economic) rationality and human reason.[37] Thus, the thesis is not so much that the German *Energiewende* itself may be an important step towards sustainable societies, but that the 'risk consciousness' behind it is a critically important social transformation. That consciousness allowed the energy turnaround to be implemented. But this does not diminish the significance of the turnaround in building sustainable societies; it rather emphasises the larger goal of sustainability.

Sustainability in an ecological sense focuses on creating and living in a society in which the integrity, resilience and productivity of Nature's Laws of Reciprocity come first.[38] One important part of this vision seems to be the satisfaction of economic desires while, ideally, not harming people and Nature. But on the other hand, it is important not to threaten people's social stability by driving them out of jobs without presenting new and realistic opportunities. The energy turnaround tries to do both. It strives to get away from a high-risk nuclear technology – an accident like Chernobyl or a catastrophe like Fukushima would be quite devastating in a densely populated country like Germany. But it also seeks to create new jobs that are attractive to people who worked in the nuclear and fossil energy industries. As mentioned above, Germany profited in this process from a strong structure of middle-sized companies that are strongly rooted in regional structures and thereby have a closer relationship to the people who live and work in their area. Thus, finding the means to take risk consciousness seriously appears favourable, while remembering the people who would probably be affected by the changes. If there is any knowledge that can be inferred from the German experience so far it might be that, on the way towards sustainable societies, all points of view need to be seriously discussed and considered.

It can still be objected that the energy turnaround is far from perfect in a sustainable sense: most of the renewable installations need rare resources mined mostly under conditions ignoring people's rights and health. Furthermore, the issue of gigantic wind turbines killing bats and birds seems the opposite of a sustainable solution. Both objections are indisputably true as far as they describe the problem. But is the conclusion, that the energy turnaround is only sustainable in a fake sense, correct? It would be right if the claimed conditions were not changeable, and/or nobody were striving to change them. Instead, particularly intense discussions take place about what can be done to avoid those effects in future. For example, Germany has funded scientific research to explore why bats and birds collide with wind turbines and what measures will help to stop this phenomenon. Astonishingly, it is never argued that staying with nuclear and fossil energies would be a better option for bats and birds. Rather, the aim is to explore what can be done to make renewable installations more closely match the expectations we have of them: to be as harmless as possible to humans and Nature. Hence, such discussions do not question the energy turnaround as a sustainable project, but instead contribute to developing the turnaround as a serious tool for creating a sustainable society.

Taking into account the welfare of people living far away, and that of bats and birds, widens our public perspective in a holistic way. Risk awareness is no longer limited to people living here and now in a closed community. Rather, the question arises as to which means are the most suitable for the government administering the turnaround to consider other entities. This question is not new, but it comes back with stronger urgency as soon as the idea of sustainable society enters public discussions. It should therefore not be a surprise that the debate on Rights of Nature continues to experience a renaissance during discussions on sustainability.[39] The German *Energiewende* deserves high marks in this process, although it can be considered just an initial method of satisfying the risk consciousness that had grown in Germany since the 1970s. But the positive experiences in Germany with the energy turnaround have encouraged tackling other subjects, such as industrial farming methods, which endanger the diversity of insects and in consequence of birds. But this is another story.

## Conclusion

Starting with the movements in the 1970s, German risk awareness and consciousness evolved that soon affected the young realms of environmental politics and law. The implementation of the Precautionary Principle obliged public institutions to consider and further develop it. A consequence was exploration of energy supplies that are, as far as possible, free of risks to humans and Nature. Therefore Germany implemented the Act of the Input of Power (*Stromeinspeisungsgesetz*) in the beginning of 1991. One aim was to encourage individuals to install renewable energy sources. Encouraged by the success of this act, the process of expanding renewable energy was taken to a higher level with the Renewable Energy Act of 2000. After the Fukushima catastrophe in 2011, this law built the legal base for the energy turnaround in which Germany faces a threefold challenge: to replace nuclear and fossil energies with renewable ones, while maintaining a strong economy.

Although the process has experienced problems due to the lack of sophisticated political planning in advance, the majority of the German population supports the turnaround.[40] Mostly, the criticism is constructive; it does not question the turnaround as such but strives to forge solutions for particular aspects that might cause fewer problems. The *Energiewende* from this point of view is highly significant in the struggle for sustainable societies rooted in a public risk consciousness that demands solutions harmless to people and Nature as much as possible. Issues, such as wind turbines causing high bat and bird mortality, provide further opportunities to design measures that take all effects into account and focus on extending concern to others beyond the immediately affected communities.

## Notes

1. Bosselmann, K. 2016. 2nd edition. *The Principle of Sustainability: Transforming Law and Governance.* London: Routledge.
2. La Follette, C. and Maser, C. 2018. *Sustainability and the Rights of Nature: An Introduction.* Boca Raton, FL: CRC Press. Introduction, xxiii.
3. This has not changed until today. The latest survey shows that over 60% of the German population still welcome the energy turnaround, 30% are undecided and only 7% reject it; see Bundesministerium für Umwelt, Naturschutz und nukleare Sicherheit. 2018. Naturbewusstsein 2017. 30. https://www.bmu.de/fileadmin/Daten_BMU/Pools/Broschueren/naturbewusstseinsstudie_2017_de_bf.pdf (accessed July 2018).
4. Compare Radkau, J. 2011. *Die Ära der Ökologie.* München: Beck-Verlag.
5. Radkau, Ibid. 18 pp.
6. At 14 October 1979 in Bonn.
7. Beck, U. 1986. *Risikogesellschaft. Auf dem Weg in eine andere Moderne.* Frankfurt am Main: Suhrkamp.
8. Similar Radkau, J. 2011. *Die Ära der Ökologie.* München: Beck-Verlag. 208, 210.
9. See Grober, U. 2010. *Die Entdeckung der Nachhaltigkeit.* München: Kunstmann. 113–120 (118).
10. Umweltprogramm der Bundesregierung. 1971. http://dipbt.bundestag.de/doc/btd/06/027/0602710.pdf and Umweltbericht der Bundesregierung. 1976. http://dipbt.bundestag.de/doc/btd/07/056/0705684.pdf (both accessed July 2018).
11. *Bundesverfassungsgericht.* Decision of 8 October 1978. BVerfGE (Kalkar) 49, 89.
12. *Bundesverfassungsgericht* Ibid. In original: *Vom Gesetzgeber im Hinblick auf seine Schutzpflichten eine Regelung zu fordern, die mit absoluter Sicherheit Grundrechtsgefährdungen ausschließt, die aus der Zulassung technischer Anlagen und ihrem Betrieb möglicherweise entstehen können, hieße die Grenzen menschlichen Erkenntnisvermögens verkennen und würde weiterhin jede staatliche Zulassung der Nutzung von Technik verbannen. Für die Gestaltung der Sozialordnung muß es insoweit bei Abschätzungen anhand praktischer Vernunft bewenden. Ungewißheiten jenseits dieser Schwelle praktischer Vernunft sind unentrinnbar und insofern als sozialadäquate Lasten von allen Bürgern zu tragen.*
13. *Gesetz über die Einspeisung vom Strom aus erneuerbaren Energien in das öffentliche Netz* (briefly: *Stromeinspeisungsgesetz*) of 7 December 1990 (BGBl. I S.2633).
14. Deutscher Bundestag. 7 September 1990. Draft *Stromeinspeisungsgesetz.* http://dip21.bundestag.de/dip21/btd/11/078/1107816.pdf (accessed July 2018).
15. *Bundesverfassungsgericht.* Decision of 9 January 1996 (2 BvL 12/95). European Court of Justice. Decision of 13th March 2001 (C-379/98).
16. Bundesministerium für Wirtschaft und Energie. February 2018. Zeitenreihe zur Entwicklung der erneuerbarer Energien in Deutschland. https://www.erneuerbare-energien.de/EE/Redaktion/DE/Downloads/zeitreihen-zur-entwicklung-der-erneuerbaren-energien-in-deutschland-1990-2017.pdf;jsessionid=19ADC78884D71E6A89DF4FF1DF74F2AD?__blob=publicationFile&v=15 (accessed July 2018).
17. Von Hirschhausen, C. 2014. The German Energiewende – An Introduction. *The Quarterly Journal of the IAEE's Energy Economics Education Foundation,* Volume 3, Number 2, 3.
18. Jacobsson, S. and Lauber, V. 2006. The politics and policy of energy system transformation – explaining the German diffusion of renewable energy technology. *Energy Politics,* Volume 34, 256–276.
19. Uekötter, F. 2007. Umweltgeschichte im 19. und 20. Jahrhundert. München: Oldenbourg, 28.
20. https://www.erneuerbare-energien.de/EE/Redaktion/DE/Dossier/eeg.html?cms_docId=71110 (accessed July 2018).
21. Von Hirschhausen, C. 2014. The German "Energiewende" – An Introduction. *The Quarterly Journal of the IAEE's Energy Economics Education Foundation,* Volume 3, Number 2, 3.
22. BDEW, 2018. Erneuerbare überholen erstmals Braun- und Steinkohle bei der Stromerzeugung. Press information: https://www.bdew.de/presse/presseinformationen/erneuerbare-ueberholen-erstmals-braun-und-steinkohle-bei-der-stromerzeugung/ and BDEW, 2017.

Erneuerbare Energien und das EEG: Zahlen, Fakten, Grafiken. 2017. Foliensatz zur BDEW-Energie-Info: https://www.bdew.de/media/documents/20170710_Foliensatz-Erneuerbare-Energien-EEG_2017.pdf (both accessed July 2018).

23. For statistics, compare https://www.unendlich-viel-energie.de/buergerenergie-bleibt-schluessel-fuer-erfolgreiche-energiewende (accessed July 2018). Democratization as one goal of the turnaround is highlighted, for example, by Hennicke, P. and Welfens, P. 2012. Energiewende nach Fukushima. Deutscher Sonderweg oder weltweites Vorbild? München: oekom verlag.
24. BMWi. 2018. Zeitreihen zur Entwicklung der erneuerbaren Energien in Deutschland. https://www.erneuerbare-energien.de/EE/Redaktion/DE/Downloads/zeitreihen-zur-entwicklung-der-erneuerbaren-energien-in-deutschland-1990-2017.pdf;jsessionid=19ADC78884D71E6A89DF4FF1DF74F2AD?__blob=publicationFile&v=15 (accessed July 2018).
25. Ministerium für Umwelt, Naturschutz und nukleare Sicherheit. 2018. Klimaschutzbericht 2017. https://www.bmu.de/fileadmin/Daten_BMU/Pools/Broschueren/klimaschutzbericht_2017_aktionsprogramm.pdf (accessed July 2018).
26. For background information on the German energy transition, see the brochure of the German Federal Foreign Office The German Energiewende, https://www.auswaertiges-amt.de/blob/610620/5d9bfec0ab35695b9db548d10c94e57d/the-german-energiewende-data.pdf (accessed July 2018).
27. Umweltbundesamt (UBA). 2018. Klimabilanz 2017: Emissionen gehen leicht zurück. Niedrigere Emissionen im Energiebereich, höhere im Verkehrssektor. Press release. https://www.umweltbundesamt.de/presse/pressemitteilungen/klimabilanz-2017-emissionen-gehen-leicht-zurueck (accessed July2018).
28. See Kübler, K. 2018. Energieeffizienz und Energieeinsparung: Politik beginnt mit dem Betrachten der Realität. In: Energiewirtschaftliche Tagesfragen 68. Jg. (2018) Heft 6.
29. Umweltbundesamt (UBA). 2018. Klimabilanz 2017: Emissionen gehen leicht zurück. Niedrigere Emissionen im Energiebereich, höhere im Verkehrssektor. https://www.umweltbundesamt.de/daten/energie/stromerzeugung-erneuerbar-konventionell#textpart 1 (accessed July 2018).
30. See Frondel, M. and Sommer, S. 2018. Diskussionspapier. Schwindende Akzeptanz für die Energiewende? Ergebnisse einer wiederholten Bürgerbefragung. rwi – Leibniz-Institut für Wirtschaftsforschung Materialien. Heft 124.
31. Jenssen, T. 2010. *Einsatz der Bioenergie in Abhängigkeit von der Raum- und Siedlungsstruktur. Wärmetechnologien zwischen technischer Machbarkeit, ökonomischer Tragfähigkeit, ökologischer Wirksamkeit und sozialer Akzeptanz.* Wiesbaden: Vieweg+Teubner Research.
32. Deutschen Rates für Landespflege (DRL). 2006. Die Auswirkungen erneuerbarer Energien auf Natur und Landschaft. Heft 79 – 2006. Schriftenreihe des Deutschen Rates für Landespflege. Link: http://www.landespflege.de/aktuelles/tagung%20EE/DRL-Stellgnahme-79_ErnEng.pdf (accessed July 2018). Moning, C. 2018. Energiewende und Naturschutz – Eine Schicksalsfrage auch für Rotmilane. In: Kühne O. and Weber, F. (eds.) *Bausteine der Energiewende. RaumFragen: Stadt – Region – Landschaft.* Wiesbaden: Springer VS.
33. Blessing, M. 2016. Planung und Genehmigung von Windenergieanlagen. Stuttgart: W. Kohlhammer GmbH. Jenssen, T. 2010. Einsatz der Bioenergie in Abhängigkeit von der Raum- und Siedlungsstruktur. Wärmetechnologien zwischen technischer Machbarkeit, ökonomischer Tragfähigkeit, ökologischer Wirksamkeit und sozialer Akzeptanz. Wiesbaden: Vieweg+Teubner Research. Zahn, A., Lustig, A. and Hammer, M. 2014. Potenzielle Auswirkungen von Windenergieanlagen auf Fledermauspopulationen. In: *Anliegen Natur.* Zeitschrift für Naturschutz und angewandte Landschaftsökologie. Heft, Volume 36, Number 1, 21–35. Hanjo Steinborn, H., Reichenbach, M., Timmermann, H. 2012. *Windkraft – Vögel – Lebensräume: Ergebnisse einer siebenjährigen Studie zum Einfluss von Windkraftanlagen und Habitatparametern auf Wiesenvögel.* Oldenburg: ARSU GmbH.
34. For example Bosselmann, K. 2011. Global Constitutionalism and the Prospect of a Global Constitution. In: Westra L., Bosselmann K., and Soskolne C. (eds.) *Globalisation and Ecological Integrity in Science and International Law.* Cambridge: Cambridge Scholars Publishing, 186.

35. See for further implications Bosselmann, K. 2015. *Earth Governance*. Cheltenham Northampton: Edward Elgar Publishing Limited.
36. Loske, R. 2015. *Politik der Zukunftsfähigkeit*. Frankfurt am Main: Fischer, 38.
37. Bosselmann, K. 2015. *Earth Governance*. Cheltenham Northampton: Edward Elgar Publishing, 13–16.
38. La Follette, C. and Maser, C. 2018. *Sustainability and the Rights of Nature: An Introduction*. Boca Raton, FL: CRC Press. Introduction, xxiii.
39. Schröter, M. W. and Bosselmann, K. 2018. Die Robbenklage im Lichte der Nachhaltigkeit. *Zeitschrift für Umweltrecht*, Volume 4, 195–205.
40. *Bundesministerium für Umwelt, Naturschutz und nukleare Sicherheit*. 2018. Naturbewusstsein 2017. 30. https://www.bmu.de/fileadmin/Daten_BMU/Pools/Broschueren/naturbewusstseinsstudie_2017_de_bf.pdf (accessed July 2018)

# 11

# *Seasonally Flooded Savannas of South America: Sustainability and the Cattle-Wildlife Mosaic*

**Almira Hoogesteijn, José Luis Febles and Rafael Hoogesteijn**

**CONTENTS**

Introduction ..... 205
Two Principal South American Savanna Ecosystems ..... 207
The Llanos ..... 207
Climatic Factors ..... 208
Soils ..... 208
Fire ..... 209
Fauna ..... 210
The Pantanal ..... 211
Climatic Factors ..... 211
Soils ..... 212
Fire ..... 212
Biology and Fauna ..... 213
Quaternary Extinctions ..... 213
The European Conquest of the New World ..... 215
Here to Stay: The Explosive Dissemination of Cattle ..... 216
Native Grass versus Non-Native Grass ..... 217
Sustainable Management of the Llanos and Pantanal: Ecological Promise ..... 219
Water ..... 220
Grazing and Overgrazing ..... 221
Introduced Grasses ..... 222
Cattle ..... 223
The Human Factor ..... 224
Solutions to Create Sustainable Savannas ..... 225
Conclusion ..... 228
Notes ..... 228
Appendix: Common and Scientific Names of Plants and Animals ..... 234

## Introduction

The basis of interesting storytelling is conflict, and every story related to environmental conservation is crammed with conflict. Many concepts have been given to portray life and its interconnectedness: Pachamama, Gaia, Land community, Animated Earth, Earth community … Although these concepts have permeated our intellect, they have not permeated our hearts yet. We still think Nature exists to serve humans; this puts us in

DOI: 10.1201/9780429505959-11

an adversarial position towards the environment. If Nature and all it encompasses is an object to profit from, then it is almost impossible to protect Nature; protection is left in hands of the market forces, and we are a long way from giving Nature the same rights as corporations: legal personhood with guardians to protect it.

In this moment of our human history, the biggest challenge ever is unfolding: meeting food demands for the exponentially growing human population, whilst reducing agriculture's environmental harm. Under proper appreciation and stewardship we portray such a possibility through two examples of flooded savanna ecosystems in South America. We support our presentation by way of a historical review of the formation of these grasslands and their biology, the introduction of European domestic animals as the main economy and how this activity can bring prosperity to all. We discuss the main threats to these ecosystems and possible solutions through the understanding of the unique ecology these ecosystems present.

Agriculture is the dominant force behind most of the environmental threats we face. The reality of the impact of the production of food is best described by the sobering appraisals brought by Foley and collaborators[1]: (1) human population increases by 75 million souls per year, the equivalent to the population of Germany; (2) one in seven people lacks access to food or is chronically malnourished; (3) the production of domestic ungulates occupies 30 million km$^2$ (11,583,065 sq. miles), the equivalent of Africa's surface; (4) 70% of the grasslands, 50% of savannas, 45% of deciduous forests and 27% of tropical forests have been cleared and converted to agriculture; (5) in the past 50 years, irrigated areas have doubled, and most of the large rivers have become small or disappeared; (6) the use of agrochemicals has increased 500%, causing water degradation, increased energy use and widespread pollution; and (7) agriculture is responsible for 30%–35% of greenhouse gas emissions (deforestation, methane emission of livestock, rice cultivation and nitrous oxide emissions from fertilised soils). In a nutshell, agriculture and animal husbandry are the most powerful and devastating forces that have been unleashed on the planet since the last glaciation.

This appraisal is echoed in an article by Machovina and collaborators.[2] In it the authors propose that, in order to conserve biodiversity, the demand of animal-based food consumption has to decrease to 10% of the weight of food in the human diet (world median of approximately 21%, versus developed countries' approximately 40%); to replace 'inefficient' ruminants (cattle, sheep, goats) with more 'efficient' monogastrics (animals with single-chambered stomachs, such as poultry and pigs) and integrated aquaculture. Livestock production would have to break away from single-product, intensive, fossil-fuel–dependent systems (e.g. the feedlot). Although the argumentation in both articles is flawless, the wide scope (Earth) of the analysis presented practically eliminates from consideration certain ecosystems that are not suitable for integrated small productions, feed crop production or the industrial rearing of poultry and pigs: flooded savannas. We describe the reason these areas should continue to be devoted to this kind of economy as the more sustainable management form.

The term 'savanna' usually refers to a biome based on herbaceous formations with or without trees, under seasonal conditions that define water availability, a rhythmic temporal range were a flooding season follows a dry season throughout a year.[3] This concept excludes human-created systems of pastures and fields. Under this definition, all savannas have common functional and structural attributes, but the ecosystems differ considerably depending on the combination of the following environmental factors: (1) fluctuations in water, (2) fluctuations in temperature, (3) soil quality, (4) fire and (5) the behaviour of herbivores (plant-eating animals).

Because sufficient ecological diversity exists to differentiate the two types, it would be incorrect to consider them one: (1) the Llanos, a savanna in the Orinoco River basin (Venezuela and Colombia) and (2) the Pantanal, a savanna in the Paraguay River basin (Brazil). However, all points presented could also apply to other flooded savanna landscapes, such as the Llanos de Moxos in Bolivia and the Rio Branco-Rupununi savannas in Guyana.

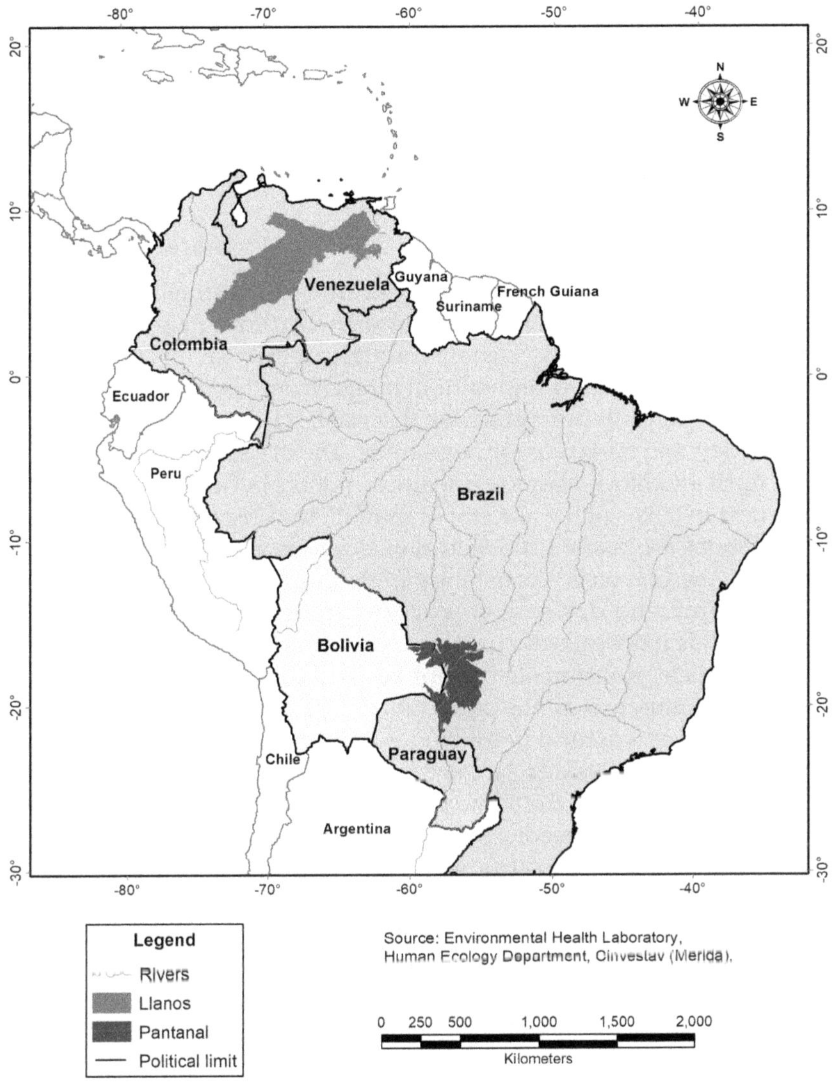


Map of South America showing the Llanos and Patanal. Map courtesy of José Luis Febles Patron, Environmental Health Laboratory, Human Ecology Department, Cinvestav, Mérida Unit.[116]

## Two Principal South American Savanna Ecosystems

### The Llanos

The Llanos are a vast alluvial floodplain, present in the intertropical belt, that occupy close to 375,787 km$^2$ (145,092.17 sq. miles) in Colombia and Venezuela. This basin is part of a subsidence zone formed at the time of the Andean mountain-building, via geological processes that folded and deformed the Earth's crust to form the mountain range. The Llanos is defined by a great variety of vegetation, which goes from grasses to dense shrubs and gallery forests, characterised by a high biodiversity but a low productivity potential. Forests always appear under the most favourable annual water budgets; they can be islands surrounded by a more or less continuous savanna matrix and/or adjacent to rivers, forming 'gallery forests'. A gallery forest grows as a corridor along rivers or wetlands and projects into otherwise sparsely treed landscapes, such as savannas, grasslands or deserts. The main economic production of the Llanos has been extensive cattle ranching.[4] 'Llano' in Spanish means 'flat', implying that the region has little elevation and no mountains; it could be translated as 'the flatlands'. Cattle-working people of the Llanos are described as 'Llaneros', which could equate to the term 'cowboy' in English.

### Climatic Factors

Grisebach was probably the first to characterise savannas by their climate.[5] These biomes are characterised by nearly constant warm climate with two well-defined seasons. The mean monthly temperatures are relatively constant throughout the year (21–28°C; 69–82°F), ranging according to latitude and elevation, and rainfall ranges from 1,000 to 2,000 mm (39–78 inches). About 80%–90% of the rain falls in a wet season that can last from 5 to 8 months.[6] The wet season usually begins in May, and rains cease in September. The dry season begins in November and ends in March. April and November are transition months between both seasons. No year is the same as the next in terms of the amount of rainfall, the onset and finish of rains and level of flooding. The Llanos are below 1,200 metres of elevation above sea level (3,937 feet).

The key factor in the Llanos is a constantly high temperature throughout the year, with alterations in the rainy and dry season. Water surplus occurs during the rainy season, while strong water deficit prevails during the dry season. Plants are especially well adapted to this alternating cycle. Rainforests cannot resist the extreme periods of drought, nor could the vegetation compete with the perennial grasses adapted to equally long periods of surplus water.[7] From the structural point of view, savannas have more or less a continuous herbaceous stratum, which tends to be very active when in presence of water during the rainy season. Grasses dry out at the same time the soils dry out until they die in the peak of the dry season. The dry herbaceous stratum is highly combustible, facilitating frequent fires. Seasonal savannas have a significant and diverse tree stratum, but when water stress is extreme, grasses become predominant.

Unlike the dry savannas of Africa and Australia, the American tropical savannas never exist in semi-arid climates; savannas disappear when annual precipitations are lower than 700–800 mm (27–31 inches).[8] The rainfall cycles are ecological 'adaptation filters' – plants and animals respond to the limiting factor of water availability, therefore the name 'seasonally flooded savanna'.

Three major ecological types of savanna have been distinguished in the Llanos according to the soil-water regime and how the vegetation functions seasonally[9]: (1) Seasonal savannas, where the dominant perennial grasses overcome an extended period of water shortage induced by rainfall seasonality, although water excess is absent or unimportant; (2) hyperseasonal savannas, where water excess and water shortage occur during each annual cycle. The topsoil becomes water saturated during the rains and dries out completely during the dry season and (3) semiseasonal savannas, which suffer from extended periods of water excess without any long period of water deficiency, thus approaching the soil-water regime of seasonal marshes (although the waterlogging may not be as pronounced and as extended as it is in marshes).[10]

### Soils

Several soil features have been postulated as major determinants of savannas: (1) soil drainage, (2) soil-water retention capacity and (3) mineral supply. Soils that evolved in the humid tropics tend to be very poor in nutrients, precisely because of the intense dissolution of those nutrients caused by water of the seasonal flooding. Leaching of minerals from the soil decreases only when the waters do not drain, and the content of nutrients is better.[11] This water saturation produces alternate periods of oxidation and reduction during the year, and the characteristic red, yellow-grey mottled colours. These soils are unfavourable for plant growth almost the whole year, remaining anaerobic during the wet season and quickly drying after the rains end.[12]

As described above, the three types of savannas (seasonal, hyperseasonal and semiseasonal) coexist in the savanna landscape. The seasonal savannas have a higher elevation with coarse, well-drained soils; hyperseasonal savannas develop on silty soils of intermediate topographic positions and semiseasonal savannas in the bottomlands with heavy clay soils.

Tropical soils in contrast to temperate soils are nutrient poor. In temperate zones the soil accumulates organic material in form of plant litter. This accumulation is possible because it is warm in relatively short periods of the year, when the fallen material decomposes. In the tropics, the soil is generally so warm and humid throughout the year that any plant litter is quickly broken down completely by the abundant decomposers (termites, fungi and bacteria). Therefore, there is little chance for build-up of organic material. Additionally, the minerals useful to plants in the soil have long ago been weathered away. When agriculture is conducted in such soils, the consequences are obvious: the few nutrients available in the soil are used up and the soil becomes quickly exhausted.[13]

Two other factors may contribute to soil impoverishment: soil erosion and fire. With heavy rainfall and open vegetation, loss of the upper soil horizon is probable, especially on steep slopes. Fire results in the accumulation of mineral elements on the soil surface, which can be lost to wind and water erosion.[14]

### *Fire*

All savannas burn repeatedly at intervals varying from a few months to several years, and the great majority burn at least once every 1 or 2 years. Fire does not affect the establishment of some woody species in the savanna; tree density is not dependent on the fires, but rather on soil quality.[15] From the point of view of vegetation structure, all savannas have a more or less continuous herbaceous stratum. This stratum function is completely seasonal; all plants are extremely active when water is available and dry out along with the soils until they die. These dead grasses are extremely combustible, favouring fires that frequently completely burn all the vegetation.[16] In present-day conditions, fires are more frequent than in any other natural system elsewhere in the world. However, we have no evidence that these savannas were different 400 to 500 years ago, before European colonisers arrived in tropical America, when they might have been subjected to a much lower fire frequency by the aboriginal population.[17] Many researchers maintained that frequent fires prevented the establishment of plants not resistant to fire. Fire of natural or human origin will act as a selective filter to which only fire-tolerant/resistant plants (pyrophytes) will survive, and those are the ones currently present in this ecosystem.[18]

Thirty years of cattle management by the third author portrays the burning activities as follows: Llaneros traditionally wait for 2 to 3 intense days of sun after the first or second rain (March-April). Cattle are evacuated from the areas to be burned, usually the areas with the lowest elevation ('esteros' and 'bajos'), and firebreaks are cleared. Fires are lit at night and always supervised. The area is kept clear of cattle after burning for at least 2 months. This system guarantees that the upper layer of dead and dry grass is burned. It produces vigorous grass regrowth, while also avoiding grazing and trampling. If cattle are allowed to enter these paddocks too soon, the grass growth is hampered and spiny/bushy vegetation takes the upper hand. Higher areas, surrounded by water, are burned at the beginning of the dry season (November/December) to eliminate spiny bushes. As they are surrounded by humid soils and water, fire does not spread to other parts of the ranch.

Calf check-up, Apure State, Venezuela. March and April are the intense dry season in the Llanos. Convenient water sources are fundamental to good husbandry, and corralling the cattle at night helps prevent rustling and feline predation. (Photograph by Rafael Hoogesteijn.)

### *Fauna*

There are 102 species of mammals in the Venezuelan Llanos; most of them bats (59 species). In the wet flooded savannas, the largest existing rodent, the capybara, occupies the large herbivore niche. Colombia has one of the richest bird diversities in the world (1,700 species); 40% of those are in the Llanos. Venezuela has approximately 1,330 bird species, and roughly 50% of them live in the Llanos. Hundreds of species of migratory birds overwinter in the Llanos, and it is a well-known area for fantastic bird aggregations around water sources, especially in the dry season. Venezuela's reptile list comprises 75 species, whilst Colombia's comprises 119 species. Two reptile species are exceptional, the largest American fluvial Arrau turtle, which reaches a weight of 50 kg (110 pounds), as well as the Orinoco Crocodile, the only crocodile restricted to a river basin.

Venezuela has 36 species of amphibians, whilst the Colombia hosts 28 species. Three hundred species of fish live in the Venezuelan Llanos; some of them have amazing physiological adaptations that allow them to extract oxygen from the air when water conditions become oxygen depleted due to high temperatures and no water circulation. There are many endangered species, mostly due to habitat destruction, illegal hunting or overharvest: the giant armadillo, giant river otter, ocelot, jaguar, tapir, Arrau turtle, Orinoco crocodile and, finally, the Amazon river dolphin, as well as several species of parrots and macaws.

*Savanna Formation Theories*

All theories related to savanna formation conclude that savannas form because of the interaction and concurrence of many factors. One theory explains that savannas are the only possible ecology within a region of tropical and dry climate, flat areas with heavy soils with no or limited drainage, that dries completely off during the dry season. Another theory states that in tropical rainy climates, rain produces intensive soil leaching. Well-drained, coarse-textured soils suffer loss of all critical elements. These impoverished soils allow the growth of plants able to live in hot, seasonal environments, with characteristic hard leaves, prone to burning. After repeated burnings a savanna becomes established and is maintained by fire.[19]

In summary, seasonally flooded savannas evolved under the simultaneous pressure of extreme seasonality, nutritionally poor soils and recurrent fires. Savannas did exist in tropical America before the arrival of the Europeans, before the development of aboriginal cultures and most probably before the arrival of humans – even before the appearance of humans on the planet.[20] Many present-day savanna species already existed in the middle Eocene; the differentiation of the distinct existing flora is evidence of the considerable age of the savanna formation.[21] The interpretation of savannas as human-induced ecosystems may be discarded, without ignoring the fact the human action may be partly responsible for the maintenance of savannas.[22]

## The Pantanal

The Pantanal is defined as a geological depression through which the Paraguay River floods and flows. It is a large wetland ecosystem of approximately 190,000 km$^2$ (73,359 sq. miles), of which 140,000 km$^2$ (54,054 sq. miles) are in Brazil, approximately 30,000 km$^2$ (11,583 sq. miles) are in Bolivia and 20,000 km$^2$ (7,722 sq. miles) are in Paraguay. Like the Llanos, the Pantanal results from the same subsidence zone formed at the time of the Andean mountain-building, which stretches into the Chaco area.[23] It receives its maximum inundation level in February, but the extent and length of the flood vary considerably from year to year. During the dry season it reverts to a grassland savanna.[23] In these systems, the main stress is the excess of water. If the excess water becomes permanent, a wetland is formed and the savanna is replaced by a marsh.

'Pantanal' means 'swampland' in Portuguese. It has areas that hardly dry out, except for a few weeks at the end of the dry season. The Pantanal is not a homogeneous swamp; it consists of a mosaic of tens of large-sized rivers and their deltas, thousands of lakes and salt pans, all interspersed with scrubby savanna and grassland, stretches of riverine forest, fragments of mountain ranges and scores of isolated rocky monadnocks.[24] (A 'monadnock' is an isolated hill or ridge or erosion-resistant rock standing conspicuously above the general level of the surrounding area.)

### *Climatic Factors*

The Pantanal has dry winters (May to September) and wet summers (October to April). The annual mean temperature is 25°C (77°F). The highest temperature usually occurs in early summer, reaching 40°C (104°F). The wet summer is due to the air masses heavy with humidity that blow from the Amazon area. Sometimes in winter, Antarctic atmospheric fronts can drop the temperature to 0°C (32°F). Precipitation is over 1,500 mm/year (59 inches.). The Pantanal receives the run-off from the surrounding rivers, which succeed in maintaining a wetland environment despite the basically semi-arid climate. The flow of the rivers decreases downstream instead of increasing, as usual.[25] The Pantanal is probably the most important region for evaporative freshwater loss on the globe. The wetlands of Pantanal owe their existence to the rivers and not to the rainfall, so the floods are very dependent on the hydrological conditions of the rivers' headwaters and surrounding areas far outside the Pantanal itself.[26]

The economic importance of the floods resides in the fact that in the plains of the Pantanal, every additional metre (39 inches) in the depth of water results in the flooding of huge areas[27]; therefore, comprehending and forecasting the floods in the Pantanal is, and will be, a very complicated task. Overall, four hydrological seasons have been defined[28]: (1) beginning of flooding ('enchente') between October and December; (2) high waters ('cheia'), between January and March; (3) receding waters ('vazante'), between April and June and (4) dry season ('estiagem'), between July and September. These seasons vary in the different parts of the Pantanal, out of phase sometimes by many weeks and even a few months.

Onset of flooding, levels of flooding and onset and finishing times vary considerably in consecutive years. Floods are classified according to their height. Ordinary floods fluctuate around a medium value of 3.8 m (12 feet) between the ranges of 3.5–4.5 m (11–15 feet). Exceptionally high floods can reach over 6 m (19 feet), and exceptionally low floods are below 3.5 m (11 feet).[29] During low flood years, husbandry expands, whereas during high floods, cattle mortality is high when the fields are flooded. In the floods of 1974, for instance, about 800,000 cattle died, either by drowning or for lack of pasture.[30]

### *Soils*

There is today a growing acceptance that the graben (a depressed block of the Earth's crust bordered by parallel faults) of the Pantanal was formed through the fragmentation of an old Cretaceous shield balanced with the uplifting of the Andes and of the Brazilian highlands. As the formation of mountains progressed, the syncline of the Pantanal sank deeper and filled up with alluvial sediments. As the Pantanal mainly consists of the alluvial fans of the main rivers, soils are formed by the action of water. There are: (1) sandy soils with 15% clay, (2) mixed soils with an approximate sand content of 15% and clay content up to 35% and (3) clay soils consisting of more than 35% clay. Sodium content may be high because of deficient flushing and usually increases in the lower reaches of the rivers, especially towards the Bolivian Chaco. More than 70% of the Pantanal is considered low fertility, in the north because it is frequently waterlogged, in the south because of the high sodium content.[31] With the exception of extremely acid soils, the amount of organic matter is the main determinant of the soil nutrient release. In wet savannas, high rainfall and an extended wet season favour increased plant production with a consequent input of organic matter into the soil.[32]

### *Fire*

Coutinho considers periodic burning a natural and even necessary event in the Pantanal. Natural fires tend to happen at the end of the dry season (August-September).[33] As in the Llanos, pantaneiros (people whose first line of income is cattle-ranching) use fire as the most effective way of replenishing pastures. However, if this practise gets out of control, it may produce wildfires of high intensity that burn accumulated turf for months. The actual fire regime for rural areas in the Pantanal is on an annual basis. In general they are not prescribed, and many of them are accidental.

Cattle roundup in flood season in the Patanal, Brazil. Pantaneiros (cowboys) drive the heard in the peak of the rainy season in the Brazilian Patanal. Ranches have to be large in order to move the herds to higher and lower areas seasonally. (Photograph by Rafael Hoogesteijn.)

### Biology and Fauna

Practically all plants in the Pantanal are capable of living in and out of water for extended periods of time. Because of the constant change of succession and the moving waters, Pantanal is an area in which few species are unique to its geographic location, but populations are large.[34] Pantanal has elements of all its surrounding regions: The Amazon, the Cerrado, the Chaco and the Atlantic Forest. Only very resistant species of these biogeographic provinces survive in the Pantanal.[35] Pantanal presents successional plant communities that have been classified as follows: (1) Campo limpo (clean grassland), grasses with no trees; (2) Campo sujo (unclean grassland), with trees up to 3 m (9 feet) in height; (3) Campo cerrado, with trees up to 4 m (13 feet); (4) Cerrado, with trees up to 6 m (20 feet) and (5) Cerradão, with trees up to 9 m (30 feet) in height. This classification, although rather artificial, clearly describes the succession from grassland to forest.[36] The Pantanal is home to at least 650 species of birds, 260 species of fish, 80 mammal species, 50 reptile species and more than 2,000 identified floral species.[37] The Pantanal's principal economic activity includes fishing, mining and tourism, but cattle-ranching has been the dominant economic activity for at least 300 years. About 95% of the Pantanal is privately owned, and 80% of the land is dedicated to cattle-ranching.[38]

## Quaternary Extinctions

As the Panama isthmus closed 3 million years ago, there was a large exchange of species known as the Great American Biotic Interchange, in which South American species migrated north (e.g. rodents, sloths, armadillos, anteaters, opossums) and North American species migrated south (e.g. cats, ferrets, skunks, bears, shrews, rabbits, horses, camels, deer, tapirs and peccaries).[39]

Ten thousand years ago, an instant in geological terms, South America had unique original fauna. There was an advanced radiation of marsupials and a group of mammals, the superorder Xenarthra, composed of sloths, armadillos and anteaters. Marsupials filled omnivorous, insectivorous and carnivorous trophic niches. The Xenarthra differentiated in insectivorous, omnivorous and herbivorous niches. Very large extinct hoofed mammals from two orders, the Litopterna and the Notungulata, filled the large herbivore niches. The arrivals of the Great American Biotic Interchange had to settle in the available free niches left by the above-mentioned orders.

What caused the nearly simultaneous extinction of about 35 varied large mammal species that lived in different habitats over such a wide geographic area? Gary Haynes presents possible hypotheses for this extinction event: (1) overkill, (2) overchill, (3) overill, (4) overgrill.[40] Were the animals hunted to extinction? Did they freeze to death during the Younger Dryas climate change? Did disease brought in by humans and their dogs kill off the native species? Or was there a cosmic impact?[41] This was not a mass extinction (it extended approximately from 10,000 to 4,000 years ago), and only affected large mammals; no other plants or animals became extinct.[42] The extinction was unequally distributed, and was stronger in South America than in North America.

There is a huge, and so far inconclusive, debate as to which model better explains the disappearance of American large mammals. For further information on the subject, we recommend an overview of the different positions.[43]

One point is clear: South American prehistory mammals occupied a wide array of the available herbivore niches. When hoofed mammals arrived from North America, their success was limited due to the widespread presence of the resident species, therefore achieving little diversification and being relegated to the forests. Since the extinction of the large herbivores, South America still lacks large hoofed grass-eating ungulates.[44] The largest ungulates are the camelids, tapirs and marsh deer.

The camelids inhabit the Andean grasslands, alpaca (55–90 kg) (121–200 pounds) and llamas (113–250 kg) (250–550 pounds) exist only as domesticated species; vicuña (45–55 kg) (100–121 pounds) and guanacos (100–120 kg) (220–265 pounds) are wild species.[45] The tapir, a browser, inhabits dense forests and can weigh up to approximately 300 kg (662 pounds). The marsh deer is found in standing waters and dense cover; also a browser, it can weigh up to 125 kg (275 pounds).[46] The only native grazer that remained was the capybara, a giant rodent restricted to habitats near rivers and lagoons.[47] Of the remaining South-American ungulates, white-tailed deer, pampas deer and marsh deer can be regarded as major components of the tropical savanna ecosystems, but they seem unable to maintain populations levels high enough to be a major force in ecosystem dynamics.[48]

This natural scarcity of large herbivores has been the seminal factor in permitting the introduced cattle/horses to exploit these grasslands.[49] All the modern Pantanal and Llanos large herbivorous species that survived the extinctions are related to savanna and forest environments (peccaries, deer, capybaras and tapirs). The total mass of tropical mammals, with the exception of the capybara, is low, especially when compared with the overall mass of large herbivore herds present in the African and Asian savannas.

Whatever the reason for extinction at the end of the Quaternary, South America no longer had large grass-eating ungulates or marsupials, which remained the case until the arrival of the European conquistadors.

## The European Conquest of the New World

> I saw neither sheep, nor goats nor any other beast, but I have been here but a short time, half a day, yet if there were any I couldn't have failed to see them ... (Christopher Columbus.)[50]

For Gold, Glory and God![51] Early Spanish exploration was motivated less by the desire to find agricultural settlements, or freedom of creed, than by the desire to extract valuables and natural resources, such as gold, silver, salt, pearls and later hides and cocoa. At the arrival of Christopher Columbus, some American civilisations were predominantly agriculturalists, but few were livestock keepers. The only domesticated species in America were dogs, turkeys, Muscovy ducks, guinea pigs, llamas and alpacas. European animals were necessary for the 'conquest'; dogs (mastiff types), oxen, horses, mules and alien microbes were part of the war machinery brought by Spaniards and Portuguese.

The settlers also felt a desperate need to make the Americas 'European'. Livestock species were brought from the Iberian Peninsula, primarily from Andalucía, Extremadura and the Canary Islands, starting with the second trip of Columbus, which departed from Cadiz in 1493. Columbus brought horses, cattle, sheep, goats and pigs in 17 caravels. It is difficult to estimate how many animals escaped, but by the middle seventeenth century, cattle were roaming free in the whole continent.[52]

European domestic animals were self-replicators. For example, pigs were released in Española (today's Dominican Republic and Haiti), and after a few years of running wild, the number was estimated as *'infinitos'*. Pigs spread from the Antilles to the mainland, and by 1514 Cuba counted more than 30,000.[53] Pigs followed the footsteps of Pizarro. The first resident Bishop, Friar Bartolomé de las Casas, a historian famous for his book *A Short Account of the Destruction of the Indies*, published in 1542, stated that all the swarms of pigs were descendants of the Columbus original.[54] Breeding herds were well established in the West Indies by 1512, in Mexico by 1520, in Peru by 1530 and in Florida by 1562. In the case of the flooded savannas, adaptation took a bit longer, but adapt they did.[55]

The introduction of domestic animals in South America was directly related to the advance of colonisation. Cattle had two advantages over pigs: more efficient coping with the tropical heat and turning rough grasses into milk, meat, hides and above all draft power. In his book *The Criollo: Spanish Cattle in the Americas*, Rouse determined that the cattle arriving from Spain had the strongest influence in the establishment of herds in the New World.[56]

The history of cattle in the Llanos is difficult to follow; it is a story rich in unreliable facts. We know for sure that Spanish stock arrived at Margarita Island (east of the Venezuelan coast) between 1525 and 1562. From there it is not clear how cattle reached the mainland, and there are three plausible scenarios. All three versions converge in the importance of a town, El Tocuyo, Lara State, founded in 1545, from which cattle expanded to the Llanos. By 1593, three important colonies mushroomed from El Tocuyo to the northwest border of the Llanos: Barinas, San Sebastian de los Reyes and Guanare. From there the colonisation of the flatlands took force and with it the ranching activity. One account tells that Cristóbal Rodríguez, a Spanish native from Córdoba, established in El Tocuyo, founded a ranch with 18 cows; 25 years later the same account states that he owned around 12,000 to 14,000 cows. Modern rancher associations recognise Cristóbal Rodríguez as the first significant commercial cattle rancher in Venezuela.

By 1646, an expedition led by Miguel Ochogavia to the south of the Apure State discovered a feral cattle population so vast that he paid 9,500 pesos, a huge sum of money for the time, in taxes to the Spanish crown (Derecho de quinto, 20% of the value of the goods) for the exploitation rights of the herd. The presence of this herd 150 years after the first arrival in Isla de Margarita is a most impressive achievement for any exotic invader. The most plausible explanation is that these feral cattle were descendants of the escapees from the Jesuit and Franciscan missions established at the left margin of the Orinoco River around the years 1531 to 1534.[57] Another influx came from Bogotá, Colombia, where in 1548 a stockman passed through with a small herd of cattle, heading in a southeast direction. By 1600, as many as 45 ranches had been founded in the Venezuelan plains. A half-century later, something like 140,000 head of cattle grazed the Llanos.[58]

The first significant introduction of bovines in Pantanal occurred in 1568, and it was rather accidental. Local ethnic communities played a key role in the establishment of bovines. From the sixteenth to the eighteenth century the Guaicurus, Paiaguás and Caiapós Native peoples actively resisted conquest and colonisation attempts by the Europeans. One of their main strategies of opposition was the theft of domestic animals; however, they were not interested in animal husbandry. Felipe de Cáceres was the first to record such an episode. He was named governor of La Asunción (Paraguay) whilst in Lima, Peru. He had to reach La Asunción by land, and for that purpose he rounded up a huge caravan. Paiaguás, and later Guaicurus Indians, harassed the caravan and stole around 730 head.

Jesuits also played an important role through the foundation of the 'Misiones' (missions), which guaranteed access to food based on European domestic animals. The main proclaimed objective of these Misiones was the protection of the local ethnic groups from the infamous 'bandeirantes', Portuguese settlers and fortune hunters, who captured the local indigenous population and sold them into the slave markets.[59] The period drama film 'The Mission' from Roland Joffé (2007) recreates a beautiful romanticised version of the Jesuit activity in the New World. Bandeirantes later became miners when gold was discovered in northern Pantanal. They established the first villages, and with them the need for a constant supply of goods and food. In the beginning, most of the transport was by river, but time and distance conspired against a steady supply of resources, and by 1736 a road was opened from Vila Boa in the state of Goias to Cuiaba, to bring supplies by oxen-cart. This road was extended from Vila Boa to São Paulo. It was along this road that the first farms dedicated to cattle ranching (fazendas) appeared, and with them the final establishment of cattle as the main economy of the area.[60]

## Here to Stay: The Explosive Dissemination of Cattle

Many of the cattle that landed in South America escaped, adapted, became feral and produced viable populations, sometimes after just a few years of introduction. The ease with which these cattle became adapted to local conditions, their survival capacity in harsh conditions and the variations that developed into breeds are astounding.[61] It has been suggested that a probable reason for this outstanding radiation is that European domestic animals found an environment initially free of devastating diseases and parasites, with very few predators and no competing large herbivores. Mammalian communities can be analysed in terms of niche occupancy, understanding niche as the status of an organism within the community, the space it lives in and what it does. Considering that bovines thrive best in exploited open habitats with their profound adaptation for grazing, the open niche left by the herbivore extinction, an extinction without replacement, facilitated the bovine radiation in the New World.[62]

The escaped animals reverted to a very different type from what the Europeans released. Cattle became feral 'Cimarrón' (Venezuela) or 'Bagual' (Brazil); these animals range farther and faster; are smaller; go farther without water in dry conditions; use shrubs as fodder; are resistant to parasites and hemo-parasites; grow slower but have a high fertility; are able to actively defend themselves from large predators, such as jaguars and pumas;[63] are long lived and are able to extract more nutrients from coarse grasses than their modern counterparts (in a nutshell: fast, lean and mean).

These animals have undergone 400 years of adaptation and natural selection without human interference. In each area, cattle evolved defined phenotypes and behaviours that facilitated their survival. We have observed that, for example, in Pantanal, the Pantaneiro Criollo cattle are able to feed on grasses that are under the water surface, sticking their heads under the water to reach their food. They are able to swim for 3 hours without touching land. In the Colombian Llanos, the calves of the Casanareño Criollo very rarely die of maggot infections in their navel, because females constantly check and lick the navel area of newborns, cleaning every infestation. Cows of both breeds eat the placenta after birth, to avoid attracting predators and carrion eaters by the odours. Calves are born small and are very active soon after birth. These features are exclusive of these breeds, and do not manifest in the original Iberian stock.

These cattle were the basis of the extensive livestock production system so characteristic of the Spanish colonies. We need to emphasise that, with the Iberian cattle, South America also inherited mediaeval Iberian ranching ways, which have changed little in time and geography. It is important to understand the extractive Nature of these enterprises. The art of handling cattle under open range conditions, by a human on horseback, and the free movement of herds from one area to the other was carried on from mediaeval Spain to the American continent.[64] In colonial times, the main economic value of the feral herds was the hide and the scarce subcutaneous fat of these animals; meat was an additional bonus, usually salted in long and thin strips but mostly left to rot. In better-established ranches, the usual method of management was a roundup (usually twice a year) to cull the herd. Selected animals were herded in large herds driven on horseback for many days, to consumption centres in *comitivas* (Brazil) or *arreos* (Venezuela).

Cattle and horses of the Llanos, as well as the Llaneros (plainsmen), were key to Venezuela's independence from Spain. They and their cattle and horses played a very important role in the independence of Colombia, Ecuador, Peru and Bolivia as well. It was the abundance of meat and horses of the Llanos that supported the war; in a similar way, Cracker cattle supported the first sound economy in the state of Florida, after the Civil War. This richness of resources is hard to fathom in environments such as the Llanos or Pantanal. However, the amount of time cattle-ranching has been the main economy of these ecosystems, and the abundant presence of wildlife, are indicative that the present regime has merit that should not be changed in favour of other economic schemes, especially not large-scale monocultures of cereals and grains.

## Native Grass versus Non-Native Grass

The total world area occupied by grazing is about 26% of the ice-free terrestrial surface of the planet. More worrying is that the total area dedicated to feed-crop production amounts to 33% of total arable land.[65] Seventy percent of previous forested land in the Amazon is occupied by pastures, and feed crops cover large parts of the remainder.[66] Deforestation goes hand in hand with the growing demand for beef. The livestock sector in Latin America has grown at an annual rate of 3.7%, which is higher than the average global rate (2.1%).

Llanero in Venezuela. Working oxen and llaneros form a partnership; the oxen help the roundups, cross watercourses and corral the cattle herds. Llaneros refuse to send old oxen to the slaughterhouse, as they are considered true working partners. (Photograph by Rafael Hoogesteijn.)

While the increase in poultry and pork satisfies an internal demand, beef production in Latin America satisfies export demands.[67]

The creation of new pasturelands is not the only problem. As the demand of meat increases around the world, so does the intensive cattle production based on exotic pasture grasses and the demand for crop products dedicated to the intensive, large-scale production of meat on a feedlot system. Between 1994 and 2004, the land devoted to growing soybeans for exportation to China more than doubled to 39 million hectares (150,580 sq. miles).[68] The demand for feed from countries with relatively abundant land (even if this land is not suitable to agriculture) is staggering and does not agree with the 'sustainability' discourse so frequently discussed in political arenas nationally and internationally. Cattle are ruminants, and as such able to convert grasses of low biological value into proteins of high biological value. To change the diet of a ruminant to the diet of a single-chambered stomach animal (i.e. a pig), based on soy and/or corn, for the sake of quick, high revenues in a feedlot system is obscene, an insult to cattle, the consumer and above all to Nature.

South America's grassy biomes have historically been considered a degraded form of human-created forest, caused by tree clearing, burning and grazing, or else a subclimax of secondary successional stages.[69] However, science has shown that the occurrence of natural savannas on previously forested areas, without changes in the original environment, has not been documented anywhere in the Neotropics.[70] Archaeological studies of pollens and other spores from the Holocene and the Upper Pleistocene show almost continuous replacement of forests and savannas, explained by climatic changes, geomorphological processes and gradual modification of soil and water conditions.[71] Sarmiento reported that no forests were converted to savannas or vice-versa during historical times.[72]

Savannas are always dominated by perennial grasses and have much smaller vegetal mass than forests (Savannas: 100–200t ha$^{-1}$ [40–80 tons per acre]; forests: 300–500t ha$^{-1}$ [121–202 tons per acre]).[73] In this way, the contents of minerals are smaller in savannas than in forests, and grasses are well adapted to nutrient shortages.

In tropical forests, there is an efficient recycling and direct transference of nutrients from decaying plants via decomposition and mycorrhizae. If forested areas are deforested, this cycle is lost. Additionally, the secondary plant formations replacing the original forests differ completely from primary savannas. These successions are equally subjected to fires and grazing and render the ecosystem fragile and less able to recover.[74] The lack of nutrients in the soil tends to facilitate the growth of hard-leaved vegetation which copes better with fire, but makes poor forage.[75]

Introduced grasses dominate these pastures, mainly from Africa with different and improved varieties like switchgrass, giant thatching grass, molasses grass and signalgrass. The diversity of grass species tends to be limited, and weeds and pests are a problem. The stability of these pastures tends to be low, and their development is dependent on constant herbivore grazing, fertilisation, soil alkalisation, use of fire and overall agronomical management. We need to insist on the difference of these two kinds of grasslands: natural and human-influenced pastures. The authors do not support the replacement of forests and jungles with pasturelands or intensive monocultures, such as African oil palm, rice, cotton, soy or sorghum.

## Sustainable Management of the Llanos and Pantanal: Ecological Promise

How can wetlands, like the Pantanal and the Llanos, survive through the progressive destruction and deterioration produced by the demands of development, agriculture and human activity? Through a brief review, we established that the Llanos and Pantanal are old grasslands biomes, not of anthropogenic origin, definitively influenced by fire and herbivory, but able to support a large mammal biomass.[76]

Today the Llanos and Pantanal present a low wild-mammal biomass (<1000 kg/km$^2$; 9 pounds/acre) when compared with tropical forests (946–4431 kg/km$^2$; 8–40 pounds/acre), but high cattle biomass (7600 kg/km$^2$; 68 pounds/acre and 3750 kg/km$^2$; 33 pounds/acre, respectively).[77] Schaller indicates that cattle and wildlife can coexist, as long as an area retains a mosaic of habitat.[78] This is especially important considering that it is possible to produce meat whilst conserving this mosaic in which South American fauna thrive. Traditional cattle-ranching has been demonstrated to have close to no impact on the macro-invertebrate populations (ants, termites, earthworms).[79]

Historically, conservation policy and practises assume that protected areas are the foundation of all efforts to protect biodiversity. Unfortunately, evidence shows there has often been a systematic downgrading, downsizing and loss of legal protection for entire national parks and other designated protected areas.[80] In fact, protected areas in most countries are subjected to land-cover degradation, leading to their not achieving their intended goal.[81] In lieu of this situation, and the huge extensions of land in private hands, cattle ranches are becoming important reservoirs of wildlife in South America.

Flooded savannas provide services that regulate the main natural cycles, such as water, carbon and nitrogen. These services include maintenance of watershed functions (infiltration, flow, soil protection), pollution removal (sequestration of carbon and chemical pollutants) and provision of habitat for wildlife. For humans they provide food, materials and water and aesthetic cultural and recreational values. So far, we have not been able to create any device able to transform grass into meat or milk; we are dependent on livestock.

Until approximately 60 years ago, the traditional extensive livestock-grazing enterprise did not employ fertilisers, pesticides, machinery, improved animals or introduced grass varieties. We find that we now face such threats as replacement of livestock production by agriculture with high rates of clearance, increased land use intensity, woody encroachment and disruption of the flood season regimes.[82]

Between these two extremes, there is a system of medium-intensity management that involves simple management (burning and controlled grazing), low-level technological inputs (genetic, reproductive and sanitary programmes, livestock record keeping, strategic feeding) that lead to notable increases in productivity in comparison to the extensive Iberian inherited system, but that also protect the environment without large environmental changes, producing enough revenue to support a stable economy.

To maintain the Llanos and Pantanal in their current productive but sustainable forms, we need to analyse the principal actors.

### Water

Hydrological patterns are the overriding factor in structuring floodplain communities. However, water cycles, more specifically rainfall floods, originate far away. For example, the Amazon controls the summer rainfall in the Pantanal; therefore, deforestation in the Amazon has deep implications for water conservation and ecosystem services in Pantanal.[83] Another problem is the severe siltation caused by the intense erosion on the sandy highlands, triggered primarily by intensive agriculture and by mining, especially on the lower Taquari River, which is the largest sheet flood fan in Pantanal. This affects over 11,000 km$^2$ (6,835 miles) of lowland, greatly changing the hydrology (i.e. from wet-and-dry to permanent wet) and consequently the fauna and flora. This rise in flood level and water table kills the riparian forest and woods and changes seasonally flooded grasslands to permanent marshes.[84]

Contamination with pesticides and heavy metals is another problem. Gold mining with mercury amalgamation has thrived since 1980, affecting the entire Pantanal basin. Mercury concentrations in the sediments were higher than the baseline world levels; however, it is important to note that only approximately 2%–8% of mercury was recovered in the sediment; the remaining mercury was lost to the atmosphere or downstream areas or stored in the biota.[85] Jaguars in the northern Pantanal have higher mercury levels in hair samples than jaguars in southern Pantanal.[86] Large-scale agricultural development of upland savanna introduces toxic chemicals into the floodplain. Laws regulating the use of agrochemicals are difficult to enforce, and persistent pesticides (including disulfan, endosulfan and thiodan) were freely used and sold[87] and some probably still are.

In 1980, Argentina, Brazil, Bolivia, Paraguay and Uruguay proposed a plan called the Paraguay-Paraná Hidrovía. Its purpose was to construct a navigable channel between the Paraguay and Parana Rivers, integrating the La Plata River Basin. By 1997, a panel of experts concluded that the engineering studies were incomplete, and the long-term hydrological implications and the environmental assessments were lacking.[88] In 1998, Brazil's Federal Environmental Agency announced that it had abandoned plans for construction activities along the Brazilian portion of the waterway. The Brazilian court system ordered a suspension of all federal government studies and constructions related to the Hidrovía. However, by 1998, eleven ports in the Paraguay River in the northern Pantanal (in the State of Mato Grosso) had been constructed for transportation interests, agribusiness and mining sectors. Construction companies continue to emphasise the potential economic gains of commercial waterways. Dredging and river straightening activities will affect the hydrology and ecology of the Pantanal.

These small projects, less subject to planning and environmental oversight, may produce worse outcomes than the Hidrovía.[89] The cumulative effect of smaller projects may be avoided only if planners, politicians, scientists and engineers adopt a large-scale, holistic perspective.[90] The Llanos have also been affected by the construction of dikes. In the 1970s a network of low dikes, the 'Modulos de Apure', were constructed. Their purpose was to control flooding during the rainy season and save water during the drought. The transformation altered the hydrologic flood/drought cycle of the savannas, as an artificial increase in flooding almost eliminated drought. The dikes reduced plant diversity by half, but the project benefitted livestock as well as aquatic and wetland fauna. Unfortunately, the programme was never thoroughly evaluated. At the moment (2019) it is paralysed, basically for lack of administration, control and maintenance.

## Grazing and Overgrazing

An important point to consider when protecting the land from the impact of cattle is that every ranch has particular biological, geographical and geomorphological conditions, and ranchers need to adapt and steward their land accordingly. It encompasses science, but it is also an art, supported by observation, learning and experience. Apparently, the bovine herds that found a nearly empty niche on the natural grasslands are not damaging to the environment when the rules of natural selection are applied; otherwise, it would not have been possible for cattle to adapt so well in the new continent.[91] Calculations of the number of native herbivores that were present before the Pleistocene indicate there was an average of 21 animals per km$^2$ (247 acres), with a variation of 15 to 50 animals per km$^2$ (247 acres), each weighing approximately 450 kg (992 pounds). This would mean there was an average of 5 hectares (12 acres) per animal, with a range of 2 to 7 hectares (5–17 acres), suggesting a large carrying capacity, which would explain why the presence of livestock did not drastically change natural habitat in South American grasslands.[92]

One seminal point, which is often overlooked, is the maximum sustainable cattle concentration – in one word: overgrazing. About 20% of the world's pastures and rangelands, with 73% of rangelands in dry areas, have been degraded to some extent through overgrazing.[93] Ranches everywhere frequently exceed the maximum, sustainable concentration of cattle. The abundance of grassland and woody vegetation is very dynamic. The public in general is usually more concerned with deforestation without understanding that although it is a huge problem, woody plant encroachment can be as damaging. The phenomenon of woody plant encroachment contrasts with the idea of relentless deforestation, because to the uneducated eye the more shrubs and trees there are, the better it is. Numerous factors – climate, fire, grazing/browsing regimes, carbon dioxide concentrations and nitrogen deposits – interact in savanna dynamics. One or several of these conditions have to change for woody plant encroachment to occur.[94] Unpalatable shrubs can replace grasses; the proliferation of trees and shrubs threatens the savanna ecosystems and the plants and animals in them.

Grazing and fire regimes and competitive interactions among plants are more dominant in humid regions. The preferential use of grasses by grazers create opportunities for woody plants to establish themselves, via reduction of competition, and persist thanks to reduction of fuel mass.[95] It is most important to understand the fine balance between controlled grazing adapted to the ecosystem, which limits shrub encroachment, and heavy livestock grazing which reduces ground cover and promotes shrub recruitment.

Overgrazing is usually associated with cattle; however, it is a problem that can rear its ugly head wherever there are hoofed mammals. Elephants and buffalos can overgraze a National Park as much as a herd of sheep can overgraze an English paddock. The vegetation changes may not produce great variation in biodiversity, but they cause great changes in community structure. The proliferation of woody plants may create a virtual monoculture, with little or no understory (e.g. morning glory, arrowleaf sida). Small animals, especially birds and arthropods, respond to the structural changes when woody cover exceeds species-specific thresholds. Populations of grassland-associated species are displaced, and communities shift from grassland-associated species to shrubland-associated species.[96] The costs of brush management are high and their effectiveness tends to be low. Prevention, and not 'cure', should be the prevailing management practise.

With increasing global scarcity of land for agriculture and increasing food demand, land use intensity has increased. There is a definitive trend of shifting pasturelands to commercial agriculture, even in lands that have been traditionally perceived as being of marginal agricultural value. Technical innovations in managing highly weathered tropical soils and the breeding of suitable crop varieties (introduced exotic grasses, African oil palm, rice, corn, sorghum, cotton and soy, amongst others) are transforming grasslands in the Llanos and in the Pantanal into immense agricultural monocultures devoid of any other type of vegetation and almost devoid of wildlife.[97]

### Introduced Grasses

Since the early 1970s, ranchers, the owners of the Llanos and Pantanal, have cleared land and planted pastures in order to increase cattle stocking rates and replace Creole cattle with Zebu cattle. Deforestation and introduced pastures may conflict with conservation; however, there must be a careful consideration of the introduction of pastures. Native grasses (more than 215 species) have a particular distribution related to topography, soil and flooding characteristics. Grass composition mostly depends on where it grows.[98] The best-quality grasses are found in the low, inundated areas available when waters recede. Cattle equally do not use all forage; feeding behaviour is very dependent on the quality of the forage. Bovines prefer to forage in the lower parts of the savanna, where smaller grasses with higher nutritional value grow. This produces areas of intense foraging and areas not foraged enough (areas with sandy soils and little organic material).

The accumulation of dry grasses in these little foraged areas enables intense fires that are difficult to control. It is therefore extremely important to facilitate cattle distribution. There are many ways to manage cattle, such as to divide pasturelands into smaller paddocks, separate herds by age and productivity category, dig water holes in specific areas and rotate cattle through the fields, thus allowing grasses to regrow. One such option is to introduce grasses in the areas where cattle do not naturally forage, because of the high tendency of plants to become woody and the low quality of pasture. These are usually the higher areas of the ranch. The ideal practise would be to plant native South American grass families (e.g. *Mesosetum spp*).[99]

Unfortunately, there is a knowledge vacuum on the management and use of native grasses. We need better understanding of such subjects as germoplasm banks, management forage characterisation, growth requirements and yields, nutritional values, forage dynamics by cattle and wildlife and association with legumes and mycorrhyzae. It is important to avoid any deforestation when planting introduced grasses as to guarantee plant diversity. Additional advantages to the minimum tillage practises are shade; browsing plants, especially in the dry season; landscape fidelity; low cost in the establishment of pastures and last, but not least, the conservation of local fauna[100]. Areas with introduced grasses should be used mainly for cattle in need of a boost, like cows that weaned a calf and will enter the next breeding season, bulls after a breeding season, first-year heifers, first calving heifers and so forth.

## Cattle

Southern Iberia was the only part of western Europe in Renaissance times in which open-range ranching was common, the technique that would come to characterise ranching in America.[101] The constant use of the horse, periodic round-ups, overland drives and eventually branding, are our Iberian heritage. Learning how to produce beef cattle in a manner that will not cause environmental degradation or habitat alteration is the main challenge for the Llanos and Pantanal.

The first step for change must be to recognise that the Spanish mediaeval inheritance, which still prevails in many ranches on the American continent, needs to be put in the past. The weekend rancher or landlord (communal or not), ignorant of even the number of livestock in his or her care, conducting occasional roundups or check-ups when in need of cash, or every blue moon, has to become history. Extensive cattle-ranching does not mean abandoning cattle to their own devices and collecting surplus animals once in a while. Cattle-ranching in flooded savannas is a well-developed, scientifically based discipline in the animal sciences.[102]

It is not the aim of this paper to describe the best practises. However, such activities as periodic check-ups and inventories with record keeping, sanitary programmes with proper vaccination and deworming schedules, reproductive programmes (cows and bulls) with well-established short breeding seasons, nutritional programmes with careful grass management and mineral supplementation, respecting the maximum sustainable herbivore concentration and careful application of prescribed burning and genetic selection applying pressure on the best traits for a flooded savanna ecosystem maximise productivity without major ecological changes to the savanna that could jeopardise the natural equilibrium. In relation to the genetic stock available, evolution has favoured floodpulse-adapted cattle breeds (Creole breeds), such as the Tucurá, the San Martinero, Romosinuano and Casanareño.

Livestock management practises need to be associated with marketing strategies that provide added value to products from the Llanos and Pantanal, such as organic-certified beef production free of mad cow disease, antibiotics and hormones. An added bonus of the coexistence of cattle with other herbivores is that the minimal habitat modification allows wildlife to be abundant.[103]

Studies conducted on a ranch in Pantanal compared wildlife biodiversity and abundance in three systems: rice paddies, cattle paddocks and forest. This study demonstrated higher diversity and abundance in cattle paddocks than in rice paddies. Intuitively this would not cause surprise; however, we cannot forget that flooded savannas' fauna is specifically adapted to flooded environments (e.g. rice paddies). Even so, wildlife diversity and abundance was higher in the cattle sector of the ranch.[104]

Such studies confirm that extensive cattle ranching can coexist with wildlife conservation, especially when land stewards prevent hunting and wildlife harassment. The presence of wildlife in ranches has an additional bonus: the development of tourism. Although experience shows that tourism represents a mixed blessing, given its limited potential and sometimes-negative impact, it remains a good incentive to protect natural areas in private properties, especially because tourism activities provide a far more persuasive argument for wildlife conservation compared to ethical or existential values.[105] There are numerous reports of the profitability of tourism and cattle ranching in the Llanos and in Pantanal in their different forms: bird-watching, cultural tourism and wildlife observation, to name a few.[106]

One of those examples is especially striking because there has been a positive shift in rural communities' attitude toward one of the most feared, vilified and persecuted species: the jaguar. A comparative study conducted by Tortato and collaborators showed that in terms of observed costs and benefits of providing safe habitat to a jaguar population, the benefits accrued from tourism far outweighed the costs of cattle losses due to predation.[107] Jaguars generated annual revenue 56 times higher than the annual, hypothetical damage caused within cattle ranches with an area equivalent to that required to support a jaguar population under conditions of frequent human contact. That people are willing to pay a fortune ($150 to $900 US dollars/day) to observe a jaguar is no surprise. What is surprising is that the local population dependent on jaguars are truly opposed to jaguar hunting, and their constant presence prevents hunting activity.[108]

Although the presence of tourist operators in the river observing wildlife in general and jaguars in particular, especially during the dry season, does not mean they need to enter private property, the situation has nevertheless brought conflict at times between tour operators and ranchers. Tour operators capitalise on a high-density jaguar population; the landowners, whose land helps to protect the jaguars, incur costs because of cattle predation. Local cattle owners do not receive a financial benefit from any level of jaguar tolerance. How could this situation be ameliorated? Eighty percent of all tourists visiting the study site were willing to donate additional funding to develop a compensation scheme. There are many logistical difficulties to overcome in such schemes. However, the seminal point is that tourists, tour operators and landowners can be on the same page, and where there is a will there is a way.[109]

## The Human Factor

In the book of Genesis it is written, 'And God gave man dominion over the animals'. But what if that was a mistranslation, and the phrase should read: 'And God gave man responsibility/stewardship over the animals'?

What people do about their ecology depends on what they think about themselves in relation to the things around them. Ecosystem management or human ecology is conditioned by beliefs about Nature and destiny – that is, by religion. Our habits and actions are dominated by the faith in perpetual progress, rooted in a Judeo-Christian teleology teaching that the reason for Nature's existence is for human use and benefit. Man named all the animals, thus establishing his dominance over them. Christianity established the dualism of man and Nature, and insists that it is God's will that man exploit Nature for his proper ends.[110]

Many ranchers approach their work as stewards of the land entrusted to them by their families and society; however, many others ascribe to a formula based on extraction, a probable backlash of our Christian and Iberian inheritance (for Gold, Glory and God …). Traditionally, the free seasonal micromigrations of herds in unfenced ranches from wet lower areas to dry higher areas ensured a substantial nutritional level during the whole year, despite the harsh conditions imposed by the drought and flooding periods. Now properties are becoming smaller and fenced in, the subdivision of the terrain due to inheritance customs. Smaller ranches limit cattle movement, and the younger generations of owners have higher economic expectations.

Pantaneiro and Llanero descendants are now educated in cities and have become professionals. They usually do not want to go back to work on the ranch of their forefathers, eating beef jerky with rice and beans, enduring the rains, mud and bugs. Ranches are sold to rich investors with no cattle ranching culture, and adverse land uses are imposed on their new properties, sometimes with disastrous results. In the northern Pantanal, many ranches were abandoned after the great floods of 1974, and these lands will probably fall victim to soybean production. The new generation of ranchers and investors clear land and plant pastures in order to increase the cattle stocking rates.

This is perceived as the optimal land use, and is enforced by government policies. It is taught in schools of agronomy, and encouraged by multinational companies that sell the technological knowhow package. But the ecological integrity of the Llanos and Pantanal, and supporting the livelihoods of people who live in these landscapes, should not be viewed as competing demands.

Another big problem ranchers face is the staggeringly low meat prices compared to the prices of supplies. A downloaded recount from the Brazilian Instituto de Economia Agrícola (Sao Paulo) describes the last 15 years, when meat prices increased 110%; however, all supplies (salt, diesel, ploughs, etc.) increased on average approximately 508%.[111]

We need to add a special section related to the actual situation in Venezuela as of 2018. All productivity has ceased under the current political circumstances. Cattle ranches have been especially affected since 2001, when every private initiative was affected by the actions against land tenure, protected by the Land and Agrarian Development Decree (Official Gazette No. 37,323, November 13, 2011). This decree, which has the force of law, states: 'Three basic levels of productivity are established: idle or uncultivated land/farm/ranch, improvable land/farm/ranch and productive land/farm/ranch. The lands qualifying as idle or uncultivated are those that do not meet the minimum production requirements; in this sense, they can be subject to agrarian intervention or expropriation and will be taxed with a tribute; this tax and the eventual intervention or expropriation, more than a punishment for unproductiveness, seek to be a means through which they are put into production'. The process of expropriation is covered in Article 37: 'Any citizen may submit a reasoned complaint to the respective Regional Land Office, when he or she becomes aware of the existence of idle or uncultivated lands'.

The minimum production requirements are nowhere established in the law, and the Official Land Registry, where the complaint is filed, delivers the final decision. This policy has led to the expropriation of productive ranches in the Llanos by cronies of the government, those with close personal or family ties to the politicians in power. Their interests and short-term benefits, and not the long-term management of the land, dictate their influences on political processes. Areas that maintained excellent wildlife populations with well-established cattle-ranching and ecotourism facilities disappeared. Private conservation actions have become nonexistent, as has any productive investment in rural areas. Oral communication with collaborators and scientists who still work in the country make it clear that if the ecological disruption and wildlife mismanagement continues, as in the last 5 years, there is a great probability that threatened species will disappear. In all of Venezuela's history there has never been such a condition of total devastation as in the last 18 years.

## Solutions to Create Sustainable Savannas

We have described how misplaced policies and social pressure affect the flooded savannas of the Llanos and Pantanal. Yet, in each place, large areas of savanna remain relatively intact with most native species still present and the habitat structure still relatively healthy. Each of these places represents an opportunity to protect or rehabilitate the savannas. Because of the demographic pressure the savannas will receive, it is of utmost importance to establish effective spatial planning and regulations, to protect the ecosystems from opportunistic agricultural expansion. We would recommend the following actions:

a. Further understand and protect the hydrology of flooded savannas. The far-reaching influence of the Brazilian Central Plateau in the case of the Pantanal is especially important. Mining, deforestation, urban development, dams, hydroelectric plants, ports, agricultural erosion and pollution all conspire against needed annual flooding with good-quality water.
b. Recognise that the soils of the flooded savannas are not conducive to long-term agriculture, even if new adapted varieties of crop plants were developed. The richness of the savanna lies in its native pastures, which have adapted to the special conditions of the savannas since before the Holocene era.
c. Produce beef cattle in a sustainable manner without causing degradation or habitat alteration. The technologies have been developed and have been proven to be applicable and effective. This does not necessarily mean the maximisation of beef production. Certain points need refining, such as the rescue of floodpulse-adapted cattle (Creole cattle), which have been replaced by the Zebu cattle. Creole cattle provide a superb foundation to breed high-yield cattle adapted to the local conditions. Experiences in the Colombian Llanos have demonstrated the economic impact of such programme.
d. Establish a research priority to study the native grasses and their natural history, as well as analysis of feeding preferences by cattle and wildlife. DNA metabar coding could enhance understanding of what species use which plant resources. The development of educational opportunities for ranchers to appreciate and recognise the problems of woody plant encroachment and how to manage it should be a priority in government and academia extension programmes.
e. Continue diversification of activities, such as ecotourism, with their multiple facets (cultural, adventure, bird-watching, wildlife watching etc.). These activities are an excellent motivation for ecosystem protection. The business of tourism seems to be insatiable, one of the few economic sectors still growing. Pantanal (and maybe the Llanos in the future, once the political situation has stabilised) may as well reap part of the economic benefits.
f. Support wildlife, which is essential to ecosystem stability. If ranchers have to live with the costs wildlife produces, then society as a whole needs to support landowners. Contrary to the current practise of compensation for losses produced by wildlife, we support the idea that ranchers should receive substantial economic support when wildlife is present and abundant on their land. For example, instead of paying a rancher for the loss of cattle to carnivore predation, ranchers should be rewarded for the presence of carnivores in their land. Such programmes should not be too difficult to put into effect, given the modern technologies available. The law, in most Latin American countries, states that wildlife is the good of the nation. As such 'the nation' should be responsible for the guardianship of 'the good'.
g. Consider using market strategies that add value for sustainable and/or organically produced meat. Pantanal and the Llanos produce a superior quality of meat, free of mad cow disease, antibiotics and growth hormones. There is an international market for the consumption of 'organic' meat. Unions of ranchers should push at the national and international level for the recognition of such quality and receive revenues for a high-quality product. Effective and severe sanitary programmes for the control of foot and mouth disease should be enforced, to render South American meat saleable in the international markets.

A pair of jaguars lift their heads from a bank of grass in the Brazilian Pantanal. Male and female jaguars will spend several days together while the female is in oestrus. (Photograph by Rafael Hoogesteijn.)

h. Pursue creation of land conservancies amongst ranchers, which would greatly benefit the conservation of species that need large territories. Such organisations could also be considered when expanding tourism opportunities for guests. Pantanal and the Llanos are a mosaic of habitats, but the relentless division of land limits their functional capacity. Local land-protection organisations could negotiate activities to maximise the characteristics of every ranch.

i. Explore the best ways to end 'business as usual' in the Llanos and Patanal. Disastrous ecological backlash will probably occur if there are no major changes to current economic practises that are so destructive to the ecology of the flooded savannas. The usual attitude that considers that science and technology will solve all problems is not the answer. Despite Copernicus and Darwin, we do not feel we are part of Nature in our hearts. As long as we continue to behave as if we are superior to Nature, our environmental problems will increase. Since most people in the Llanos and in Pantanal have a Christian orientation, it would be advisable to ponder the teachings of Francis of Assisi, who believed in the virtue of humility, not only individually but also for man as a species. We need to believe in the democracy of all creatures.[112]

j. Last, but not least, we need to research ways to meet the challenges of population expansion. The profound challenges of hunger and environmental conservation will detonate in the tropics, because major projected population growth will occur in tropical nations.[113] This includes the Llanos and the Pantanal, Therefore, the wisest long-range action to protect Nature must be an aggressive and effective family planning programme. An estimated 214 million women of reproductive age in nonindustrialized countries want to avoid pregnancy but are not using a modern contraceptive method for many reasons. This needs to change, especially if women do not want more children or want to delay the next child. The gap between women's reproductive intentions and contraception behaviour needs to be eliminated, and doing so should be a priority in any health-related government programme.[114]

## Conclusion

De Chant calculated if every person in the world lived like the average United States citizen, we would need 4.1 more planets.[115] We don't have more planets; we only have this one. In the 1970s, the fur trade nearly drove spotted South American cats to extinction. However, Nature is incredibly resilient. When people understood that it was morally reprehensible to dress in spotted cat fur coats, and legal actions were taken against the fur trade, the cat populations rebounded. This story fills us with inspiration and hope.

Death starts when we lose hope. Young people are apathetic, depressed and full of rage because we have compromised their future. However, history has showed time and again that humans are able to overcome everything: wars, disease outbreaks, natural disasters, poverty. To survive we need to push our mental barriers. How can we respect Nature when half of humanity is fighting to survive and the other half is driven by greed? We suggest that the secret lies in establishing a goal. We propose this goal should be a wise way of producing food. The means to achieve the goal is work, work towards the well-being of Nature, based on our knowledge of history, biology and animal husbandry. The message: That we can make a better world for all living beings and thrive. Change will happen when the heart is in tune with the brain, when knowledge is in tune with compassion.

Only better land management will produce better lives and then, maybe Nature will be respected. In this chapter we presented alternatives of meat production that we demonstrated are in synchrony with flooded savanna ecosystems. We know that when production is in harmony with Nature, there is such a surplus that people are even willing to spare land to facilitate conservation. We should not forget we are intelligent, with a heart and a brain capable of generous selfless acts, rapturous art and technology.

## Notes

1. Foley, J.A., Ramankutty, N., Brauman, K.A. and others. 2011. 'Solutions for a cultivated planet'. *Nature*, 478:337–342.
2. Machovina, B., Feeley, K.J. and Ripple, W.J. 2015. Biodiversity conservation: The key is reducing meat consumption. *Science of the Total Environment*, 536:419–431.
3. Sarmiento, G. 1990. 'Ecología comparada de ecosistemas de sabanas en América del Sur'. pp. 15–56. In: *Las sabanas Americanas, aspectos de su biogeografía, ecología y manejo.* (G. Sarmiento, ed) Venezuela: Fondo Editorial Acta Científica Venezolana, Caracas.
4. Serna-Isaza, R.A. 2001. 'Teledetección para la cartografía de la vegetación de sabana'. pp. 81–96. In: *Agroecología y biodiversidad de las sabanas en los Llanos Orientales de Colombia.* (G. Rippstein, G. Escobar and F. Motta, eds) Centro Internacional de Agricultura Tropical, Cali, Colombia.
5. Grisebach, A.H. 1872. *Die Vegetation der Erde nach ihrer klimatischen Anordnung.* Vol. 2. Engelmann, Leipzig.

6. Sarmiento, G. 1996. 'Ecología de pastizales y sabanas en América Latina'. pp. 15–24. In: *Biodiversidad y Funcionamiento de Pastizales y Sabanas en América Latina.* (G. Sarmiento and M. Cabido eds) Venezuela: Merida. Cyted-Cielat.
7. Sarmiento, G. and Monasterios, M. 1975. 'A critical consideration of the environmental conditions associated with the occurrence of savanna ecosystems in tropical America'. pp. 223–250. In: *Tropical Ecology Systems.* (E. Medina editor). Springer. New York.
8. Sarmiento, G. 1990. 'Ecología comparada de ecosistemas de sabanas en América del Sur'. *Op. cit.*
9. Sarmiento, G. and Monasterios, M. 1975. 'A critical consideration of the environmental conditions associated with the occurrence of savanna ecosystems in tropical America'. *Op. cit.*
10. Sarmiento, G. 1992. 'A conceptual model relating environmental factors and vegetation formations in the lowlands of tropical South America'. pp. 583–601. In: *Nature and Dynamics of Forest-Savanna Boundaries.* (P.A. Proctor, Furley, J. and J.A. Ratter, eds) Chapman & Hall, London.
11. Sarmiento, G. 1996. 'Ecología de pastizales y sabanas en América Latina'. *Op. cit.*
12. Sarmiento, G. and Monasterios, M. 1975. 'A critical consideration of the environmental conditions associated with the occurrence of savanna ecosystems in tropical America'. *Op. cit.*
13. Pearson, D.L. and Beletsky, L. 2002. *Brazil, Amazon and Pantanal, the Ecotravellers' Wildife Guide.* Academic Press, London.
14. Sarmiento, G. and Monasterios, M. 1975. 'A critical consideration of the environmental conditions associated with the occurrence of savanna ecosystems in tropical America'. *Op. cit.*
15. Mooney, H.A. and Gulmon, S.L. 1979. 'Environmental and evolutionary constraints on the photosynthetic characteristics of higher plants'. pp. 316–337. In: *Plant Population Biology.* (O.T. Solbrig, S. Jain, G.B. Johnson and P.H. Raven, eds) Columbia University Press, New York.
16. Sarmiento, G. 1996. 'Ecología de pastizales y sabanas en América Latina'. *Op. cit.*
17. Sarmiento, G. and M. Monasterios. 1975. 'A critical consideration of the environmental conditions associated with the occurrence of savanna ecosystems in tropical America'. *Op. cit.*
18. (a) Tamayo, F. 1964. 'Ensayo en la clasificación de las sabanas de Venezuela': Universidad Central de Venezuela, Caracas; and (b) Vareschi, V. 'La quema como factor ecológico de los Llanos'. *Boletín de la Sociedad Venezolana de Ciencias Naturales,* 101:9–26.
19. (a) Sarmiento, G. 1984. *The Ecology of Neotropical Savannas.* Cambridge, MA: Harvard University Press; (b) Sarmiento, G. 1990. 'Ecología comparada de ecosistemas de sabanas en América del Sur'. *Op. cit.*; (c) Sarmiento, G. 1992. 'A conceptual model relating environmental factors and vegetation formations in the lowlands of tropical South America'. *Op. cit.*; (d) Sarmiento, G. 1996. 'Ecología de pastizales y sabanas en América Latina'. *Op. cit.*; (e) Sarmiento, G. and Monasterios, M. 1975. 'A critical consideration of the environmental conditions associated with the occurrence of savanna ecosystems in tropical America'. *Op. cit.*
20. van Hammen, T. 1983. 'The paleo ecology and paleo geography of savannas'. pp. 79–108. In: *Ecosystems of the World.* (F. Bourlière, ed) Amsterdam.
21. van Hammen, T. 1972. *Historia de la vegetación y el medio ambiente del Norte Sudamericano.* I Congreso Latinoamericano de Botánica, Mexico.
22. Sarmiento, G. and M. Monasterios. 1975. 'A critical consideration of the environmental conditions associated with the occurrence of savanna ecosystems in tropical America'. *Op. cit.*
23. Dubs, B. 1994. 'Differentiation of woodland and wet savanna habitats in the Pantanal of Mato Grosso, Brazil'. Vol. 1, *The Botany of Mato Grosso*: Betrona-Verlag.
24. Pearson, D.L. and L. Beletsky. 2002. *Brazil, Amazon and Pantanal, the Ecotravellers' Wildife Guide. Op. cit.*
25. Por, F.D. 1995. 'The Pantanal of Mato Grosso (Brazil) world's largest wetlands'. (H.J. Dumont and M.J.A. Werger, eds) Vol. 73, *Monographiae Biologicae,* Springer.
26. Klammer, G. 1982. 'Die Palaeowüste des Pantanal von Mato Grosso und die pleistozene Klimageschichte der brasilianischen Randtropen'. *Zeitschrift für Geomorphologie,* 26:393–416.
27. Por, F.D. 1995. 'The Pantanal of Mato Grosso (Brazil) world's largest wetlands'. *Op. cit.*
28. Ibid.
29. Silva, C.I. 1990. *Influencia da variaçião do nivel de agua sobre a estrutura e funcionamento de uma area alagavel do Pantanal Matogrossense* (Pantanal de Barao de Melgarço, MT).

30. Adamoli, J.A. 1986a. 'A dinamica da Inundações no Pantanal'. Anais do I Simpósio sobre Recursos Naturais e Sócio-Econômicos do Pantanal., Corumbá, MGS.
31. Sucksdorff, A. 1989. 'Desenvolvimento e Preservação. II'. Visão Nacional. Anais do I Congresso Internacional sobre Conservação do Pantanal, Campo Grande.
32. Por, F.D. 1995. 'The Pantanal of Mato Grosso (Brazil) world's largest wetlands'. *Op. cit.*
33. Solbrig, O.T. 1996. 'The diversity of the savanna ecosystems'. pp. 1–27. In: *Biodiversity and Savanna Ecosystem Process, A Global Perspective.* (O.T. Solbrig, E. Medina and J.F. Silva, eds) Springer, Berlin.
34. Coutinho, L.M. 1990. 'Fire in the Ecology of the Brazilian Cerrado'. pp. 82–105. In: *Fire in the Tropical Biota.* (J.G. Goldammer, ed) Springer Verlag, Berlin-Heidelberg.
35. Por, F.D. 1995. 'The Pantanal of Mato Grosso (Brazil) world's largest wetlands'. *Op. cit.*
36. Adamoli, J.A. 1986b. 'Fitogeografia do Pantanal. Anais do I Simpósio sobre Recursos Naturais e Sócio-Econômicos do Pantanal', Corumbá.
37. Coutinho, L.M. 1990. 'Fire in the ecology of the Brazilian Cerrado'. *Op. cit.*
38. Seidl, A.F., Vila-de-Silva, J.S. and Steffens-Moraes, A. 2001. 'Cattle ranching and deforestation in the Brazilian Pantanal'. *Ecological Economics*, 36:413–425.
39. Ibid.
40. Woodburne, M. 2010. 'The great American Biotic Interchange: Dispersals, tectonics, climate, sea level and holding pens'. *Journal of Mammalian Evolution*, 17:245–264.
41. Haynes, G. 2014. 'North American megafauna extinction: Climate or overhunting?' pp. 5382–5388. In: *Encyclopedia of Global Archaeology.* (C. Smith, ed) Springer, New York.
42. Fariña, R.A., Vizcaíno, S.F. and De Iuliis, G. 2013. *Extinction.* (J.O. Farlow, ed) Indiana University Press, Bloomington, IN.
43. Cione, A.L., Tonni, E.P. and Soibelzon, L. 2009. 'Did humans cause the late Pleistocene-early Holocene mammalian extinctions in South America in a context of shrinking open areas?' pp. 125–144. In: *American megafaunal extinctions at the end of the Pleistocene.* (G. Haynes, ed) Springer, Dordrecht.
44. Fariña, R.A., Vizcaíno, S.F. and De Iuliis, G. 2013. *Extinction. Op. cit.*
45. Fowler, M. 1998. *Medicine and Surgery of South American Camelids.* 2nd ed. Iowa State University Press, Ames, IA.
46. Redford, K. and Eisenberg, J.F. 1992. *Mammals of the Neotropics.* Vol. 2. The University of Chicago Press, Chicago.
47. Ojasti, J. 1991. 'Human exploitation of capybara'. pp. 236–252. In: *Neotropical Wildlife Use and Conservation.* (J.G. Robinson and K.H. Redford, eds) The University of Chicago Press, Chicago.
48. Ojasti, J. 1983. 'Ungulates and large rodents of South America'. *Ecosystems of the World*, 13:427–439.
49. Hoogesteijn, R. and Chapman, C. 1997. 'Large ranches as conservation tools in the Venezuelan Llanos'. *Oryx*, 31:274–284.
50. Christopher Columbus quote in: Crosby, A.W. 1973. *The Columbian Exchange: Biological and Cultural Consequences of 1492*, 2nd ed. Greenwood Press, Westport, CT.
51. Crosby, A.W. 2004. *Ecological Imperialism, the Biological Expansion of Europe 900–1900.* Cambridge University Press, Cambridge, UK.
52. Ibid.
53. Crosby, A.W. 1973. *The Columbian Exchange: Biological and Cultural Consequences of 1492. Op. cit.*
54. Ibid.
55. Crosby, A.W. 2004. *Ecological Imperialism, the Biological Expansion of Europe 900–1900. Op. cit.*
56. Rouse, J.E. 1978. *The Criollo: Spanish Cattle in the Americas.* University of Oklahoma Press, Norman, OK.
57. Gomez-Pernía, O.G. 2012. *Nuestra carne, origen, cualidades y culinaria de la carne bovina venezolana.* 2nd ed. Ediciones Grupo Tei C.A., Soluciones Gráficas, Caracas.
58. Crosby, A.W. 1973. *The Columbian Exchange: Biological and Cultural Consequences of 1492. Op. cit.*
59. Madeiros-Mazza, M.C., Silva-Mazza, C.A., Bezerra-Sereneo, J.R. and Oliveira-Pellegrin, A.O. 1994. 'Etnobiologia e conservação do bovino pantaneiro'. Corumbá, Mato Grosso do Sul: EMBRAPA Centro de Pesquisa Agropecuaria do Pantanal.

60. Rouse, J.E. 1978. *The Criollo: Spanish Cattle in the Americas. Op. cit.*
61. Ibid.
62. (a) Vizcaino, S.F., Cassini, G.H., Toledo, N. and Bargo, M.S. 2012. 'On the evolution of large size in mammalian herbivores of Cenozoic faunas of South America'. pp. 76–101. In: *Bones, Clones and Biomes, the History and Geography of Recent Neotropical Mammals.* (B.D. Patterson and L.P. Costa, eds); (b) Bocherens, H. 2018. 'The rise of the anthroposphere since 50,000 years: An ecological replacement of megaherbivores by humans in terrestrial ecosystems?' *Frontiers in Ecological Evolution,* 6 (3). doi: 10.3389/fevo.2018.00003.
63. Hoogesteijn, R., Payán, E., Valderrama-Vázquez, C.A., Tortato, F.R. and Hoogesteijn, A.L. 2016. 'Comportamiento del ganado criollo Sanmartinero y Pantaneiro: La experiencia brasileña y colombiana'. pp. 193–208. In: *Conflictos entre felinos y humanos en América Latina.* (C. Castaño-Uribe, C.A. Lasso, R. Hoogesteijn, A. Díaz-Pulido and E. Payán, eds) Bogotá, D.C., Colombia: Instituto de Investigación de Recursos Biológicos Alexander von Humboldt.
64. Rouse, J.E. 1978. *The Criollo: Spanish Cattle in the Americas. Op. cit.*
65. FAO. 2016. *Livestock Production in Latin America and the Caribbean.* http://www.fao.org/americas/perspectivas/produccion-pecuaria/en/ (accessed 21 June 2018.)
66. (a) Machovina, B., Feeley, K.J. and Ripple, W.J. 2015. 'Biodiversity conservation: The key is reducing meat consumption'. *Op. cit;* (b) FAO. 2016. *Livestock Production in Latin America and the Caribbean. Op. cit.*
67. FAO. 2016. *Livestock Production in Latin America and the Caribbean. Op. cit.*
68. Ibid.
69. Lehman, C.E.R. and Parr, C.L. 2016. 'Tropical grassy biomes: linking ecology, human use and conservation in *Philosophical Transactions Royal Society B,* 371. doi: 10.1098/rstb.2016.0329.
70. Sarmiento, G. 1992. 'A conceptual model relating environmental factors and vegetation formations in the lowlands of tropical South America'. *Op. cit.*
71. van Hammen, T. 1983. 'The paleo ecology and paleo geography of savannas'. *Op. cit.*
72. Sarmiento, G. 1992. 'A conceptual model relating environmental factors and vegetation formations in the lowlands of tropical South America'. *Op. cit.*
73. Sarmiento, G. 1984. 'The ecology of neotropical savannas'. *Op. cit.*
74. Sarmiento, G. 1992. 'A conceptual model relating environmental factors and vegetation formations in the lowlands of tropical South America'. *Op. cit.*
75. Rizzini, C.T. 1979. *Tratado de Fitogeografia do Brasil.* 2nd ed. Editora Interciencias, São Paulo.
76. Lehman, C.E.R. and C.L. Parr. 2016. 'Tropical grassy biomes: Linking ecology, human use and conservation'. *Op. cit.*
77. (a) Eisenberg, J.F., O'Connell, M.A. and August, P.V. 1979. 'Density, productivity and distribution of mammals in two Venezuelan habitats'. pp. 187–207. In: *Vertebrate Ecology in the Northern Neotropics.* (J.F. Eisenberg, ed) Smithsonian Institution, Washington, DC; (b) Schaller, G.B. 1983. 'Mammals and their biomass on a Brazilian ranch'. *Arquivos de Zoologia,* 25:1–36.
78. Schaller, G.B. 1983. 'Mammals and their biomass on a Brazilian ranch'. *Op. cit.*
79. Decaens, T., Jiménez, J.J., Rangel, A.F., Cepeda, A., Moreno, A.G. and Lavelle, P. 2001. 'La macrofauna del suelo en la sabana bien drenada de los Llanos Orientales'. pp. 111–137. In: *Agroecología y biodiversidad de las sabanas en los Llanos Orientales de Colombia.* (G. Rippstein, G. Escobar and F. Motta, eds) Centro Internacional de Agricultura Tropical, Cali, Colombia.
80. Mascia, M.B. and Pailler, S. 2011. 'Protected area downgrading, downsizeing, and degazettement (PADDD) and its conservation implications'. *Conservation Letters,* 4:9–20.
81. Leicher, C., Touval, J., Hess, S.M., Boucher, T.M. and Reymondin. L. 2013. 'Land and forest degradation inside protected areas in Latin America'. *Diversity and Distributions,* 5:779–795.
82. Lehman, C.E.R. and Parr, C.l. 2016. 'Tropical grassy biomes: Linking ecology, human use and conservation'. *Op. cit.*
83. Bergier, I., Assine, M.L., McGlue, M.M., Alho, C.J.R., Silva, A., Guerreiro, R.L. and Carvalho, J.C. 2017. 'Amazon rainforest modulation of water security in Pantanal wetland'. *Science of the Total Environment,* 1116–1125. doi: 10.1016/j.scitotenv.2017.11.163.

84. Pott, A. and Pott, V.J. 2004. 'Features and conservation of the Brazilian Pantanal wetland'. *Wetland Ecology and Management*, 12:547–552.
85. Leady, B.S. 2013. 'Historic patterns of deposition and biomagnification of mercury in selected wetland systems'. Doctor of Philosophy degree in Biology, University of Toledo, Toledo, OH.
86. May-Júnior, J. A., Carvalho-Junior, M.R., Frescura, V.L.A. and others. 2014. 'Mercury analysis in hair of free-ranging jaguars (*Panthera onca*) in Pantanal, Brazil'. Toxilatin 2014, 1st Latin American Congress of Clinical and Laboratorial Toxicology. Technological Development for the Advances of Toxicology and Promotion of Health., Porto Alegre, RS, Brazil. 27 to 30 April 2014.
87. Alho, C.J.R. and Vieira, L.M. 1997. 'Fish and wildlife resources in the Pantanal wetlands of Brazil and potential disturbances from the release of environmental contaminants'. *Environmental Toxicology and Chemistry*, 16:71–74.
88. Gottgens, J.F. Perry, J.E., Fortney, R.H., Meyer, J.E. and others. 2001. 'The Paraguay-Paraná hidrovía: Protecting the Pantanal with lessons from the past'. *BioScience*, 51:301–308.
89. Ibid.
90. Odum, W.E. 1982. 'Environmental degradation and the tyranny of small decisions'. *BioScience*, 32:728–729.
91. (a) Pott, A. and Pott, V.J. 'Features and conservation of the Brazilian Pantanal wetland'. *Op. cit*; (b) Rouse, J.E. 1978. *The Criollo: Spanish Cattle in the Americas. Op. cit.*
92. Guevara, S. 2012. 'Introduction'. pp. 125–130. In: *Grazing Systems and Biodiversity in Latin American Areas: Colombia, Chile and Mexico.* (S. Guevara and J. Laborde, eds) Revista de la Sociedad Española para el estudio de los pastos, Madrid, Spain.
93. FAO. 2016. *Livestock Production in Latin America and the Caribbean. Op. cit.*
94. Archer, S.R., Andersen, E.M., Predick, K.I., Schwinning, S. and others. 2017. 'Woody plant encroachment: Causes and consequences'. pp. 25–84. In: *Rangeland Systems.* (D.D. Briske, ed) Springer.
95. Ibid.
96. Sirami, C., Seymour, C., Midgley, G. and Barnard, P. 2009. 'The impact of shrub encroachment on savanna bird diversity from local to regional scale'. *Diversity and Distributions*, 15:948–957.
97. Lehman, C.E.R. and Parr, C.L. 2016. 'Tropical grassy biomes: Linking ecology, human use and conservation'. *Op. cit.*
98. Santos, S.A., Araujo-Crispin, S.M., Comastri-Filho, J.A., Pott, A. and Cardoso, E.L. 2005. 'Substitução de pastagem nativa de baixo valor nutritivo por forrageiras de melhor qualidade no Pantanal'. *Circular Técnica EMBRAPA Pantanal*, 62:5.
99. Ibid.
100. (a) Hoogesteijn, A.L., Lemos-Monteiro, J. and Hoogesteijn, R. 2010. 'El arado ecológico: una alternativa sustentable para la introducción de pasturas en las sabanas inundables neotropicales'. pp. 41–71. In: XXV *Cursillo Sobre Bovinos De Carne.* (R. Romero, J. Salomón, J. De Venanzi and M. Arias, eds) Universidad Central de Venezuela, Facultad de Ciencias Veterinarias, Maracay, Venezuela; (b) Santos, S.A., S.M. Araujo-Crispin, J.A. Comastri-Filho, A. Pott and E.L. Cardoso. 'Substitução de pastagem nativa de baixo valor nutritivo por forrageiras de melhor qualidade no Pantanal'. *Op. Cit.*
101. Crosby, A.W. 1973. *The Columbian Exchange: Biological and Cultural Consequences of 1492. Op. cit.*
102. Plasse, D. and Salom, R. 1985. *Ganadería de Carne en Venezuela.* Italgrafica SRL. Caracas, Venezuela
103. Hoogesteijn, R. and Chapman, C. 'Large ranches as conservation tools in the Venezuelan Llanos'. *Op. Cit.*
104. Teribele, R. 2007. 'Comparações entre taxas de encontro de mamíferos de médio e grande porte em focagens noturnas, em dois períodos sazonais, na Fazenda San Francisco' (Pantanal, Miranda, Mato Grosso do Sul). Masters of Ecology, Universidade Federal de Moto Grosso do Sul.
105. Chardonet, P.H., des-Clers, B., Fischer, J., Gerhold, R., Jori, F. and Lamarque, F. 2002. 'The value of wildlife'. *Revue Scientifique et Technique International Office of Epizootics*, 21:15–51.
106. (a) Hoogesteijn, A.L. and Hoogesteijn, R. 2010. 'Cattle ranching and biodiversity conservation as allies in South America's flooded savannas'. *Great Plains Research*, 20:37–50; (b) Hoogesteijn, R., Hoogesteijn, A.L., Tortato, F.R., Rampim, L.E. et al. 2015. 'Conservación de jaguares (*Panthera onca*)

fuera de áreas protegidas: Turismo de observación de jaguares en propiedades privadas del Pantanal, Brasil'. pp. 259–271. In: *Conservación de grandes vertebrados en áreas no protegidas de Colombia, Venezuela y Brasil.* (E. Payan, C.A. Lasso and C. Castaño-Uribe, eds). Instituto de Investigación de Recursos Biológicos Alexander von Humboldt, Bogotá, Colombia; (c) Hoogesteijn, R., Hoogesteijn, A.L. and González-Fernández, A. 2005. 'Ganadería y ecoturismo, dos actividades productivas, compatibles y sustentables en hatos de sabana inundable'. pp. 23–77. In: XX *Cursillo sobre Bovinos de Carne.* (R. R. Romero, J. Salomón and J. De-Venanzi, eds) Universidad Central de Venezuela. Facultad de Ciencias Veterinarias, Maracay, Venezuela.

107. Tortato, F.R., Izzo, T.J., Hoogesteijn, R. and Peres. C. A. 2017. 'The numbers of the beast: Valuation of jaguars (*Panthera onca*) tourism and cattle predation in the Brazilian Pantanal'. *Global Ecology and Conservation*, 11:106–114.
108. Ibid.
109. Ibid.
110. White, L. 1967. 'The historical roots of our ecologic crisis'. *Science*, 155:1203–1207.
111. Instituto de Economía Agrícola, Banco de Dados. http://www.iea.agricultura.sp.gov.br/out/Bancodedaos2.html (Accessed 28 August 2018).
112. White, L. 'The historical roots of our ecologic crisis'. *Op. cit.*
113. Laurance, W.F., Sayer, J. and Crassman, G. 2013. 'Agricultural expansion and its impacts on tropical nature'. *Trends in Ecology and Evolution*, 29:107–116.
114. WHO. 2018. *Sexual and Reproductive Health*. The University of Chicago Press, Chicago. http://www.who.int/reproductivehealth/topics/family_planning/fp-global-handbook/en/ (Accessed 19 June 2018).
115. De Chant, T. 2011. Global Footprint Network: National Footprint Accounts. https://persquaremile.com/2012/08/08if-the-worlds-population_lived_like/ (Accessed 28 August 2018).
116. Map shape file from: Olson, D. M., Dinerstein, E., Wikramanayake, E.D., Burgess, N. D., Powell, G. V. N., Underwood, E. C., D'Amico, J. A., Itoua, I., Strand, H. E., Morrison, J. C., Loucks, C. J., Allnutt T. F., Ricketts, T. H., Kura, Y., Lamoreaux, J. F., Wettengel, W. W., Hedao, P. and Kassem, K.R. 2001. 'Terrestrial ecoregions of the world: A new map of life on Earth.' *Bioscience*, 51(11):933–938.

## Appendix: Common and Scientific Names of Plants and Animals

*PLANTS*

**GRASSES**

| | |
|---|---|
| Corn (a.k.a. maize) | *Zea mays* |
| Koronivia grass | *Brachiaria humidicola* |
| Rice | *Oryza* spp. |
| Sorghum | Poaceae |
| Surinam grass (a.k.a. signal grass) | *Brachiaria humidicola* |
| Switchgrass | *Panicum* |

**FORBES**

| | |
|---|---|
| Arrowleaf sida | *Sida rhombifolia* |
| Cotton | *Gossypium* spp. |
| Morning glory | *Ipomea crassicaulis* |
| Soy | *Glycine max* |

**TREES**

| | |
|---|---|
| African oil palm | *Elaeis guineensis* |

*INVERTEBRATES*

**VIRUSES**

| | |
|---|---|
| Foot and mouth disease | *Aphthae epizooticae* |

**WORMS**

| | |
|---|---|
| Earthworms | Annelida |

**INSECTS**

| | |
|---|---|
| Ants | Formicidae |
| Bugs | Hemiptera |
| Termites | Isoptera |

*VERTEBRATES*

**REPTILES**

| | |
|---|---|
| Arrau turtle | *Podocnemis expansa* |
| Orinoco crocodile | *Crocodylus intermedius* |

**BIRDS**

| | |
|---|---|
| Macaws | Psittaciformes |
| Muscovy duck | *Cairina moschata* |
| Parrots | Psittaciformes |
| Domestic turkey | *Meleagris gallopavo* |

**MAMMALS**

| | |
|---|---|
| African (a.k.a. Cape) buffalo | *Syncerus caffer* |
| Alpaca | *Lama pacos* |
| Amazon river dolphin | *Inia geoffrensis* |
| Anteaters | Vermilingua |
| Armadillos | Cingulata |
| Bats | Chiroptera |

| | |
|---|---|
| Bears | Ursidae |
| Camels | Camelidae |
| Capybara | *Hydrochoerus hydrochaeris* |
| Creole cattle | *Bos taurus* |
| Domestic cattle (including oxen) | *Bos taurus* |
| Deer | Cervidae |
| Domestic cat | *Felis catus* |
| Domestic cattle (including oxen) | *Bos taurus* |
| Domestic dog | *Canis lupus familiaris* |
| Domestic goat | *Capra aegagrus* |
| Domestic horse | *Equus ferus caballus* |
| Domestic pig | *Sus scrofa domesticus* |
| Domestic rabbit | *Oryctolagus* spp. |
| Domestic sheep | *Ovis aries* |
| Elephants | Elephantidae |
| Ferrets | Mustelidae |
| Giant armadillo | *Priodontes maximus* |
| Giant river otter | *Pteronura brasiliensis* |
| Glyptodonts | Glyptodontinae |
| Guanaco | *Llamas guanicoe* |
| Guinea pig | *Cavia porcellus* |
| Jaguar | *Pantera onca* |
| Llama | *Llamas glama* |
| Marsh deer (a.k.a. swamp deer) | *Blastoceros dichotomus* |
| Marsupials | Marsupialia |
| Mule | *Equus* spp. |
| Ocelot | *Leopardus pardalis* |
| Opossums | Didelphidae |
| Pampas deer | *Ozotocerus bezoarticus* |
| Peccaries | Tayassuidae |
| Rodents | Rodentia |
| Shrews | Soricidae |
| Skunks | Mephitidae |
| Sloths | Pilosa |
| Spanish cattle | *Bos taurus turdetanus* |
| South American tapir (a.k.a. Brazilian tapir) | *Tapirus terrestres* |
| Tapirs | *Tapirus* spp. |
| Vicuña | *Vicugna vicugna* |
| White-tailed deer | *Odocoileus virginianus* |
| Zebu cattle | *Bos indicus* |

# 12

# *Ocean Rights: The Baltic Sea and World Ocean Health*

**Michelle Bender**

**CONTENTS**

World Ocean Health ..... 237
European Union Waters and The Baltic Sea ..... 239
Legal and Institutional Frameworks of the European Union and the Baltic Sea Region ..... 240
Ocean Rights: A Systemic Solution ..... 243
What if Fisheries Policies Evolved to Include the Rights of the Ocean? ..... 245
How Would Rights of Nature Affect the Criteria for Decision-Making? ..... 246
The Movement Is Growing ..... 248
United Nations (UN) ..... 248
International Union for the Conservation of Nature (IUCN) ..... 249
National Oceanic and Atmospheric Administration (NOAA) ..... 249
Food and Agriculture Organization (FAO) ..... 249
Shifting out of the Dated Governance Paradigm ..... 250
Conclusion ..... 250
Notes ..... 251

It should be obvious that we are a part of nature, not apart from nature, and that what we do to the living world we do to ourselves. We must protect the ocean as if our lives depend on it – because they do.

**Dr. Sylvia Earle**[1]

## World Ocean Health

The ocean. A blue unknown. Even though it comprises 70% of the planet, most of its internal workings and functions escape the notice of most humans. What is really happening underneath the ocean surface?

At its peak in 1996, an estimated 130 million tonnes of fish were pulled out of the ocean worldwide, only decreasing by 0.38 million tonnes each year.[2] As a result, 10% of large predatory fish, such as tuna, swordfish and marlin, remain,[3] and 30% of shark species are threatened with extinction.[4] And even though there has been an international moratorium on commercial whaling since 1986, some populations of the world's largest species have

DOI: 10.1201/9780429505959-12

yet to recover, such as the Western North Pacific grey whale, estimated to have fewer than 100 individuals left.[5]

Similarly, about 27% of coral reefs (and half the Great Barrier Reef off the Northeastern coast of Australia) worldwide have already been lost to ocean acidification and other climate factors such as warmer sea temperature and sea level rise.[6] Though coral reefs cover less than 1% of the ocean, they provide a home to 25% of known marine species.[7]

Humans also generate vast amounts of pollution that finds its way into the ocean, including plastic and carbon dioxide.[8] According to a study originating from the UC Santa Barbara's National Center for Ecological Analysis and Synthesis (NCEAS), every year 8 million metric tonnes of plastic ends up in the oceans – the same amount as five grocery bags filled with plastic for every foot of coastline in the world.[9] By 2025, it is estimated that the amount will have doubled, leading to more plastic than fish in the sea (in tonnes).[10] Plastic debris kills around 100,000 marine mammals, as well as millions of birds and fishes, annually though ingestion and entanglement.[11]

The oceans absorb one-quarter of all human-created carbon emissions. This leads to acidification, as it lowers the pH of the surface waters.[12] When the chemistry in the ocean is not working, marine ecosystems and the coastal economies, which depend on the ecosystems, don't work either. For example, as acidity levels in the ocean rise, the carbonate levels go down.[13] Calcium carbonate is required in order for reefs and shellfish to build and maintain their skeletons, and so acidification threatens the existence of these creatures. As many fish, seabirds and marine mammals depend on shellfish as food for their survival, acidification creates a threat to the whole marine ecosystem.[14]

Whale Flipper. (Photograph by Michelle Bender.)

Decline in ocean health does not just affect the ocean. One billion people worldwide rely on fish as their primary source of protein[15] and fishing is 'central to the livelihood … of 200 million people, especially in the developing world'.[16] As a result, the decline in marine biodiversity threatens millions of peoples' livelihoods and survival. For example, when the Newfoundland cod fishery closed in 1992, 40,000 jobs disappeared (10,000 being fishermen[17]) and entire communities virtually vanished.[18]

## European Union Waters and the Baltic Sea

The European Union (EU) is considered the fourth largest producer of fish and seafood in the world, and '[a]lmost half of Europe's population lives within 50 kilometers of the sea and regularly uses its resources'.[19]

The Baltic Sea is the youngest, yet most polluted, sea in the world.[20] It contains a uniquely biodiverse ecosystem, with 100 fish species and 4,450 species of macroalgae, plankton and zoobenthos (seabed animal), all adapted to its brackish waters.[21] Positioned in between Sweden, Finland, Russia, Germany and five other countries,[22] it is also an isolated sea, taking approximately 30 years for the waters to get fully exchanged.[23]

Around 85 million inhabitants directly depend on the functions and health of this water basin with limited exchange to the open ocean.[24] Human activities have affected the Baltic for thousands of years, both locally via land-use changes[25] and regionally, for example through the mining and fishing industries.[26] Over the past few decades, several disaster reports have been released concerning the state of the Baltic Sea.[27] Overfishing, toxic algal blooming and dead zones have emerged as the main threats to the health of the Sea.[28]

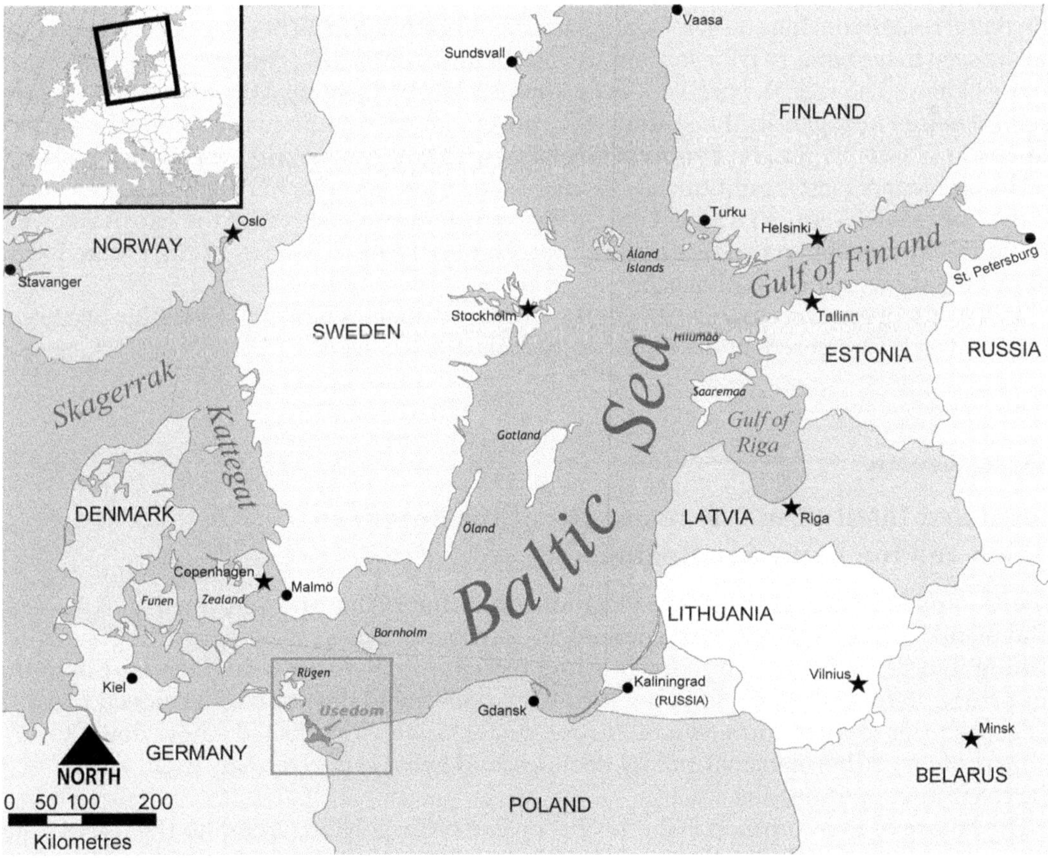


The Baltic Sea and the surrounding countries. https://commons.wikimedia.org/wiki/File:Baltic_Sea_map_Usedom_location.png

Studies show that 75% of the most important fishing areas in the Baltic Sea are either overexploited or fully exploited.[29] Fisheries in the Baltic Sea target both marine and freshwater species;[30] major industries include cod, herring and sprat.[31]For example, the Kattegat Baltic cod populations have decreased by more than 90% since 1981.[32] And recent data trends show that there have never been so few large cod caught,[33] a sign also exhibited before the Newfoundland cod-fishing collapse.[34] Decline in fish populations impact the entire food web, sending impacts up and down the chain of interdependent species, known as a trophic cascade.

Eutrophication is considered the threat most severely impacting the Baltic Sea.[35] High levels of nitrogen and phosphorus from land increases primary production in the Sea, reducing water quality and creating 'dead zones' or areas depleted of oxygen and life.[36–38] Parts of the Baltic have long suffered from low oxygen levels, but a recent study shows that oxygen loss in coastal areas over the past century is unprecedented compared to the last 1,500 years.[39] Human-induced pollution from fertilisers and sewage run-off is the main driver of recent oxygen loss in the Baltic, according to researchers.[40] In fact, the Baltic Marine Environment Protection Commission (HELCOM) found that 97% of the surface area in the Baltic Sea, from the Kattegatt to the inner bays, is eutrophied.[41] Such ecosystem changes severely impact the survival of many important species, such as cod. Identified as a major threat affecting cod populations,[42] eutrophication disrupts the 'balance of temperature, salinity and oxygen' that is 'critical for cod at different stages in their development'.[43]

Additionally, increased nutrients and temperatures have led to recurrent algal blooms in the Baltic Sea, rendering the sea's waters toxic.[44] Negative effects of Cyanobacteria blooms can be seen in humans, marine mammals and shellfish, impacting recreation and fisheries.[45] These blooms also feed the problem of eutrophication, by creating a positive feedback loop that increases nitrogen in the system, the movement of organic matter from the surface downwards and oxygen consumption by plants, to the detriment of the marine ecosystem.[46]

The problem of plastic pollution is also increasing quite severely in Atlantic coastal waters of the European Union. In a recent study, 95% of fulmars (a seabird) of the North Sea were found to have plastic in their stomachs.[47] And a horrifying 712 items of litter were found on just a 100-metre stretch of beach.[48]

As a result of human activity, 18% of the identified species who call the Baltic Sea home are listed as 'endangered' and 43% as 'vulnerable'.[49]

## Legal and Institutional Frameworks of the European Union and the Baltic Sea Region

The Common Fisheries Policy was introduced in the 1970s, and reformed in 2014. This framework guides fishery management in European Union waters, creating rules for fishing in order to 'foster a dynamic fishing industry and ensure a fair standard of living for fishing communities'.[50] However, the Policy does not appear to have achieved many of its goals: prior to the reform 88% of European stocks were overfished – now down to 40% – but profit for fishermen continually declines, and bycatch is extremely high, with 40% of the catch discarded dead at sea.

The failure of the Common Fisheries Policy can be largely attributed to the use of Total Allowable Catch based on a theoretical concept of Maximum Sustainable Yield. The Total Allowable Catches are 'on average 48% higher than those advised by scientists', with industry representatives holding two-thirds of the seats of the advisory committees that negotiate fishing limits. For example, the International Council for the Exploration of the Sea determined that the total allowed cod catch should be lowered by 28%, but the European Union ministers decided to only lower it by 8%.[51]

Additionally, according to the reform, the Maximum Sustained Yield is defined as 'the highest possible annual catch that can be sustained over time, by keeping the stock at the level producing maximum growth'.[52] It regulates according to a hypothetical equilibrium between the exploited population and the fishing activity. A focus on the short-term, maximum amount of fish we can take out of the system represents a deeply rooted problem in our worldview and legal systems.

The Maximum Sustained Yield metric is known to be 'devilishly hard to predict accurately',[53] and that does not take into account other factors affecting fish (such as other predation effects, unforeseen events and pollutants).[54] In the Baltic Sea environment, Maximum Sustained Yield does not take changes in density-dependent growth into account, which has recently resulted in drastic reductions in the Eastern Baltic cod stock.[55] Additionally, this metric focuses on 'mortality to be achieved'[56] rather than population and ecosystem health.

Due to the well-known problem of the tragedy of the commons,[57] and the many countries' policies affecting the Baltic Sea, governance has been challenging. Political frameworks must be agreed upon and implemented by nine countries. If one country continues to pollute and fish unsustainably, the Baltic Sea continues to suffer, which in turn diminishes the sea for all the other countries that also depend on it – a classic instance of the tragedy of the commons.

Boats in the Port of Miiduranna, Estonia, September 2016. Photograph by Pjotr Mahhonin/Wikimedia. https://commons.wikimedia.org/wiki/File:EK-0903_Kotkas_EK-9222_Narvia_FIN-145-O_Baltic_EK-1003_Viktory_Miiduranna_10_September_2016.jpg

The overall objective of the Baltic sea fisheries is to ensure economically, environmentally and socially sustainable fisheries and use of resources in alignment with an ecosystem-based approach.[58] However, an historical analysis of agreed Total Allowable Catches for all European Union waters between 2001 and 2016 shows that 7 out of every 10 allowable catch limits were set above the scientifically advisable level.

Additional frameworks addressing the Baltic Sea's problems include the Convention on the Protection of the Marine Environment of the Baltic Sea Area, the European Union Strategy for the Baltic Sea Region, the Water Framework Directive and Marine Strategy Framework Directive. The governing body for the Convention on the Protection of the Marine Environment of the Baltic Sea Area of 1992, the Baltic Marine Environment Protection Commission (HELCOM), contains representatives from all nine Baltic Sea countries and the European Union. HELCOM creates and implements common environmental actions, provides data and scientific information and monitors the measures to protect the Baltic Sea.[59]

The Convention on the Protection of the Marine Environment of the Baltic Sea Area seeks to reduce land-based pollution and aims to protect and enhance the environment of the Baltic Sea, through ecosystem-based management and implementation of the Precautionary and Polluter-pays Principles. The Precautionary Principle states that whenever a significant amount of uncertainty exists, actions must be precautionary ('when in doubt, err on the side of conservation'[60]). The Polluter-pays Principle requires that those who pollute and cause harm to the ecosystem must pay for the damage caused and subsequent restoration. It is therefore the contracting parties' obligation to implement the best available technology and environmental practises to reduce, prevent and eliminate pollution that not only harms human health but also the marine ecosystem and its living resources.[61]

The Convention has seen some success in the past 40 years, including reducing the frequency of oil spills, reducing industrial heavy-metal pollution, reducing the concentration of dioxins in wildlife, and recovering seal and eagle populations. Remaining problems include high phosphorous and cyanobacteria blooms, and fish catches that are not tailored to ecosystem needs. Perhaps most troubling for the Sea's ecosystem integrity, 'interconnected problems [are] still managed apart [and] fisheries [are] handled by a different convention'.[62]

The European Parliament's Marine Strategy Framework Directive is the European Union's framework for developing and implementing a strategy to conserve the marine environment. It requires Member States to take the necessary measures to achieve or maintain good environmental health in the marine environment by the target date of 2020. Implementing the Directive has proven to be a challenging task. The last status report, in 2014, confirmed that European seas are doing poorly environmentally.[63] For example, '88 percent [of stocks] in the Mediterranean and Black Seas are still overfished and the situation is improving only slowly'.[64]

One roadblock to successful implementation of the Marine Strategy Framework Directive is the requirement for each Member State to create a programme of cost-effective measures to attain the best health for marine ecosystems. The Directive requires the 'integration of the value of marine ecosystem services ... so that policies are designed to achieve sustainable management of these ecosystems'.[65] However, even if these 'services' attempt to take into account the cost of environmental degradation, it is unlikely that we understand the linked social and ecological dynamics well enough to make an accurate estimate. Additionally, one cannot put a price on a healthy, stable and functioning ecosystem and the benefits it provides, not only to humans, but the entire Earth community.

Another roadblock to successful implementation of both the Convention and Directive lies in its mandate to use an ecosystemic analysis for resource problems. The ecosystem-based management approach requires that humans consider the cumulative impacts and links between living and nonliving resources, and weigh human activities in the context of the broader ecological surroundings.[66] Defined by the U.S. National Oceanic and Atmospheric Administration as 'a holistic way of managing fisheries and marine resources',

Port City of Klaipeda, Lithuania. (Photograph by Mantas Volungevicius/Wikimedia. https://commons.wikimedia.org/wiki/File:Quite_summer_evening_in_the_port_city_Klaipeda.jpg)

ecosystem-based management strives to 'maintain ecosystems in a healthy, productive, and resilient condition *so they can provide the services humans want and need* (emphasis added)'.[67] But holistic management of human impacts on nonhuman systems requires a broader horizon than this.[68] A truly holistic framework that takes into account interactions among all species, and human impacts from land on the marine an environment, has yet to be fully developed and implemented.

Finally, current European legal and policy frameworks aim to achieve 'sustainable use' of the marine environment. The main concern with this goal is the dominance of anthropocentric language, and the representation of Nature as having mainly instrumental value. Nevertheless, the concept of sustainable development was created on human-centred values, focusing on support for human needs and preventing the loss of biodiversity for human benefit. As is currently interpreted, it reinforces the idea that humans are above and separate from Nature.[69]

Indeed, the language of Sustainable Development Goal 14 to 'conserve and sustainably use the oceans, seas and marine resources for sustainable development' equates the oceans value to human wants and needs, failing to mention conserving the ocean for use and enjoyment of the species that live there, and the web of life that depends on it. It is now imperative, in order to prevent the crisis now evident,[70] to evolve our perceptions and values, transitioning to language and law that represents the interconnectedness of ecological processes, that sees humans as inseparably embedded within Nature, and listens to the sea.[71]

## Ocean Rights: A Systemic Solution

A Rights of Nature paradigm places the ocean's rights to exist and flourish on a level playing field with the rights of other living beings (i.e., humans). Only this framework can provide the changes needed to turn our current ocean-use paradigm – common in the Baltic Sea region, and in many other fishing cultures worldwide – into one that creates a long-term healthy ecosystem in which humans participate, but do not own. Rather than treating the ocean as property and a *resource*, we treat the ocean as an entity, our *source* of life. This requires humans to 'converge towards a more holistic approach that balances both human well-being and ecological well-being'[72] by making protection and restoration a legal responsibility.

By recognising legal rights for the ocean, we move beyond the traditional model of perpetual economic growth and development, linear progress and a mechanistic worldview consisting of separate parts. We require that decisions take into account the costs and benefits to the whole Earth community, not just the human population. It now becomes our legal responsibility to recognise, respect and protect the ocean's rights and prohibit activities that will violate them.

The ocean, as a legal entity, has the following rights:

*TO LIFE*: The right to maintain the integrity of living systems and natural processes that sustain the ocean and Earth as a whole, and capacities and conditions for regeneration. All species of the ocean, plants, animals and microorganisms, have the right to life. The right to have critical and significant areas set aside for the continuation of cycles and processes where no human activity may occur (no take zones). The ocean has a right to life in perpetuity, and for humans to ensure that the pursuit of human well-being contributes to the well-being of the ocean now and into the future.

*TO HEALTH AND WELL-BEING*: Where health is defined in terms of the ocean's own well-being and in relation to its natural state. The right to live free from torture or cruel treatment by human beings and to exist in its natural state and habitat. The right to be free from contamination, pollution (including noise and plastic) and toxic or radioactive waste.

*TO THE DIVERSITY OF LIFE*: The ocean has the right to biodiversity and to evolve. It has the right to the differentiation and variety of beings that make up the ocean. It has the right to not have its beings genetically altered or structurally modified in an artificial way, including in any way that threatens their existence, functioning or future potential.

*TO WATER*: The right to water as a source of life. The right to preserve the functionality of the water cycle, its existence in the quantity and quality needed to sustain living and nonliving systems, and its protection from pollution for the reproduction of the life of the ocean and all its components. This includes the right to maintain ocean temperature and chemical composition (carbon dioxide proportions) at a level which protects the ocean's integrity or vital and healthy functioning.

*TO CLEAN AIR*: The right to preserve the quality and composition of air, and the functionality of the carbon cycle, for sustaining living and nonliving systems and its protection from pollution, for the reproduction of the life of the ocean and all its components.

*TO EQUILIBRIUM*: The right to maintenance or restoration of the interrelationship, interdependence, complementarity and functionality of the components of the ocean in a balanced way for the continuation of its vital cycles and processes. The ocean has a right to live in harmony with humans and exhibit normal form and function.

*TO RESTORATION*: The right to timely and full restoration of impacts by direct or indirect human activities.

*TO REPRESENTATION*: The right to recognition everywhere before the law, during decision making about activities that may impact the ocean and its rights, and before the commencement of activities which may impact the ocean and its rights.

In order to implement a holistic framework that achieves effective and equitable governance,[73] guides humanity to live in harmony with Nature,[74] ensures the proper functioning of natural systems and recognises the interconnectedness and complexities of marine ecosystems,[75] our laws must properly place humans within the larger Earth community. Rights of Nature is one way to do so, by placing us on the same level as all other life, not above it. This would also give humans the opportunity to retool our technologies and consumption levels to fit within Nature's limitations.

## What if Fisheries Policies Evolved to Include the Rights of the Ocean?

Rights of Nature in fishery policy would create an environmentally sustainable economy – an eco-economy – that requires that the principles of ecology establish the framework for economic policy and that economists and ecologists work together to fashion the new economy. Ecologists understand that all economic activity, indeed all life, depends on the Earth's ecosystem – the complex of individual species living together, interacting with one another and their physical habitats.[76]

Rather than focusing on mortality and aiming to prevent collapse, Rights of Nature shifts our approach towards aiming to maintain healthy and thriving populations. Fishing levels in European fisheries would be based on what allows the population and ecosystem to 'maintain normal form and function';[77] a level of health that is determined based on the fish population to maintain its vital role now and into the future, rather than focusing on human utility and traditional cost-benefit analysis.

Rights of Nature in fishery policy would require decisions (such as quotas) to consider 'no human activity' as a reasonable alternative, employ a nonconsumptive model for weighing alternatives, deploy the best available and least disruptive technology, and apply the precautionary principle. Also, recall that advisory committees currently determine the level of fishing allowed, and that two-thirds of these committee members are industry representatives. Rights of Nature would place 'guardians' on these committees charged with representing the fish populations and ecosystems themselves, to ensure their interests are represented in decisions and enable the shift to alternative livelihoods fashioned within Nature's limits.

As we are seeing fish populations decline, we also see greater effort required to catch fewer fish.[78] We must not let the time needed to receive benefits from a regulatory decision lead to prolonging protective efforts. For example, a no-take fishing policy is often seen as 'inefficient' when looking at it in the short-term because fishermen lose their benefits, and it takes years to rebuild ecosystem health and see 'spillover' or recruitment effects.[79] Rights of Nature requires that in addition to considerations of efficiency and fishing effort, decision-making also include the rights of future generations (future generations should inherit the world in no worse a condition than we received it) and the rights of the species to remain healthy.

This may then lead to determining that the 'optimum allocation' of fishing privileges is *no* allocation. In fact, reducing fishing effort, or closing an area to fishing, has proven to rebuild fisheries. For example, fish populations of Apo Island in the Philippines have tripled since the marine protected area was created, leading to a 50% increase in catch per unit of effort for fishermen.[80]

Additionally, it is important for decisions regarding fisheries to be science-based, and grasp the wholeness of the ecosystem as fully as possible.[81] Decisions must move past merely assigning a dollar value to fisheries, and where possible evaluate the decision alternatives from a nonconsumer approach.[82] For example, managers in the North Pacific Fishery Management Council in the Bering Sea, Alaska, determine allowable catch of pollock by managing human interactions as predators within the food web, rather than outside the system.[83] Scientists recommend biologically safe harvest levels using these models, and as a result, the Alaska pollock fishery quotas are never set higher than the acceptable biological catch[84] and the Bering Sea fishery is considered one of 'the best-managed fisheries in the world'.[85]

Recall that scientists advised the committee for Baltic Sea cod to lower the allowable catch by 28%, but the catch was only reduced by 8%.[86] With a Rights of Nature framework, the committee would have been legally required to enforce the limits informed by the best available science, perhaps preventing the collapse. Rights of Nature also represents the ability to address the gaps in the current policies, by managing the interconnected problems together, rather than apart, and creating standards for land-based pollution, such as phosphorus. For example, traditional standards for industrial discharge are created based on human health standards, but using a rights of nature framework requires these standards to be developed based on what allows the downstream ecosystem, and all of its parts to remain healthy.

## How Would Rights of Nature Affect the Criteria for Decision-Making?

Fishery policymaking can use a weighted score to help assess alternatives. For example, the Global Ocean Refuge System (GLORES) evaluation employs a classification system based on the number of fishing gears allowed, their ecological impact, the types of bottom exploitation and aquaculture allowed and the regulations relating to recreational boating. In the weighted scores, the lower impact of fishing, the higher the score.[87] Assigning scores to attributes such as type of fishing gear, type of activity (extractive = lower score, tourism = higher score) and impact of activity (purse seines = lower score) can help provide a total score to help assess alternatives and make sound decisions.

Take the Baltic Sea cod fishery, where the majority of cod are caught through 'bottom trawling',[88] a method which is highly destructive to the seafloor ecosystem, and nonselective, killing huge numbers of discarded fish. Because of its indiscriminate destructiveness, bottom trawling is banned in various locations worldwide.[89] Rights of Nature would assign use of this fishing gear a low score, translating to a policy that bans trawling or greatly reduces its use. Some believe that the ban on trawling since the 1930s has lead to the cod population of Öresund ('Sound' between Sweden and Denmark) being much healthier than that of the Baltic Sea, with greater population size and average size of individuals.[90] The ban also led to a transition to tourism, where tourism brings in 80 million more Swedish Krona than the large-scale cod fishery did.[91]

Additionally, Rights of Nature requires that the Precautionary Principle provide the foundation for decision-making, because it is preventive rather than reactive. Four elements of the principle can be identified: (a) level of damage, (b) scientific criterion, (c) remedy and (d) burden of proof. The precautionary principle calls for early prevention to avoid harm before it occurs, and relieve serious and irreversible damage to marine ecosystems. While there is no single definition of the precautionary principle, and its multiple competing formulations are highly contested, it nevertheless seeks to transfer the burden of proof. Instead of policymakers having to prove that an action is potentially harmful, the burden is on those who wish to pursue the allegedly harmful action to demonstrate sufficient evidence of safety.[92] These elements provide the basis for a minimum-harm threshold so that 'only those threats that present a genuinely harmful outcome will allow the principle to come into effect'.[93] This criterion deals with the problem of how to make a decision when there is a lack of complete scientific information regarding serious damage.[94]

As previously noted, under a Rights of Nature management framework, the committees determining allowable catches in European waters would consist of 'guardians' to represent the species and ecosystems interests. Per the precedent set by New Zealand law,[95] guardians have a legal responsibility to protect and act on behalf of the marine ecosystem. It is the duty of the management body to protect the integrity and diversity of the marine ecosystem, and to defend the area from activities that may harm the ecosystem and its inhabitants.

Guardians for the ecosystem can participate in any legal process that may affect its health (particularly 'appearing before national legislative and rule-making bodies to help clarify ocean impacts of proposed actions'),[96] develop or review any relevant guidelines, monitor the health of the ecosystem, monitor compliance with applicable laws and treaties and represent the ecosystem in disputes. Using a holistic and rights-based approach, the new fisheries management committee, now composed of guardians, would determine the allowance and extent of fishing and other activities that potentially disrupt healthy ecosystem functioning. The Rights of Nature framework in the context of fisheries requires a determination of the following:

1. Impacts of the proposed action
2. Whether the impacts violate the ecosystem's rights, and, if so, to what extent
3. The alternatives and their impacts
4. Choice of the alternative that fulfils the highest environmental and human needs beyond economic considerations

At the top of the list for evaluating alternatives is 'alternative livelihoods' that would cause fewer environmental impacts, such as education or ecotourism. The benefits of tourism have been proven to be 'closely linked to the quality of the environment'.[97] The healthier the ocean and coastal environment is, the more benefits to the local communities. Ecotourism offers a solution to replace livelihoods disrupted by fisheries and their subsequent rules and regulations. Research has found that for heavily exploited fisheries, developing a nonextractive activity, such as ecotourism, may help to overcome the dilemma between the need for long-term resource conservation and the immediate necessity to provide jobs and income to the local population.[98] Ecotourism is an alternative that balances the rights and needs of both humans and the ecosystem.

In theory, applying the Rights of Nature to the Common Fisheries Policy that governs European Union waters, and other ocean laws and policies, would create a future with healthy fish populations, plentiful and safe to eat. Our decisions would be guided by science, precaution and holistic evaluation of entire ecosystem (including human) impacts and benefits. Policy would strive for maintaining and restoring fish populations and ecosystem health, rather than causing a slow decline to the point of irreparable harm. It would force us to look to the future, and ensure we leave the ocean no worse than we inherited it. The defining question of this paradigm is: What does the ocean want and need?

## The Movement Is Growing

Experts, governments and organisations worldwide agree that such a shift to holism and alternative forms of management are needed in order to once again have flourishing oceans that sustain all the species – and also allow humans to fish. Here we survey a few of the organisations worldwide calling for systemic change in our relationship with the ocean.

### United Nations (UN)

In 2017, the UN hosted the first Ocean Conference, which brought together governments, stakeholders, businesses and civil society representatives worldwide to discuss reversing the decline in the health of the oceans for people, the planet and future economies.[99] At the conference, the Prime Minister of the Cook Islands, Henry Puna, spoke about the need to recognise the ocean's rights.[100]

Wall painting in Galapagos: 'Ecosystems are important to our world'. (Photograph by Michelle Bender.)

Additionally, the United Nations in promoting sustainable development for the oceans and seas states that '[h]uman well-being cannot be achieved without the protection and conservation of the Earth's ecosystem. To maintain the quality of life that the oceans have provided to humankind, while sustaining the integrity of their ecosystems, a change will be required in how humans view, manage and use oceans, seas and marine resources'.[101]

## International Union for the Conservation of Nature (IUCN)

In 2011, a joint workshop between the IUCN, the International Programme on the State of the Ocean and the World Commission on Protected Areas concluded that current approaches to manage human activities are inadequate. They issued a call for systemic change, warning: 'maintain[ing] the goods and services [the ocean] has provided to humankind for millennia demands change in how we view, manage, govern and use marine ecosystems'. Additionally, the experts concluded we must rapidly adopt a holistic approach to the many wide-ranging activities that impinge on marine ecosystems.[102]

The IUCN in 2017 produced a guideline document for large-scale marine protected areas.[103] The framework notes: '[t]he key is for all players to commit to effective and equitable governance and management that seeks to conserve biodiversity in parallel with influencing, for the better, the economic, social and political drivers that affect ecosystem management, nature-based livelihoods, and the rights and responsibilities for nature'.[104] It further requires human activities to be managed holistically, and strongly recommends use of a holistic management model that is based on the relationship between nature, culture and the human dimension.[105]

## National Oceanic and Atmospheric Administration (NOAA)

The United States National Oceanic and Atmospheric Administration, a leader in ocean research and management worldwide, acknowledges that 'virtually all commercial fishing involves harvesting of a magnitude that is well beyond being fully sustainable',[106] leading to escalating failures of conventional fisheries management strategies.[107] NOAA stresses that recognition of ecological interconnectedness and complexity is crucial to managing marine ecosystems.[108] The need for holism is highlighted throughout NOAA's work, pinpointed as essential to sustainability in ocean activities. In multiple reports, NOAA noted that a holistic approach is distinct from current approaches,[109] and to achieve the needed holism, we must reject and replace 'many (but not all) of the processes upon which conventional management depends'.[110]

## Food and Agriculture Organization (FAO)

The International Food and Agriculture Organization highlights the failures of traditional fishing methods. FAO listed over 70% of the world's fish species as either fully exploited or depleted,[111] and therefore unlikely to rebound to healthy populations. In providing technical guidelines for responsible fisheries, FAO recognises the need to improve current fisheries management,[112] highlighting the use of marine protected areas and a holistic approach to do so.[113] However, FAO notes that marine protected areas must merge two converging paradigms: ecosystem management and traditional fisheries management.[114] Sustainable development can be achieved only if the two converge, become more holistic and balance human and ecological well-being.[115]

## Shifting out of the Dated Governance Paradigm

In sum, experts worldwide recognise the need to shift our approach to ocean conservation and management. Earth Law Center, a non-profit based in the United States, leads the transition towards holistic, rights-based ocean governance. Earth Law Center launched the initiative in 2017 at the United Nations Ocean Conference, with an initiative calling on stakeholders and governments to incorporate Rights of Nature into ocean protection efforts. The initiative (as of 2018) has gained 74 signatories from 34 countries in support of a paradigm shift.

To show how the Rights of Nature framework can be turned from theory into practise, Earth Law Center created the Earth Law Framework for Marine Protected Areas.[116] This framework serves as a guideline for adopting this approach into the creation and management of marine protected areas, to ensure these areas are fully and effectively protected. The most essential part – and difference – between the Earth Law structure for governance and those currently in use is the requirement that law recognise the ocean as an entity subject to rights that maintain its integrity, and that of all its species.

- There are a few more locations around the world where the concept of 'ocean rights' is beginning to take hold:
- The Environmental Defenders Office of Northern Queensland, Australia, began in 2014 to explore what legal personality would mean for the Great Barrier Reef.[117]
- The Galapagos Marine Reserve of Ecuador includes the Rights of Nature as a guiding principle for management; a successful case on illegal shark finning marked the first crime against nature in the Galapagos.[118]
- A commitment was also made at the conference by the French Research Institute for Development (based in Nouméa, New Caledonia), the Secretariat of the Pacific Regional Environment Programme, the non-profit Conservation International, and the International Centre of Comparative Environmental Law to create a Convention on the Rights of the Pacific Ocean to be signed by Pacific Island nations by 2020.[119] The first meeting to recognise the rights of the Pacific Ocean took place in November 2018.
- A campaign began in late 2017 to gain legal rights for the Salish Sea, which lies in Washington, United States, and Canada.
- A campaign began in 2018 to gain legal rights for the Whale and Dolphin Sanctuary in Uruguay.

## Conclusion

The European Union and Baltic Sea countries have taken great strides to protect the ocean and its integrity. The Common Fisheries Policy, Convention on the Protection of the Marine Environment of the Baltic Sea Area and the Marine Strategy Framework Directive all aim to conserve and protect the ocean. But there is only so much our current laws can do. In a legal system where the ocean is regarded as property, our economic system will continue to drive decision-making and ocean health will continue to decline. Until we undertake a fundamental change in our legal system and our relationship with the ocean, we cannot hope to change this downward spiral, in the Baltic or other oceans of the world.

Overfishing, pollution and habitat destruction have led to a global decline in marine biodiversity of 49%, roughly half of what it was 50 years ago.[120] The rapid decline in ocean health can be largely attributed to a lack of understanding of our relationship with the ocean and the majority's worldview of the ocean as an endless resource. It is in fact, a finite and beautiful entity that contains up to 80% of all life on Earth.[121] Human well-being is entirely dependent on the well-being of the ocean: 50% of the oxygen we breathe, the regulation of the water cycle and carbon cycle and a source of food and livelihoods for millions of people. We can only hope to reverse the tide of destruction by adopting the Rights of Nature paradigm. Only this framework provides the shift to holism desperately needed in order to maintain healthy and thriving marine ecosystems.

## Notes

1. Earth Law Center. Press Release of 29 April 2018. https://www.earthlawcenter.org/elc-in-the-news/2018/4/earth-defenders-launch-first-ocean-rights-framework-in-the-world-on-earth-day.
2. Pauly, D. and Zeller, D. 2016. Catch reconstructions reveal that global marine fisheries catches are higher than reported and declining. *Nat. Commun.* 7: 10244 doi: 10.1038/ncomms10244.
3. Myers, R.A. and Worm, B. 2003. Rapid worldwide depletion of predatory fish communities, *Nature*. 423(May): 280–283. www.nature.com/nature/journal/v423/n6937/abs/nature01610.html.
4. Cronin, A.M. 2018. Time to get real: There aren't plenty of fish in the sea and it's our fault. *One Green Planet*. https://www.onegreenplanet.org/environment/there-arent-plenty-of-fish-in-the-sea/ (accessed 26 July 2018); Platt, J.R. 2014. 30 Percent of sharks, rays and related species at risk of extinction. *Scientific American*. https://blogs.scientificamerican.com/extinction-countdown/30-percent-of-sharks-rays-and-related-species-at-risk-of-extinction/ (accessed 26 July 2018).
5. National Oceanic and Atmospheric Administration, Gray Whale (*Eschrichtius robustus*), NOAA Fisheries. http://www.nmfs.noaa.gov/pr/species/mammals/whales/gray-whale.html#population (accessed 26 July 2018).
6. World Wildlife Fund, Fast Facts: Why Coral Reefs are Important to People, at: www.wwf.panda.org/about_our_earth/blue_planet/coasts/coral_reefs/coral_facts (accessed 26 July 2018).
7. Lallanilla, M. 2013. What are Coral Reefs? Live Science, https://www.livescience.com/40276-coral-reefs.html (accessed 27 July 2018).
8. Dancheck, M. 2018. Ocean Pollution: The Dirty Facts. National Resource Defense Council. https://www.nrdc.org/stories/ocean-pollution-dirty-facts (accessed 27 July 2018).
9. Guem, C.L. 2018. When the Mermaids Cry: The Great Plastic Tide. Coastal Care. http://plastic-pollution.org. (accessed 26 July 2018).
10. World Economic Forum. 2016. *The New Plastics Economy: Rethinking the Future of Plastics*. Industry Agenda, http://www3.weforum.org/docs/WEF_The_New_Plastics_Economy.pdf.
11. National Oceanic and Atmospheric Administration (NOAA). 2018. Plastics, NOAA Marine Debris Program, https://marinedebris.noaa.gov/info/plastic.html.
12. Dancheck, M. 2018. *Ocean Pollution: The Dirty Facts*. National Resource Defense Council. https://www.nrdc.org/stories/ocean-pollution-dirty-facts (accessed 27 July 2018).
13. Id.
14. Id.
15. Institut de Recherche Pour le Developpement. Towards Responsible and Sustainble Fisheries. http://www.suds-en-ligne.ird.fr/ecosys/ang_ecosys/intro1.htm (accessed 10 October 2018).

16. Nutall, N. 2018. Overfishing: A Threat to Marine Biodiversity. United Nations Environment Programme (UNEP). http://www.unis.unvienna.org/documents/unis/ten_stories/09fisheries.pdf. (accessed 18 July 2018).
17. White, C. 2018. How Lobsters Became Victims fo the Tragedy of the Commons. Green Biz. https://www.greenbiz.com/article/how-lobsters-became-victims-tragedy-commons (accessed 15 July 2018).
18. Rinaldi, L. 2014. Whole communities disappeared: Bill Broderick, a Newfoundland fisherman, is waiting for the cod to return. *Maclean's*. https://www.macleans.ca/news/canada/history-of-cod-fishery-in-newfoundland/ (accessed 22 July 2018).
19. European Commission. 2011. Seas for Life: Protected – Sustainable – Shared European Seas by 2020. https://publications.europa.eu/en/publication-detail/-/publication/ff3c7a4d-7ce5-4427-b9c3-8ed90d58a4a0 (accessed 27 July 2018).
20. Fonselius, S.H., 1972. On biogenic elements and organic matter in the Baltic. *Ambio Spec. Rep.*, 1: 29–36.
21. Brackish waters are those that are a mixture of salt and freshwater.
22. Baltic Sea – Map and Details. *World Atlas*. https://www.worldatlas.com/aatlas/infopage/balticsea.htm (accessed 27 July 2018).
23. Stigebrandt, A. 2001. Physical Oceanography of the Baltic Sea. A Systems Analysis of the Baltic Sea, pp. 19–74.
24. Andrén, E., Telford, R.J. and Jonsson, P. 2017. Reconstructing the history of eutrophication and quantifying total nitrogen reference conditions in Bothnian Sea coastal waters, *Estuarine, Coastal and Shelf Science*. 198: 320–328, doi: 10.1016/j.ecss.2016.07.015.
25. Andrén, E., Telford, R.J. and Jonsson, P. 2017. Reconstructing the history of eutrophication and quantifying total nitrogen reference conditions in Bothnian Sea coastal waters, *Estuarine, Coastal and Shelf Science*. 198: 320–328, doi: 10.1016/j.ecss.2016.07.015.
26. Bignert, A., Nyberg, E., Greyerz, E., Brännvall, M.-L., Sundqvist, K., Wiberg, K. and Haglund, P. 2005. 'Miljögifter i Biota', Gruppen för miljögiftsforskning, Naturhistoriska riksmuseet, Länsstyrelsen i Gävleborg, Kemiska institutionen, Miljökemi, Umeå Universitet.
27. First Version of the 'State of the Baltic Sea' Report. 2017. http://stateofthebalticsea.helcom.fi (accessed 1 July 2018).
28. Id.
29. Id.
30. Id.
31. ICES Fisheries Overview. 2017. Baltic Sea Ecoregion. doi: 10.17895/ices.pub.3053. http://www.ices.dk/sites/pub/Publication%20Reports/Advice/2017/2017/Baltic_Sea_Ecoregion_Fisheries_Overview.pdf.
32. HELCOM, Red List Fish and Lamprey Species Expert Group 2013, p. 12. http://www.helcom.fi/Red%20List%20Species%20Information%20Sheet/HELCOM%20Red%20List%20Gadus%20morhua.pdf.
33. Hamrén, H. 2017. Large-sized cod at a historic low in the Baltic Sea, Baltic Sea Centre, Stockholm University, https://balticeye.org/en/fisheries/fewer-large-cod/ (accessed 4 July 2018).
34. Cutlip, K.A. 2016. New angle on Baltic Sea Cod = Upheaval for the fishery, *Global Fishing Watch*. http://globalfishingwatch.org/news-views/a-new-angle-on-baltic-sea-cod-means-upheaval-for-the-fishery/ (accessed 17 July 2018).
35. Nixon, S.W. 1995. Coastal marine eutrophication: A definition, social causes, and future concerns. *Ophelia*. 41: 199–219.
36. Rosenberg R., Elmgren R., Fleischer S., Jonsson P., Persson G. and Dahlin H., 1990. Marine eutrophication case studies in Sweden, *Ambio*. 19: 102–108; Norkko, J., Gammal, J., Hewitt, J.E., Josefson, A.B., Carstensen, J. and Norkko, A. 2015. Seafloor ecosystem function relationships: In situ patterns of change across gradients of increasing hypoxic stress. *Ecosystems*. 18: 1424–1439.
37. Bonsdorff, E, Blomqvist, E.M, Mattila, J. and Norkko, A. 1997. Long-term changes and coastal eutrophication. Examples from the Aland Islands and the Archipelago Sea, northern Baltic Sea. *Oceanologica Acta*. 20(1): 319–329.

38. First Version of the 'State of the Baltic Sea' Report. 2017. http://stateofthebalticsea.helcom.fi (accessed 21 July 2018).
39. Oxygen loss in the coastal Baltic Sea is 'unprecedentedly severe'. 2018. *Sciencedaily*. https://www.sciencedaily.com/releases/2018/07/180705084213.htm.
40. Id.
41. State of The Baltic Sea – Second HELCOM Holistic Assessment 2011–2016. 2018. http://stateofthebalticsea.helcom.fi/ (accessed 26 July 2018).
42. Species Information Sheet. 2018. *Helcom.Fi*. http://www.helcom.fi/Red%20List%20Species%20Information%20Sheet/HELCOM%20Red%20List%20Gadus%20morhua.pdf.
43. Cutlip, K. 2018. Baltic Sea Cod quota slashed. *Global Fishing Watch*. http://globalfishingwatch.org/news-views/a-new-angle-on-baltic-sea-cod-means-upheaval-for-the-fishery/ (accessed 26 July 2018).
44. Coldeway, D. 2018. Beautiful but toxic algae swirl in satellite view of Baltic Sea. *NBC News*. https://www.nbcnews.com/science/environment/beautiful-toxic-algae-swirl-stunning-satellite-view-baltic-sea-n422011 (accessed 26 July 2018).
45. Funkey, C.P., Conley, D.J., Reuss, N.S., Humborg, C., Jilbert, T. and Slomp, C.P. 2014. Hypoxia sustains cyanobacteria blooms in the Baltic Sea. *Environmental Science & Technology*. 48(5): 2598–2602. doi: 10.1021/es404395a. https://pubs.acs.org/doi/full/10.1021/es404395a.
46. Vahtera, E., Conley, D.J., Gustafsson, B.G., Kuosa, H., Pitkanen, H., Savchuk, O.P., Tamminen, T., Viitasalo, M., Voss, M., Wasmund, N., and Wulff, F. Internal ecosystem feedbacks enhance nitrogen-fixing cyanobacteria blooms and complicate management in the Baltic Sea. *Ambio*. 2007: 362–3186–194.
47. Seas For Life: Protected – Sustainable – Shared European Seas By 2020. 2018. *Publications.Europa.Eu*. https://publications.europa.eu/en/publication-detail/-/publication/ff3c7a4d-7ce5-4427-b9c3-8ed90d58a4a0.
48. Report from the Commission to the Council And the European Parliament. The First Phase of Implementation of The Marine Strategy Framework Directive. 2018. *Eur-Lex.Europa.Eu*. https://eur-lex.europa.eu/legal-content/EN/TXT/PDF/?uri=CELEX:52014DC0097&from=EN.
49. Red List Of Species – HELCOM. 2018. *Helcom.Fi*. http://www.helcom.fi/baltic-sea-trends/biodiversity/red-list-of-species/.
50. European Commission, The Common Fisheries Policy (CFP). https://ec.europa.eu/fisheries/cfp_en.
51. Hamren, H. 2018. Torsken Förlorare I Kvotsättning. 2018. *Baltic Eye*. https://balticeye.org/sv/hallbart-fiske/fiskekvoter-2018/.
52. Common Fisheries Policy Reform: Getting MSY Right. 2018. *Awsassets.Panda.Org*. http://awsassets.panda.org/downloads/wwf_msy_oct2011_final.pdf.
53. Wilder, R.J. 1998. *Listening to the Sea: The Politics of Improving Environmental Protection*, University of Pittsburgh Press. p. 95.
54. Stolpe, N. 2018. Optimum Yield In Fisheries Is Far From Optimum. *Fishnet-Usa.Com*. http://www.fishnet-usa.com/maximum_sustainable_yield.htm.
55. Elmgren, R., Blenckner, T. and Andersson, A. 2015. Baltic Sea management: Successes and failures. *Ambio*. 44(Suppl 3), 335–344. http://doi.org/10.1007/s13280-015-0653-9.
56. CFP Reform – Maximum Sustainable Yield. 2018. *Ec.Europa.Eu*. https://ec.europa.eu/fisheries/sites/fisheries/files/docs/body/msy_en.pdf.
57. Hardin, G. 1968. The tragedy of the commons. *Science*. 162(3859): 1243–1248, doi: 10.1126/science.162.3859.1243. http://science.sciencemag.org/content/162/3859/1243.full (accessed 26 July 2018).
58. First Version of the 'State of the Baltic Sea' Report. 2017. http://stateofthebalticsea.helcom.fi (accessed 21 July 2018).
59. About Us – HELCOM. 2018. *Helcom.Fi*. http://www.helcom.fi/about-us.
60. Marc Mangel et al. 1996. Principles for the conservation of wild living resources. *Ecological Applications*. 6(2): 354, available at: http://www.jstor.org/stable/2269369.
61. Convention on the Protection of the Marine Environment. Article 3.

62. Elmgren, R., Blenckner, T., and Andersson, A. 2015. Baltic Sea management: Successes and failures. *Ambio.* 44(Suppl 3): 335–344. http://doi.org/10.1007/s13280-015-0653-9.
63. Report from the Commission to the Council and the European Parliament. 2018. The First Phase of Implementation of the Marine Strategy Framework Directive. p. 3. *Eur-Lex.Europa.Eu.* https://eur-lex.europa.eu/legal-content/EN/TXT/PDF/?uri=CELEX:52014DC0097&from=EN.
64. Id.
65. Know the Sea Guidelines. 2018. *Msfd.Eu.* http://www.msfd.eu/knowseas/guidelines/4-C-B-Guideline.pdf.
66. Framework for the National System of Marine Protected Areas of the United States of America. 2015. *National Marine Protected Areas Center,* 27. http://marineprotectedareas.noaa.gov/nationalsystem/framework/final-mpa-framework-0315.pdf.
67. Ecosystem-Based Fisheries Management. 2017. *NOAA Fisheries.* https://www.fisheries.noaa.gov/insight/ecosystem-basedfisheries-management.
68. Fowler, C.W., Belgrano, A. and Casini, M. 2013. Holistic Fisheries Management: Combining Macroecology, Ecology, and Evolutionary Biology. 2013. *Marine Fisheries Review.* Scientific Publications Office, National Marine Fisheries Service, NOAA. 75: 1–2.
69. Campagna, C., Guevara, D. and Bernard Le Boeuf. 2017. Sustainable development as deus ex machina. *Biological Conservation.* 209: 54–61, pp. 58–59. https://doi.org/10.1016/j.biocon.2017.01.016.
70. Id.
71. Wilder, R.J. 1998. *Listening to the Sea: The Politics of Improving Environmental Protection.* University of Pittsburgh Press, p. 189, 203.
72. Fisheries Management-2: The Ecosystem Approach to Fisheries, Food and Agriculture. 2003. *Organization of the United Nations.* http://www.fao.org/3/ay4470e.html.
73. IUCN Programme 2017–2020: Approved by the IUCN World Conservation Congress, September 2016, Target 15, p. 35, at: http://bit.ly/2kkHWCo
74. UN Resolution 71/232: 'Harmony with Nature'. *United Nations.* http://www.un.org/ga/search/view_doc.asp?symbol=A/RES/71/232
75. Fowler, C.W., Belgrano, A. and Casini, M. 2013. Holistic Fisheries Management: Combining Macroecology, Ecology, and Evolutionary Biology. *Marine Fisheries Review* (Scientific Publications Office, National Marine Fisheries Service, NOAA). 75, 1–2. http://aquaticcommons.org/14550/1/mfr751-21.pdf.
76. Brown, L.R. 2002. Building an environmentally sustainable economy. *Mother Earth News.* https://www.motherearthnews.com/nature-andenvironment/environmentally-sustainable-economyzmaz02fmzgoe.
77. Costanza, R. and M. Mageau. What is a healthy ecosystem. 1999. *Aquatic Ecology.* 33: 105–115, 105. Healthy as a state in which the ocean can maintain its 'normal form and function' and therefore has the ability to 'maintain its structure (organization) and function (vigor) over time in the face of external stress (resilience)'.
78. World Ocean Review. 2010. Classic Approaches to Fisheries Management. http://www.fao.org/3/ai3720e.pdf (accessed 24 July 2018).
79. Dahlgren, C. 2014. Review of the Benefits of No-Take Zones: A Report to the Wildlife Conservation Society. https://appliedecology.cals.ncsu.edu/absci/wpcontent/uploads/Review-of-the-Benefits-of-No-TakeZones_Final.pdf (accessed 24 July 2018).
80. Environment Foundation. 2017. Environment Guide, Benefits of Marine Protected Areas. http://www.environmentguide.org.nz/issues/marine/marine-protected-areas/ (accessed 24 July 2018).
81. Center for Environmental Policy, University of Florida. 2018. Emergy Systems, https://cep.ees.ufl.edu/emergy/resources/templates.shtml (accessed 24 July 2018).
82. Angelo, M.J. 2008. Harnessing the power of science in environmental law: Why we should, why we don't, and how we can, 86 *Tex. L. Rev.*:1527. https://scholarship.law.ufl.edu/cgi/viewcontent.cgi?article=1034&context=facultypub.

83. NOAA Fisheries, New Comprehensive Bering Sea Climate Change Study to Focus on Fish and Fishing and Provide Insights for Management in a Changing Marine Environment, AFSC News, August 2015, https://www.afsc.noaa.gov/News/BS_climate-change-study.htm; Alaska Fisheries Science Center, Assessment of the Pollock Stock in the Aleutian Islands, 2013, https://www.afsc.noaa.gov/REFM/Docs/2013/AIpollock.pdf.
84. Genuine Alaska Pollock, A Model for Sustainable Fisheries Management, http://www.alaskapollock.org/sustainabilityMan.html.
85. NOAA Fisheries, Alaska Pollock, https://www.fisheries.noaa.gov/species/alaska-pollock.
86. Hamren, H. 2018. Torsken Förlorare I Kvotsättning. 2018. *Baltic Eye*. https://balticeye.org/sv/hallbart-fiske/fiskekvoter-2018/.
87. Marine Conservation Institute. 2018. Global Ocean Refuge System Criteria. Seattle, WA. https://globaloceanrefuge.org/wp-content/uploads/2018/04/GLORES_2018_Criteria_web.pdf (accessed 24 July 2018).
88. Hamrén, H. 2017. Large-Sized Cod at a Historic Low in the Baltic Sea, Baltic Sea Centre, Stockholm University, https://balticeye.org/en/fisheries/fewer-large-cod/ (accessed 4 July 2018).
89. Stiles, M.L., Stockbridge, J., Lande, M. and M.F. Hirshfield. 2010. Impacts of Bottom Trawling on Fisheries, Tourism, and the Marine Environment. *Oceana*. http://oceana.org/sites/default/files/reports/Trawling_BZ_10may10_toAudrey.pdf (accessed 24 July 2018).
90. Hamrén, H. 2017. Large-Sized Cod at a Historic Low in the Baltic Sea, Baltic Sea Centre, Stockholm University. https://balticeye.org/en/fisheries/fewer-large-cod/ (accessed 18 July 2018).
91. Ida Martenssen, Fisheries Brief No. 7: Who Is Entitled to the Fish? BalticSea2020, http://balticsea2020.org/english/press-room/438-fisheries-brief-no-7-who-is-entitled-to-the-fish (accessed 10 October 2018).
92. Mariqueo-Russell, A. 2017. Rights of Nature and the Precautionary Principle. Can nature have rights? Legal and political insights. RCC Perspectives: *Transformations in Environment and Society*. 6: 21–27. doi: 10.5282/rcc/8211.
93. Id.
94. Wang, R. 2011. The precautionary principle in maritime affairs. *WMU Journal of Maritime Affairs*. 10: 143. doi: 10.1007/s13437-011-0009-7.
95. Bender, M. 2018. *Earth Law Framework for Marine Protected Areas: Advancing Holistic Systems and Rights-Based Approach to Ocean Governance*. pp. 57–59. https://static1.squarespace.com/static/55914fd1e4b01fb0b851a814/t/5adca14b352f538288f4ea67/1524408668126/Final+Draft+3.pdf. In its originating statute, Te Urewera is declared a legal entity, and has all the rights, powers, duties and liabilities of a legal person. '[T]he rights, powers, and duties of Te Urewera must be exercised and performed on behalf of, and in the name of, Te Urewera … by Te Urewera Board'. The management body, the Te Urewera Board, is therefore 'responsible for protecting the entity and its rights'.
96. Stone, C. 2010. Should trees have standing?: *Law, Morality, and the Environment*, 101 3rd ed.
97. Russi D., Pantzar M., Kettunen M., Gitti G., Mutafoglu K., Kotulak M. and P. Brink. 2016. Socio-Economic Benefits of the EU Marine Protected Areas. Report prepared by the Institute for European Environmental Policy (IEEP) for DG Environment. p. 27. http://ec.europa.eu/environment/nature/natura2000/marine/docs/Socio%20-Economic%20Benefits%20of%20EU%20MPAs.pdf.
98. International League of Conservation Photographers. 2016. Changing Planet. National Geographic. https://blog.nationalgeographic.org/2016/10/17/marineecotourism-the-wealth-of-the-oceans-goes-well-beyondfisheries/ (accessed 18 July 2018).
99. United Nations. 2017. The Ocean Conference. https://oceanconference.un.org/about (accessed 18 July 2018).
100. Moore, R. 2017. Puna's Plea for Rights of the Ocean. Cook Islands News. http://www.sprep.org/biodiversity-ecosystems-management/qrights-of-the-ocean-need-to-be-exploredq-cook-islands-prime-minister (accessed 24 July 2018).

101. United Nations. 2017. Sustainable Development Platform. https://sustainabledevelopment.un.org/topics/oceanandseas (accessed 24 July 2018).
102. Rogers, A.D. and D. Laffoley. 2011. International Earth system expert workshop on ocean stresses and impacts. Summary report. *IPSO Oxford*, 18 pp.
103. Lewis, N., Day, J.C., Wilhelm, A. et al., 2017. Large-Scale Marine Protected Areas: Guidelines for Design and Management. Best Practice Protected Area Guidelines Series, No. 26, Gland, Switzerland, https://portals.iucn.org/library/node/46933.
104. Id. at 4.
105. Id. at 18.
106. Fowler, C.W., Redekopp, R.D., Vissar, V. and Oppenheimer, J. 2014. Pattern-based control rules for fisheries management. U.S. Dep. Commer., NOAA Tech. Memo. NMFS-AFSC-268. p. 1. https://www.afsc.noaa.gov/publications/afsc-tm/noaa-tm-afsc-268.pdf.
107. Id.
108. Id. at 2.
109. Fowler, C.W., Belgrano, A. and M. Casini. 2013. *Holistic Fisheries Management: Combining Macroecology, Ecology, and Evolutionary Biology, Marine Fisheries Review*. Scientific Publications Office, National Marine Fisheries Service, NOAA., 75(1–2): 1. http://aquaticcommons.org/14550/1/mfr751-21.pdf.
110. Fowler, C.W., Redekopp, R.D., Vissar, V. and Oppenheimer, J. 2014. Pattern-based control rules for fisheries management. U.S. Dep. Commer., NOAA Tech. Memo. NMFS-AFSC-268. p. 2, https://www.afsc.noaa.gov/publications/afsc-tm/noaa-tm-afsc-268.pdf.
111. Nutall, N. Overfishing: A Threat to Marine Biodiversity. United Nations Environment Programme. http://www.un.org/events/tenstories/06/story.asp?storyID=800 (accessed 24 July 2018).
112. Food and Agriculture Organization of the United Nations. 2003. Fisheries Management-2. The Ecosystem Approach to Fisheries. http://www.fao.org/3/a-y4470e.html.
113. Food and Agriculture Organization of the United Nations. 2011. Fisheries management-4. Marine protected areas and fisheries. FAO Technical Guidelines for Responsible Fisheries. No. 4, Suppl. 4. Rome. http://www.fao.org/docrep/015/i2090e/i2090e00.htm.
114. Food and Agriculture Organization of the United Nations. 2003. Fisheries Management-2. The Ecosystem Approach to Fisheries. http://www.fao.org/3/a-y4470e.html.
115. Id.
116. Bender, M. 2018. *Earth Law Framework for Marine Protected Areas: Advancing Holistic Systems and Rights-Based Approach to Ocean Governance*. https://static1.squarespace.com/static/55914fd1e4b01fb0b851a814/t/5adca14b352f538288f4ea67/1524408668126/Final+Draft+3.pdf (accessed 24 July 2018).
117. Environmental Defenders Office of Northern Queensland. 2014. Legal personality for great barrier reef. *Chain Reaction*. 120: 40. https://search.informit.com.au/documentSummary;dn=170948694928592;res=IELHSS (accessed 24 July 2018).
118. Kauffman, C.M. and Martin, P.L. 2016. Testing Ecuador's Rights of Nature: Why Some Lawsuits Succeed and Others Fail. https://static1.squarespace.com/static/55914fd1e4b01fb0b851a814/t/5748568c8259b5e5a34ae6bf/1464358541319/Kauffman++Martin+16+Testing+Ecuadors+RoN+Laws.pdf (accessed 24 July 2018).
119. Reinert, M. 2017. Legal Rights of the Pacific Ocean Pushed Forward. *Sci Dev Net*. https://www.scidev.net/asia-pacific/environment/news/legal-rights-of-pacific-ocean-pushed-forward.html (accessed 24 July 2018).
120. Living Planet Index, Living Blue Planet Report. http://www.livingplanetindex.org/projects?main_page_project=BluePlanetReport&home_flag=1 (accessed 24 July 2018).
121. Mora, C. et al. 2011. How many species are there on earth and in the ocean? *PLoS Biology*. 9: 8 https://www.ncbi.nlm.nih.gov/pmc/articles/PMC3160336/; Marine Bio. 2017. Little Known Facts About The Ocean http://www.marinebio.org/marinebio/facts/ (accessed 24 July 2018).

# Section III

# Rights of Nature in the Law

# 13

# *A River Is Born: New Zealand Confers Legal Personhood on the Whanganui River to Protect It and Its Native People*

**The Honorable Christopher Finlayson**

**CONTENTS**

Summary ........ 259
Introduction ........ 260
Background ........ 261
Treaty of Waitangi ........ 261
The Waitangi Tribunal ........ 263
The Claims Settlement Process ........ 264
Components of a Settlement ........ 266
The Whanganui River Settlement ........ 268
The Solution ........ 271
Conclusion ........ 274
Acknowledgments ........ 275
Notes ........ 275

## Summary

For over a millennium, the Māori, New Zealand's indigenous people, inhabited the banks of the Whanganui River. Uninformed of Western ideas of property ownership, Māori considered the river a sacred ancestor and the source of their *'ora'*, or environmental, social, cultural, economic, physical and spiritual well-being.

As Europeans began to settle in the country, the British Crown proposed a treaty with the Māori. In 1840, the Māori signed the treaty, which had three components: Māori became British subjects; the Crown agreed to safeguard the lands, forests and other treasures of Māori and sovereignty over New Zealand was assumed by the British Crown. Over time, the British government and subsequently the New Zealand government breached parts of the treaty. In the case of the Whanganui River, for example, it was damaged by the removal of minerals and traditional structures from the river and the diversion of the headwaters of the river for hydroelectric production. These and other acts degraded the river and injured its people.

One hundred and seventy-seven years later, the New Zealand parliament conferred personhood on the Whanganui River, using a novel legal theory that was in alignment with the ancient beliefs of the Māori who lived alongside the river. This legislation recognises the interdependence of a river and its people and resolves longstanding grievances by the Māori people against the Crown.

DOI: 10.1201/9780429505959-13

## Introduction

The purpose of this chapter is to explain the settlement entered in 2014 between the Crown (New Zealand Government) and Te Ati Haunui a Paparangi, the Whanganui Iwi (the tribe of indigenous Māori people) who have lived along the banks of the Whanganui River for many centuries. The settlement was given effect by legislation passed by the New Zealand Parliament in 2017, titled the 'Whanganui River Claims Settlement Act, Te Awa Tupua'.[1]

The settlement recognised the Whanganui River as a legal person. Before outlining the terms of the settlement and the legislation, it is necessary to provide some background to New Zealand and how the Crown and Māori have addressed historical issues and grievances, such as confiscation of land by forcible seizure or dubious contracts.

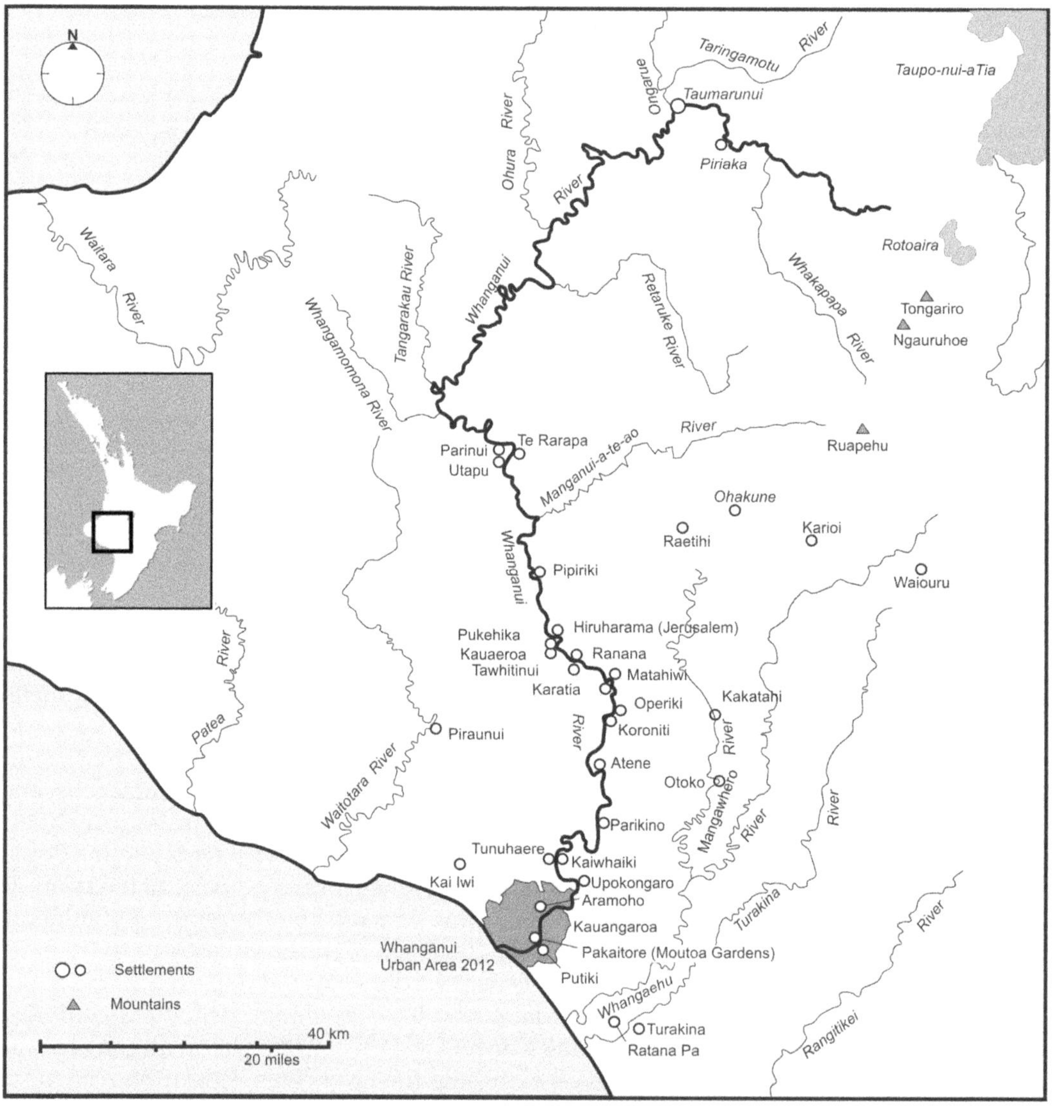


Map of Whanganui River and locations of the hapū. (Map courtesy of Andrew A. Young.)

## Background

Māori first came to New Zealand about 1,000 years ago. They are divided into tribes or *iwi*, each of which consists of several subtribes or *hapū*.[2] For hundreds of years, the hapū of Whanganui lived alongside the Whanganui River, which flows from the central North Island down to the Tasman Sea. At 280 km, it is New Zealand's longest navigable river. The hapū were known as the River People, who lived in a narrow margin along the banks of the River. At one stage there were 140 river settlements.

Three women wearing cloaks and taniko headbands sitting and standing on a flax mat above the Whanganui River. (Photograph by James Ingram McDonald in 1921. Courtesy of the National Library of New Zealand.)

## Treaty of Waitangi

In the early nineteenth century, Europeans began visiting New Zealand on a regular basis and, in 1840, to facilitate immigration in NZ of European people, the British government negotiated a treaty with Māori. This treaty, which recognised indigenous rights, was known as the Treaty of Waitangi. It was a very short document, which said:[3]

> HER MAJESTY VICTORIA Queen of the United Kingdom of Great Britain and Ireland regarding with Her Royal Favour the Native Chiefs and Tribes of New Zealand and anxious to protect their just Rights and Property and to secure to them the enjoyment of Peace and Good Order has deemed it necessary in consequence of the great number of Her Majesty's Subjects who have already settled in New Zealand and the rapid extension of Emigration both from Europe and Australia which is still in progress to constitute and appoint a functionary properly authorized to treat with the Aborigines of New Zealand for the recognition of Her Majesty's Sovereign authority over the whole or any part of those islands – Her Majesty therefore being desirous to establish a settled form of Civil Government with a view to avert the evil consequences which must result from the absence of the necessary Laws and Institutions alike to the native population and to Her subjects has been graciously pleased to empower and to authorize me William Hobson a Captain in Her Majesty's Royal Navy Consul and Lieutenant Governor of

such parts of New Zealand as may be or hereafter shall be ceded to her Majesty to invite the confederated and independent Chiefs of New Zealand to concur in the following Articles and Conditions.

**Article the First**

The Chiefs of the Confederation of the United Tribes of New Zealand and the separate and independent Chiefs who have not become members of the Confederation cede to Her Majesty the Queen of England absolutely and without reservation all the rights and powers of Sovereignty which the said Confederation or Individual Chiefs respectively exercise or possess, or may be supposed to exercise or to possess over their respective Territories as the sole Sovereigns thereof.

**Article the Second**

Her Majesty the Queen of England confirms and guarantees to the Chiefs and Tribes of New Zealand and to the respective families and individuals thereof the full exclusive and undisturbed possession of their Lands and Estates Forests Fisheries and other properties which they may collectively or individually possess so long as it is their wish and desire to retain the same in their possession; but the Chiefs of the United Tribes and the individual Chiefs yield to Her Majesty the exclusive right of Preemption over such lands as the proprietors thereof may be disposed to alienate at such prices as may be agreed upon between the respective Proprietors and persons appointed by Her Majesty to treat with them in that behalf.

Unidentified Māori girl in a waka. (Photograph by Frank J. Denton, c. 1900. Courtesy of the National Library of New Zealand.)

**Article the Third**

> In consideration thereof, Her Majesty the Queen of England extends to the Natives of New Zealand Her royal protection and imparts to them all the Rights and Privileges of British Subjects.

Despite an original good-faith intent by the Crown, there has been a vigorous debate over the meaning and effect of the Treaty, and whether the English version, which relied on common-law principles of property ownership, differs in meaning from the Māori version, which was based on native title, grounded in beliefs of river ancestry. Scholars have asked whether the Treaty has status in international law, and whether, apart from the Treaty, Māori customary law has protection at common law. However, any analysis of these arguments is beyond the scope of this chapter.[4]

## The Waitangi Tribunal

In the early 1970s, there was a growing awareness of Māori ethos and the need to address longstanding historical grievances.[5] In 1975, Parliament passed the Treaty of Waitangi Act, the preamble of which explains its purpose:[6]

> Whereas on the 6th day of February 1840 a Treaty was entered into at Waitangi between Her late Majesty Queen Victoria and the Māori people of New Zealand: And whereas the text of the Treaty in the English language differs from the text of the Treaty in the Māori language: And whereas it is desirable that a Tribunal be established to make recommendations on claims relating to the practical application of the principles of the Treaty and, for that purpose, to determine its meaning and effect and whether certain matters are inconsistent with those principles.

Initially, the jurisdiction of the Tribunal was limited to contemporary grievances, but was enlarged in 1985 by an amendment to the Act[7] to include claims by any Māori, or group of Māori, who were or were likely to be, prejudicially affected by any act done, proposed or omitted by or on behalf of the Crown since the 6th day of February 1840.

The Act and amendment transformed the Crown/Māori relationship in the years following its passage. Although Māori claims related to the Whanganui River date as far back as 1883, many remained unsettled. Since the mid-1980s, Māori, either individually or in groups, made claims to the Tribunal, which produced many reports recommending action by the Crown to remedy its past failures.[8]

The first reports, released in the late 1980s and early 1990s, challenged the government, led at that time by Jim Bolger,[9] to respond. In 1992, the government set aside $NZ 1 billion for settlement of all historical claims.[10] Mr. Bolger appointed a Minister for Treaty of Waitangi Negotiations whose job it was to negotiate and conclude the settlements of all historical grievances.[11] The first major settlement was concluded in the mid-1990s with the Waikato Tainui iwi, in respect of land confiscation (or raupatu) during the Māori Land Wars.[12] However, this settlement only related to land claims, specifically excluding claims by the Waikato Māori to the rivers and harbours within their territory. So, the settlement was not full and final but partial.

Unidentified Māori group outside a meeting-house in Tieke, upper Whanganui River, circa 1910. (Photographer unknown. Courtesy of the National Library of New Zealand.)

The Waikato River claim was settled in 2010;[13] the Waikato Harbour claims are yet to be resolved. Settlements with many other iwi around New Zealand have been concluded since the mid 1990s.[14] The goal was to achieve just and durable settlements with Māori to properly address past failings, and then move the country forward. Sixty-one percent of claims have been settled thus far, with general support across the political divide. The public generally regards these settlements as fair.

## The Claims Settlement Process

The process has been reviewed and refined several times since the first negotiations in the early 1990s. The current approach is as follows:

a. *Deed of mandate*: This is the first and vital stage. The Crown needs to know with whom it is negotiating and must therefore recognise the mandate of an entity endorsed by the claimant community to represent that community in negotiations with the Crown. This can be a difficult (and litigious) process especially if, for example, a dissenting faction is opposed to one group representing the negotiating iwi. Sometimes the issues in this stage cannot be resolved.

b. *Terms of negotiation*: The next stage is to be very clear about the way in which the negotiation will be conducted. The terms will set out the 'ground rules' and objectives for the negotiations.

c. *Agreement in principle*: After sometimes-lengthy negotiations, the parties will agree the essential elements of a settlement and will sign an agreement in principle or heads of agreement. This is a nonbinding agreement which outlines, at a high level, all redress proposed to settle the claims. Technical and drafting details are agreed during the deed of settlement and legislation phases. Some agreements in principle are skeletal, whilst others are very detailed and approach the level of detail required in a deed of settlement. It all depends on the circumstances.

d. *Initialling a deed of settlement*: An initialled deed of settlement sets out in technical detail the historical claims and the redress agreed between the Crown and mandated body. For the purposes of the Crown's internal accountancy, this is the point at which the value of the settlement is counted against multiyear Treaty settlements appropriation. Interest on the settlement figure usually commences from this date, although some settlements have started to accrue interest once the agreement in principle is signed.

Elders and other representatives of Whanganui River tribes, with their counsel, D. G. B. Morison, in Wellington for a hearing of the Māori Appellate Court of their claim to ownership of the bed of the Whanganui River. Back row, from left to right: Kaiwhare Kiriona, Tanginoa Tapa, Tekiira Peina, Tonga Tume, Hohepa Hekenui, Henare Keremeneta. Middle row: Te Rama Whanarere, Hekenui Whakarake, D.G.B Morison, Titi Tihi, Tonga Awhikau. Front row: Taka-te-iwa Anderson, Kahukiwi Whakareke. (Photograph taken at the Dominion Museum in 1945, by an unidentified photographer. Courtesy of the National Library of New Zealand.)

e. *Ratification*: During ratification, the claimant community has the opportunity to vote on the final Crown offer as set out in the initial deed of settlement. It will also usually vote on the proposed postsettlement governance entity that will receive, hold and manage settlement redress on their behalf. This stage can also be very litigious, as dissenters sometimes try to stop a settlement proceeding. Some recent settlements have been delayed because of ratification problems.
f. *Signing a deed of settlement*: The mandated body, the postsettlement governance entity and the Crown sign the final deed of settlement once the Crown is satisfied that the claimant community has accepted, by ratifying, the deed of settlement as concluding all their historical claims.
g. *Legislation*: The Crown's office responsible for drafting statutes drafts a settlement bill for introduction to Parliament. This legislation must be agreed to by the negotiating group. It gives effect to the ratified deed of settlement and authorises settlement redress to transfer to the ratified postsettlement governance entity.

Most negotiations are led by Chief Crown negotiators appointed by the Minister from both the private and public sectors.[15] At key points in a negotiation, the Minister will be available to conduct a face-to-face negotiation. This can be very useful if there is a stalemate in negotiations.

## Components of a Settlement

What are the components of a settlement?

First (and arguably most importantly), an acknowledgment of historical acts or omissions by the Crown which may have resulted in the erosion of traditional tribal structures and practises, loss of tribal lands, loss of access to forests and waterways, food sources and sacred places.

Second, an apology for the Crown's actions. Apologies are not pro forma; they are carefully drafted by Crown and Māori historians. They can be the source of intense debate between the parties.

Third, commercial redress. This is not *restitutio in integrum* or full compensation. The country could never afford to pay such large sums. Nor is it notional, as is often alleged by some people. It is redress that augments existing iwi assets.[16] An important aspect of commercial redress is a right of first refusal to purchase surplus property owned by the Crown for a period following settlement. In urban areas this can be a very valuable part of the settlement package.

Fourth, the return to the iwi of culturally significant land now in the ownership of the Crown. This will include, for example, ancient burial sites and places where villages once stood. Originally Crown policy provided for the return of small and significant sites, but the policy has been applied more liberally in recent years.[17]

Fifth, an agreement will provide for a postsettlement relationship between the Crown and the settling iwi. These settlements are not 'commercial deals', as they are sometimes inaccurately and derisively called; they are important historical settlements which will hopefully transform the relationship between Crown and Māori in the future. A detailed and meticulous approach is taken to settlements because they are designed to last forever.

As has been said, these settlements are to be full and final. However, the greatest risk to finality is that the Crown forget its obligations in settlements. There are currently around 7,000 ongoing commitments to date. That is why a Post-Settlement Commitments Unit has been established to work within the Crown to safeguard the durability of settlements. New Zealand history is littered with examples of agreements between Crown and Māori which have been undermined or ignored because of Crown actions.

The final component of a settlement may be an agreement to give iwi an opportunity to co-govern a natural resource like a river or lake. The extent of the co-governance mechanism will depend on the intensity of the relationship between the settling iwi and the resource.[18] There is no 'one size fits all' approach to this aspect of a settlement. Each presents unique circumstances. For example, the agreements negotiated with Whanganui River are very different to those negotiated with other iwi over the Waipa[19] or Rangitaiki[20] Rivers, or other agreements for rivers in an area of the east coast of New Zealand's North Island called Hawke's Bay.[21] The reason for that will be explained later in this chapter.

Although successful, the negotiating process is far from perfect. A former United Nations Special Rapporteur, James Anaya, visited New Zealand in 2010 to look at the way this country negotiated historical settlements with Māori. He noted the various faults with the settlement process but concluded that 'the Treaty settlement process in New Zealand, despite evident shortcomings, is one of the most important examples in the world of an effort to address historical and ongoing grievances of indigenous peoples, and settlements already achieved have provided significant benefits in several cases. However, steps could be taken to strengthen this process'.[22]

Poutama Te ture, his wife and others smoking eels at Koriniti. (Photograph by James Ingram MacDonald in 1921. Courtesy of the National Library of New Zealand.)

## The Whanganui River Settlement

When I first became involved in negotiations over the Whanganui River, many factors persuaded me of the urgent need for a just and durable settlement with the local Māori:

1. The length of association between the Māori of Whanganui and the River. This was nowhere better expressed than in the opening paragraph of the executive summary of the Waitangi Tribunal's 1999 Report. There the Tribunal summarised the historic link:

    > For nearly a millennium, the Atihaunui hapū have held the Whanganui River. They were known as the river people, for uniquely amongst the rivers of New Zealand, the Whanganui River winds through a precipitous terrain that confined most of the large Atihaunui population to a narrow margin along its banks. There were, last century, some 140 river pa (fortified villages) and many large, carved houses that tell of substantial and permanent settlements. The river was central to Atihaunui lives, their source of food, their single highway, their spiritual mentor. It was the aortic artery of the Atihaunui identity. It is the main focal point for the Atihaunui people, whether living there or away. Numerous marae still line its shores.[23]

2. The many claims about the River go back almost a century. These claims were brought in the general courts of New Zealand as well as in the Māori Land Court. (The Māori Land Court has jurisdiction to determine claims brought in relation to Māori land which, originally, being 100% of the land in New Zealand, now makes up 11% of the North Island and a very small amount of the South Island.) In its interim report on the River, the Tribunal summarised the position:

    > Rarely has a Māori river claim been so persistently maintained as that of the Whanganui people. Uniquely in the annals of Māori settlement, the country's longest navigable river is home to just one iwi, the Atihau-a-Paparangi. It has been described as the aortic artery, the central bloodline of that one heart.
    >
    > The Atihau-a-Paparangi claim to the authority of the river has continued unabated from when it was first put into question. The tribal concern is evidenced by numerous petitions to Parliament from 1887. In addition, legal proceedings were commenced as early as 1938, in the Māori land Court, on an application for the investigation of the title to the riverbed. From there the action passed to the Māori Appellate Court in 1944, the Māori Land Court again in 1945, the Supreme Court in 1949, to a further petition and the appointment of a Royal Commission in 1950, to a reference to the Court of Appeal in 1953, to a reference to the Māori appellate Court in 1958 and to a decision of the Court of Appeal in 1962. This may represent one of the longest legal proceedings in Māori claims history, yet in all those proceedings, it is claimed, the principles of the Treaty of Waitangi had no direct bearing. Nor did the matter rest there, for the court hearings were followed by further petitions and investigations, and in more recent times, Atihau-a-Paparangi were again involved in the Catchment Board inquiry on minimum river flows in 1988 and in the Planning Tribunal and High Court hearings on the same matters in 1989, 1990 and 1992'.[24]

3. The claim was based on the Treaty of Waitangi but also on the Common Law of England, inherited by New Zealand. The Treaty of Waitangi affirmed Māori property rights. Of any reading of Article 2, it was at least arguable that rivers came within the phrase 'other properties' possessed by Māori as set out in Article 2 of the Treaty.

The legal recognition of prior property interests of an indigenous people (often referred to as native title) was a long-standing principle of English law, recognised by New Zealand courts as early as 1847.[25] New Zealand courts had determined, as a matter of law, that the prior ownership of Māori extended to rivers. That ownership was affected by legislation in 1903[26] which provided that the beds of all navigable rivers 'shall remain and shall be deemed to have been always vested in the Crown'. As so often happened with legislation that expropriated the property interests of Māori, it was rushed through the Parliament without any debate and no consultation.

4. The denigration of the River by the Crown. The way in which the River had been used by the Crown over the years for various public purposes including transportation, harbour works, waste disposal, flood control, tourism, power generation and scenery preservation led to water pollution and loss of use of the River. Some of this occurred even before the bed of the River was taken under the 1903 legislation. This happened without any consultation or adequate consultation with Māori. As so often happened in New Zealand's history, Māori were bystanders as their property was either taken or harmed. For example, in 1958, a Crown order in Council (like an executive order in the United States) authorised the diversion of water from the Whanganui River into a proposed massive water scheme. The Crown failed to consult Māori, who opposed this scheme and contended that the diversion and the reduced flow damaged the Whanganui River and adversely affected their cultural and spiritual values.
5. The effect that lack of respect for Māori property rights had on the Whanganui people. The Māori association with the River had either been severed or spoilt by new developments. James K Baxter, New Zealand's bard, who lived and died in the river settlement of Jerusalem, wrote the following sonnet:[27]

> The brown river, te taniwha, flows on
> Between his banks – he could even be on my side
>
> I suspect, if there is a side – there are still notches worn
> In the cliffs downstream where they used to shove
>
> The big canoes up; and just last week some men
> Floated a ridge-pole down from an old pa
>
> For the museum – he can also be
> A brutal lover; they say he sucked under
>
> A young girl once, and the place at the river-bend is named
> After her tears – I accept that – I wait for
>
> The taniwha in the heart to rise – when will that happen?
> Is He dead or alive? A car goes by on the road
>
> With an enormous slogan advertising
> Rides for tourists on the jetboat at Pipiriki

This sense of loss was referred to by witness after witness in the Tribunal hearings. Tariana Turia[28] talked about how her adopted mother spoke of her relationship with the River as if it was an integral part of her life:

> Her life was about the river as their sustenance. It provided for them physically, spiritually, and culturally. It was also their place of recreation. I never understood

> this as a teenager, because my experience of the river down here at Putiki was that we were not allowed to swim or eat kai from the river as it was polluted.

Other witnesses in the tribunal spoke about the harmful effect of Crown actions on the children of the river.

6. The length of the claim. This was referred to by James Anaya in his report to the United Nations in 2011.[29] He commented that the Tribunal had reported as far back as 1999. Prior to the 2008 election of the author's party to the government, the Crown negotiated with the Whanganui Iwi in 2003–2004, but only in response to the iwi's reoccupation of a contentious parcel of land (Moutoa Gardens, see map) in central Whanganui city. The protest intended to draw attention to the lack of negotiation since the Waitangi Tribunal Report, exemplifying the point made below that any redress was on the Crown's terms and timetable. Thus, it was only in September 2009 that Te Ati Haunui a Paparangi (the Whanganui Iwi) entered into settlement negotiations with the Crown over the River. This illustrates a weakness of the New Zealand Treaty Settlement process. It is the Crown who determines when it will enter into negotiations. Save in a few respects, which are irrelevant to this situation, recommendations of the Tribunal are not binding. The Crown often only acts on a report when it is ready to do so. It is in many respects a very one-sided approach.
7. The need to protect Māori property rights. Many European New Zealanders, or Pakeha, have a deep understanding of their own property rights but seem to regard Māori property rights as either illegitimate or incidental. Once something was taken from Māori, whether legally or illegally, many people thought that the passage of time meant it should be regarded as forming part of the public estate, and that public access to the outdoors should not in any way be compromised, let alone by Māori claims.[30]

   In 2002, the New Zealand Court of Appeal had found that Māori had the right to go to court to try to prove customary title in the foreshore and seabed.[31] Following a public outcry (and all the usual lamentations about 'activist judges'), the Labour Government passed legislation that prevented court action and extinguished any uninvestigated customary title.[32] When the National government took office, one of its first steps was to review that legislation and restore the litigation rights of Māori.[33] Once again, those claiming public estate via passage of time strongly opposed any reform and were adamant that the beaches were for all and that customary title in the foreshore and seabed was an unthinkable proposition. So often in New Zealand's history, attempts by Māori to protect their property rights have been met with spurious incantations about public access and public property. These issues were sure to be litigated in any settlement with Māori.
8. The unbreakable connection between Whanganui Māori and the River. From a European point of view, this concept may appear to be bizarre, even animist, but it was a deep-seated and genuine expression of belief by countless generations of Māori who had lived along the River. As one of the ancient sayings (or *pepeha*) of the Whanganui people provides:

> *E rere kau mai te awa nui*
> *nei, Mai te kahui maunga ki*

*Tangaroa Ko au te Awa*
*Ko te Awa ko au*

The River flows from the mountain to the sea
I am the river
The river is me

The interdependence of the river and its people was described by Niko Tangaroa, a Māori elder:

> *Ka mate te Awa, ka mate tatou te Iwi*
>
> The river and the land and its people are inseparable. And so if one is affected the other is affected also. The river is the heartbeat, the pulse of our people … [if the river] dies, we die as a people.

In conclusion, I was convinced that the Māori of the Whanganui River had a powerful case based on well-established property rights and Treaty principles, and that this matter cried out for a just and durable settlement as quickly as possible. In my 9 years as a Minister I learned a lot about iwi and their aspirations. I began to understand that when it came to property rights, Māori were more often than not correct in their contentions. It is one of New Zealand's tragedies that Māori property rights were not protected like European property rights and that the rule of law did not on occasion apply to everyone, but only to the majority. Negotiating a just and durable settlement was going to be a challenge.

## The Solution

As Chief Crown Negotiator, Dr. John Wood's first negotiation was with a tribe called Ngai Tuhoe. Many chapters could be written on the struggles of the Tuhoe people since the nineteenth century. Very briefly, their lands had been effectively taken by the Crown and in 1954 turned into a national park without any consultation, let alone consent, of the tribe. Dr. Wood and his team negotiated a settlement with Ngai Tuhoe that involved taking the national park out of Crown ownership and vesting the land in itself.[34] This area, known as Te Urewera, was to be initially governed by a board made up of equal numbers of Tuhoe and Crown representatives and then, after a period, it was to be governed by a Tuhoe majority.[35]

In the course of thinking about whether that precedent could be employed here, I was referred to some North American writings that discussed the idea of legal personality for natural resources.[36] I was attracted by this approach, as I knew that the corporate form developed by Anglo-American law could be a very flexible tool, not only for companies, but also for incorporated societies and charitable trusts. Legal rights are self-evidently not the same as human rights. A legal person does not have to be a human being. Recognising the legal personality of a natural resource was not an outlandish idea but could provide the basis for a just and durable settlement. I saw no reason in principle why these legal rights could not be appropriate for Nature.

Dame Tariana Turia and the Honorable Christopher Finlayson engaged in hongi at the signing of the Whanganui River Claims Settlement Agreement, August 6, 2014. (Photograph by Bevan Conley. Courtesy of the Whanganui Chronicle.)

In 2014, the New Zealand Cabinet accepted the proposal that the River be given the status of a legal person. There were other aspects to the redress, including various acknowledgements, an apology and payment of redress. The two key parts of the Whanganui River Claims Settlement Act established the framework for the river and set out redress for the Whanganui River iwi. The first part consists of the following:

a. The River was recognised as an indivisible and living whole, comprising the Whanganui River from the mountains to the sea, incorporating all its physical and metaphysical elements.[37] Some people have looked at that and expressed surprise that this kind of language is contained in a statute. Their concerns are misplaced. It is a much more coherent and realistic way of looking at a river as opposed to an unreal and almost bizarre segmentation of a river into a riverbed, the water column and the airspace above the water. That was the way water bodies in New Zealand tended to be analysed in the past, including for the purposes of Treaty settlements. When looked at objectively, the River is a single entity.
b. The River is a legal person and has all the rights, powers, duties and liabilities of a legal person.[38] The legal effect of the declaration is that any person exercising or performing a function in certain specified statutes (for example, the Resource Management Act 1991, which is the primary planning statute) must recognise and provide for the status of the River and the intrinsic values that represent the River.[39]
c. The River is to be represented by an office called Te Pou Tupua.[40] This is its human face. Te Pou Tupua is a joint appointment by the Crown and the iwi of Te Awa Tupua. One person is nominated by the Crown, and one person is nominated by the iwi, but neither owes a duty it its nominator. The appointment holds all powers

reasonably necessary to achieve its purpose, laid out in the statute as including the following functions:

> To speak and act for Te Awa Tupua;
> To uphold its status and the intrinsic values;
> To promote and protect the health and wellbeing of the River;
> At all times to act in the interests of Te Awa Tupua and consistently with the intrinsic values.

Very importantly, the human face may engage with any agency or decision-maker to assist it to understand, apply and implement the status and the intrinsic values of the River. They may also participate in any statutory process affecting the River in which they would be entitled to participate under any legislation.[41]

The first two appointees to Te Pou Tupua were Turama Hawira, a senior leader and tribal historian appointed by the Whanganui iwi, and Dame Tariana Turia, a witness in the Waitangi Tribunal hearings over the Whanganui River. First elected as a Labour Member of Parliament, she served as a Minister in the early 2000s but left her Party in the aftermath of the Foreshore and Seabed controversy. She then founded the Māori Party, which was in a confidence-and-supply relationship[42] with the national government between 2008 and 2017. She exemplifies what American President John F. Kennedy once said: 'It takes courage to stand up to your enemies and even more courage to stand up to your friends'. No New Zealander has a greater understanding of the aspirations of the Whanganui iwi.

The two representatives of the River have only recently been appointed (2018) and are just embarking on their tasks. One thing is certain – perfection will not be achieved any time soon. The environmental degradation of the River will not be reversed for many years. For example, the consents to take water for the hydroelectric scheme referred to above will continue for at least another decade. Change will be incremental, not revolutionary. Critical to the success of this project will be the development of close working relationships with the Regional Council and the relevant district councils, particularly the Whanganui District Council.

Four years ago, the idea that the Whanganui River would become a legal person in 2017 would have been regarded as a fantasy. Yet in a decade, local government has become very supportive of Treaty settlements, even if there are legitimate concerns about the costs to them. They particularly see the implementation of settlements as part of their core duties. They are committed to make them work. Building good relationships with communities is an important first task for the human face of the River.

In my experience, acceptance of new realities comes reasonably quickly, provided the preparatory work is done well. For example, those who have engaged with communities and advocacy groups (like those who contend for public areas and recreational groups) will find the task much easier than those who immediately after settlement seek to 'lay down' the law. A self-evident statement perhaps, but one that is lost on some iwi in recent times. At least one major settlement of which I am aware could fail to achieve its potential because the settling iwi have failed to engage with their community and have consequently alienated the community.

d. A strategy document must be developed directed to the future environmental, social, cultural and economic health and wellbeing of the River. This document will identify issues and provide a strategy. Decision-makers in legislation affecting the River will be required to have particular regard for the River strategy.[43]

e. A strategy group will be created comprising representatives of iwi, cultural and local government, and other groups with interests in the River, including the primary sector, recreational and environmental groups. This will provide a great opportunity for all intended parties to work together to devise a long-term strategy.[44]
f. The Crown agreed to establish a $NZ 30 million fund. This will support the health and wellbeing of the River.

The second part of the Whanganui River Claims Settlement Act sets out redress for the Whanganui River iwi, which consists of the following:

a. Whanganui iwi redress. There is an agreed historical account, Crown acknowledgments and an apology for its acts or omission, including its failure to protect the interests of the iwi.[45]
b. Whanganui iwi are recognised as having an interest in, and responsibility to, the River, for the first time in legislation. That interest is greater than, and separate from, the interests of the public generally. They will have the statutory rights to lodge submissions and participate in hearings related to the River.
c. Payment of fundamental redress of $NZ 80 million to Whanganui iwi in recognition of all claims and to assist their work to advance the future health and well-being of the River and its people.
d. Other cultural redress. An example is the recognition of traditional customary activities which can now be undertaken without the need for consent or other authorisation under certain statutes.[46]

## Conclusion

There are a number of reasons why I am confident that this redress will be successful.

First, both the River and the Whanganui iwi are adequately funded to do their job properly. It would have been counterproductive to establish these new structures and not have them properly funded.[47]

Second, this is not standalone legislation. Every effort has been made to ensure that the new regime for the River is aligned with related resource legislation administered by regional and local government. Early signs indicate a willingness on all sides to work together to address the environmental problems faced by the River.

Finally, several of the new structures mandate collaboration between previously antagonistic parties. So, the energy company (whose predecessor established the scheme which damaged the River) will have to talk to the environmental and recreational groups. There is every incentive for everyone to work together to get ahead without the need for time-consuming and expensive legislation.

The international attention to the novelty of this settlement may be greater than that received at home. But if all players commit to the settlement, I think that the new arrangements will be successful. After years of misery, watching the degradation of the River and being unable to do anything, Whanganui iwi now have the clearance to advance great things for their River and the region.

## Acknowledgments

The author would like to thank Mr. Gerrard Albert, who provided essential advice and editing for this paper. Mr. Albert was a lead negotiator for the Whanganui Iwi. He now leads the Te Kopuka group, which is composed of representatives from across the River's governance and stakeholder spectrum. Te Kopuka will formulate the strategy necessary to realise the potential of the Te Awa Tupua agreement protecting the Whanganui River. Mr. Albert has an intimate understanding of the customs, traditional values and etiquette of the Whanganui Iwi in relating to the River.

The author would like to thank Ms. Lynne Reid for her invaluable help in organising and editing this paper. Lynne L. Reid is a criminal defence attorney practising in the federal courts in the Eastern District of North Carolina and the Fourth Circuit Court of Appeals. She resides in Chapel Hill, North Carolina, with her husband Andrew A. Young, a native of Taranaki, New Zealand.

## Notes

1. Te Awa Tupua (Whanganui River Claims Settlement) Act, Public Act 2017 No. 7, 20 March, 2017. Te Awa Tupua is a Maori expression referring to the river with ancestral power as a whole integrated entity from the mountains to the sea.
2. In Māoridom and New Zealand, a hapū ('subtribe', or 'clan') functions as 'the basic political unit within Māori society'. The iwi (tribe) is made up of various hapū and may have up to several hundred members and trace back to the waka (canoe) on which their founding ancestor arrived. For example, the Waikato tribes trace their descent from the *Tainui* waka. War, migration and family infighting also caused new iwi or hapū to emerge. Iwi and hapū are often named after an ancestor. For example, Ngāpuhi means 'the people of Puhi' and Te Uri-o-Rātā means 'the descendants of Rātā'. Sometimes tribal names came from an important event that involved their ancestor, such as a battle. Each hapū is made up of whānau (extended families). Whānau included much-respected elders, adults, children and grandchildren. Everyone helped each other, working for the group and caring for each other's children and the elderly. See https://teara.govt.nz/en/tribal-organisation
3. Found in Schedule 1 of the Treaty of Waitangi Act 1975.
4. There is a mass of material on this subject. For a fair external summary of the position and the arguments, see Anaya, James. 2011. 'Report of the Special Rapporteur on the rights of indigenous peoples', United Nations General Assembly 2011, A/HRC/18/xx/Addy
5. The Māori Land Wars were an example of such historical grievances. Triggered by tensions over disputed land purchases, they took place from 1845 to 1872 between the New Zealand government and the Māori. They escalated dramatically from 1860 as the government became convinced it was facing united Māori resistance to further land

sales and a refusal to acknowledge Crown sovereignty. At the peak of hostilities in the 1860s, 18,000 British troops, supported by artillery, cavalry and local militia, battled about 4,000 Māori warriors in what became a gross imbalance of manpower and weaponry. Large areas of land were confiscated from the Māori by the government under the New Zealand Settlements Act in 1863, purportedly as punishment for rebellion. In reality, land was confiscated from 'loyal' and 'rebel' tribes alike. See https://en.wikipedia.org/wiki/New_Zealand_Wars.

6. Treaty of Waitangi Act 1975.
7. Treaty of Waitangi Amendment Act 1985.
8. See, for example, the report on which the Whanganui claim was negotiated – WAI 167, *The Whanganui River Report* (1999). This document was a history of the settlement, use, ownership and control of the Whanganui River, commissioned by The Treaty of Waitangi Tribunal, authored by Thomas Bennion. He drew heavily upon the research of David Young, who published the authoritative *Woven by Water: Histories from the Whanganui River* a few years later (Huia Publishers, 1998). It described 150 years of Maori/Pakeha relationship. It is clear that some developments, such as the establishment of a steamboat service, were embraced by many Maori, who saw it as facilitating their livelihood via trade, and even sought to become investors in the company. But the subsequent clearing of the river to prepare for the steamboat service, and removal of traditional eel and lamprey traps, was oblivious to customary Maori use and claim to the river, as were many of the public works to alter the bed of the river. Transactions involving land alongside the river were varied; some claimed to the high-water mark, some to the low-water mark and some to the center of the river, but universally ignored Maori claim to the river. There were also many bad-faith actions: land confiscations, shady land deals, diversion of water, all without the consultation or consent of Maori and which resulted in damage to the river and injury to the well-being of the Maori along the river. The establishment of the native land court, which served mainly to acquire Maori land, was not a sincere instrument to dispense justice. There were further insults to the Maori culture that led to grievances.
9. Rt. Hon James Bolger ONZ, Prime Minister of New Zealand 1990–1997; subsequently Ambassador to the United States, 1998–2002. His autobiography *A View from the Top: My Seven Years as Prime Minister*, Viking Press, 1998, has a very interesting account of his stewardship of the early Treaty Settlements; see pp. 176–184. An excellent summary.
10. The national government of Jim Bolger settled on this sum, in the knowledge that it was never going to be adequate, and the figure was abandoned by the late 1990s. The settlements are currently worth about $NZ 2.5 billion, and the final cash figure will be around $NZ 3.5 billion. Cash was not the only commercial component – right of first refusal of government property was also valuable compensation, and now constitutes the bulk of the postsettlement value to the iwi. See footnote 17.
11. The author held this position from 2008 to 2017, as well as being Attorney-General.
12. Waikato-Tainui Raupatu Claims Settlement Act 1995: see particularly section 8 for the meaning of the phrase 'Raupatu Claims'. See also footnote 6 above, on the Maori Land Wars.
13. Waikato-Tainui Raupatu Claims (Waikato River) Settlement Act 2010.
14. A full list of settlement legislation is contained in schedule 3 of the Treaty of Waitangi Act 1975. Once an historic claim has been settled, the Waitangi Tribunal has no jurisdiction to inquire or further inquire into a claim – see section 6.
15. For example, the Chief Crown Negotiator for the Whanganui River Claims was John Wood CNZM, QSO, a retired diplomat who served as New Zealand's Ambassador to Iran, Turkey, Pakistan and twice in the United States (1994–1998 and 2002–2006). He is also a former Deputy Secretary of Foreign Affairs, and is currently Chancellor of the University of Canterbury. He was recommended to the author by New Zealand's current Ambassador to the United States, Hon. Tim Groser, who described John as a brilliant and innovative negotiator. Groser was completely correct. The Whanganui River Settlement was a result of his very clever and principled negotiation.

16. See further http://www.chapmantripp.com/news/crown-m%C4%81ori-relationship-evolves-as-$50b-industry-grows, where the Māori economy is estimated to be worth approximately $NZ 50 billion, of which $NZ 6 billion comes from Treaty Settlements.
17. See, for example, part 2 of the Ngati Pahauwera Claims Settlements Act 2012, which sets out the cultural redress negotiated with the iwi.
18. For example, some iwi will be satisfied with an advisory body, while others seek a full co-governance arrangement.
19. Nga Wai o Maniapoto (Waipa River) Act 2012, a full co-governance and co-management regime for the Waipa River, the King country.
20. Ngati Whare Claims Settlement Act 2012; see especially sections 108–120, which provide for the establishment of the Rangitaiki Rivers Forum, the purpose of which is the protection of the wellbeing of the Rangitaiki River in the Bay of Plenty.
21. Hawkes Bay Regional Planning Committee Act of 2015.
22. Anaya, James. 2011. 'Report of the Special Rapporteur on the rights of indigenous peoples', United Nations General Assembly 2011, A/HRC/18/xx/Addy, p. 22, paragraph 67.
23. Report of the Waitangi Tribunal Report (Wai 167) into the Whanganui River.
24. Interim Report of the Waitangi Tribunal on Wai 167, which can be found in Appendix IV of the Tribunal's 1999 Report on the Whanganui River Claim, p. 379.
25. *R v Symonds* (1847) NZPCC 387.
26. Coal Mines Amendment Act 1903.
27. From 'Jerusalem Sonnets, 1969'. *Collected Poems of James K Baxter*, Ed John Weir. Oxford University Press, 1980. The James K. Baxter Trust is gratefully acknowledged for permission to cite this poem. Maori mythology holds that a taniwha, a supernatural being inhabiting a waterway, guarded and protected it and the hapu. The taniwha, male or female, could sometimes take a lover. In this passage, te taniwha is accused of dereliction of his protective duty. Pipiriki is a Whanganui River settlement popular with tourists. The jetboat, a New Zealand invention, by virtue of its power requirement and shallow draft, is particularly noisy and invasive.
28. See Report of the Waitangi Tribunal (Wai 167) into the Whanganui River Claim, 1999, pp. 80 and 84.
29. Anaya, James. 2011. 'Report of the Special Rapporteur on the rights of indigenous peoples', United Nations General Assembly 2011, A/HRC/18/xx/Addy.
30. For example, the Federated Mountain Clubs of New Zealand, represented by Dr. Hugh Barr, had given evidence in the Tribunal and had been adamant that anything short of public ownership of the Whanganui River would be unacceptable, a position also held during the Foreshore and Seabed controversy of the early 2000's. See Barr, Hugh. *The Gathering Storm over the Foreshore and Seabed:Why It Must Remain in Crown Ownership*. Wellington [N.Z.]: Tross Pub, 2010.
31. *Ngati Apa v Attorney General* [2003] 3NZLR 643
32. Foreshore and Seabed Act 2004.
33. Marine Coastal Act 2011.
34. Te Urewera Act 2014.
35. The reader might assume that the Tuhoe negotiations over Te Urewera led to the decision to accord the Whanganui River legal person status. In fact, the opposite was the case. Whanganui Iwi provided an outline of its settlement aspirations to the author in 2009. The WAI 168 Whanganui River Claim Outline for Settlement described the establishment of a unique title in relation to the Whanganui River. The Outline for Settlement described the purpose and function of the Awa Tupua Title. The title would provide for legal expression of the relationship of the Iwi with the River that is not otherwise capable of legal recognition. The Awa Tupua Title was thus considered central to highest-level recognition of the River as a unique entity, the River's needs, the River-Iwi relationship, and Restoration of the Iwi-Crown relationship. From workshops held in 2010, 'legal personality' was agreed as the most appropriate legal vehicle whereby the River, including its headwaters and tributaries, could be recognised in statute as a single entity with accordant 'rights'. These 'rights' would be consistent with the overarching Principles and Vision set out in the Outline for Settlement document. It was following the Whanganui initiative, that it was proposed to Ngai Tuhoe as a solution to the 'tenure' impasse over Te Urewera National Park.

36. For example: Stone, Christopher D. 1974. *Should Trees Have Standing? Toward Legal Rights for Natural Objects.* William Kaufman, Inc.
37. Te Awa Tupua (Whanganui Claims Settlement) Act 2017, Sections 12 and 13.
38. Ibid, Section 14.
39. Ibid, Section 15.
40. Ibid, Subpart 3.
41. Ibid, Section 19.
42. A confidence-and-supply agreement is one whereby a party, or independent members of parliament, will support the government in motions of confidence and appropriation or budget (supply) votes, by either voting in favour or abstaining.
43. Te Awa Tupua (Whanganui Claims Settlement) Act 2017, Sections 35–38.
44. Ibid, Sections 29–32.
45. Ibid, Sections 69 and 70.
46. Ibid, Subpart 3.
47. Although the Iwi received 80 million dollars redress from the Crown, the funding for a postsettlement body and implement the Te Awa Tupua framework is less than a million dollars; its adequacy is uncertain.

# 14

# *The Rights of Nature in Ecuador: An Overview of the New Environmental Paradigm*

**Hugo Echeverria and Francisco José Bustamante Romo Leroux**

**CONTENTS**

The Rights of Nature: Origins ... 281
The Rights of Nature: Content ... 283
Constitutional Case Law Relating to Rights of Nature in Ecuador ... 284
Selection Criteria ... 284
Rights of Nature Case Law: An Ongoing Effort ... 285
Selected Cases ... 285
The Case of the Biodigesters ... 285
Mangroves Case ... 286
Unauthorised Mining Case ... 287
Vilcabamba River Case ... 289
Binding Jurisprudence on the Constitutional Rights of Nature Language ... 290
Conclusion ... 290
Notes ... 291

The Rights of Nature is undoubtedly the most significant legal step forward on major environmental issues brought about by the passage of Ecuador's 2008 Constitution.[1] The new language, for the first time giving Nature legal rights, gave rise to great expectations, in particular the possibility of transforming the traditional – and worldwide – paradigm in which Nature is understood to be a mere object available for human use. Many hoped that the view of human beings as the owners, de facto keepers and ultimate end purpose of all-natural resources would be challenged by this emerging environmental paradigm.

However, the expectations created thus far may prove to be ill justified. In practise, there is still no certainty about the effectiveness of the Rights of Nature language. Despite the new Constitution having been in force for 10 years, the State and civil society have progressively brought legal action to ensure the protection of these rights.

Although there are several developments, the State has yet to fully implement the institutional and legal framework required to ensure the adequate protection of the Rights of Nature. For example, there are no courts specialising in Rights of Nature. This is a pending task, considering that Ecuador was the first country ever to have constitutionally recognised Nature as a legal subject.[2]

In the public policy domain, there have – so far – been few actions aimed at promoting the Rights of Nature; in contrast, mining policy and the exploitation of nonrenewable natural resources remain priorities in the State's agenda to produce economic income. The Yasuní ITT initiative failed miserably. 'First World' countries were censured for not contributing with the necessary funds (equivalent to part of the income that Ecuador would have received from

DOI: 10.1201/9780429505959-14

hydrocarbon exploitation) to keep the oil below ground and prevent oil exploitation within Yasuní National Park. This led the government to make the 'difficult' decision to exploit oil within this Protected Area, which is internationally recognised for its mega-biodiversity, in addition to being the home of indigenous communities living in voluntary isolation.

Yasuni National Park, Ecuador. (Photo Wikimedia Commons/Agencia Latinoamericana de Información (ALAI), 2010. https://commons.wikimedia.org/wiki/File:Amazon%C3%ADa_Ecuador.jpg)

On the legal front, the current Environment Act – in force since April 2018 – was designed to protect the environment under the current model of sustainable development. It sets an environmental impact process which a project must go through in order to obtain an environmental permit. It also defines the jurisdiction of every governmental entity involved in environmental protection as well as administrative sanctions, legally protected area regime and so on.

However, it does not fully address Rights of Nature. The Act does not develop the content of these rights nor does it set any specific institutional and policy frameworks for implementing it. For example, it does not define 'integral respect for Nature', a key constitutional concept underlying Rights of Nature.

As a result, the work of constitutional judges takes on special significance under Ecuador's legal system, since their work goes beyond being mere enforcers of the law. On the contrary, in the absence of legislation implementing the constitutional level, constitutional judges are responsible for gradually setting out the scope and content of constitutional rights – including the Rights of Nature – through the development of case law and by directly enforcing constitutional provisions.

This analysis is divided into two sections. An opening theoretical section examines the Rights of Nature, their origin, content and key legal effects. With these at hand, the essay goes on to examine the most important cases settled through Ecuadoran courts on the Rights of Nature, in order to assess their effectiveness, and explain how they have been understood over these past 10 years of the 2008 Constitution.

## The Rights of Nature: Origins

Any attempt to establish the 'origin' of the Rights of Nature concept is a difficult task. However, in order to better understand its inception, it must be examined from a historical, political and social perspective. From an exclusively traditional legal standpoint, it would be impossible to do so, since according to Romano-Germanic legal systems, only humans may possess the characteristics of a legal subject: capacity, dignity, subjective rights and equality.[3]

As general background, Ecuador lived through an acute political and social crisis in the mid-2000s. A blend of general fatigue with the political class and distrust of democratic institutions led to a coalition of various social sectors (including indigenous communities, Afro-Ecuadorans and Montubio people). This was unprecedented in Ecuador. Social demands and the primacy of human beings above capital and private interests, as well as the need to establish a new relationship between human beings and Nature, were the major discourses of this new socio-political momentum.

One proposal the new coalition put forth was the need to introduce a new Constitution to help overhaul power relations, as well as the economic and social order. Among the proposed changes would be protecting the rights of indigenous peoples who had historically been excluded from the formation of the State and society since the Americas were colonised, without acknowledgment and respect of their values, customs, culture and worldviews.[4]

Discussion of indigenous rights first introduced the idea of granting rights to Nature, because in the ancient worldview of Andean peoples, the Pachamama (Mother Earth) is alive, and human beings are but a small component of this great living 'whole'. According to the indigenous worldview:

> Mother Earth is the centerpiece, understood as a mother who protects her daughters and sons, to whom it provides the necessary space, food, and the required – cosmic, physical, emotional, spiritual, identity-forming, cultural, and existential – elements in order to survive. It is the body of Nature that receives and gives the seed of life in its infinite expressions (…) there is no division between man-woman (or culture) and Nature. The balance, development, and survival of society depend on this harmonious relationship of embeddedness.[5]

Therefore, the recognition of rights in favour of Nature arose in the context of asserting the worldview of the indigenous Andean peoples, and the formation of a social, economic and political practise based on a qualitative relationship with Nature. This principle of the ancient worldview of Andean peoples is known in Ecuador as *Sumak Kawsay*, which translates as 'good living' in English. *Sumak Kawsay* thus establishes itself upon existing political-philosophical foundations to recognise the existence of rights in favour of other beings that do not belong to the human species.

This principle, *Sumak Kawsay*, is based on four subprinciples that enable us to arrive at a more precise understanding of the term. These subprinciples are relatedness, correspondence, complementarity and reciprocity.

The principle of relatedness, according to Ávila Santamaría:

> has to do with a holistic view of life. Everything is related, linked, connected to each other (…) The upshot of this principle is that Nature needs the beings that inhabit it, and those beings in turn would be unable to survive without Nature. Moreover, human beings are not in Nature, nor does Nature shelter human beings, but rather human beings are themselves Nature.[6]

This means that all elements that are part of Nature (or rather, *are* Nature) have a connection between them, so that any imbalance in any one of them directly affects the others.

The principle of correspondence, according to Walsh, refers to qualitative links established among the different beings that *are* Nature, insofar as all of Nature's elements are related to each other: '... reality's different aspects or components are in harmonious correspondence. It is not about (...) an analogous-proportional correlation, or a causal nexus, but relational bonds of a qualitative, symbolic, celebratory, ritual, and affective kind'.[7]

The principle of complementarity holds that all beings that are part of this great 'whole' coexist and complement each other, which departs from a dualistic view of reality. Given the above, Western concepts such as good and evil would not be antagonistic, but rather two notions that complement each other to form a 'whole', whereby: '... heaven and Earth, sun and moon, light and dark, truth and falsehood, male and female, Nature and humans cannot be mutually exclusive, but must instead necessarily complement each other for their assertion as a superior and integral entity'.[8]

Finally, the principle of reciprocity refers to the duties that all the elements of Pachamama have towards each one of themselves. For Walsh, this principle enables the establishment of reciprocal relationships among all beings and elements that make up Pachamama: 'This principle is founded on the necessary interrelationship between heavenly, earthly, and present worlds, between human beings and Nature'.[9]

*Sumak Kawsay* is therefore the joint expression and permanent interaction of these four subprinciples, as replicated in all spheres of life and which are also based 'on balanced, harmonious, equitable, and solidarity relations between humans and Nature, on the dignity of every human being and the necessary interrelationship between beings, knowledge, cultures, rationales and thinking, acting and living logics'.[10] *Sumak Kawsay* amounts to the principle or philosophical basis that guides social, economic and political structures, as well as the relationship and interaction with the natural environment.

*Sumak Kawsay*, as a concept, comes up repeatedly in the text of the Ecuadoran Constitution. In its preamble, for instance, the Constitution refers to *Sumak Kawsay* as the basis and purpose for building a form of coexistence between humans and Nature, which differs from the traditional relationship.[11] Similarly, under its provisions, the Constitution enshrines *Sumak Kawsay* as an objective of the right to a healthy and balanced environment.[12]

Likewise, the model of development[13] must seek the enjoyment of constitutional rights in a way that ensures *Sumak Kawsay* becomes the purpose and goal of the State.[14] In this regard, respect and protection of the Rights of Nature is a means to *Sumak Kawsay*. Another constitutional obligation for the State in the context of *Sumak Kawsay* is the promotion of and respect for ancestral knowledge,[15] which in practise translates into the source for understanding the inception of the Rights of Nature.

Thus, *Sumak Kawsay* amounts to an overarching constitutional principle about many kinds of relations. In addition, it became both a foundation and a goal for the State in building a new model of society that acknowledged the need to enter a different relationship with Nature for the purpose of social and economic development.

While Sumak Kawsay is the fundamental pillar to Rights of Nature, Echeverría argues that there are other pillars contributing to this emerging paradigm, namely the development of Ecuadoran Environmental Constitutionalism, which has included Nature's related topics since the 1990s, as well as International Environmental Law. Principle 1 of the 1982 World Charter for Nature proclaims respect for Nature as a general principle. As article 71 of the Constitution States, this principle lies at the centre of the Ecuadoran recognition of rights to Nature.[16]

## The Rights of Nature: Content

Now that the context leading up to the Constitutional recognition of Nature as a legal subject has been addressed, it is necessary to examine the content and scope of Nature's rights in order to establish how they are to be understood and interpreted.

We start by pointing out something that seems obvious, but it is important to state: when referring to the Rights of Nature, we acknowledge the emergence of a new subject or rights. These rights are not human rights, but rights recognised to Nature. Moreover, it is worth noting that Ecuadoran environmental constitutionalism has been influenced by two trends: on the one hand, a traditional anthropocentric approach, acknowledging the right of human beings to live in a healthy environment, to participate in environmental decision-making, to have access to environmental information and justice; and a contemporary, biocentric approach. As a result, under the Ecuadoran constitutional framework, two visions of Nature coexist: on the one hand, as an object of protection under the law, and second, as a subject of rights.

In this context, the rights to which Nature is entitled under the Ecuadoran Constitution are: (a) respect for her existence; (b) maintenance and regeneration of her life cycles, structure, functions and evolutionary processes[17] and (c) restoration.[18] Nevertheless, while the constitutional recognition of the Rights of Nature has been regarded as a step forward, it is no less true that there are doubts regarding how public officials, both administrative and judicial, should correctly interpret these rights. This especially applies to judges, since they are the chief guarantors and protectors of Constitutional rights.

What is to be understood by 'respect for the existence of Nature'? What elements, whether acts or omissions, may be considered violations to Nature's existence? It nevertheless appears there are helpful insights into the content and scope of any act that would be contrary to the Rights of Nature, or to the maintenance and regeneration of her life cycles, structure, functions and evolutionary processes. That is, any impairment or prejudice relating to this right could be verified by means of scientific or technical parameters. Such violation might even be visible at first sight.

As regards the right of Nature to be restored, one could argue that this is the least abstract of the rights. Restoration would involve a restitution of the affected ecosystem. In a comprehensive interpretation of the Constitution, the right to restoration could be linked to the quality of the life cycles, structure, functions and evolutionary processes of Nature.

In the absence of regulations below the constitutional level, fully setting out the scope and content of rights to Nature, the judges' task acquires greater relevance, since they are responsible for building it through their judicial decisions, which is a scenario related to the common law tradition. This task may be hindered by the civil law tradition roots of the Ecuadoran legal system, in which rules understood as legal provisions are the primary normative source to address legal matters, including Rights of Nature. But the problem remains the same: Rights of Nature lacks legal provisions that would help defining and explaining its scope.

Therefore, constitutional jurisprudence becomes the most appropriate mechanism for the further development of Nature's rights, given that the judges, in the absence of legal provisions, are compelled to apply constitutional principles, rules and values to resolve a given legal dispute. Thus judges, basing their decisions on the facts of each individual legal case and the relevant constitutional principles, progressively construct the content of the Rights of Nature.[19]

In addition to the duty of applying principles and constitutional values to build the content and scope of the Rights of Nature, the Constitution establishes criteria for the implementation of constitutional rights, which must always be observed by any public authority in exercise of their duties, namely:

a. All constitutional rights are enforceable, regardless of whether they are traditionally known as civil or political rights, or of an economic, social or cultural nature.[20]
b. No legal regulation (below constitutional level) can restrict the contents of rights or constitutional guarantees. This criterion must, above all, be observed by those agencies having regulatory authority.[21]
c. Rights and constitutional guarantees must be applied in their most favourable terms, meaning that public, administrative or legal servants must always interpret and apply them in keeping with the meaning that is most favourable to their effective force.[22]
d. All constitutional rights, including Rights of Nature, are inalienable, indivisible, unwaivable, interdependent and of equal hierarchy. Therefore, the Rights of Nature are at the same level of importance as any other constitutional right to which a person is entitled.[23]
e. The foremost duty of the State, and that of its bodies, is to ensure respect for and compliance with constitutional rights.[24]

Where there are no legal provisions that expand upon the content of rights, constitutional judges must always consider the application of principles, and hence these play a major role in providing guidance to agents of justice in developing the content of rights. As a result, the Rights of Nature and their construction via case law will rely on citizens as their main overseers and guarantors. Citizens must be empowered and encouraged to make use of case law mechanisms to protect these rights, thus requesting judges to build content on a case-by-case basis.

## Constitutional Case Law Relating to Rights of Nature in Ecuador

In Ecuador, there are no official statistics on the number of Rights of Nature cases. However, there are data on a variety of constitutional, criminal, civil and even administrative actions throughout the country whose purpose has been the safeguarding of the Rights of Nature.[25] These actions relate to the protection of ecosystems and species, such as mangroves and sharks. But actions have also been filed to protect water and rivers, including the Vilcabamba River.

### Selection Criteria

From the body of case law, we have selected four cases that were admitted and resolved by the Constitutional Court of Ecuador between 2007 and 2018. These cases were selected based on two criteria: (a) they were resolved by the country's highest body responsible for the control, interpretation and administration of constitutional justice and (b) they provide a constitutional approach for interpretation of the Rights of Nature.

As regards the first criterion, the highest body responsible for the administration of constitutional justice is the Constitutional Court, which exercises national jurisdiction, and has among its powers the admission and resolution of actions filed by citizens or State bodies, whose purpose is to safeguard constitutional rights, including the Rights of Nature. These powers enable the Constitutional Court to examine the content and constitutional adherence of judicial rulings issued by lower courts, as well as to ensure their compliance.[26]

These cases were also selected because they present an interpretation of the Rights of Nature based on constitutional law,[27] which supersedes their interpretation based on environmental law, under which these new rights tend to be examined.

## Rights of Nature Case Law: An Ongoing Effort

Before discussing each of the cases, it is worth mentioning that Ecuadoran constitutional case law does not yet provide background elements regarding the legal content of the Rights of Nature. As it has been noted, approaching the subject matter is marked by the difficulty of defining and characterising these new rights: What are the Rights of Nature? How are they different from environmental rights? What are their legal effects in practise? These are the questions that have yet to find answers in case law.[28]

To some extent, this approach – which could be described as preliminary – is explained by the complexity of the subject matter, which entails a paradigm shift. But it is also explained by the short amount of time that has passed: a decade is very little time to develop the legal content of any constitutional right. Let us recall that the Universal Declaration of Human Rights dates to 1948, and 70 years on, its content continues to be developed.

Nevertheless, Ecuadoran constitutional case law has provided two important notions: (a) being a subject of rights – an attribution previously reserved for persons – can be extended to Nature, and (b) the State's duty of guardianship – previously applied to human rights – can be extended to the Rights of Nature. As will be seen below, these two notions stand out in each of the selected cases.

## Selected Cases

The selected cases refer to the protection of water, mangrove forest and a river on the one hand, and the impacts of mining on Nature on the other. Citizens filed two of the cases, while the other two were filed by State bodies. These cases not only reflect balance in terms of the actors, but also the implementation of judicial guarantees whose purpose is the effective and immediate protection of the rights enshrined in the Constitution.[29]

### *The Case of the Biodigesters*[30]

This case dates to 2007 and was the first to be resolved from a Rights of Nature perspective. It concerned the installation of six biodigesters in an industrial hog farm. The plaintiffs, members of an ethnic group whose territory borders the farm, complained about the environmental management of the agro-industrial enterprise, particularly air pollution through the emission of odours, as well as water pollution resulting from the discharge of organic waste into the surrounding bodies of water. However, their specific request referred solely to suspending the installation of the biodigesters.

Given that the constitutional action was filed in 2007 – prior to the entry into force of the 2008 Constitution, which recognises Nature's entitlement to rights – the Constitutional

Court studied the case under a framework of environmental rights, and rejected the action, arguing that the installation of biodigesters did not violate the right of people to live in a healthy environment. The court argued that on the contrary, biodigesters are operations that fall under the Clean Development Mechanism of the Kyoto Protocol, since they replace the use of fossil fuels with biogas recycled from animal manure.

The Constitutional Court nevertheless acknowledged the plaintiffs' concern about the environmental management of the agro-industrial enterprise in question, in particular the water pollution caused by the discharge of organic waste, which had been reported to several authorities without an effective response. In order to respond to this concern, the Constitutional Court decided to apply the Constitution of 2008, current at the time of the ruling on the case (2009), with specific reference to the Rights of Nature.

The Court, invoking the principle of comprehensiveness, established the need to '... consider all the elements of the case, and the parties involved, among which was Nature (...) since living coexistence depends on water as one of its elements (components of Nature), not just that of humans, but also that of other living species'.[31]

Accordingly, the Court emphasised the State's duty of guardianship, which was applied for the very first time to protect the Rights of Nature:

> For this reason, we cannot set aside the content of the protection of the Rights of Nature, as set out in articles 71 and 72 of the Supreme Law of the State, which sets out that Nature has the right to have its existence fully respected, together with the maintenance and regeneration of its life cycles, structure, functions, and evolutionary processes, as well as the right to be restored when its natural systems are affected. Even more so when, according to the same foundational Charter, as set out in articles 3 and 277, it is the State's duty to protect natural assets and guarantee the rights of individuals, communities, and Nature... It is the duty of this Court, as the guardian of compliance with constitutional mandates, to substantiate the will of the People, in that our Foundational Charter grants rights to Nature.[32]

The Constitutional Court also decided to set up a Commission (made up of representatives of the plaintiffs, supervisory bodies and the defendant) in order to monitor the operation of the biodigesters, as well as waste management and other aspects relating to the industrial farm's environmental management.[33]

### *Mangroves Case*[34]

This case dates to 2012, and addressed the continued operation of a shrimp plant within a natural area protected by the Ecuadoran State. The ecological reserve is home to the world's tallest mangroves, and for this reason was designated a Ramsar site under the Convention on Wetlands of International Importance.[35] By this Convention, member parties must designate significant important wetlands to be part of a List of wetlands – Ramsar sites – which obliges the State to implement special measures towards their protection.

The context of this case was the eviction of the shrimp plant from the protected area. This decision, made by the Ecuadoran environmental authority, was challenged by the owner of the shrimp plant, who argued that his right to property and to work should prevail. Although the environmental authority demonstrated, using satellite evidence, that the shrimp plant had been installed in the area *after* it was declared a protected area, the lower courts resolved the case from the perspective of the constitutional rights to property and work, thereby ruling in favour of the shrimp plant owner. The lower courts did not consider the constitutional Rights of Nature.

The case was referred to the Constitutional Court of Ecuador, which in 2015 examined the case from the comprehensive perspective of constitutional rights, including the Rights of Nature. In this context, the Court held that the Rights of Nature:

> ... amount to one of the most interesting and relevant innovations of the current Constitution, as it departs from the traditional notion of 'nature-object' – which views nature as property, and approaches its protection solely through the right of people to enjoy a healthy environment – and gives way to a notion that recognizes that nature is entitled to rights of its own. The novelty therefore consists of a paradigm shift, based on which Nature, as a living being, is considered to be a subject entitled to rights.[36]

The Court remarked that this case '... must have considered the potential impacts arising in nature from the production process involved in the aquaculture of shrimps (...) especially (...) when such activity is carried out within an area that has been declared a nature reserve ...'[37]

The Court noted the absence of a comprehensive legal examination of the case by the courts below, pointing out that this not consistent with Constitutional Law, thus breaching the requirements of reasonableness and judicial logic needed to issue a judgment. Accordingly, the Constitutional Court ordered a review of the case from the appeal stage.

This decision put into practise the State's role as guarantor of Rights to Nature, which had already been stated by the Court in the case of the biodigesters. However, in this case, the Court went one step further, by clarifying the guardianship role of judicial bodies:

> ... the constitutionally recognized character of the Rights of Nature implies the obligation of the State to ensure their effective enjoyment, and judicial bodies specifically have the task of ensuring their guardianship and protection, in cases brought to their attention, where these may have been undermined.[38]

### *Unauthorised Mining Case*[39]

This case, also dating back to 2012, addressed unauthorised mining activities. The context of this case was the suspension of mining work by the Ecuadoran mining authority. The suspension arose because the volume of operations and the investment conditions were found to be different from those declared by the beneficiary of the concession.

Authorisation had been granted to carry out small-scale mining activities, that is, using manual tools. However, the beneficiary hired and used mining machinery, which due to its environmental impact failed to comply with the terms of small-scale mining. In addition to the suspension of activities, the machinery was confiscated by order of the authority.

This decision was challenged by the beneficiary of the concession and by the owner of the machinery, who argued that their right to work and property should prevail. The mining authority was able to demonstrate that the mining activity carried out was not consistent with the authorisation, since material was extracted using not manual tools but an excavator, and its volume was so large that it was transported in six dump trucks. Nevertheless, the lower court failed to examine the case from the perspective of the effects on Nature, and instead resolved the case from the perspective of the right to property and work, thereby ruling in favour of the beneficiary of the mining concession and the owner of the machinery: The court lifted the suspension of the mining activity and returned the machinery to the owner.

Considering these circumstances, the case was referred to the Constitutional Court of Ecuador, which in 2015 examined it from the comprehensive perspective of constitutional rights, including the Rights of Nature. In this context, the Court held that the Constitution:

> … is inclined towards the Biocentric perspective of a 'nature-society' relationship that acknowledges nature as a living being, and as a giver of life, and therefore bases the respect owed to it by humans in valuing it as holder of rights beyond its utility to people.[40]

The Court thus ruled – for the first time – on the complex relation between the Rights of Nature and the economic system, by pointing out that:

> … we emphasize the importance of the dynamic and balanced relationship between society, the State, and the market, with Nature (…) the Rights of Nature evidently encompass both social relations and each of the elements of the country's economic system, so as to ensure that production and consumption do not become predatory processes but are instead geared towards respecting the existence, maintenance and regeneration of its elements. In this order of ideas, if we take as our reference the articles of the Constitution that address the Rights of Nature, as well as those that govern economic, socio-cultural, and environmental systems, the mention of Nature and each of its components in the Constitution clearly corresponds to a holder of rights, whose respect must come before any individual economic interest.[41]

The Court also ruled on a specific topic: the importance of the truthfulness of information declared, and its role in guaranteeing the Rights of Nature. Given that the information upon which the mining activity had been authorised was not declared in good faith, the State arrived at a decision of required environmental protections that were at odds with actual events.

This ruling is significant in that the Court established '… a direct relationship between the information provided to the State to obtain the environmental permit with the Rights of Nature, since it enabled the establishment of more appropriate measures to ensure complete respect for Nature …'[42]

Against this background and reiterating that the State '… must ensure the effective protection of the Rights of Nature …',[43] the Constitutional Court ordered the assessment and quantification of the environmental damage caused by the unauthorised mining activity in order to proceed to its restoration. To date, this has been the only decision to concretely and correctly refer to the right of Nature to be restored. This is because case law tends to confuse restoration, which is a constitutional right of Nature, with reparation, which is an obligation that arises from environmental damage. Restoration has also been confused with the human rights institution of integral reparation, which includes economic compensation. This has been particularly evident in penal cases, including a 2015 case regarding the illegal traffic in marine iguanas, which are endemic to the Galapagos Islands. Instead of issuing a restoration order, a local judge ordered reparation of biodiversity, by means of economic compensation. Given the complexity of biodiversity restoration, especially on species reintroduction to their habitat, restoration requires other means, namely in situ research and other scientific and technical responses.[44]

Marine iguanas are endemic to the Galápagos Islands. (Photo Wikimedia Commons/Lt. Elizabeth Crapo, U.S. National Oceanic and Atmospheric Administration. https://commons.wikimedia.org/wiki/File:Marine_Iguana_On_Rock_In_Galapagos_Islands.jpg)

### *Vilcabamba River Case*[45]

The case dates to 2010, and dealt with the effects of large deposits of debris on the banks of the Vilcabamba River, produced by nearby road improvement works. The Vilcabamba River supplies water to the town of Vilcabamba, which is known for the longevity of its inhabitants. Moreover, it flows through a valley of the same name, which is nationally known as a sacred place for Native cultures.

Even though their riverside property was directly impacted, the plaintiffs – residents of the location that was most affected – did not file the action on account of the violation of their property rights, but on account of the violation of the Rights of Nature.

Their action was the first time these new rights were implemented by citizens on behalf of Nature. Following a dismissal in the first instance, a Court of Appeals ruled in favour: Nature's constitutional rights had been violated. The Court of Appeals ordered measures to remedy the environmental damage on the banks of the Vilcabamba river, which were not promptly carried out. In 2012, the plaintiffs went before the Constitutional Court to challenge the failure to comply with the remedial measures.

In 2018, the Constitutional Court confirmed that the measures required by the Court of Appeals were complied with, and declared the case closed. The significance of this lies not so much in the content of the judgment itself, but in the fact that this was a pioneering case with regard to the protection of rivers, which was replicated in subsequent years in other countries with different legal systems, such as India and Colombia: as with the Vilcabamba River, the Ganges River now enjoys theoretical legal subject status, as granted by courts of India, while the Constitutional Court of Colombia has recognised the Atrato River as an entity subject of rights. The latter is of significance because the Colombian Constitution does not recognise rights for Nature. Yet, the Constitutional Court, acknowledging the ineffectiveness of the environmental protections to the river – greatly degraded by decades of mining activities – adopted a new *biocultural* approach that focuses on the biological as well as cultural context of the Atrato river basin, which crosses the Chocó, one of the most biodiverse regions on Earth. The Court Stated:

> … it is necessary to advance the interpretation of the applicable law and the forms of protection of fundamental rights and their subjects, due to the great degree of degradation and threat in which it found the Atrato River basin.[46]

The Court further noted:

> … environmental justice must be applied beyond the human stage and must allow nature to be subject to rights. Under this understanding is that the Chamber considers it necessary to take a step forward in the jurisprudence towards the constitutional protection of one of our most important sources of biodiversity.[47]

Returning to Ecuadoran case law, it is worth taking into consideration, with reference to the Vilcabamba ruling, as well as the ruling on unauthorised mining, that the Constitutional Court has adopted the biocentric approach of Rights to Nature, with no objections. Although the Court has not yet provided a substantial reasoning on this topic, it has accepted its incorporation in the Ecuadoran legal system. This is an important step towards recognition and enforcement of these rights.

## Binding Jurisprudence on the Constitutional Rights of Nature Language

In February 2019, a New Constitutional Court took office in Ecuador. In May 2019, the Court selected an important case for its docket regarding the role of the State in controlling environmental management of a hydropower plant. For years, the rural community of San Pablo de Amalí, directly affected by the plant, has been litigating in courts and tribunals for enforcement of human and Nature's rights. Early this year, a constitutional action was filed in this case. According to Ecuadoran Constitutional Law, the Constitutional Court can select cases that are novel or relevant to the protection of constitutional rights. The Court's decision in such cases becomes universal binding jurisprudence, not merely affecting the parties to the litigation. In addition, the Court's decisions provide the substantive content of constitutional rights. Hence, it is very important for environmental protection in Ecuador that the Court selected a case needing constitutional interpretation of the Nature's Rights language. The Court has already announced it will address standards and limits regarding the exploitation of resources and its impacts on Nature.[48]

## Conclusion

The recognition of Rights of Nature constitutes an important legal development for many reasons, including the acceptance of a new paradigm aiming at building a different relationship with Nature. This is a difficult goal, one that requires time to achieve. Ten years are not enough to reach final conclusions.

Nevertheless, 10 years provide an important timeframe for two preliminary conclusions. The first relates to the impact of Rights of Nature in the Ecuadoran legal system, and the second relates to its global impact.

The impact of Rights of Nature in the Ecuadoran legal system is yet to be fully assessed. To date, the scope and substantial content of these rights remain legislatively undefined. The leading national doctrine has been cautious about these new rights, acknowledging a symbolic effect derived from their recognition, but without real significance. Judges, on the other hand, have been the ones enforcing these rights. Nevertheless, key questions remain, such as what Rights of Nature are, what the difference is between Rights of Nature and environmental rights and how to articulate Rights of Nature with human rights and sustainable development. These are yet to be addressed.

As previously mentioned, the development of rights takes time. It is not possible to expect the Rights of Nature to have fully evolved after just 10 years, that is, that they be understood and interpreted in the most appropriate, effective and favourable way to ensure the effective enjoyment of rights. The development of the Rights of Nature almost entirety depends on civil society, which is a key actor, both in ethical and constitutional terms, in protecting and guaranteeing the Rights of Nature, by calling upon the various State bodies to safeguard them.

In contrast, the impact of Rights to Nature outside of Ecuador has been of importance to the extent that, after only 10 years, the Interamerican Court on Human Rights acknowledges a *trend* towards recognition of rights for Nature. In other parts of the world, like India and New Zealand, Rights of Nature are also advancing relatively fast.

But, more important than the impact of Ecuador's recognition of Rights of Nature on current legal developments is its contribution to revisiting the almost-forgotten debate on the human-Nature legal relationship, first advanced in the 1970s by Professor Christopher Stone[49] in the United States and in the 1980s by the United Nations worldwide, with the adoption of the World Charter for Nature.

Forty years later, Rights of Nature are alive again.

## Notes

1. Constitution of the Republic of Ecuador. 2008. Published in Official Record No. 449 (20 October 2008)
2. Ibid. Article 10: 'Persons, communities, peoples, nations, and communities are bearers of rights and shall enjoy the rights guaranteed to them in the Constitution and in international instruments. *Nature shall be the subject of those rights that the Constitution recognizes for it'.*
3. Ávila Santamaría, Ramiro. 2012. *Los derechos y sus garantías: ensayos críticos.* Quito: Constitutional Court for the Transitional Period, 1st reprint.
4. Walsh, Catherine. 2009. *Interculturalidad, Estado y Sociedad: Luchas (de) coloniales de nuestra época.* Quito: Universidad Andina Simón Bolívar – Ediciones Abya-Yala. First Edition.
5. Ibid.
6. Ávila Santamaría, Ramiro. 2012. *Los derechos y sus garantías: ensayos críticos.* Quito: Constitutional Court for the Transitional Period, 1st reprint.
7. Walsh, Catherine. 2009. *Interculturalidad, Estado y Sociedad: Luchas (de) coloniales de nuestra época.* Quito: Universidad Andina Simón Bolívar – Ediciones Abya-Yala, First Edition.
8. Ibid.
9. Ibid.
10. Ibid, p. 219.
11. Constitution of the Republic of Ecuador. 2008. Preamble: 'We hereby decide to build: A new form of citizen coexistence, diversely and in harmony with Nature, in order to achieve good living, i.e. *sumak kawsay* ...'

12. Ibid. Article 14, first paragraph: 'The right of the population to live in a healthy and ecologically balanced environment that guarantees sustainability and the good living (*sumak kawsay*) is recognized ...'
13. Ibid. Article 275: 'The development structure is the organized, sustainable, and dynamic group of economic, political, socio-cultural, and environmental systems that underpin the achievement of the good living (*sumak kawsay*). The state shall plan the development of the country to assure the exercise of rights, the achievement of the objectives of the development structure, and the principles enshrined in the Constitution. Planning shall aspire to social and territorial equity, promote cooperation, and be participatory, decentralized, deconcentrated, and transparent. The good living shall require persons, communities, peoples, and nationalities to effectively exercise their rights and fulfill their responsibilities within the framework of interculturality, respect for their diversity, and harmonious coexistence with Nature'.
14. Ibid. Article 277: 'The general duties of the state in order to achieve the good living shall be: 1. To guarantee the rights of people, communities, and Nature ...'
15. Ibid. Article 387, Section 2: 'To promote the generation and production of knowledge, to foster scientific and technological research, and to upgrade ancestral wisdom to thus contribute to the achievement of the good way of living (*sumak kawsay*)'.
16. Echeverría, Hugo. 2017. Rights of Nature: The Ecuadoran case. *Revista ESMAT. Escola Superior da Magistratura Tocantinense.* V9, n13. Palmas: ESMAT.
17. Constitution of the Republic of Ecuador. 2008. Article 71: 'Nature, or Pachamama, where life is reproduced and occurs, has the right to integral respect for its existence and for the maintenance and regeneration of its life cycles, structure, functions, and evolutionary processes.

    'All persons, communities, peoples, and nations can call upon public authorities to enforce the Rights of Nature. To enforce and interpret these rights, the principles set forth in the Constitution shall be observed, as appropriate.

    The state shall give incentives to natural persons and legal entities and to communities to protect Nature and to promote respect for all the elements comprising an ecosystem'.
18. Ibid. Article 72: 'Nature has the right to be restored. This restoration shall be apart from the obligation of the state and natural persons or legal entities to compensate individuals and communities that depend on affected natural systems'.
19. Constitution of the Republic of Ecuador. 2008. Article 11, Section 8, first paragraph: 'The contents of rights shall be developed progressively by means of standards, case law, and public policies. The state shall generate and guarantee the conditions needed for their full recognition and exercise ...'
20. Ibid. Section 3.
21. Ibid. Section 4.
22. Ibid. Section 5.
23. Ibid. Section 6.
24. Ibid. Section 9.
25. The Global Alliance for the Rights of Nature, Ecuador chapter, has made available a compilation of cases on the subject here: http://www.derechosdelanaturaleza.org.ec/casos-en-ecuador/
26. The rules on the jurisdiction and powers of the Constitutional Court are provided for in Chapter II, Title XI of the Constitution of the Republic of Ecuador.
27. Article 3 of the Jurisdictional Guarantees and Constitutional Control Act provides that constitutional provisions must be interpreted in accordance with the meaning that best fits the Constitution in its entirety; and, in the event of doubt, these must be interpreted in accordance with the meaning that best favours the full enjoyment of rights.
28. Some argue that the Rights of Nature do not offer any different elements to those offered by environmental law for the protection of the environment. Accordingly, these rights are deemed to play a symbolic rather than real role. Ecuadoran lawyer Ricardo Crespo Plaza, a pioneer in the study of environmental law, has strongly supported this approach:

    Crespo Plaza, Ricardo. 2009. 'La naturaleza como sujeto de derechos. Símbolo o realidad jurídica', Iuris Dictio 8, no. 12: 31–37, http://revistas.usfq.edu.ec/index.php/iurisdictio/article/view/685/757 (accessed 22 October 2018).

29. This is the legal definition of judicial guarantees, as provided for in Article 6 of the Judicial Guarantees and Constitutional Control Act.
30. Constitutional Court for the Transitional Period. First Chamber. No. 0567-08-RA. Official Record. Special Edition. No. 23: 8 December 2009.
31. Ibid, p. 7.
32. Ibid, p. 7.
33. Ibid, p. 8.
34. Constitutional Court of Ecuador, Ruling No. 166-15-SEP-CC, Case No. 0507-12-EP, Official Record 575, 28 August 2015.
35. Convention on Wetlands of International Importance, Ramsar – Iran. 1971. 'Ramsar Sites Around the World', https://www.ramsar.org/sites-countries/ramsar-sites-around-the-world (accessed 22 October 2018).
36. Constitutional Court of Ecuador, Ruling No. 166-15-SEP-CC, Case No. 0507-12-EP, Official Record 575, 28 August 2015, Page No. 154
37. Ibid, p. 156.
38. Ibid, p. 155.
39. Constitutional Court of Ecuador, Ruling No. 218-15-SEP-CC, Case No. 1281-12-EP, Official Record 629; 17 November 2015.
40. Ibid. p. 193.
41. Ibid. p. 194.
42. Bustamante, Francisco. 2016. 'Los derechos de la naturaleza en la jurisprudencia constitucional ecuatoriana'. *Ecuadoran Observatory of Constitutional Justice.* http://observatoriojusticiaconstitucional.uasb.edu.ec/articulistas/-/asset_publisher/6iE7o2o3Gu0e/content/los-derechos-de-la-naturaleza-en-la-jurisprudencia-constitucional-ecuatoriana?inheritRedirect=true
43. Ibid, p. 195.
44. Multicompetent Judge of Santa Cruz, Galapagos Province. Case No. 20332-2015 00616 www.funcionjudicial.gob.ec
45. Constitutional Court of Ecuador Ruling No. 012-18-SIS-CC. Case No. 0032-12-IS, Official Record Constitutional Edition 53; 28 June 2015.
46. Constitutional Court of Colombia. Ruling No. T622/16. Finding 9.28. Page 142. http://www.corteconstitucional.gov.co/relatoria/2016/T-622-16.htm
47. Ibid., Finding 9.31, p. 144. For an initial analysis, see Echeverria, Hugo. June 2017. 'The Case of the Atrato River: A Legal View on the Rights of Nature'. Located at: http://ow.ly/d/6pb8 (accessed 24 January 2019).
48. Constitutional Court of Ecuador. Selection Chamber. 6 May 2019. Case No. 502-19-JP.
49. Stone, Christopher. 1972. *Should Trees Have Standing? Toward Legal Rights for Natural Objects*, 45 S. Cal. L. Rev. 450.

# 15

# *The Godavari Marble Case and Rights of Nature Discourse in Nepal*

**Jony Mainaly**

**CONTENTS**

Introduction ........ 295
Background ........ 296
Environmental Problems in Nepal ........ 297
The Prefederal Environmental Protection Regime in Nepal ........ 298
Environmental Federalism in Nepal ........ 301
Legacy of the Godavari Marble 1 Case ........ 302
Godavari Marble 2 Case ........ 303
Facts of the Case ........ 304
Issues ........ 304
Holding ........ 305
Underpinnings of Rights of Nature in the Case ........ 305
Analysis and Conclusion ........ 306
Notes ........ 307

## Introduction

A void in alternative thinking with respect to the planetary crisis emanating from environmental degradation and natural resource exploitation has been gradually closing, via the Rights of Nature discourse, though on a limited scale worldwide. Nevertheless, discussion exploring Rights of Nature is emerging globally on academic frontiers, and a few countries are giving the philosophical underpinnings constitutional and legal effect. The recognition of this new idea, that Nature has its own intrinsic value, above and beyond merely serving humans, is gaining momentum, though gradually.

As it is considered to be outside the current legal and economic paradigm, many countries are hesitant to adopt this principle into their legal systems. Nonetheless, gradual acknowledgment of the idea is taking shape. Unlike Bolivia and Ecuador, the Rights of Nature are not explicitly mentioned in the laws or constitution of Nepal. This paper outlines and analyses a case relating to a marble quarrying industry in Nepal, where the Supreme Court of Nepal handed down a decision that, for the first time, opened a Rights of Nature discussion in Nepal's judicial system. Shedding light on the historical development of the environmental regime in pre- and postfederal governance structure, this case analysis specifically outlines the concepts of Rights of Nature in this particular decision, and explores the future possibilities the case provides.

DOI: 10.1201/9780429505959-15

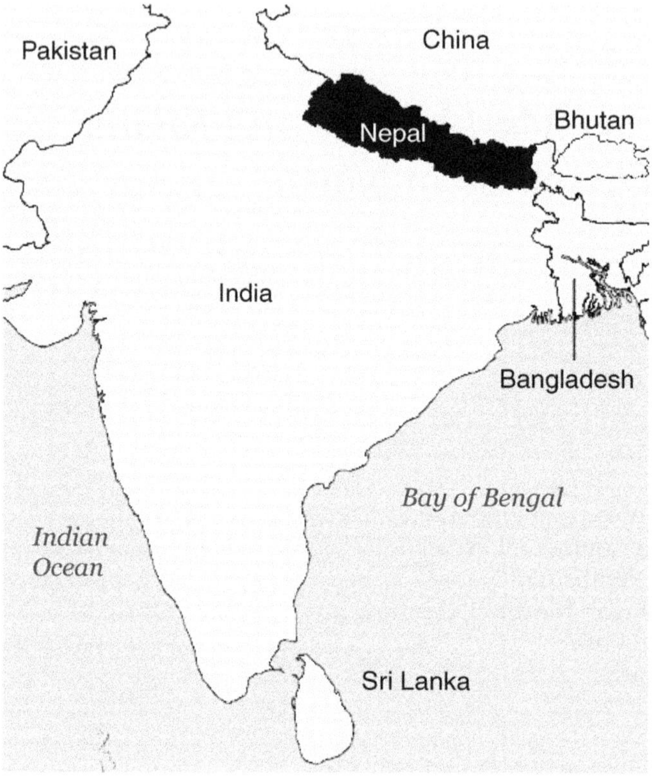


Map of Nepal. (Map courtesy of the United States Centers for Disease Control and Prevention.) https://www.cdc.gov/immigrantrefugeehealth/images/profiles/color_nepal_bhutan-580.gif

## Background

Nepal is a mountainous and landlocked country in South Asia situated between China and India. Lying in the lap of the Himalayas, Nepal borders China in the north and India in the east, west and south. The total area of Nepal is 174,181 km$^2$, which is, comparatively, a little bigger in the area occupied by the U.S. state of Tennessee. (109,247 km$^2$). About 850 km long (east-west) and 200 km wide (south-north), Nepal is an elongated rectangle in shape. Nepal consists of three major geographic regions: plains (Terai), hills (Pahad) and mountains (Himal). The southern plains, hills in the middle and mountains in the northern part of the country occupy, respectively, 17%, 65% and 16% of Nepal's total land area.

The Terai region encompasses a wide range of fertile land for cultivation, as well as dense forests containing a wide variety of flora and fauna. The Pahad covers the largest part of the country. The mountain region of Nepal boasts 8 of the 10 tallest mountains in the world, including the world's highest, Mount Everest. Within these three broad physiographic variations, at least five climatic zones corresponding to the latitudes are found in the country, ranging from tropical to arctic. Nepal occupies only about 0.1% of the Earth, but it contains 3.2% of the flora and 1.1% of the fauna. Some 284 species of flowering plants and160 of species of animals are endemic to Nepal.[1] Nepal ranks 25th in the world in terms of biodiversity abundance.[2] Possessing around 6,000 rivers and streams, Nepal accounts for 2.27% of the world's freshwater resources.[3] Within an area comparable to one of the smaller states of the United States, Nepal contains huge geographic diversity that makes it a very ecologically rich country.

However, two opposing realities often clash in Nepal. It is one of the richest countries in the world in terms of ecological resources, and it is one of the poorest countries in economic terms.[4] Home to about 30 million people, Nepal has a predominantly rural economy, with 65.6% of the total population dependent on agriculture.[5] Its socio-economic and human development indicators are weak. Nepal ranks 149th in the Human Development Index (HDI), with a 0.574 HDI value.[6] With a rich natural resource base, the country is still fighting to address poverty. Nepal is an example of a country abundantly endowed with natural resources, but poor in economic and environmental conditions.

With high geographic and climatic variation in a small region, an abundance of natural resources and biological diversity and sociocultural diversity backed by its transition to a democratic and peaceful country, Nepal is poised to transition to a socio-economically developed nation. Such transition is possible, presuming Nepal prioritises sustainable management and equitable distribution of these resources, in harmony with Nature. As natural resources are its foundation, it is important that Nepal prioritise their integrity and sanctity going forward.

With the promulgation of the Constitution of Nepal in 2015, the country is now the Federal Democratic Republic of Nepal. The internal insurgency that gripped the country for more than a decade was finally settled in 2006, leading to formulation of the law of the land from the Constituent Assembly. Via the 2015 Constitution, Nepal switched from a unitary government to a three-tiered federal governance structure. In the new governing framework, the lawmaking power vests in all tiers of government. In terms of natural resources and environment, all three tiers of government have their respective functions, as stipulated in the schedules of the constitution.

There are several legal, regulatory and judicial interventions emerging to address seemingly intractable environmental problems. The Supreme Court of Nepal has played a very progressive role in grappling with environmental issues, especially during the period of legal vacuum in Nepal in the 1990s, with verdicts ranging across issues such as river cleanup, pollution fines and environmental standards compliance.[7] The Supreme Court's role in environmental jurisprudence continues to be strong and wide ranging in more recent cases as well.

## Environmental Problems in Nepal

Environmental challenges are increasingly becoming more serious in Nepal. The intersection of weak environment management and regulation coupled with urbanisation, population growth, improper disposal of solid waste, haphazard discharge of industrial effluents in the river systems and vehicular emissions is creating a new era of ecological crisis. Unregulated infrastructure development, particularly road construction, has created a major problem of dust pollution. Unregulated brick kilns operating in the vicinity of Kathmandu Valley are another prominent cause of air pollution in the region. Dust pollution and vehicular emissions are major pollutants affecting urban areas of the country.[8] Nepal is not a large industrialised nation. As the population is predominantly rural, the indoor burning of wood, dung and similar biomass is one of the major sources of indoor air pollution.[9] Many environmental problems in Nepal are due to improper management and regulation, rather than quantity of discharge. Nevertheless, Nepal ranks in the top range of 'most polluted countries', as indicated by the Environmental Performance Index, which shows that Nepal was ranked 176th in environmental performance in 2018.[10]

All these problems increase the constant, and tragic, loss of Nepal's tremendous biodiversity. River pollution continuously endangers the biological life in the nation's many river systems. Air pollution, haphazard urbanisation, unsystematic solid waste discharge and water pollution have made many urban areas of Nepal unlivable.

Namche Bazaar, Nepal. (Photograph by Kalle Kortelainen.) https://unsplash.com/photos/6F-uGWod7Xk

## The Prefederal Environmental Protection Regime in Nepal

The Constitution of the Kingdom of Nepal 1990 had not provided an explicit right to a clean and healthy environment as a fundamental right. However, in its chapter on 'Directive Principles and State Policies', it laid out the state's responsibility to prioritise to the 'protection of the environment and also to the prevention of its further damage due to physical development activities by increasing the awareness of the general public about environmental cleanliness, and [the] state shall also make arrangements for the protection of the rare wildlife, forests and vegetation'.[11] The marked difference in the provisions contained in the fundamental rights chapter and the provisions in the Directive Principles chapter is especially highlighted by Article 24 of the 1990 Constitution that states that the principles and policies contained in the Directive Principles are not enforceable in any court. Though the principles and policies formed important aspects of government activity, the nonjusticiable nature of the provisions made environmental considerations unenforceable in courts of law.

Nevertheless, the Supreme Court of Nepal, with its landmark decisions, progressively interpreted the provisions of the Directive Principles chapter to give it constitutional effect. The role of the Supreme Court of Nepal, therefore, was significant in shaping environmental jurisprudence in the country. It would not be an overstatement that environmental jurisprudence has principally advanced via the proactive role of the Supreme Court of Nepal. The Supreme Court's environmental foresight dates back to the first significant case that was filed against Godavari Marble Industry Pvt. Ltd. (hereinafter Godavari Marble 1), filed by Surya Prasad Dhungel[12] against a marble-quarrying industry in the postdemocracy restoration period in Nepal.

In this case, the Supreme Court of Nepal ruled on matters related to the legal standing of nongovernmental organisations and environmentally concerned individuals to file public interest lawsuits in the Supreme Court and the nature of judicial orders allowed in the name of the government, in the absence of specific legal obligation. The court's verdict in recognising a clean and healthy environment as a precondition to exercise the fundamental right to live a dignified life (though discussed only in the context of legal standing) paved the way for further elaborating the link in subsequent lawsuits. The court's verdict is considered instrumental because it laid the foundation for several subsequent public interest lawsuits seeking redress for environmental harms and ensuring environmental protection, thus establishing and reinforcing environmental jurisprudence in the country.

After the court gave the government its directive order in the Godavari Marble 1 case to regulate environmental problems, the Parliament of Nepal approved the Environment Protection Act of 1997 and the Environment Protection Rule of 1997 as a comprehensive legal and regulatory response to environmental degradation. Based on the foundation set in the Godavari Marble 1 case, the verdicts of many other subsequent lawsuits determined that the right to environmental protection is a fundamental right of citizens.

The Environment Protection Act of 1997, an umbrella law, was promulgated with the objectives of maintaining a clean and healthy environment, protecting the environment, minimising adverse impacts of environmental degradation while achieving sustainable development and striking a balance between economic development and environmental protection. The Act, and the Rule devised to enforce the Act, required an Environment Impact Assessment (EIA)[13] for development projects. It also provided for the prevention and control of pollution. Apart from the substantive provisions to achieve the Act's objectives, the law also created the framework to carry out these functions. These include creating the position of Environmental Inspector to undertake pollution prevention and control,[14] establishment and operation of the Environment Protection Fund[15] and appointment of the Environment Protection Council.[16] The Act also integrated the 'polluter pays' principle by providing compensation for loss or damage as a result of actions carried out in contravention of the law.[17]

These new provisions were all progressive aspects of the 1997 statute. But implementation has faltered because the new rules and requirements were unfamiliar to the concerned agencies, being outside their regular norms of operation. This led to failure to enforce the law. The lack of a robust and high-quality Environmental Impact Assessment process, as well as failure of effective monitoring or evaluation, has meant there is no effective implementation of the environmental protection provisions.[18]

The Supreme Court of Nepal has handed down several decisions in environmental protection cases since passage of the 1997 law. Though there are some evident successes, such as closure of polluting factories and orders to appoint statutorily required Environmental Inspectors, effective enforcement of judicial decisions is largely lacking. For example, in the 2003 case of *Bhojraj Ayer vs. Office of Prime Minister and Council of Ministers and others*, the Supreme Court of Nepal[19] imposed a 0.5-paisa fine as a pollution levy and created a pollution control fund to finance pollution mitigation activities. However, as of this date (2019), the fund has not been created, and no monies have been disbursed for pollution reduction and control. Similarly, several cases against Bagmati River pollution have been filed in the Supreme Court of Nepal, followed by contempt of court petitions against the government's inaction in enforcing the rulings of the Court. However, 17 years since the initial Bagmati River pollution case,[20] the plight of the Bagmati remains more or less the same.

The Pisang Valley, Nepal. (Photograph by Jerome Sallerin.) https://unsplash.com/photos/VwswcJbffr0

Further continuing the legacy of the Godavari Marble 1 case, the Supreme Court explicitly recognised the right to live in a clean and healthy environment as a precondition to the right to a dignified life in the case of Yogi Naraharinath.[21] This case gave enforceability to the 1990 constitution's Directive Principles chapter by stating that the Court could not remain silent in matters provided in the constitution when the provisions were not given consideration in state governance. This case established the link between fundamental rights and the Government's Directives and Principles issued under the 1990 constitution. Continuing with such progressive decisions, the Supreme Court of Nepal has issued orders to limit the discharge and effluents in rivers and water bodies;[22] implemented the 'polluter pays' principle by imposing a tax on every litre of petrol;[23] required protection of groundwater from pollution;[24] protected historical and cultural heritages;[25] set pollution standards;[26] laid out the protection of national parks, wetlands and protected areas, recognising the sensitivity of the areas and the need for their protection;[27] restricted operation of polluting vehicles[28] and required equitable distribution of environmental services and intergenerational equity[29] and control of visual pollution.[30] These judicial decisions paved the way for the progressive advancement of environmental protection in the country during the pre-Federalist period, despite the uneven enforcement.

As discussed above, the Constitution of Kingdom of Nepal 1990 had no explicit environmental protection provision in the Fundamental Rights section, but had nonenforceable statements in the Directive Principles chapter. But the Supreme Court, by establishing a linkage between directive principles and fundamental rights via its holdings in several lawsuits, essentially created the fundamental right to live in a pollution-free environment under the 1990 constitution. Nepal subsequently formulated and implemented an Interim Constitution in 2007, which for the first time specified right to live in a healthy environment as a fundamental right.[31]

## Environmental Federalism in Nepal

In 2015, Nepal promulgated a new Constitution through the Constituent Assembly. Departing from the unitary system, the Constitution adopted a federal governance structure. It devised three spheres – federal, provincial (state) and local. Discussing both shared and self-rule jurisdiction, the new constitution specifies the powers of these three spheres, which are detailed in different Schedules of the Constitution. Schedules 5, 6 and 8 articulate the exclusive powers of federal, provincial and local governments, respectively. Schedules 7 and 9 provide for the concurrent functions of federal and province; and federal, provincial and local governments, respectively. Environment protection and management fall under the exclusive jurisdiction of all tiers of government, as well as under their shared functions. Although such overlapping jurisdiction may provide avenues for inaction in environmental protection, the provisions themselves indicate that the 2015 Constitution of Nepal has envisioned a strong role for all three tiers of government in environmental regulation.

Apart from that, the new constitution provides for the right to live in a clean and healthy environment.[32] More progressive than the Interim Constitution of Nepal 2007, the 2015 Constitution also created a right to compensation against environmental pollution and degradation in Article 30(2), and constitutionalised the principle of sustainable development in Article 30(3) with the requirement to adopt laws that balance environmental considerations and development pursuits. These provisions marked a sharp difference from that of the then-existing environmental regime.

Since the 2015 Constitution of Nepal came into effect, the Supreme Court of Nepal has made few interpretations of the provisions relating to clean and healthy environment specified in Article 30. However, in the case of *Amita Gautam Poudel vs. Prime Minister and Office of Council of Ministers and others,*[33] the Supreme Court handed down a progressive decision covering a wide array of environmental issues. The case was filed in the aftermath of the devastating 2015 earthquake and concerned the reconstruction activities that had direct bearing on the renovation of historically and culturally significant structures. Additionally, the blatant disregard of environmental concerns while carrying out development activities in the postearthquake environment resulted in air pollution due to reconstruction work, lack of systematic waste management and lack of intergovernmental coordination in carrying out development projects.

These problems all had implications for the right to live in a clean and healthy environment. The Supreme Court of Nepal handed down a verdict in line with the requirements of the new Constitution's Article 30. The Court, in its order of Mandamus, stressed the government's obligation to provide free medical treatment to individuals affected by a polluted environment resulting from its failure to reduce environmental harms. The Justices focused strongly on the new provision of the right to compensation under Article 30(2) of the Constitution. This was one of the first cases where the Court relied upon the new right to compensation language. The Court also ruled that reconstruction of ruined historical and cultural heritage sites have significance not only for the present generation, but for future generations as well, which must guide reconstruction activities. The Court thus reiterated the intergenerational equity principle that it had already propounded in several past decisions.

The Court, importantly, also ruled that mere passage of laws is not enough to address environmental problems; effective implementation of laws is required to achieve the desired outcomes. This is a general ruling with significant bearing on enforcement of environmental laws and rules, which is a serious issue in Nepal. The Court's ruling also directed the government to establish a coordinating body to address the problem of lack of intergovernmental/agency coordination in solving environmental challenges. In a nutshell, the decision has set an important precedent for the progressive interpretation of Article 30 of the 2015 Constitution.

## Legacy of the Godavari Marble 1 Case

Before discussing the second case of Godavari Marble (the present case for analysis), it is important to shed light on the legacy of Godavari Marble 1 for three reasons. First, the facts and issues of both lawsuits were the same and brought against the same industry. Second, the Godavari Marble 1 case played a significant role in fashioning environmental jurisprudence in Nepal by shaping the future actions of the government in devising legislative and regulatory mechanisms to ensure environmental protection. Finally, it was a revolutionary step in understanding the right to a clean environment, initiating the state directives and policies that led to the current environmental law regime via subsequent public interest litigation. Based on the foundation set in this case, the verdicts of many other subsequent lawsuits were able to determine that the right to environmental protection is a fundamental right of citizens, because in the state's principles and policies it was defined as a precondition to achieving the explicitly provided fundamental right to live a dignified life.[34] The decision of the court revolutionised the environmental outlook in Nepal. This case, which was a far-reaching milestone in shaping the discourse of environmental jurisprudence, faced a glorious fate until the second case was filed.

The Godavari Marble 1 case was decided in 1996. As one of the major fallouts, the Supreme Court of Nepal handed down a direct order to the Government of Nepal to devise adequate legislative measures for environmental protection to comply with its ruling and to implement the Mines and Minerals Act, which had been approved in 1985 – but not enforced. Subsequently, the Nepalese assembly promulgated the Environment Protection Act of 1997, and the government devised the Environment Protection Rule of 1997.

The Environment Protection Act created a broad legislative framework for the government to regulate the environment. As envisaged by the Rio Declaration[35] on integration of sustainable development in national legislation, the Act provided for the objective to promote conservation while achieving economic development. The overarching provisions of the Act and the Rule fashioned a procedure to carry out environmental impact assessment, control pollution and designate protected areas, all of which provided abundant space for the government to achieve sustainability. They set out institutional and financial arrangements to achieve the twin goals of sustainability and economic development. The Act also provided for integration of the 'polluter pays' principle by imposing liability on the polluters. Despite gaps and failures in enforcement, the Act was an important milestone in Nepal's effort to solve environmental problems.

As discussed above, the Court's holding demonstrates that Godavari Marble 1 was one of the drivers influencing enactment of these two important environmental protection directives. However, Godavari Marble's operations were unaffected after the promulgation of the Environment Protection Act. On the contrary, the Government of Nepal in 1997 permitted Godavari to continue quarrying operations for 10 years, from 2001 until 2011, without requiring the industry to implement the new legislative requirements.

The closed Godavari Marble industry in the Phulchowki hills, Lalitpur District of the Kathmandu Valley, after the decision of the Supreme Court of Nepal. (Photograph by Jony Mainaly.)

## Godavari Marble 2 Case

Godavari Marble Industry is a marble quarry located in the Lalitpur District in Kathmandu Valley, which began operation in 1966. Godavari Marble is situated in the heart of the Phulchowki hills in the Godavari area, which is considered a living museum of rich biodiversity. The area is home to at least 571 species of flowers, 300 species of butterflies, 254 species of birds and 80 species of trees.[36] The Godavari region also has a natural water recharge system, providing rare habitat for wildlife and vegetation. Apart from its biological riches, the area also has high religious and cultural importance. The presence of Pancha Dhara (five streams), Nav Dhara (nine streams) and Godavari Kun (pond) in the Godavari area makes the region one of enormous religious importance.

The Godavari area is home to a large and popular religious festival every 12 years, and many temples of various Gods and Goddesses, and Buddhist monasteries, are sited there.[37] The Government of Nepal and the United Nations Educational, Scientific and Cultural Organization (UNESCO) had jointly prepared a report that provided for the protection of this area because of such religious and cultural importance, apart from its natural and biological richness.[38] In Godavari Marble 1, the Supreme Court handed down a directive to the government to devise legislation that would regulate the quarry appropriately.

## Facts of the Case

The Godavari Marble industry had been in operation since 1966. In 2001, the Government of Nepal granted approval to Godavari Marble to continue operating until 2011 in the biodiverse and fragile Godavari region, even after the Supreme Court's 1996 ruling in Godavari I and promulgation of the Environment Protection Act in 1997. Advocate Prakashmani Sharma then filed a public interest lawsuit in the Supreme Court of Nepal against the decision, stating that the government's decision was unlawful and was carried out with 'malafide intention'. The petitioner claimed that the important influence the Godavari Marble 1 case had in devising the 1997 Environment Protection Act was disregarded when the government allowed the industry to continue operations.

The Division Bench of the Supreme Court rendered a split decision in 2008, and the case went before the Full Bench. In 2010, while the case was still before the Supreme Court, the Department of Mines and Geology extended Godavari Marble's permission to continue quarrying for an additional 10 years, effective from the 2011 expiration date, until 2021. The Department gave this approval before the completion of Godavari's prior approved period, and while the second case was still before the Full Bench of the Supreme Court. Following this, Sharma again filed a writ of petition, asking the Court to quash the Department's extension of the approval, arguing that it was arbitrary and unlawful.

Pursuant to Article 32 and 107(2) of the Interim Constitution of Nepal 2007, Sharma and the other petitioners requested the Court to: (a) quash the January 2011 decision of the Department of Mines and Geology that granted Godavari permission to operate for an additional 10 years; (b) issue an order requiring the Government of Nepal to establish the Godavari region as an environmentally protected area where mineral activities are prohibited; (c) implement a 'polluter pays' principle, after ascertaining damages suffered by the families in the area, and force the quarry to pay compensation; (d) hold the quarry accountable, requiring them to pay the full cost of restoring the Godavari region and (e) require the quarry to reimburse the legal expenses of the lawyers bringing the public interest litigation. Sharma's major argument to the court was the government's failure to abide by new legal provisions mandating environmental protection in biologically diverse, ecologically sensitive and religiously and historically important areas like Godavari.

### Issues

The major dispute in the case was whether Godavari Marble's operations should be allowed under Nepal's constitutional and legal environmental protections. The court identified the following issues to be adjudged and resolved:

- The evolving dynamics of environmental justice and its applicability in the case
- Jurisprudential norms determined by the Supreme Court of Nepal in matters of environmental justice
- Significance of the Godavari area in biological diversity, as well as religious and cultural importance, and whether the industry caused adverse impact in sustaining the balance
- Whether the industry's activities are compatible with legal and constitutional provisions of environmental protection in Nepal

In coming to its decision, the Court also examined international standards, norms and practices in maintaining the balance between environmental protection, justice and economic development. While examining the balance, the court also examined the significance of the Phulchowki-Godavari area from biological, environmental, religious and cultural perspectives.

### Holding

The Supreme Court, in its verdict, ruled as follows. The Court:

- Closed the Godavari marble quarry
- Prohibited future mines and mineral operations in the Godavari area
- Ordered the government of Nepal to form a committee of representatives from the Ministry of Forests, Ministry of Environment, the Department of Mines and Minerals and at least two other experts to identify the accurate and actual adverse impacts on the ground and the damage caused to the environment and biological diversity as a result of the mining
- Ordered the government to designate the Godavari area as a protected area/reserve/park within 1 year of receiving the order

## Underpinnings of Rights of Nature in the Case

The Supreme Court of Nepal for the first time based its verdict on a holistic recognition of all species on the planet. This recognition of apportioning legal rights to Nature is emerging globally in the concept of Rights of Nature. While discussing the rights of Nature, the Supreme Court frequently referenced Eastern philosophy and its deep reverence towards all aspects of Nature.

Referring to ancient scriptures and Eastern philosophy, the case highlighted the interactions and interdependence of Nature with human beings and other life forms. The five elements (air, water, land, sky and sun), referred to as 'Pancha Tatva' in Eastern philosophy,[39] have great significance in ensuring, maintaining and sustaining life on Earth. The Court in its verdict discussed in great detail the importance of the law of Nature and its importance in Eastern philosophy. The Court stated that the survival of human beings is based on the survival of the natural components, and not vice-versa. Environmental protection, in this sense, requires keeping these elements intact in their natural form, and not distorting their qualities or their balance. Human beings are creatures of Nature; however, they are degrading it to serve their vested interests, disregarding Nature's prerequisites and limits.

The Supreme Court commented on the supposed supremacy of human beings over Nature. It pointed out that all life forms, human beings, vegetation and animals are dependent on Nature; both their existence and the quality of their existence depend on Nature's own quality. The Court further opined that Nature has its own rules and punishes disobedience of those rules. The self-proclaimed victory of humans over Nature is not sustainable, because Nature also has limits.

By basing their decision on Nature's need to exist for its own sake, the Justices made a strong pronouncement in the ongoing debate about the intrinsic versus instrumental value of Nature.[40] The Court made it clear that its decision arose from understanding that all components of Nature serve their own distinct purpose and have their own intrinsic value. Nature is not at the service of addressing humans' unfettered needs, but rather is an 'end-in-itself'.[41]

The case, uniquely in Nepalese jurisprudence, recognised that Nature has its own rules, and that any species not respecting those rules threatens the overall balance of Nature. The ruling explicitly stated that human beings are the disobedient ones in the family of Nature and are threatening the balance of Nature, thereby triggering a major ecological crisis. Dangerous human activities are no longer limited to the Earth's surface, but have now penetrated into the womb of the Earth, in the name of extracting oil, gas and other minerals.

Eastern philosophy worships Nature, its elements and other life forms on Earth. Hindu philosophy, which finds its base in ancient scriptures such as the Vedas, contains no separation between the Divine and Nature. It recognises the sanctity and sacredness of Nature,[42] and describes the *pancha tatwa* (also *panchabhut*) as the five elements that form both the human body and the universe, inferring that the basic elements of human life and Nature are the same.[43] Human development activities are creating grave pressure on Nature. Hindu philosophy holds that the ecological crisis stems from a spiritual crisis,[44] because the components of Nature are closely intertwined with the existence and well-being of its different forms, creating fundamental balance. The court decided, in this context, that such equilibrium may not be undermined by development imperatives; having Nature function well is equally important as the achievement of development needs.

Focusing on intergenerational equity, the court recognised that Nature's health is critical not only for the present, but also for all generations to come. Explaining its reasons for closing down the Godavari quarry, the court boldly stated that the biodiverse and culturally rich Godavari area needed protection even if its rocks held gold or diamonds, let alone marble. With that statement, the court made clear that it based its decision on the primacy of Nature's needs over development needs.

## Analysis and Conclusion

The constitutional, legal and judicial recourse in terms of environmental conservation has been progressive in Nepal. The decisions of the Supreme Court, from the earlier case of Godavari Marble 1 (1996) that led the promulgation of laws, to the decision in the case of Godavari Marble 2 (2016) that tapped in the emerging discourse on Rights of Nature, is testament to the judicial commitment to environmental sustainability in the country.

The exercise of judicial power in integrating environmental protections has been instrumental in Nepal from the beginning. Its progressive role in the 1996 (Godavari Marble 1 case) marked an important initial shift. Its decision in the recent Godavari Marble 2 case has set another milestone in Nepal's environmental regime with its discourse on Rights of Nature. The Court has opened room for further discourse by its statement that even if 'gold or diamonds' were available in biologically rich and culturally and religiously important sites like Godavari, their exploitation must be prohibited, as Nature has its own interest and right to survive and thrive. The decision also highlights the need to recognise the importance of letting ecosystems thrive in their own right, limiting human interference with the survival of Nature.

The Godavari 2 case is important for two broad reasons. First, the court closed down the marble quarrying industry that was operating in an ecologically sensitive area. After some 15 years of contested battles in the second case alone, the Supreme Court resolved the issue by an order closing Godavari Marble and prohibiting any further mines or mining activity in the area. The verdict settled the fight between environmentalists and the industry in favour of environmental conservation. Second, the verdict opened opportunities for alternative thinking on sustainability by providing scope for a Rights of Nature discourse. Apart from the case's primary fallout, the Supreme Court of Nepal also discussed several important legal dimensions that have the potential to shape the future course of environmental protection in the country, especially in exploration of a Rights of Nature framework for environmental protection.

The federal governance structure under the 2015 Constitution of Nepal that constitutionalised the right to live in a clean and healthy environment and the right to compensation and sustainable development opens a new approach to defining these rights. The interpretation of the right to compensation in the case of Amita Gautam Poudel points towards greater protection against environmental harms and violations. Similarly, the Supreme Court's progressive role in requiring the government to provide effective interagency coordination to protect individuals' right to live in a clean and healthy environment points towards a continuing effort to focus on individual environmental needs.

Decisions as early as that of Godavari Marble 1, and the recent verdicts in Godavari Marble 2 and the Amita Gautam Poudel case (the most recent, as of 2019, based on the new constitution), provide a hopeful future for an environmentally cautious development imperative in the country. The more progressive interpretation of Article 30(3), on balancing the rights to environmental protection and development, should focus future lawsuits on ensuring that the Rights of Nature are recognised and strengthened, along with protecting the legitimate interests of human beings.

Tenacious development needs of poor countries provide both opportunities and challenges for sustainable development. The literature supports the possibility of achieving development needs, as well addressing environmental conservation. However, existing development narratives that are capitalist in nature need redefinition.[45] Rethinking development modes provides scope for changing the outlook to recognising the intrinsic value of Nature, where Nature and its important components, as well as human needs, can survive and thrive together. The Godavari Marble 2 case is a pointer in that direction. The case has laid down a foundational principle that no human encroachment is permissible in areas that are rich in natural, cultural and religious treasures, even for the sake of immense economic value. The Court's holding provides the means for examination of priorities in the context of Nature's right to exist and human beings' responsibility to limit their unfettered interests.

## Notes

1. Ministry of Forests and Soil Conservation. 2014. 'Nepal Fifth National Report to Convention on Biological Diversity.' Government of Nepal. Available at https://www.cbd.int/doc/world/np/np-nr-05-en.pdf
2. As cited in ADB. 2014. 'Country Environment Note Nepal.' Asian Development Bank.
3. ADB. 2014. 'Country Environment Note Nepal.' Asian Development Bank.
4. Jha. Sri, Gopal. 2007. 'A Brief Appraisal of Existing Main Environmental Issues in Nepal and Potential Intervention to Solve the Perceived Problems.' Banko Janakari. Vol. 17, No. 1.

5. Ministry of Forests and Environment. 2018. '25 Years of Achievement on Biodiversity Conservation in Nepal.' Government of Nepal. Available at http://mofe.gov.np/downloadfile/25%20Years%20of%20Achievements%20on%20Biodiversity%20Conservation%20In%20Nepal_1530603709.pdf
6. UNDP. 2018. 'Human Development Indices and Indicators: 2018 Statistical Update: Briefing Note for Countries on the 2018 Statistical Update, Nepal.' United Nations Development Program. Available at http://hdr.undp.org/sites/all/themes/hdr_theme/country-notes/NPL.pdf
7. See, for example, *Bhojraj Ayer vs. Office of Prime Minister and Council of Ministers and others.* NKP 2061, No. 10. Decision No. 7453.
8. ADB. 2014. 'Country Environment Note Nepal.' Asian Development Bank.
9. More than 80% of the rural population is dependent on burning biomass.
10. Yale Center for Environmental Law & Policy & Center for International Earth Science Information Network. 2018. '2018 Environment Performance Index.' Yale University and Columbia University. Available at: https://epi.envirocenter.yale.edu/downloads/epi2018policymakerssummaryv01.pdf
11. Article 26 (4), State Policies. The Constitution of the Kingdom of Nepal.
12. *Surya Prasad Dhungel vs. Godavari Marble Industry and others,* 2052 (1996) Swarna Subhajanmotsab Biseshanka, p. 169.
13. Section 3-6, Environment Protection Act, 1997, Nepal.
14. Section 8, Environment Protection Act, 1997, Nepal.
15. Section 13, Environment Protection Act, 1997, Nepal.
16. Section 14, Environment Protection Act, 1997, Nepal.
17. Section 17, Environment Protection Act, 1997, Nepal.
18. Bhatt. P.R. and Khanal, S. 2009. 'Environmental Impact Assessment System in Nepal – An Overview of Policy, Legal Instruments and Process.' *Kathmandu University Journal of Science, Engineering and Technology* Vol. 5, No. II.
19. *Bhojraj Ayer vs. Office of Prime Minister and Council of Ministers and others* NKP 2061, No. 10. Decision No. 7453
20. *Bhojraj Ayer vs. Prime Minister and Office of Council of Minister* (Writ No. 99 of the year 2061 [Nepali]; 2004-2005 [English]), Date of Decision: 2066/12/15 [Nepali], 3/28/2010 [English].
21. *Naraharinath, Y. vs. Office of Prime Minister and others.* NKP, 2053. No 1. Decision No. 6127.
22. *Bhandari, S. vs. Shree Distillary Pvt. Ltd.* Decision number 3259, Decision year 1997; *Shatrughan Prasad Gupta and others vs. Everest Paper Mills Pvt. Ltd and others.* NKP 2061, No 8. Decision no. 7427; *Ramesh Kumar Singh vs. Ramagya Shah Kalwar.* NKP 2067. No. 10. Decision No. 8490.
23. *Bhojraj Ayer vs. Office of Prime Minister and Council of Ministers and others.* NKP 2061, No. 10. Decision No. 7453.
24. *Sharma, P. and others on behalf of Pro-Public vs. HMG, Ministry of Local Development and others.* (Writ No. 3440 of 2000. Decision year 2002).
25. *Sharma, P. vs. Council of Ministers and others.* NKP. (1997) Vol. 39, No. 6.
26. *Bhojraj Ayer on behalf of Pro Public vs. HMG, Ministry of Water Resources and others.* Writ No 3305 of 2001.
27. *Khanal, D. vs. Hon'ble Prime Minister, Office of Council of Ministers and others.* NKP. 2063. No. 5. Decision No. 7695; *Devkota, N.P. vs. Hon'ble Prime Minister and the Office of Council of Ministers and others.* NKP. 2067. No. 12. Decision No. 8521; *Acharya, R.K. vs. Office of the Prime Minister and Council of Ministers and others.* NKP. 2070. No. 1. Decision No. 8942.
28. *Tuluman Lama on behalf of Kathmandu Valley Tempo Profession Union vs. HMG and others.*
29. *Pokharel, K. vs. Office of the Prime Minister and Council of Ministers and others.* NKP. 2070. No. 4. Decision No. 8995; *Poudel, S.P. and others vs. Hon'ble Prime Minister and Office of Prime Minister and Council of Ministers and others.* NKP. 2070. No. 7. Decision No. 9030.
30. *Shrestha, P.B. vs. Office of Kathmandu Metropolitan City and others.* NKP. 2073. Vol. 58, No. 4.
31. Article 16. Interim Constitution of Nepal, 2007.
32. Article 30 (1). The Constitution of Nepal of 2015.
33. NKP. 2074. Vol. 59, No. 6.
34. *Naraharinath, Y. vs. Office of Prime Minister and others.* NKP, 2053. No 1. Decision No. 6127.

35. United Nations. 1992. 'Rio Declaration on Environment and Development.' United Nations Conference on Environment and Development. Available at http://www.unesco.org/education/pdf/RIO_E.PDF
36. As cited in the petition of Godavari Marble Case.
37. Petition of the Godavari Marble 2 Case.
38. As cited in the petition of the Godavari Marble 2 Case.
39. (a) Shrestha, T.B. 2003. 'Education for the Sustainability of Biological Diversity.' Page 79. Available at: https://pub.iges.or.jp/system/files/publication_documents/pub/policyreport/749/999-report9_full.pdf#page=84; (b) Saraswati, S. 2007. 'Purifying the Five Elements of Our Being.' *Yoga Journal.* Available at: https://www.yogajournal.com/teach/purifying-the-five-elements-of-our-being. Obiter dictum of the case provides discussion on these five elements.
40. Piccolo, J.J. 2017. 'Intrinsic Values in Nature: Objective Good or Simply Half of an Unhelpful Dichotomy?' *Journal for Nature Conservation* Vol. 37, pp. 8–11.
41. Ibid.
42. Frawley, D. (Vamadeva Shastri). 2017. 'Hindu View of Nature.' American Institute of Vedic Studies. First published in Hindu Voice UK. Available at: https://www.vedanet.com/hindu-view-of-nature/ (Accessed 28 March 2019).
43. Shree, B.C. (editor in chief). 2052 (1996). 'Dharma Bigyan (Science of Religion).' Shree Khaptad Aashram Prakashan Samiti.
44. Swami Tripurari, B.V. 1995. 'Ancient Wisdom for Modern Ignorance.' Clarion Call Publishing, Eugene, OR.
45. Magdoff, F. and Foster Bellamy, J. 2011. *What Every Environmentalist Needs to Know about Capitalism*. Monthly Review Press. New York, NY.

# 16

# *Nature's Rights: Why the European Union Needs a Paradigm Shift in Law to Achieve Its 2050 Vision*

**Mumta Ito**

**CONTENTS**

Why Is the European Union Environmental Law and Policy Failing? ... 311
The Need for a Systemic Approach ... 313
Challenges in Implementation of European Union Environmental Law ... 314
Systemic Problems with Environmental Law ... 315
Neoliberalism, Economic Growth and Structure of Law That Works to Support It ... 318
Green Economy ... 319
How Law Created These Conditions ... 321
The Way Forward ... 323
Notes ... 327

We are far beyond the limitations of the planet and we have to face the simple fact that the approach that we had before to set up specific environmental legal texts is simply not working. It's not that we don't have the best or the proper regulation in the system but perhaps the problem is with the system itself. Clearly the current decision making structures and the structure of environmental law can manage certain externalities, as mentioned, and some effects of production and consumption – but it's not really challenging the basis of the problem. Forty years ago it was already on the table – do we create an overly complex un-understandable system of environmental law – or do we simply give rights to Nature?

**Benedek Jávor**
*Member of the European Parliament, Co-Vice Chair of the Environment Committee and Professor of EU Environmental Law.*[1]

## Why Is the European Union Environmental Law and Policy Failing?

The European Union (EU) has one of the most extensive environmental laws of any international organisation – over 500 directives, regulations and decisions.[2] So why is it that the state of Nature in Europe is worse than ever and still in decline?

EU environmental policy is set out in the 7th Environmental Action Programme,[3] which describes the following vision, along with certain key objectives and enablers:

> In 2050, we live well, within the planet's ecological limits. Our prosperity and healthy environment stem from an innovative, circular economy where nothing is wasted and

DOI: 10.1201/9780429505959-16

> where natural resources are managed sustainably, and biodiversity is protected, valued and restored in ways that enhance our society's resilience. Our low-carbon growth has long been decoupled from resource use, setting the pace for a safe and sustainable global society.

A mid-term review of the 7th Environmental Action Programme was carried out and the results were published by the European Environment Agency in the State of the Environment Report 2015.[4] The overall the outlook was very bleak. In particular, the study found:

- Sixty percent of protected species and 77% of habitat types are in unfavourable conservation status and Europe is not on track to meet its overall target of halting biodiversity loss by 2020.
- Climate change impacts are projected to intensify and the underlying drivers of biodiversity loss are expected to persist.
- Loss of soil functions, land degradation and climate change remain major concerns.
- Projected reductions of greenhouse gas emissions are not sufficient to bring the EU onto a pathway towards its 2050 target of reducing emissions by 80%–95%.
- Air and noise pollution continue to cause serious health impacts, particularly in urban areas.
- Health impacts resulting from climate change are expected to worsen.

According to the report, 'the systemic and transboundary nature of many long-term environmental challenges are significant obstacles to achieving the EU's 2050 vision of living well within the limits of the planet'.[5] It went on to acknowledge that 'Europe's success in responding to these challenges will depend greatly on how effectively it implements existing environmental policies and takes necessary additional steps to formulate integrated

European Parliament. Nature's Rights: The Missing Piece of the Puzzle Conference, 2017. (Photograph courtesy of Nature's Rights.)

approaches to today's environmental and health challenges'.[6] At the 2nd Evaluation Workshop in 2018,[7] the European Environment Bureau reiterated concerns about the 'major implementation deficit' of EU environmental law resulting in eroding citizens' confidence in the rule of law and in institutions due to 'insufficient action' and 'policy dissonance'. The Summary Report[8] from the event called for the need to take a systemic approach.

## The Need for a Systemic Approach

The 7th Environmental Action Programme vision is not just an environmental one; it is 'inseparable from its broader economic and societal context'[9]. Unsustainable production and consumption patterns not only undermine ecosystem resilience but also have 'direct and indirect implications for health and living standards',[10] which, in turn, affects our economy and well-being. Our economies are approaching the ecological limits within which they are embedded. This is further pressured by the growing population and increasing middle classes estimated to be 9 billion and 5 billion worldwide by 2050, respectively,[11] resulting in increased competition for resources and demand on ecosystems. It is clear from the planetary boundaries discussion[12] that the planet's ecological limits cannot sustain the economic growth upon which our consumption and production patterns rely.

The increasing globalisation of environmental drivers, trends and impacts adds an additional layer of complexity when it comes to governance. Better understanding of systemic challenges and their time dimension has, in recent years, led to the framing of global environmental issues in terms of tipping points, limits and gaps. However, overall, societies, economies, finance systems, political ideologies and knowledge systems fail to acknowledge or incorporate seriously the idea of planetary boundaries or limits. A study was done in 2017[13] to examine whether in planetary boundaries terms the EU is 'living well within the limits of the planet'. It was found, based on past and current data, that the EU far exceeded its per-capita share of the climate change, nitrogen and phosphorus flows – two of the three critical planetary boundaries that are far above safe limits globally.

Notably, the study did not include biodiversity (which is the third and most critical boundary), as it does 'not yet have quantified control variables, making them impossible to measure'.[14] The midterm review of the European Union Biodiversity Strategy to 2020 identified the key threats to biodiversity as 'habitat loss (in particular through urban sprawl, agricultural intensification, land abandonment, and intensively managed forests), pollution, over-exploitation (in particular fisheries), invasive alien species and climate change'.[15] Of all of the planetary boundaries, biodiversity is in the most precarious position – far more precarious than climate change[16] – and it defies a quantitative and reductionist approach, as the causes of the sixth mass extinction are systemic.

According to the midterm review, 'It is clear that tomorrow's economic performance will depend on making environmental concerns a fundamental part of our economic and social policies, rather than merely regarding nature protection as an "add-on"'.[17] Article 3 of the Treaty on European Union envisages such integration, but in practise it has not yet been achieved. Achieving the 7th Environmental Action Programme vision therefore involves 'going beyond economic efficiency and optimisation strategies to embracing society-wide changes'.[18] This means that the traditional incremental approach is not enough; the mobility, agriculture, energy, urban development and other core systems of provision will have to be radically redesigned to operate in a way that maintains the resilience and integrity of the global natural systems that support and constitute the web of life. In order for that to happen, we need a radically different structure of law.

## Challenges in Implementation of European Union Environmental Law

Law forms the framework for the execution of policy, so the failure of European Union environmental policy must also be seen in the light of the structure of law that is used to implement it. EU environmental law faces severe implementation challenges. For several years running,[19] the Environment policy area has occupied the largest number of infringement proceedings.[20] The European Parliament's midterm review of the 7th Environmental Action Programme states that the root causes of common implementation problems include 'ineffective coordination between authorities in Member States, a lack of administrative capacity and financing, and policy incoherence'.[21] However, these are merely effects of a deeper root cause that stems from continuing to apply an outdated paradigm of law developed in the sixteenth–eighteenth centuries[22] to solve our twenty-first–century problems.

Over the last 40 years in Europe, environmental law has evolved through two main opposing phases: the 'regulatory phase' and 'deregulatory phase'.[23] The regulatory phase tried to protect Nature through legislation aimed at managing negative environmental externalities of economic activities. This has resulted in an emergency response approach, lack of medium- to long-term vision and continued support of the unsustainable mainstream economic model (infinite growth on a finite planet). The deregulatory trend tries to revise existing environmental legislation in order to streamline, simplify and reduce it, without questioning the means of managing the negative externalities of 'business as usual'. It does not address the underlying root causes of environmental degradation. This has led to a decrease in the level of environmental protection and further support of the mainstream economic model.

This is exacerbated by incoherence between European Union environmental law and other policy initiatives, such as the continuation of subsidies in the agricultural, meat/dairy and fossil fuel industries, as well as inconsistency between EU's Sustainable Development Agenda and the 2020 Agenda for Growth and Jobs. This incoherence includes economic objectives and economic concerns as legitimate derogations from fulfilling obligations under the EU's Environmental Directives and Regulations (e.g. Habitats,[24] Birds[25] and Water[26] directives). The foregoing aligns with the growing recognition worldwide that environmental laws premised on regulating the use of Nature are, and have been so far, unable to protect it sufficiently – because they arise out of the same paradigm that created the problem in the first place.

Furthermore, access to justice on environmental issues at the European Union level is poor. At present, all decisions affecting the environment and applicable in the Member States are made at an EU level, which means the focus of any challenge is also at an EU level, since national courts cannot declare the acts of EU institutions invalid.[27] A litigant who wishes to challenge the validity of an EU act must either bring a challenge in the Court of Justice of the European Union under article 263 of the Treaty on the Functioning of the European Union ('Treaty')[28] or persuade a national court to refer the matter to the Court of Justice under article 267 of the Treaty.

The European Union is a party to the Aarhus Convention[29] and has adopted Regulation 1376/2006 in order to implement and apply it to the EU institutions. However, there are considerable shortcomings. Over the last few years, the Court of Justice has consolidated its case law in claims brought by nongovernmental organisations, clarifying that members of the public have no standing under Article 263 of the Treaty[30] to challenge acts and omissions of EU institutions that are not addressed directly to them. As a result, in 2017 the Aarhus Convention Compliance Committee concluded that the EU is in breach of Article 9(3) and 9(4) of the Convention relating to the provision of access to justice, due to its failure to provide members of the public with access to the EU courts. It recommended amending the Aarhus Regulation or adopting new legislation, in the absence of a change in jurisprudence of the Court of Justice of the European Union.[31] The EU has postponed the decision on this until 2021.

Public participation is needed to keep institutions in check and accountable for mistakes, corruption or passivity, especially where Member States are unwilling to prosecute. Enforceability of substantive laws is a core feature of effective environmental governance[32] and remains a challenge in the European Union.

## Systemic Problems with Environmental Law

At the root of the many interconnected ecological, economic and social crises we face today is a legal system based on an obsolete worldview. Traditionally science and law (or jurisprudence) have always developed in parallel, influencing each other.[33] However, in the past few decades, science has moved away from seeing the world as a machine best understood by analysing its discrete parts to seeing the world as a dynamic and fluid interconnected community of life best understood by thinking in terms of patterns and relationships. Science also acknowledges that Nature sustains life through ecological principles that are generative rather than extractive. However, law remains stuck in the mechanistic worldview of the scientific revolution of the sixteenth–eighteenth centuries, failing to evolve and keep pace with current scientific knowledge, particularly in the field of systems science, complexity science, quantum physics and ecology.

At the root of the implementation challenge is the fundamental mismatch between a fragmented, mechanistic, reductionist, top-down, fixed, quantitative and outdated system of law – with the holistic, dynamic, multidimensional and unpredictable nature of complex adaptive systems such as Nature and human societies (which are a subsystem of Nature). If law is to be part of the solution, a radically different whole-systems approach is needed to overcome the following issues:

a. Many environmental problems are invisible. By the time they become visible it may be too late to stop the damage,[34] which is challenging for a reductionist approach to deal with, as we can see clearly from the failure of the endangered species listing system to prevent the mass extinctions we are witnessing today. In the time it takes to do the scientific studies to list a species, it is already too late. Listing individual species also ignores the fact that ecosystems are an integrated whole, where the loss of an unlisted species (e.g. bees[35]) may have huge ramifications for the entire system.[36]

b. Many environmental problems are complex and marked by uncertainties and controversies about cause and effect relationships. This is exacerbated by a great lack of knowledge about how various organisms, including human beings, react in the short and long term to cumulative ecological stress. Environmental policy

Not a single species of bee is protected in European Union law despite the systemic bee/pesticide crisis. (Photograph by John Severns/Wikimedia. https://commons.wikimedia.org/wiki/File:European_honey_bee_extracts_nectar.jpg)

and law must often address uncertainties and moving targets, which is difficult to do with a reductionist approach that insists on quantitative rather than qualitative measures.

c. Many environmental problems are long term, and future generations and other species – who do not have a defined legal status – will be the victims. Again, climate change and loss of biodiversity are clear examples. The role of law – as it is today – is to regulate present conflicts between persons (including legal persons) and the interests of today. Law and legal institutions are not designed or equipped to regulate intergenerational conflicts and secure intergenerational justice. This presents a profound challenge for environmental policy and law in its present form.
d. Many environmental problems are the cumulative effect of many actions and decisions that, by themselves, look harmless. Currently most decisions (especially planning decisions) are made on a case-by-case basis with no overarching requirement to look at cumulative or systemic implications for the whole.
e. Many environmental problems are cross-sectoral both in their causes and their effects. The combination of activities in many different sectors create effects, which may then affect different parts of society. At the same time, institutional organisation and legislation are mainly developed along the lines of economic sectors, and are therefore too fragmented and specialised to deal consistently with the problem. Climate change and reduction of biological diversity are clear examples of this, as they need a systemic approach.
f. The most serious environmental problems cross administrative borders: boundaries between local communities, counties and regions and also between states. The environmental effects of an activity may appear far away from its source. Completely different constituents may, on the one hand, get the benefits and, on the other hand, incur the costs of environmental degradation. Environmental issues cannot, therefore, just be dealt with at the local, regional or Member State level. There needs to be a more integrated and holistic approach with greater collaboration.

g. Environmental law is fragmented both vertically and horizontally. It is defined narrowly as the body of law that has environmental protection as its main objective, ignoring the fact that most of the main sectors of economic and social activity have environmental impacts. These sectors are regulated by their own laws and policies that operate in silos isolated from one other. Within each sector decision-making authority is distributed among state, regional and local authorities. Due to this fragmentation, different acts may be aimed at the protection of different elements of the ecosystem, such as the water and the species; or they aim to regulate different activities within that ecosystem, such as shipping, building, aquaculture and fishing; or they may have an unintended effect on the ecosystem, such as the adverse effects from industry, trade and transport. This is perhaps complicated by the fact that these different legal acts are implemented under different administrative sectors with different interests, tools and traditions, and by the fact that the legal acts contain a degree of discretion or flexibility to aid in administrative decision-making.

h. The legislation and policies offer the authorities broad discretionary power to balance environmental concerns and other (often conflicting) social goals and considerations, such as the objectives of economic growth; employment; increased production of food, energy and goods and improved transport, and against such principles as cost-effectiveness and local self-rule. This is not just political, but is also a result of how environmental law is construed.[37] Inconsistency and lack of coherence in environmental law (when looking at all laws that affect an ecosystem rather than just laws that have environmental protection as their main objective) allows for diverging value judgments by administrative bodies on different aspects of the same ecosystem. As the legislation and policies are generally formed in an open style, this leaves the executive a broad discretionary power in how the provisions are applied. Broad discretionary power coupled with vertical and horizontal fragmentation makes it very difficult to take a holistic ecosystem approach.

i. Despite the European Union having well-formulated policy objectives, integration of environmental law (in its broadest sense) has not yet happened, because it requires changes in the paradigm, objective and priorities of important sectors and authorities at different levels.[38] Often, environmental issues are perceived as a 'problem' that works against the success criteria of a sector authority. However, sector authorities usually know little about environmental objectives and problems, and see them as irrelevant to their primary mission. This narrow, symptomatic thinking and application of law works against achieving systemic solutions.

j. Environmental legislation establishes institutions, systems and procedural rules, and it lays down certain general objectives and principles to be observed. However, when it comes to the actual protection of the environment, the legal core is neither very precise nor very 'hard'. This is not just in the European Union; the soft, discretionary character of many international environmental treaties is well known.

k. Developments in international environmental law have been influenced by several principles, such as 'sustainable development'[39] and the 'precautionary principle',[40] which could counteract some problems of fragmentation and broad discretionary power, if they are understood and applied consistently by the authorities.

Unfortunately, this has not yet happened. Whilst the principle of sustainable development has been incorporated in European Union policy, there is no definition of sustainable development or guidance on implementation in EU law.[41] It is, therefore, no surprise that a 2009 analysis in Norway showed that inclusion of sustainable development or 'sustainability' as an objective, in reality does not mean anything very different from 'business as usual'.[42] Similarly, the precautionary principle has been enshrined in Article 191(2) of the Treaty. However, in practise the principle has been observed to be highly malleable and performs many functions.[43] Accordingly, application of the principle to create systemic change in major policy areas is difficult to identify.

1. One of the biggest issues in environmental law worldwide is legal standing[44] (i.e. the entitlement of a party to demonstrate to a court sufficient connection to support participation in the case), which is a fundamental problem of access to justice. This procedural dimension is often not given sufficient attention, though it leads to failure of the justice system.[45] Standing for judicial review in the European Union has for years remained severely restricted. European courts continue to interpret access to justice criteria so rigidly that it precludes challenges, particularly by nongovernmental organisations, against the decisions of EU institutions, agencies and bodies. Courts hold that standing is only available to individuals who are personally affected by a situation, over and above the way that situation may affect anyone else.[46] In an anthropocentric perspective, an injury to Nature is an injury to everyone and therefore an injury to no one – as it does not individually concern any one person, but many people. The concept of individual and direct concern is not only incompatible with the nature of ecological damage but also contradictory to the Aarhus Convention rights.[47] This problem arises because Nature is merely an object of the law – property or fair game – and as such incapable of sustaining any harm itself. If ecosystems and species were subjects of the law with legal personality and rights, this problem of Nature's standing would vanish.

In order to implement European Union environmental policy, fundamental structural changes must be made to the law to bring it into the twenty-first century. This is even more important because the current structure of law is responsible for facilitating the concentration of wealth, power and the extractive economics that are at the root of the interconnected crises.

## Neoliberalism, Economic Growth and Structure of Law That Works to Support It

Like much of the world today, the European Union pursues a neoliberal economic model, based on an ideal of infinite growth on a finite planet. Environmental law and policy don't tackle this fundamental mismatch in societal objectives. This becomes clearer through an overview of the burgeoning problems, and the historical and philosophical trends that created the dangerous system controlling the European Union's current economy and laws.

### Green Economy

The 7th Environmental Action Programme relies heavily on creating a 'green economy' based on decarbonisation and increasing resource efficiency, as a way to continue growing the economy in terms of Gross Domestic Product, whilst achieving the goal of 'living well within the means of the planet'. However, according to critics, not everything that is 'green' and efficient is also environmentally sustainable and socially equitable (such as biogas and genetic engineering). Critics also point out that more growth and consumption are not sustainable through increased use of green technology.[48] Ensuring that the direction of technological development is consistent with environmental objectives (such as renewable energy and organic farming) is better – though still facing limits when applied on a large scale – because these activities need resources, all of which come from Nature, which has physical limits. There is also the 'rebound effect', in which greater efficiency leads to lower prices, which leads to increased consumption. Neither strategies of efficiency or consistency will be able to achieve their objectives unless accompanied by the 'principle of sufficiency', which is currently absent from all concepts of a green economy.

Introducing a principle of sufficiency into the green economy will require a move away from Gross Domestic Product as the sole measure of well-being and progress. The gross domestic product concept was first introduced in 1934, during the Great Depression, by Simon Kuznets, who cautioned against equating GDP growth with economic or social well-being.[49] Yet is has become the sole measure of progress ever since. The gross domestic product is the combined transactions (private consumption, investment, government consumption and total imports) in a country. Nonmarket transactions, such as family work, unpaid care, neighbourliness and volunteering are not a part of the calculation. Nor does the Gross Domestic Product measure quality of life or well-being; it does not take into account the risks and financial burdens of overconsumption, such as pollution, environmental damage, social stress, increasing inequalities and public health risk factors. As Nobel Prize winning economist Joseph Stiglitz said, 'GDP tells you nothing about sustainability'.[50] Yet this outdated measure forms the underlying basis for law and policymaking aiming to improve societal progress.

Most conventional economists and policymakers now endorse the idea that growth of the gross domestic product can be 'decoupled' from environmental impacts[51] – that the economy can grow without using more resources and exacerbating environmental problems. However, critics say that what people are observing (and labelling) as decoupling is only partly due to genuine efficiency gains, but all too often at the cost of biophysical effectiveness and long-term social–environmental sustainability. The rest is a combination of three illusory effects: (a) substitution (replacing one energy source with another, such as replacing fossil fuels with renewables), (b) financialisation (growth through the financial sector, creating 'phantom wealth') and (c) cost-shifting (moving resource-intensive modes of production away from the point of consumption, to other countries which bear the brunt of the impacts).[52] The decoupling illusion props up gross domestic product growth as a measure of well-being, though it is both outdated and limited. Instead, we need to redefine what is meant by 'living well' in a twenty-first–century context. This requires the creation of alternative indicators of happiness and well-being, which in turn requires a move towards qualitative, rather than purely quantitative, evaluation. Much of what makes human beings happy is the quality of our relationships, including our relationship with the rest of Nature.

The green economy also relies heavily on the concept of 'natural capital', turning Nature into capital to reconcile capitalist growth with environmental protection. In this way, proponents assert, conservation can be expressed in a language that economists, policymakers and CEOs understand.[53] However, critics point out that this strategy is not just self defeating; it is a dangerous illusion that masks the way capitalist growth undermines conservation itself.[54] The Natural capital proposal is based on two flawed assumptions: (a) that Nature can become capital-providing services, and (b) that this could be the basis of a sustainable economy.

The move from Nature to natural capital is problematic due to the fundamental mismatch between the homogenous and unlimited quality of monetary units and the finite, complex, qualitative and heterogeneous qualities of complex adaptive systems like Nature.[55] The assumption that human, financial and natural capital can be made equivalent and exchanged is therefore fundamentally flawed, especially when planetary boundaries are imposing physical limitations on how much monetary capital can be exchanged into nature capital without sending us all into oblivion.

The other assumption, that it can form the basis of a sustainable society, is also being proven false. Recent research shows that markets for natural capital and ecosystem services are mostly failing, and that rather than being true markets, they are subsidies in disguise.[56] In contrast, investments in unsustainable economic activities have continued unabated because these are much more profitable, and hence a much better form of capital.

Another disconcerting trend is that programmes built on natural capital are usually geared towards offsetting the destruction of Nature, which becomes the main source of the money needed for investing in conservation. In the logic of natural capital, investments in unsustainable economic activities are therefore 'compensated' for by equal investments in sustainable activities. This practise, which in theory should lead to no net loss of – or better yet, net positive impact on – Nature and biodiversity, leads to a contradiction: that Nature can only be conserved if it is first destroyed.[57]

A further consequence of the natural capital approach is the financialisation of Nature. Here the components and functions of Nature, including biodiversity, are priced according to their utility and assigned an economic value, which forms the basis for the creation of financial instruments that can be traded on the primary and secondary capital markets. The instruments are acquired by corporations to offset their overuse, degradation or pollution of the environment, and they can further profit from trading them. Pollution permits, natural capital bonds, biodiversity banks and offsetting already exist. Essential prerequisites for financialising nature are: pricing Nature, characterising Nature's functions as 'ecosystem services', and redefining Nature as 'natural capital'. This approach has several drawbacks that could seriously accelerate the rate of ecosystem destruction.[58] This extension of property rights increases the commodification and privatisation of Nature and leads to further concentration of power into corporate hands.

In all green economy scenarios, political, social, economic and cultural rights are largely left out of the picture. In the absence of political action to prevent it, there is a clear and alarming tendency towards concentration of power.

An example of this is in the agricultural and food industry. Global food production and marketing is controlled by a few large agricultural monopolies.[59] Production of fertilisers, pesticides, seed and genetically modified seed is largely concentrated in the hands of a few conglomerates – the same ones that control the global food market. The powerful seed, fertiliser and pesticide lobby is intent on securing market power in this area for itself. Its representatives exert increasing influence on policymaking everywhere in the world.

Another example of innovation leading to an increasing concentration of power comes with the bio-economy, which is now part of European Union policy.[60] The bio-economy fosters techniques such as synthetic biology and nanotechnology, which transform living 'biomass' into fuels, chemicals and power. However, this seemingly 'green' switch from fossil fuels to plant-based production is now threatening biodiversity, fuelling land grabs in the global South, where 86% of all biomass is located, and enabling new corporate claims on Nature.[61]

Underpinning all of this concentration of power is property rights. This is why access to intellectual property rights is part of the repertoire of economic negotiations and of innumerable bilateral trade agreements between industrialised and developing (i.e. nonindustrialised) countries. Small farmers and rural workers rarely have the power to defend their interests and rights against the conditions imposed by global corporations. None of the green economy strategy papers or policies – from the Organisation for Economic Cooperation and Development to the United Nations Environment Programme to the European Union – tackle the issues of power and distribution of resources. In the political sphere, and among the general public, there is little awareness that the problem even exists.

A fundamental failing of environmental law and policy today is its inability to deal with the extraordinarily unequal power relations that exist in today's world, and the interests at play in the operation of this destructive global economic system. Law has progressively facilitated a situation where the capacity of existing political systems to establish regulations and restrictions to the free operation of the markets – even when a large majority of the population call for them – is seriously limited by the political and financial power of the corporations. For example, although the European Commission ('Commission') is tasked with the role of competition watchdog, its track record for enforcing merger rules to date confirms a strong proconcentration stance. The recent mega-mergers between the agro-chemical giants Bayer and Monsanto, Dow and DuPont, and ChemChina and Syngenta are an example that will lead to unseen economic concentration in the markets for seeds and pesticides and other chemical inputs, which will not only affect the future of biodiversity, wildlife and the conditions under which farmers produce their crops, but also the lives and food choices of billions of people around the world.[62] The Commission has legitimised its proconcentration stance by referring to synergy effects, such as lower costs and thus lower prices for consumers, product innovation and the displacement of inefficient management structures.[63] However, so far smaller and less competitive companies have suffered, whilst the transnationals have profited. The Commission's policy of consolidation of economic power into ever-fewer transnational corporations therefore works against both people and Nature.

## How Law Created These Conditions

The progressive evolution of law in Europe – from the holism in ancient Greece to the mechanistic and institutionalised form of modern law that we use today – has played a powerful role in creating and legalising the destruction of Nature, as well as the concentration of power and inequality that are at the heart of our interconnected crises.

Prior to the scientific revolution of the sixteenth and seventeenth centuries, philosophers in Europe saw the world as a living organism rather than a machine – an ordered harmonious whole (a macrocosm) whose general properties are reflected in its parts (microcosm). These ideas are also an integral part of systems science today and can be seen in action in examples such as DNA, which is found in each cell but contains the code of life for the whole. The scientific revolution put an end to the concept of systemic holism by introducing the mechanistic worldview. During this time, mathematical regularities observed in Nature got solidified into 'laws of nature' that were deemed to be absolute and fixed. Just as the scientists fragmented the universe into an aggregate of separate parts (often described as 'independent variables'), legal scientists fragmented the legal order from a holistic system of customary laws adapted to the practical requirements of human relationships into an aggregate of component parts governed by strict natural laws of individual reason.[64]

Private ownership of land became the most important legal concept dividing the whole into parts, governed by what became the sovereign state. This created the intellectual foundation for a dramatic shift away from commons-based folk institutions into legally formalised concentrated private property and eventually into capital. This shift in law provided the legal justification for the (violent) private enclosure of the commons, colonialism, slavery, industrialisation and capitalism – progressively concentrating wealth and power in the hands of the few, and requiring progressive increases in the extraction, accumulation and mobilisation of natural and human resources, later to be understood as 'capital'.[65]

Ownership and sovereignty became the two organising principles of law that are still with us today. Law developed into a system of relationships among sovereign individuals who could exercise their sovereignty over private property, which was seen as an extension of themselves.[66] If there is any legal limit to this sovereignty, it could only derive from the presence of other subjects who had similar formal legal rights.

During this period, distributive justice was completely eliminated from Western legal and political thought.[67] Absolute property rights, immune from possible redistributive plans and from any concern about the commons, are also the basis of legal support for our current disastrous model of development and concentration of wealth and power.

The rational, scientific way of thinking spread from science into law, and then into economics. Once bound by their duties toward one another, people were now defined instead by their individual property rights. Today the legacy of these ideas is an unexamined faith in a mechanistic, top-down rule of law, which opened the way for the conception of corporations as 'legal persons' with limited liability. This has led to widespread expansion and exploitation of Nature, natural 'resources' and people worldwide. 'Natural law' was used as a justification to acquire private property that was unused as 'empty' even if it was under foreign jurisdiction.[68] The Dutch East India Company – a precursor to our present day multinational – was the first example of a private ownership structure that claimed legal advantages even over states. This created a very favourable playing field for corporations venturing into transnational activities. Very similar arguments today are the foundation of the World Trade Organisation – no public power can limit the corporate right to roam the globe to acquire control over natural or human resources.[69]

The law operated to allow the commons to be freely transferred into private hands. However, moving resources from private ownership back to the public could only be done under strict judicial scrutiny. This process has produced a constant and irreversible transfer of public resources into a few private hands.

After the collapse of the Soviet Union, philosophies of 'legal realism' deemed law a social science rather than a 'natural science', which made it subservient to economics. This led to political processes producing law by design to facilitate market efficiency. The predictability of law then paved the way for more efficient business activity, which led to the global capitalism that we see today, where the entire world is governed by essentially the same rules and institutions. These structures are equally unsustainable and extractive, whether in private decision-making by corporations, or short-term decision-making by governments. At the centre of global capitalism is a network of financial channels, which the law allows to develop outside of any ethical framework, supported by free-trade rules designed to support corporate growth. This depletes the Earth, and ultimately separates people. Analyses by scholars and community leaders reveal a multitude of interconnected, harmful consequences of the capitalist freedom of extraction generated by property law, including increasing social inequity and social exclusion, a breakdown of democracy, more rapid and extensive deterioration of Nature and increasing poverty and alienation.[70]

Three centuries of transforming commons into capital have weakened governments and the public interest, allowing dominance by strong corporate actors. Private corporations are not accountable for violations of international human rights, in contrast to legal limits on governments and their officials. Although corporations are the most powerful actors in international law – and capable of determining its content – the lack of an ecological vision of the legal system makes them invisible and shields them from responsibility. The Western legal tradition developed to protect the private economic agent against government, when government was strong and property owners weak. It has now produced a major imbalance protecting private vested interests against the public. People are more affected today by the decisions of corporations than the decisions of governments – yet only governments can be legitimised in democratic terms, whilst corporations are discussed only in terms of economic efficiency, highlighting a need for fundamental systemic change in European Union law and governance.

The notion of law itself as a commons, in Europe and beyond, has been eradicated by its progressive professionalisation and institutionalisation, which took law away from communities – meaning that communities no longer had any control over their legal structures, disempowering people. The only choice available today in law is between the interests of private property and state sovereignty, and the only political choice is between more or less government or more or less property. The law has progressively become a tool to support this constriction of political agency. The current legal structure in Europe (and arguably reflected in many countries around the world), built on the institution of private property, is based on the concentration of power and exclusion[71] as described in the examples of the agro-chemical mergers above.[72] Based on a mechanistic vision of the relationship between humans and the rest of Nature, property law is still perhaps the most powerful institution of exclusion, individualisation and competitive accumulation.

The tight structural connection between private property and unsustainable practises of short- term extraction explains the difficulty of exiting the mechanistic trap.[73] Mechanisms of liability embedded in corporate charters force managers to pursue shareholder value and the bottom line. Property law and freedom of contract also protect even the most systemically irrational behaviours such as monoculture for fuel production and the highly destructive form of mining known as hydraulic fracturing. Any attempt to put public interest over private ownership faces a high burden of proof because scores of economists, jurists and pundits rush to protect the sanctity of private property and the unlimited freedom of an owner. Above all, property law protects corporations.

The current dominant legal order is designed for extractive rather than generative purposes. To do so, it has progressively separated itself from politics and economics – areas where law can serve human needs. Moving from one mechanistic alternative to another is likely to fail. In order to solve our systemic problems, we need something that will take us out of the mechanistic trap,[74] whilst providing a powerful counterbalance to property rights.

## The Way Forward

Just as alternative models of economics are being proposed that are more holistic, it is equally important that law evolve to meet the challenges of our time and become part of the solution. We have already looked at how the structure of law in the European Union contributes to the problem. It is not so much any individual provision of law, but rather the structural basis upon which these laws are founded, that gave rise to the problem. To move us into a sustainable future, a few key pieces of legislation that start to shift the structural frame – such as that presented in the Draft Directive for the Rights of Nature – are needed. Slavery ended as a mainstream model when slaves got legal personality and rights. Similarly, in order to put an end to the systemic root cause of climate change and the sixth mass extinction, we need Nature's rights within an ecological framework of law.

The first step in the journey towards a whole-systems approach is for the law to include the interests of all living beings as subjects of the law rather than objects or property. The current legal construction of Nature as an object of the law is the reason it is legal to have an economy based on infinite growth, an agricultural system that poisons the Earth and people and an energy system that depletes Nature faster than she can replenish. It is therefore imperative to redress the imbalance and ensure that Nature is represented in the European system of law as a stakeholder in its own right, given legal personality and rights in the Treaty along with an implementing framework set out in a Directive. This will act as a powerful counterbalance against corporate and property rights[75] and start transforming some of the plentiful capital now available back into commons available to Nature itself and people as part of Nature, without having to destroy Nature first.

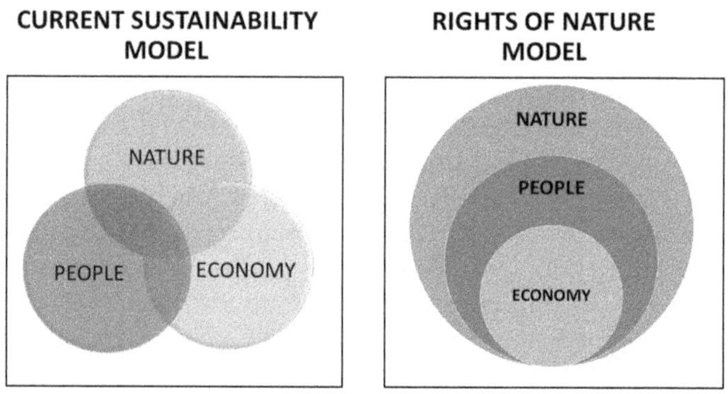


Sustainability models. (Diagram by Mumta Ito.)

It is important to change the mainstream model of sustainability to a more systemic view. The Current Sustainability Model in the diagram, which is the dominant model for sustainability in the European Union and elsewhere, represents a mechanistic worldview where the assumption is that each circle can operate independently of the others. In reality, the only circle that can operate independently is Nature, as the others are dependent on Nature for their existence. The Rights of Nature Model (three concentric circles) is therefore more accurate, as it represents a nested natural hierarchy of systems – Nature, Human Society and Economy.

European Union law doesn't expressly recognise a human right to a healthy environment, although there is reason to believe that it is supported in the jurisprudence of the European Court of Human Rights.[76] However, the human right to a healthy environment has not been successful so far in halting the onward destruction of Nature that has accompanied capital accumulation over the centuries – as the midterm review of the 7th Environment Programme illustrates – because human rights are conceived of in a vacuum. Aligning law with the systems perspective of modern science and ecology would involve a reframing of the notion of rights to reflect the whole system and the interrelationships between the parts. For this, we need to include the overarching context of our existence on this planet – the fundamental basis from which all other rights emanate: the Rights of Nature, because without Nature we cannot exist.

This is in stark contrast to the situation that we see in the world today (shown on the left in the Hierarchy of Rights diagram) which has led to the domination and concentration of wealth and power in the hands of the few.[77] Interlinked multinational banks and corporations determine economic activity throughout the European market. Their vast lobbying power has an overwhelming influence on the European Commission and Council that remains unregulated by law. Currently, human rights are in the realm of public law aimed at the relationship between governments and individuals; property and corporate rights are exercised in the realm of private law between individuals. There is no horizontal enforcement of human rights in private law, making corporations effectively immune to human rights abuses, although this is increasingly questioned by legal scholars.[78] By contrast, corporations, as legal persons, enjoy the protection of human rights in addition to property and corporate rights, leading to the current state of affairs in which corporations are shielded from the consequences of their destructive extractive activities worldwide.

A natural hierarchy of rights that derives from the natural hierarchy of systems – Nature's rights, Human rights and Economic rights – is shown in the Hierarchy of Rights diagram. In this model, the rights operate in synergy with each other, as competing rights would undermine the well-being and integrity of the whole. This model of nested rights brings a unifying overarching framework to the balancing of interests and weighing of divergent values, overcoming the existing fragmentation and imbalance. Economic rights that currently undermine human rights and Nature's rights destabilise the whole system, and would no longer be in the public interest.

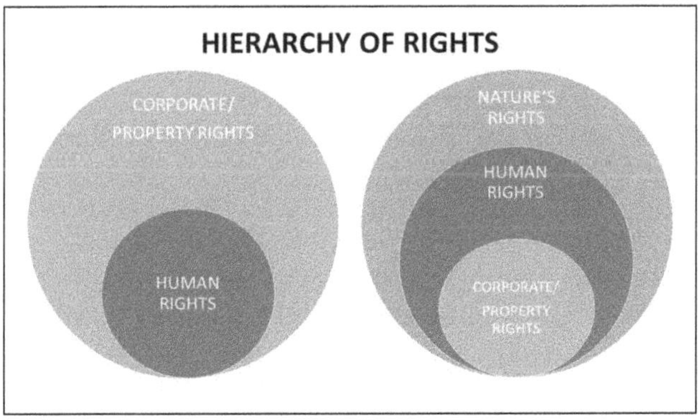


Hierarchy of Rights. (Diagram by Mumta Ito.)

To develop this idea further into a workable model, the planetary boundaries and United Nations Sustainable Development Goals (which the European Union has committed to implementing via adoption of the 2030 Agenda for Sustainable Development)[79] can then be mapped onto the three concentric circles, identifying a clear and safe operating space for humanity that takes all three layers of rights into account. This is partially depicted in the diagram below, which shows how the biosphere forms the fundamental basis for all of the UN Sustainable Development Goals.

This establishes the basis for policies integrating Nature's rights, Human rights and Economic rights (corporate and property rights) into a coherent and unified whole within a legal framework that supports that integration. Operationalising the planetary boundaries

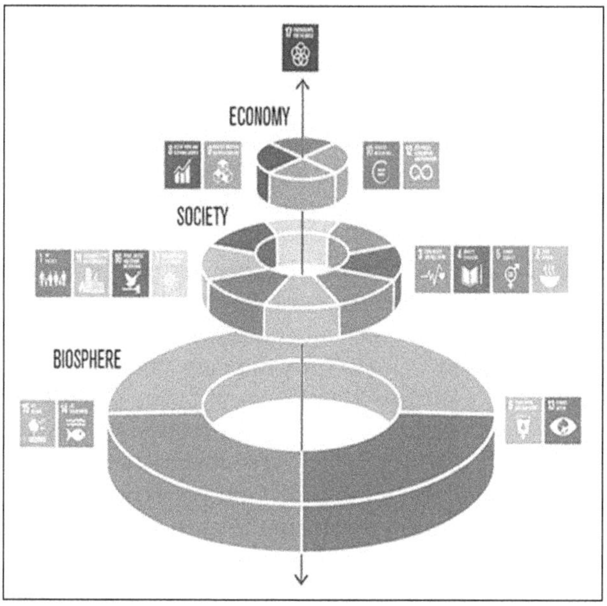


Biosphere model. (Diagram by Azote, for Stockholm Resilience Centre, Stockholm University.)

and Sustainable Development Goals at Member State levels, and identifying gaps, would form the basis of the implementation of the Nature's rights framework. However, this has to be done within a holistic and ecological legal framework, based on the principles of Earth Jurisprudence. The purpose of this framework is to embed these principles in all aspects of human lives and society. Earth Jurisprudence, which has been endorsed by the United Nations Harmony with Nature Programme[80] as a way of achieving the Sustainable Development Goals, can be distilled into the following key principles:

*Wholeness.* The Earth is a living being, a single Earth Community webbed together through interdependent relationships. All life is sacred, with inherent value, and the Earth has her thresholds and limits. The well-being of each member of the Earth Community is dependent on the well-being of the Earth as a whole.

*Universal principles.* The Earth is part of the universe, which is ordered and operates according to its own principles that govern all life, including human beings. We need to discover these principles and comply with them, for our own well-being and for the well-being of the whole.

*Duty of Care.* Earth Jurisprudence is a living law, a way of life, guided by moral responsibilities. We have a duty of care to all present and future members of the Earth Community to contribute to its integrity and well-being. If we create imbalance, then we cause disorder in the Earth's dynamic equilibrium, which we have a duty to restore.

*Nature's Rights.* The Earth and all of the Earth Community have three inherent rights: the right to exist, the right to habitat and the right to fulfil their role in the ever-renewing processes of life.

*Mutual Enhancement.* Relationships within the Earth Community are reciprocal, a cycle of giving and receiving. Our role is to participate and contribute to the health and resilience of the Earth Community. What does not enhance the whole will ultimately not enhance human life either.

*Resilience.* All healthy living systems have the ability to grow, evolve and adapt to change and disturbance, without losing inner coherence. By complying with the laws that maintain life's health and vitality, we strengthen Earth Community resilience as well as our own. To learn from Nature and understand its laws, we must become eco-literate and engage other ways of knowing, especially feeling, sensing and intuition.

Practising this approach to law requires us to make decisions that prioritise the long-term ecological interests of the whole and of future generations over short-term gain. These principles and ideas form the underlying basis of a Nature's Rights Draft European Union Directive,[81] prepared by the author and reviewed by several networks of legal experts, to form the basis of a Nature's Rights European Citizens Initiative, using participatory democracy to put the Rights of Nature on the EU legislative agenda. The proposal was presented to the European Parliament at a conference organised by the Nature's Rights organisation,[82] and co-hosted by the Vice-Chairs of the Environment Committee of the European Parliament, MEPs Benedek Jávor and Pavel Poc. Keynote speakers included the Executive Director of the European Environment Agency, Professor Hans Bruyninckx, and the Director of International Union for the Conservation of Nature Europe, Luc Bas. The event was attended by over 120 key stakeholders from key European institutions and civil society, and very well received.[83] Nature's Rights is now working with the European Economic and Social Council and the International Union for the Conservation of Nature to further raise awareness and to propose a Climate Rights Bill based on the Nature's Rights Draft European Union Directive.

If the European Union adopts a Nature's rights legal framework within an ecological legal context, it will herald the start of a new era, bringing law up to date with modern science, healing the societal disconnect from the rest of Nature – and each other. The ramifications will be profound. But if Europe is to achieve the 2050 vision set out in the 7th Environmental Action Programme, the law must make a dramatic shift into the twenty-first century, so that humanity can flourish in harmony with the rest of Nature for generations to come.

## Notes

1. (a) 'Nature's Rights – the Missing Piece of the Puzzle'. Conference in European Parliament cohosted by the nongovernmental organisation Nature's Rights and European Union MEPs on 29th March 2017 – https://youtu.be/OOSjW_02tjo; (b) UN Harmony with Nature Programme, Conference Report, 'Nature's Rights: The Missing Piece of the Puzzle'.
2. Jordan, A.J. and Adelle, C. (eds.) 2012. *Environmental Policy in the European Union: Contexts, Actors and Policy Dynamics*, Third Edition, London and Sterling, VA: Earthscan.
3. Decision No 1386/2013/EU of the European Parliament and of the Council of 20 November 2013 on a General Union Environment Action Programme to 2020 'Living Well, within the Limits of Our Planet' Text with EEA relevance https://eur-lex.europa.eu/legal-content/EN/TXT/?uri=CELEX:32013D1386
4. https://www.eea.europa.eu/soer
5. Ibid., p. 3.
6. Ibid., p. 3.
7. http://ec.europa.eu/environment/action-programme/pdf/PPT%20A%20-%20EEB.pdf

8. http://ec.europa.eu/environment/action-programme/pdf/7EAP%20-%20Second%20 Workshop_Summary_final.pdf
9. Ibid., p. 3.
10. Ibid., p. 3.
11. Kharas, H. 2010. The Emerging Middle Class in Developing Countries. OECD Development Centre, Working Paper No. 285.
12. Rockström et al. 2009. 'Planetary Boundaries: Exploring the Safe Operating Space for Humanity'. *Ecology and Society* 14(2): 32. [online] URL: http://www.ecologyandsociety.org/vol14/iss2/art32/
13. Hoff, H., Häyhä, T., Cornell, S. and Lucas, P. 2017. *Bringing EU Policy into Line with the Planetary Boundaries*, SEI, PBL Netherlands Environmental Assessment Agency and Stockholm Resilience Centre.
14. Ibid., p. 23.
15. European Commission. 2015. *The Mid-Term Review of the EU Biodiversity Strategy to 2020*. COM (2015) 0478 final.
16. https://www.stockholmresilience.org/research/planetary-boundaries/planetary-boundaries/about-the-research/the-nine-planetary-boundaries.html
17. Ibid., p. 3.
18. Ibid., p. 3.
19. European Commission. 2017. *Monitoring the Application of European Union Law: 2016 Annual Report*. COM (2017) 370 final.
20. European Commission. 2018. *Monitoring the Application of Union Law: 2017 Annual Report*. COM (2018).
21. European Parliament. 2017. *Report on the Implementation of the 7th Environment Action Programme 2017/2030* (INI).
22. Capra, F. and Mattei, U. 2015. *The Ecology of Law: Toward a Legal System in Tune with Nature and Community*. First Edition, Oakland, CA: Berrett-Koehler.,
23. Montini, M. 2017. 'The Double Failure of Environmental Regulation and Deregulation and the Need for Ecological Law' (2017) 26 *The Italian Yearbook of International Law Online* 265, p. 283.
24. Council Directive 92/43/EEC on the conservation of natural habitats and of wild fauna and flora, OJ 1992 L 206.
25. Directive 2009/147/EC of the European Parliament and of the Council on the conservation of wild birds, OJ 2010 L 20.
26. Directive 2000/60/EC of the European Parliament and of the Council establishing a framework for Community action in the field of water policy, OJ 2000 L 327.
27. Foto-Frost v Hauptzollamt Lübeck-Ost (C-314/85) 1987. http://curia.europa.eu/juris/liste.jsf?language=en&jur=C,T,F&num=314/85&td=AL
28. OJ C 326, 26.10.2012, pp. 47–390 https://eur-lex.europa.eu/legal-content/EN/TXT/?uri=celex%3A12012E%2FTXT
29. The United Nations Economic Commission for Europe (UNECE) Convention on Access to Information, Public Participation in Decision-Making and Access to Justice in Environmental Matters. http://www.unece.org/env/pp/treatytext.html
30. 'Consolidated version of the Treaty on the Functioning of the European Union – Part Six: Institutional and Financial Provisions – Title I: Institutional Provisions – Chapter 1: The institutions – Section 5: The Court of Justice of the European Union – Article 263 (ex Article 230 TEC)' *Official Journal* 115, 09/05/2008, pp. 0162–0162.
31. Findings and recommendations of the Aarhus Convention Compliance Committee concerning compliance by the European Union with the Aarhus Convention (ACCC/C/2008/32(EU)) – 17 March 2017.
32. Fulton, S. and Wolfson, S. 2014. 'Strengthening National Environmental Governance to Promote Sustainable Development'. In: Percival, R.V., Lin, J. and Piermattei, W. (eds.) *Global Environmental Law at a Crossroads*, Cheltenham, Gloucestershire, UK: Edward Elgar, p. 15.

33. Capra, F. and Mattei, U. *The Ecology of Law: Toward a Legal System in Tune with Nature and Community*. First Edition. Berrett-Koehler. Oakland, CA.
34. Beck, U. 1986. *Risikogesellschaft: Auf dem Weg in eine andere Moderne*. Frankfurt: Suhrkamp Verlag. English edition: *Risk Society. Towards a New Modernity*, London: Sage, 1992 (last edn. 2004).
35. Nieto, A. and others. 2014. *European Red List of Bees*. European Commission, p. 6. http://bookshop.europa.eu/uri?target=EUB:NOTICE:KH0714078:EN:HTML accessed 23 September 2018.
36. Ito, M. 2016. 'Rights of Nature – Why Do We Need It?' *Permaculture Design*, 99. Ecological Restoration, pp. 48–51.
37. Bugge, H.C. 2010. 'Environmental Law's Fragmentation and Discretionary Decision-Making. A Critical Reflection on the Case of Norway'. In: *Law and Economics*. Oslo, Norway: Cappelen Damm Akademisk. Kapittel 4., pp. 55–75.
38. Ibid., p. 33.
39. World Commission on Environment and Development (WCED). 1987. *Our Common Future*, Chapter 2, Para. 1.
40. United Nations Conference on Environment and Development (UNCED). 1992. Rio Declaration on Environment and Development, Principle 15.
41. Krämer, L. 2008. 'Sustainable Development in EC Law'. In: Bugge, H.C. and Voigt, C. (eds.) *Sustainable Development in International and National Law*, pp. 377–379. Amersterdam, Netherlands: Europa Law Publishing.
42. Jerkø, M. 2009. 'Det norske formålet 'bærekraftig utvikling' (The Norwegian Objective of 'Sustainable Development'), Tidsskrift for Rettsvitenskap nr. 3/2009s, pp. 354–388.
43. Scotford, E. 2017. *Environmental Principles and the Evolution of Environmental Law*, Oxford: Hart.
44. Fulton, S. and Wolfson, S. 2014. 'Strengthening National Environmental Governance to Promote Sustainable Development'. In: Percival, R.V., Lin, J. and Piermattei, W. (eds.) *Global Environmental Law at a Crossroads*, Cheltenham, Gloucestershire, UK: Edward Elgar, p.15.
45. European Commission, Access to Justice in Environmental Matters (27 April 2016).
46. *Plaumann v Commission* (C-25/62) [1963] ECR, 95.
47. ACCC/C/2008/32 (Part I) concerning compliance by the European Union, adopted on 14 April 2011 ('2011 ACCC Part I Findings and Recommendations').
48. Unmüßig, B., Sachs, W. and Fatheuer, T. 2012. *Critique of the Green Economy toward Social and Environmental Equity*. Volume 22 (English edition), in the Publication Series on Ecology, edited by the Heinrich Böll Foundation.
49. Kuznets, S. 1962. 'Inventive Activity: Problems of Definition and Measurement, The Rate and Direction of Inventive Activity: Economic and Social Factors'. In: *National Bureau of Economic Research, The Rate and Direction of Inventive Activity: Economic and Social Factors*, Princeton, NJ: Princeton University Press, pp. 19–52.
50. Nobel Prize-winning economist Joseph Stiglitz proposes alternatives to gross domestic product as a measurement of national economic success, 2008 (publ. 2010) https://www.youtube.com/watch?v=QUaJMNtW6GA
51. Hatfield-Dodds, S. et al. 2015. *Nature* 527, pp. 49–53 (5 November 2015).
52. 'The Decoupling Delusion: Rethinking Growth and Sustainability'. March 12, 2017, https://theconversation.com/the-decoupling-delusion-rethinking-growth-and-sustainability-71996
53. 'When It Comes to Natural Capital, It's Easy to Forget That We're on the Same Team' 09/16/2016, *Huffington Post*, https://www.huffingtonpost.com/natural-capital-coalition-/when-it-comes-to-natural_b_12043244.html?guccounter=1
54. Büscher, B. and Fletcher, R. 2016. Nature Is Priceless, Which Is Why Turning It into 'Natural Capital' Is Wrong. *The Conversation*. https://theconversation.com/nature-is-priceless-which-is-why-turning-it-into-natural-capital-is-wrong-65189
55. Ibid 53.
56. Dempsey, J. and Chiu Suarez, D. 2016. 'Nature and Society Arrested Development? The Promises and Paradoxes of 'Selling Nature to Save It'. *Annals of the American Association of Geographers* 106 (3), pp. 653–671.Published online: 6 Apr 2016.

57. Ibid., p. 51.
58. Ito, M. and Montini, M. 2019. "Nature's Rights and Earth Jurisprudence – A New Ecologically Based Paradigm for Environmental Law, Part III". In: *The Right to Nature: Social Movements, Environmental Justice and Neoliberal Natures*, Abingdon-on-Thames, Oxfordshire, UK: Routledge, pp. 231–233.
59. 'Agropoly – A Handful of Corporations Control World Food Production'. 2013. English Edition. Berne Declaration (DB) & EcoNexus. https://www.econexus.info/sites/econexus/files/Agropoly_Econexus_BerneDeclaration.pdf
60. EU. 2018. A Sustainable Bioeconomy for Europe: Strengthening the Connection between Economy, Society and the Environment. Updated Bioeconomy Strategy. https://ec.europa.eu/research/bioeconomy/pdf/ec_bioeconomy_strategy_2018.pdf#view=fit&pagemode=none
61. Ibid., p. 55.
62. 'Too Big to Control? The Politics of Mega-Mergers and Why the EU Is Not Stopping Them;. June 21, 2017. The Power of Lobbies. https://corporateeurope.org/power-lobbies/2017/06/too-big-control
63. Ibid., p. 62.
64. Ibid., p. 33.
65. Marx, K. 1867. *1992 Capital: A Critique of Political Economy* Volume 1. London: Penguin Classics.
66. Macpherson, C.B. 1962. *The Political Theory of Possessive Individualism: Hobbes to Locke*. Oxford: Clarendon Press.
67. Ibid., p. 33.
68. Ibid., p. 33.
69. Hertz, N. 2001. *The Silent Takeover: Global Capitalism and the Death of Democracy*. New York: Harper Business.
70. Picketty, T. 2014. *Capital in the Twenty First Century*. Translated by Arthur Goldhammer. Cambridge, MA: Harvard University Press.
71. Ibid., p. 33.
72. Ibid., p. 62.
73. Ibid., p. 33.
74. Ibid., p. 33.
75. Ito, M. 2017. 'Nature's Rights – A New Paradigm for Environmental Protection' https://theecologist.org/2017/may/09/natures-rights-new-paradigm-environmental-protection
76. Lassalle, D., Maquil, F. and Raducu Pelin, I. DATE. *Mapping Human Rights Obligations Relating to the Enjoyment of a Safe, Clean, Healthy and Sustainable Environment*. Universite de Geneve Global Studies Institute. Individual Report on the European Convention on Human Rights and the European Union. Report No. 14. Prepared for the Independent Expert on the Issue of Human Rights Obligations Relating to the Enjoyment of a Safe, Clean, Healthy, and Sustainable Environment.
77. Ibid., p. 62.
78. Friedmann, D. and Barak-Erez, D. 2002. *Human Rights in Private Law*. Oxgord, UK and Portland, OR: Hart Publishing.
79. https://ec.europa.eu/europeaid/policies/european-development-policy/2030-agenda-sustainable-development_en (accessed 9 February 2019).
80. http://www.harmonywithnatureun.org/
81. http://www.natures-rights.org/ECI-DraftDirective-Draft.pdf
82. http://www.natures-rights.org
83. (a) UN Harmony with Nature Programme, Conference Report, 'Nature's Rights: The Missing Piece of the Puzzle', http://files.harmonywithnatureun.org/uploads/upload52.pdf (accessed 20 January 2019); (b) Ibid., p. 1.

# 17

# *Nature's Rights through Lawmaking in the United States*

**Lindsey Schromen-Wawrin and Michelle Amelia Newman**[1]

**CONTENTS**

From Property to Person: What Rights of Nature Means in the Legal System ..... 332
Grassroots System Change: Local Democracy versus the Corporate State ..... 334
The Courts and System Change ..... 334
How Courts Prohibit Lawmaking Today ..... 335
Corporate Constitutional Rights ..... 335
Dormant Commerce Clause ..... 336
Contracts Clause ..... 336
Dillon's Rule ..... 336
Pre-Emption as a Ceiling on Rights Protections ..... 336
Nature as Property ..... 336
Recognising Rights of Nature Requires System Change ..... 337
Protecting Watersheds through Ecosystem Rights by Local Lawmaking ..... 337
A Bill of Rights for Lake Erie ..... 338
The Highland Township Ordinance and Home Rule Charter ..... 340
Lincoln County, Oregon, Ordinance against Aerial Spraying ..... 342
Rights for the Salish Sea, Washington State and British Columbia ..... 346
Conclusion ..... 349
Notes ..... 349
Appendix: Nature's Rights through Lawmaking in the United States ..... 355

This chapter looks at developing ecosystem rights, frequently known as Rights of Nature, in the United States through legislative lawmaking. Social movements for system change usually start from the grassroots, developing new policies through working for local, then state, and ultimately federal laws that institutionalise a social movement's goals. When a social movement seeks deep structural change, like the Rights of Nature, that grassroots policy work fuels – and is itself fuelled by – cultural changes that create societal recognition for the legitimacy of the new system, and the illegitimacy of the old.[2]

The movement for ecosystem rights is beginning that process in the United States. This chapter looks at some of the early local laws and the emerging social recognition of the legitimacy of ecosystem rights.

DOI: 10.1201/9780429505959-17

## From Property to Person: What Rights of Nature Means in the Legal System

The western legal system generally divides things into two categories: persons and property. Persons can have rights. Property, which includes Nature, cannot. But turning Nature from 'property' to 'person' is just the first step. A legal 'person' is often defined as having the ability to contract, own property, sue and be sued. Legal persons can also have rights, but not every legal person has the same rights. But categorically, 'property' does not have rights. Thus, the framework of 'person' versus 'property' is useful for understanding the legal transformation needed for ecosystem rights. But the details about what rights an ecosystem has, and what those rights mean, are not inherently answered simply by putting Nature in the 'person' category.

Historically, not all human beings have been 'persons'. The law classified enslaved people as property.[3] Women also had to fight for their rights – first by fighting to become legal persons. Meanwhile, the classification of 'person' has not been limited only to human beings. Corporations and other business entities were 'persons'. long before enslaved people were granted that legal title, and corporations fought for and gained rights as legal persons long before many human beings had rights.[4]

Western law has always considered the Earth and ecosystems to be property. The concept of Rights of Nature includes the rights of nonhuman species, who are also classified as property in the law today. There has been a longer fight for personhood and legal rights for animals than for ecosystems, and thus this chapter focuses on the more recent development of Rights of Nature for ecosystems.[5] The idea that it could be any different has only recently entered Western thought.

In 1949, ecologist Aldo Leopold advocated for ecosystem rights in his essay 'The Land Ethic'. Leopold illustrated the violence in Western society's treatment of human beings as property, then challenged us to think about the Earth as having rights:

> When god-like Odysseus returned from the wars in Troy, he hanged all on one rope a dozen slave-girls of his household whom he suspected of misbehavior during his absence.
>
> This hanging involved no question of propriety. The girls were property. The disposal of property was then, as now, a matter of expediency, not of right and wrong.
>
> ....
>
> ... There is as yet no ethic dealing with man's relation to land and to the animals and plants which grow upon it. Land, like Odysseus' slave-girls, is still property. The land-relation is still strictly economic, entailing privileges but not obligations.[6]

Leopold, however, was an ecologist, not a lawyer. So even though he conceptualised land as a rights-bearing entity, his ideas did not immediately reach into legal thinking.

It was not until the early 1970s when the idea of ecosystems having rights formally entered Western legal thought. In 1972, property law Professor Christopher Stone wrote a law review article titled 'Should Trees Have Standing? – Toward Legal Rights for Natural Objects'. Professor Stone spent the first several pages of his article framing the 'unthinkable' idea that trees could have rights. He wrote:

> Throughout legal history, each successive extension of rights to some new entity has been theretofore, a bit unthinkable. We are inclined to suppose the rightlessness of rightless 'things' to be a decree of Nature, not a legal convention acting in support of some status quo....

> The fact is, that each time there is a movement to confer rights onto some new 'entity', the proposal is bound to sound odd or frightening or laughable. This is partly because until the rightless thing receives its rights, we cannot see it as anything but a *thing* for the use of 'us' – those who are holding rights at the time.... There is something of a seamless web involved: there will be resistance to giving the thing 'rights' until it can be seen and valued for itself; yet, it is hard to see it and value it for itself until we can bring ourselves to give it 'rights' – which is almost inevitably going to sound inconceivable to a large group of people.
>
> The reason for this little discourse on the unthinkable, the reader must know by now, if only from the title of the paper. I am quite seriously proposing that we give legal rights to forests, oceans, rivers and other so-called 'natural objects' in the environment – indeed, to the natural environment as a whole.[7]

Professor Stone's paper influenced United States Supreme Court Justice Douglas's dissenting opinion in *Sierra Club v. Morton*, where Justice Douglas advocated that ecosystems should be able to sue in their own right:

> The critical question of 'standing' would be simplified and also put neatly in focus if we fashioned a federal rule that allowed environmental issues to be litigated before federal agencies or federal courts in the name of the inanimate object about to be despoiled, defaced, or invaded by roads and bulldozers and where injury is the subject of public outrage. Contemporary public concern for protecting nature's ecological equilibrium should lead to the conferral of standing upon environmental objects to sue for their own preservation....
>
> Inanimate objects are sometimes parties in litigation. A ship has a legal personality, a fiction found useful for maritime purposes. The corporation sole – a creature of ecclesiastical law – is an acceptable adversary and large fortunes ride on its cases. The ordinary corporation is a 'person' for purposes of the adjudicatory processes, whether it represents proprietary, spiritual, aesthetic, or charitable causes.
>
> So it should be as respects valleys, alpine meadows, rivers, lakes, estuaries, beaches, ridges, groves of trees, swampland, or even air that feels the destructive pressures of modern technology and modern life. The river, for example, is the living symbol of all the life it sustains or nourishes – fish, aquatic insects, water ouzels, otter, fisher, deer, elk, bear, and all other animals, including man, who are dependent on it or who enjoy it for its sight, its sound, or its life. The river as plaintiff speaks for the ecological unit of life that is part of it. Those people who have a meaningful relation to that body of water – whether it be a fisherman, a canoeist, a zoologist, or a logger – must be able to speak for the values which the river represents and which are threatened with destruction.
>
> ....
>
> Those who have that intimate relation with the inanimate object about to be injured, polluted, or otherwise despoiled are its legitimate spokesmen.[8]

During the next several years, there were a few lawyers who represented ecosystems or parts of Nature, such as a river, a beach, a national monument, a commons and a tree, as named parties in court,[9] but Professor Stone and Justice Douglas's idea did not develop beyond these efforts and their rather ambiguous results. Instead, beginning in the 1990s, lawsuits to protect the environment became mired in Justice Scalia's limitations on legal issues over which a court can exercise its judicial authority, which severely limited the ability of environmental plaintiffs to even get into court to argue their cases.[10] Plaintiffs' standing, rather than legal rights, became the critical question for environmental protection litigation.

The courts did not take the opportunity to develop ecosystem rights on their own merit. If anything, the U.S. Supreme Court and lower courts have gone on a detour in the decades since Professor Stone and Justice Douglas thought the unthinkable. Rather than explore the implications of legal rights for the Earth itself, the courts have focused on which people and organisations (which 'persons') can bring lawsuits to prevent which environmental harms. The legal issue of 'standing' became a complicated barrier for environmental protection litigants, even though Justice Douglas had pointed out that ecosystem rights would simplify the issue.

## Grassroots System Change: Local Democracy versus the Corporate State

### The Courts and System Change

We should not be surprised that the courts did not pick up the idea of ecosystem rights, or are resistant to the idea of Rights of Nature generally. United States Courts are inherently conservative and backwards-looking and do not frequently lead innovative legal change. In 1835, Alexis de Tocqueville opined on the role of lawyers and judges in society:

> In America there are no nobles or literary men, and the people are apt to mistrust the wealthy; lawyers consequently form the highest political class and the most cultivated portion of society. They have therefore nothing to gain by innovation, which adds a conservative interest to their natural taste for public order. If I were asked where I place the American aristocracy, I should reply without hesitation that it is not among the rich, who are united by no common tie, but that it occupies the judicial bench and bar.[11]

Social movements that have sought to fundamentally change who is granted legal personhood and who or what is considered property are frequently at odds with the opinions of courts. The most famous example is probably *Dred Scott v. Sandford*.[12] In that case, an enslaved person brought a lawsuit to attempt to gain his freedom. In 1857, United States Supreme Court Chief Justice Taney wrote an infamous opinion for the Supreme Court concluding that black people could never be citizens of the United States.

The opinions of the United States Supreme Court become the definitive interpretation of the United States Constitution, at least as long as they remain good precedent, and thus can close legal debate on an issue. While we are trained to think of the courts as the least political branch of government, their opinions can actually be the most authoritative tools of political power. Regardless, political debates carry on, thanks to the tireless work of advocates and those most impacted by injustice. For example, 5 months after the *Dred Scott* opinion, abolitionist Fredrick Douglass gave a speech defending struggle, agitation and resistance: 'The limits of tyrants are prescribed by the endurance of those whom they oppress'.[13] And Abraham Lincoln, then a Senator, talking about the legal implications of *Dred Scott*, gave his 'House Divided' speech the following year: 'I believe this government cannot endure, permanently half slave and half free. … It will become all one thing, or all the other'.[14]

Notably, courts have the power to decide whether a law is constitutional, and thus whether it is legal. During the populist and progressive eras at the end of the nineteenth and beginning of the twentieth centuries, in response to the social harms of industrialisation, state and local governments enacted many laws to protect people from corporate exploitation. In response, in 1905, the United States Supreme Court reasoned that people had a constitutional right to choose how to contract with others (a 'liberty of contract'), and thus held a New York state law that prohibited employing a person in a bakery for more than 60 hours a week to be unconstitutional. This decision is *Lochner v. New York*:

> The statute necessarily interferes with the right of contract between the employer and employees, concerning the number of hours in which the latter may labor in the bakery of the employer. The general right to make a contract in relation to his business is part of the liberty of the individual protected by the Fourteenth Amendment of the Federal Constitution. Under that provision no State can deprive any person of life, liberty or property without due process of law. The right to purchase or to sell labor is part of the liberty protected by this amendment, unless there are circumstances which exclude the right.[15]

*Lochner v. New York* is the namesake for what is today known as the '*Lochner* Era', the several decades at the beginning of the twentieth century when the United States Supreme Court used the 'liberty of contract' theory to strike down state and local labour and employment laws enacted by legislatures and pertaining to health, safety and welfare. By constitutionalising 'liberty of contract', the Court took away the legislative branch's ability to protect the people.

Thus, the courts can lock social movements seeking system change out of the policy world. On the one hand, courts can define who is a person and who is property, and enforce those boundaries in the interest of the status quo. On the other hand, courts can prohibit lawmaking by constitutionalising a right (such as 'liberty of contract'), thus preventing state and local governments from redefining persons and property or recognising additional rights for persons. As a result, systemic social movements are squeezed out of any opportunity for 'legal' change.

The consequences of courts stifling social change are often terrible. *Dred Scott* led to the Civil War and Reconstruction: it took a United States Constitutional amendment to undo the *Dred Scott* holding that black people are not citizens. The first sentence of the Fourteenth Amendment (proposed by Congress in 1866 and ratified by the states in 1868) expressly reverses the *Dred Scott* holding that black people are not citizens. The Fourteenth Amendment, Section 1, begins with the sentence: 'All persons born or naturalized in the United States, and subject to the jurisdiction thereof, are citizens of the United States and the State wherein they reside.'[16] The *Lochner* Era was a bloody time for the labour movement, and one of the ultimate constitutional crises in United States history, where President Roosevelt, proposing adding more Supreme Court Justices, declared: 'We have, therefore, reached the point as a Nation where we must take action to save the Constitution from the Court, and the Court from itself'.[17] The Court responded with the 'switch in time that saved nine', issuing opinions on new cases that overruled previous *Lochner* Era opinions, signalling its departure from the 'liberty of contract' theory. For example, in *West Coast Hotel Company v. Parrish* in 1937, the Supreme Court upheld Washington State's minimum wage law, ruling it constitutional and therefore legal.[18]

## How Courts Prohibit Lawmaking Today

When systemic social movements attempt to recognise ecosystem rights through lawmaking, they run up against this legal system that allows the courts to define who has rights and to prohibit lawmaking. There are at least six legal doctrines that limit Rights of Nature lawmaking in the United States today:

### *Corporate Constitutional Rights*

The courts have given corporations constitutional rights under most key provisions of the United States Constitution (First Amendment speech, Fourth Amendment privacy, Fifth

Amendment takings, Fourteenth Amendment equal protection and due process). These corporate constitutional rights allow corporations to invalidate laws that infringe on a corporation's 'rights'.

#### *Dormant Commerce Clause*

The courts say that because the United States Constitution's commerce clause gives Congress authority to regulate interstate commerce, the states (and by extension local governments) have no power to pass laws that interfere with interstate commerce. This interpretation prohibits local lawmaking to favour locally produced goods and services, or prohibit importing wastes, and other measures that are needed for environmental and economic sustainability.

#### *Contracts Clause*

The courts interpret the United States Constitution's Contracts Clause to protect corporate charters, and hold that once a state permits an entity to incorporate (which creates the entity in the eyes of the law), the state and the corporation are effectively contracting parties on equal footing. The state may not interfere with the corporation as if the state were in any way superior to the corporation.

#### *Dillon's Rule*

The courts interpret local governments (and by extension, 'the people of local governments', who really have no legal existence whatsoever) to be 'children of the state'. Local governments can therefore only pass laws that the state has authorised, and the state retains full control over the local government. It even retains the ability to destroy the local government (e.g. Michigan's 'emergency manager' law, whereby the state can nullify local authority and appoint an emergency manager to serve for 18 months or more). Note the sharp contrast here with the contracts clause protection that business corporations receive, even though they are arguably also 'state creatures'.

#### *Pre-Emption as a Ceiling on Rights Protections*

The courts interpret constitutional rights to be limited to those found in the state and federal constitutions. A local law – including the local charter, which is a constitution – that conflicts with state or federal law is usually trumped, and nullified, even when that local law provides greater protection for people's rights than the state or federal law. Compatible with Dillon's Rule, courts often require state or federal laws to explicitly authorise local lawmaking that is more protective.

#### *Nature as Property*

The courts view the Earth and ecosystems as governed by property law, including the property owners' right to destroy their property. Environmental law, which is a subset of property law, accepted this premise and attempted to provide 'reasonable' regulation of property use without actually infringing on the property owner's right to use (and exploit) the property.

### Recognising Rights of Nature Requires System Change

These six legal doctrines are currently part of the United States legal system, and work to prevent lawmaking that favours Rights of Nature. Advocating for the Rights of Nature requires simultaneously advocating against these legal doctrines. There is no secret back-door to systemic change that can avoid confronting these legal doctrines. There is no magic legal argument that will make the courts recognise ecosystem rights. As Frederick Douglass recognised in the fight against slavery, this will be a struggle.

Once again, historical examples help to illuminate the issue. In 1860, enslaved people – also known as property – were considered 48.3% of the total wealth of the South.[19] One of the four main arguments for slavery was the economic devastation that would result from emancipation.[20] During and after the Civil War, slave owners were compensated for government's taking of their property.[21] A constitutional amendment was needed to exempt emancipation from the former slave owners' constitutional right to just compensation for the taking of their 'property'.[22]

In 2012, the Carbon Tracker Initiative calculated that fossil fuel companies and petro-states hold five times as much carbon in known coal, oil and gas reserves than can be burned if the Earth has a chance of avoiding catastrophic climate change.[23] Keeping that carbon in the ground means corporations would be giving up $20 trillion in assets.[24] And, if the government causes that lost property, those corporations would be owed just compensation for the 'taking'. Just as the South justified slavery for economic reasons, our society today uses the same justification for not moving more quickly towards sustainable energy. Just as former slave owners had a constitutional right to be compensated for emancipation of their 'property', oil and gas industries will demand payment for 'their' hydrocarbons (fossil fuels) that must be kept in the ground. Since implementing Rights of Nature laws will involve some elements of climate justice, fossil fuel corporations will have to give up their property. They probably have a fiduciary duty to their shareholders to demand compensation for that financial loss.

Fredrick Douglass, in his 1857 speech, reminded us that system change does not come easily:

> If there is no struggle there is no progress. Those who profess to favor freedom and yet deprecate agitation are men who want crops without plowing up the ground; they want rain without thunder and lightning. They want the ocean without the awful roar of its many waters....
>
> Power concedes nothing without a demand. It never did and it never will. Find out just what any people will quietly submit to and you have found out the exact measure of injustice and wrong which will be imposed upon them, and these will continue till they are resisted with either words or blows, or with both. The limits of tyrants are prescribed by the endurance of those whom they oppress.[25]

Today, if we want to live in a Rights of Nature–based society, we must challenge the legal doctrines designed to prevent system change. We must demand that the law accord ecosystems the right to exist, flourish, evolve and regenerate, and elevate those rights over now-dominant corporate interests.

## Protecting Watersheds through Ecosystem Rights by Local Lawmaking

Fortunately, the path forward is already being forged. Several dozen communities in the United States have already codified Rights of Nature into law, or are working to do so. Here, we will look at several communities as case studies.

First, residents of Toledo, Ohio, approved a ballot proposal in February 2019 to protect the rights of Lake Erie from regular toxic algal blooms that threaten the drinking water of nearly half a million residents.

Second, Highland Township, in the Allegheny Mountains of western Pennsylvania, recognised Rights of Nature to protect the community's water from a fracking waste injection well, and its residents have been bravely battling to enforce those rights ever since.

Third, Lincoln County, on the Oregon coast, passed a ballot measure in 2017 to protect the community from aerial pesticide spraying and grant rights to the local watershed to flourish and maintain its ecological integrity.

Finally, in the Salish Sea of Washington State and British Columbia, communities are considering Rights of Nature lawmaking, attempting to stop the complex and interconnected harms that are leading to ecosystem collapse, most visibly illustrated by the population decline of resident orca whales.

## A Bill of Rights for Lake Erie

In the summer of 2014, in Toledo, Ohio, residents were faced with a historic water crisis on the shores of Lake Erie bordering the city. A combination of agricultural waste runoff from industrial farms and increasing temperatures from climate change converged to cause a significant bloom of cyanobacteria (blue-green algae) in the lake, Toledo's source of drinking water. Residents of Toledo have never forgotten the moment: due to the toxic algal bloom, the city was forced to shut down public water sources and nearly a half of a million people could not use the water from any faucets for two-and-a-half days. The loss of public drinking water set off a chain of negative effects: restaurants and businesses in the area shut down completely during the crisis; dialysis patients, among other medical patients, could not get treatments in the city; Toledo firefighters set up water stations to distribute water donated and imported from other regions and Lake Erie's western waters turned pea green and smelled like sewage.[26] The toxic algal blooms continue annually, with dead zones remaining in the wake of the blooms and other threats to humans still being studied.[27]

Scientific research identified the major cause of the 2014 toxic algal bloom: phosphorus-heavy fertiliser and animal manure from industrial farming and feed operations in northwest Ohio that runs off as waste into the Maumee River, which drains into Lake Erie. Because Lake Erie is relatively shallow, and warm in the summer, this waste runoff feeds the explosive growth of toxic algae.[28] As one journalist put it, 'the Western Lake Erie Basin has a phosphorus problem', and the industrial agriculture industry concentrated in the region, growing mainly soy and corn, is the biggest contributor to that problem.[29]

Agriculture, however, is an especially large and strong industry in the state, and changes to current practises never materialised. Terrified residents vowed such a crisis would never happen again, but then watched the years pass with no meaningful legislative or regulatory changes. Meanwhile, the harmful algal blooms have continued in Lake Erie annually, affecting businesses and residents that rely on the Lake's waters for their livelihoods and sustenance.

Toledoans for Safe Water, a grassroots organisation, working with local environmental lawyers and other organisations and activists, proposed a new way of protecting Lake Erie from the continuing threat of algal blooms – one that did not rely on sluggish government agencies and did not cater to the agricultural lobby. Relying on principles of self-governance, organisers proposed a Lake Erie Bill of Rights that would enshrine the Lake's right to exist and thrive, as well as the community's right to access clean and safe drinking water, and right to local self-government.[30] The Bill of Rights combines these correlative rights – for the

Lake Erie ecosystem itself and for Toledo's human residents – with the right of any resident of Toledo, or the city itself, to sue on behalf of Lake Erie, as guardians, to protect the Lake's right to survive and be free from pollution and harm.[31] The initiative describes the rights of the Lake Erie ecosystem: 'Lake Erie, and the Lake Erie watershed, possess the right to exist, flourish, and naturally evolve'. The rights of Toledoans are described as 'the right to a clean and healthy environment, which shall include the right to a clean and healthy Lake Erie and Lake Erie ecosystem'. The initiative also includes a collective and individual right to self-government in their community. The rights secured by the law are to be self-executing and enforceable against both public and private actors. Thus, the proposed initiative creates the framework of legal protection and standing, and allows for the local government to build policy around those rights.[32]

This little girl was born in a Toledo, Ohio, hospital during the 2014 Lake Erie toxic algal bloom, when not even hospitals were able to use municipal water drawn from Lake Erie. (Photo by Tish O'Dell.)

In 2018, organisers of the initiative collected over 10,000 signatures on a petition for the amendment to be submitted to a city-wide vote. Ohio requires 10% of registered voters who voted in the last gubernatorial election to validly sign a ballot initiative before the proposal will be qualified for the ballot.[33] Organisers canvassed Toledo throughout the year to collect enough signatures to submit the ballot initiative to county officials by mid-July 2018, but legal challenges delayed its appearance on the ballot for the November 2018 general election. In December 2018, the Toledo City Council determined that an adequate number of petition signatures had been collected for the issue to be submitted to electors, and in January 2019, the Ohio Supreme Court rejected a challenge to block a vote on the measure. Thus, the proposed amendment was submitted to Toledo citizens for a popular vote on the protection of Lake Erie, as well as their own health, in a special election in late February 2019. It passed by over 60% of the vote, amending the city charter to codify the right of Lake Erie to flourish and live, and Toledoans' right to a healthy environment, as local constitutional rights.[34]

As this initiative has grown into an organising movement to protect Lake Erie's right to exist, as opposed to regulating (or failing to regulate) a steady stream of pollution, organisers with Toledoans for Safe Water have noticed a feeling of engagement and empowerment around the residents' rediscovering their potential to change the law to protect themselves and their natural environment. Residents viscerally understand the threat that the annual and increasing algal blooms present to Lake Erie and every living being that relies on it. The proposed Lake Erie Bill of Rights presents a lasting, local solution to the long history of pollution and degradation of Lake Erie. And despite legal challenges the law now faces – brought by corporate attorneys working for industrial agricultural firms – organisers describe much of the amendment's value in the change it represents: local people taking collective action despite a legal system that denies their rights.

## The Highland Township Ordinance and Home Rule Charter

In contrast to the nascent Lake Erie Bill of Rights success, the small rural community of James City in Highland Township in northwest Pennsylvania has spent over half a decade fighting a proposed injection well for wastewater used in the extraction of shale gas. When they realised that traditional avenues, such as petitioning the regulatory agency or working with government officials, would fail, the township residents turned to Rights of Nature and home rule principles.

James City is an unincorporated community in Highland Township composed of a few hundred residents. The city, and Highland Township, lie within Allegheny National Forest, which is crisscrossed with oil and gas infrastructure corridors. Oil and gas is a major industry in this region, and the area has seen a boom in hydraulic fracturing ('fracking') from development of the gas and oil-rich Marcellus Shale underlying the Appalachian regions of western Pennsylvania, West Virginia and eastern Ohio.[35]

In June 2012, Texas-based oil and gas company Seneca Resources Corporation applied to the regional Environmental Protection Agency (EPA) office for a permit to construct and operate an injection disposal well for hydraulic fracturing wastewater in James City. Fracking waste consists of a blend of toxic, hazardous and sometimes radioactive materials, yet it is exempt from the requirements of several important environmental statutes and is only lightly regulated.[36] The injection well Seneca proposed was hazardously close to the municipality's pristine water source, Crystal Spring. The township municipal authority draws drinking water for James City residents from Crystal Spring, which is located at the base of a hill outside the city upon which the injection well would be built, about 3,000 feet away from the spring. In addition to threatening James City's only usable water supply, Seneca's proposed injection well would increase traffic and noise in the area, heighten dangerous conditions on the often-icy roads, pollute the fresh air and disturb the natural beauty and peaceful quiet that residents love about their environment.

Due to the fracking boom, communities, and particularly rural communities, across Pennsylvania have been increasingly threatened by injection disposal wells. Hydraulic fracturing comes with little transparency,[37] endangers drinking water and local aquifers[38] and is even linked to earthquakes near fracking waste disposal operations throughout the country.[39] In addition, Seneca had a long history of permit violations.[40] The almost inevitable risk that an underground toxic waste leak would contaminate James City's only water supply was enough to rally nearly the whole Township in fighting against the proposed well. The well is so unpopular with the community that even some local oil and gas workers opposed it.

Highland Township launched its first salvo in early 2013, when the local government adopted an ordinance entitled 'Highland Township's Community Rights and Protection from Injection Wells Ordinance'.[41] The ordinance recognised and protected the rights of Highland Township residents to clean water and air, natural communities and ecosystems, including wetlands, streams, rivers, aquifers and other systems.[42] The ordinance also specifically prohibited injection wells from disposing wastewater from shale gas extraction within the township, as wastewater injection wells would violate those rights.[43] And, like the Lake Erie Bill of Rights, the ordinance provided that the Township or any of its residents would have standing to enforce violations of the ordinance.[44]

The fight to enforce this ordinance, and thereby protect their own health and environment, has been complicated and often harrowing for the Highland residents. In 2014, the regional EPA office approved a federal permit for Seneca's waste injection well, and then denied an appeal of its decision. The township sent EPA a letter stating that the permit was invalid under its ordinance.[45] In early 2015, Seneca went on the offensive, suing Highland Township in federal court and seeking nullification of the ordinance as a violation of its constitutional right to build the well in the township.

Though Seneca had its federal permit, it could not commence work on the injection well without a well permit from Pennsylvania's Department of Environmental Protection (DEP). In light of the ordinance and the Seneca lawsuit, DEP determined it would not review Seneca's permit application before the federal court decided the validity of the ordinance.[46]

Because the Ordinance had granted rights to ecosystems within the township, the Crystal Spring ecosystem attempted in August 2015 to intervene in defence of the ordinance that recognised its rights. This was only the second time that an ecosystem in the United States had attempted intervene in a lawsuit to defend its own rights to exist and flourish.[47] Representatives described the ecosystem as encompassing the spring itself, 'as well as the surrounding hillside and riparian forests, soils, and bedrocks, through which water flows (and is filtered) to the water source at Crystal Spring', and the residents of James City themselves, who rely on Crystal Spring as their sole source of drinking water.[48] The magistrate judge did not permit Crystal Spring ecosystem to intervene, however, finding that its rights were already adequately represented in the litigation by the township.[49]

Over time, under pressure from Seneca, a new Board of Supervisors for Highland Township decided to rescind the ordinance in August 2016[50] and enter into a consent decree with Seneca, effectively ending the first court case. The ecosystem and its co-interveners attempted to intervene again, now that no one was representing its interests in the case. But the magistrate judge again denied the request and, on appeal, the Third Circuit affirmed the magistrate judge's order because the ordinance had been rescinded by that time.[51] Moreover, because the ecosystem had proposed to intervene with other entities (a nonprofit, grassroots organisation formed to advocate for the Ordinance and the municipal water authority), the Third Circuit noted that it did not need to decide the question of whether the ecosystem had standing to sue in its own right.[52]

Meanwhile, Highland Township's residents were still fighting. Shortly after the Board of Supervisors revoked the ordinance, the township residents voted to pass a Home Rule Charter pursuant to Pennsylvania's Home Rule Charter and Optional Plans Law.[53] Drafted by a citizen-elected government study commission, the charter again included provisions for community and ecosystem rights, a prohibition on hydraulic fracturing wastewater injection wells and the ability of residents to enact laws through a democratic initiative process if residents determined the township supervisors had failed to adequately represent them.[54]

Seneca sued the township again, this time over the Charter,[55] and the Pennsylvania Department of Environmental Protection also sued the township. Late in 2017, the same magistrate judge determined that portions of the Home Rule Charter were invalid, and left other portions in place. In particular, the judge declared invalid the township's ban on injection wells, found state and federal law pre-empted other provisions relating to oil and gas activity within the township and determined that the Charter's stripping of 'corporate personhood' violated Seneca's constitutional rights.[56] The judge again denied Crystal Spring ecosystem's attempt to intervene, this time finding it had not alleged more than a generalised threat of future harm.[57]

Yet the Highland residents' fight to save their environment carries on. In May 2018, a Township election again changed the composition of the Board of Supervisors, which may continue to battle Seneca. At its core, the story of Highland Township is one of a working-class community's struggle against a large and exploitative oil and gas company, and the stakes are higher than court cases or corporate threats. The residents of Highland Township seek to protect their small town, its natural environment and only water supply, buried deep in a National Forest, for generations to come.

## Lincoln County, Oregon, Ordinance against Aerial Spraying

The people of Lincoln County, Oregon, recognised ecosystem rights and the violation of those rights caused by aerial pesticide spraying.

Lincoln County is a rural county on the Pacific Coast. It is the historic territory of the Tillamook, Siletz, Yaquina, Alsea and Suislaw indigenous peoples. Non-native settlers moved into the coastal region beginning in the mid-nineteenth century. Today the human population lives mostly on the coast in small cities and towns that historically were founded for timber extraction and paper production, as well as fishing.

The Pacific Ocean is immediately west of Lincoln County. Heading east from the coast, the land rises into the low mountains of the Coast Range. From these mountains, short rivers flow to estuaries along the coast.

Middle Siletz River watershed, Lincoln County, Oregon. Between 2000 and 2013, more than 42% of the watershed was clear-cut.[58] (Photo by Rio Davidson.)

The land is covered in the dense conifer forests common in the Pacific Northwest of North America. However, twentieth-century industrialised logging has taken a heavy toll on the landscape, and it continues to do so. To manage invasive plants such as scotch broom, which can suppress tree seedlings, the timber industry sprays pesticides from aeroplanes or helicopters during the reforestation process. Recognising the hazards of broadcast pesticide application, the United States Forest Service and Bureau of Land Management prohibited aerial pesticide application on its lands in 1984.[59] However, aerial spraying continues on private and state lands.

Oregon state law provides no buffers from aerial spraying around homes and schools. There are only small buffers required around streams.

The Oregon legislature enacted the Forest Practices Act in 1971 to consolidate regulatory power over private timberland practises into the hands of the state-wide Board of Forestry.[60] Rather than provide minimum standards for protecting public health and ecosystems, the Forest Practices Act purports to set a ceiling on industry regulations. This prevents local governments from taking action to protect their people and ecosystems from harm by the timber industry.

Seeing the harm to themselves and their home environment, people in Lincoln County organised to protect themselves through passage of an initiative to regulate aerial spraying in Lincoln County. As in Toledo, Ohio, they used the progressive era tool of initiative lawmaking.

A hundred years ago, people throughout the United States saw how corporate power could control a state's legislature. With the state legislature as the only lawmaking body – and frequently under corporate control – the people were powerless to make law to protect themselves. So the populist and progressive social movements of the time attempted to redesigned the legal system to reserve the power of direct democracy. 'Home Rule' was the response to Dillon's Rule. The powers of initiative, referendum and recall gave the people the ability to pass their own laws, call for a direct vote on their representatives' actions and get new representatives if the existing 'representation' wasn't adequate.

People in Lincoln County used the initiative process to propose a community bill of rights, a county law that recognised human and ecosystem rights and prohibited certain corporate practises that violated those rights. They called the draft ordinance the 'Freedom from Aerially Sprayed Pesticides Ordinance of Lincoln County',[61] and the preamble to the ordinance made the law's purpose clear:

> At a time when the rights of corporations are being protected at the expense of the people's rights, we must reaffirm what the Oregon Constitution guarantees: 'that all [people], when they form a social compact are equal in right: that all power is inherent in the people, and all free governments are founded on their authority, and instituted for their peace, safety, and happiness; and they have at all times a right to alter, reform, or abolish the government in such manner as they may think proper'.
>
> We assert that our right of local community self-government to claim and protect our constitutionally-guaranteed right to safety is inherent, fundamental, and inalienable, and that – as the United States Declaration of Independence proclaimed – governments derive their just powers from the consent of the governed.
>
> The practice of aerial spraying of pesticides on Lincoln County's forests is causing serious chemical contamination of our county's air, people, wildlife, ecosystems, and watersheds, as well as terminal degradation of our soil. A large number of pesticides being used, among them 2,4-D, glyphosate (Roundup), and atrazine, have been proven harmful to both humans and the environment.

> It is important to note that only a small fraction of the approximately 12,000 pesticides registered for use in Oregon, including those being used in aerial spraying, have ever been tested for safety. None have been tested for the effects of repeated exposure over time, or for the effects of their routine application in combination with other pesticides and chemicals.
>
> Lincoln County's most populated communities are located at the base of our life-giving watersheds. However, clear-cut logging operations and the aerial spraying of pesticides before and after replanting are taking place upstream in these watersheds, exposing people, ecosystems, and natural communities to chemical contamination.
>
> Current laws and regulations permit and protect the practice of aerial spraying of pesticides, threaten our public health, violate our constitutionally-guaranteed right to safety, and interfere with our right of local community self-government. The risks from toxic trespass from aerial spraying of pesticides, due to the failure of our federal, state, and local governments to protect us, are therefore no longer acceptable.
>
> Our fundamental right to clean air, water, and soil not contaminated by aerial spraying of pesticides cannot be achieved when that right is routinely overridden by corporate minorities claiming legal powers to engage in that contamination. Nor can sustainability of any kind be achieved within a system of preemption which enables corporate decision-makers to wield state governmental power to override local self-government, and when the state itself restricts the county's lawmaking powers in ways that prevent the people of the county from protecting the health, safety and welfare of people and natural communities from such harms as aerial spraying of pesticides.
>
> The people of Lincoln County believe that aerial spraying of pesticides is not necessary, because the task of eliminating weeds and competing vegetation after tree planting can also be accomplished by manual action, without the use of any pesticides, however applied, without contaminating the environment, and without endangering the safety of all life in it, while at the same time creating many direly needed jobs for our community. Increasingly mechanized logging, including the practice of aerial spraying of pesticides, utilized by corporations extraneous to our communities, no longer makes any sizable contribution to our local economy, but nevertheless is now carried out by these corporations that wield tremendous power over decision-making in our county.[62]

The proposed ordinance recognised a number of rights for people and ecosystems in Lincoln County, including a right to be free from toxic trespass; right to clean air, water and soil; right to rural preservation; right of local community self-government and right to assert the right of self-government. Some of these rights extended to 'natural communities and ecosystems within the County'. The proposed ordinance also provided 'It shall be unlawful for any corporation to violate any right secured by this Ordinance', and 'It shall be unlawful for any corporation to engage in aerial spraying of pesticides within Lincoln County'.

In the enforcement section, the proposed ordinance gave citizen standing to 'any resident of Lincoln County' to enforce the rights and prohibitions. The proposed ordinance also provided a mechanism for ecosystems to defend their rights:

> Any action brought by either a resident of Lincoln County or by Lincoln County to enforce or defend the rights of ecosystems or natural communities secured by this Ordinance shall bring that action in the name of the ecosystem or natural community in a court possessing jurisdiction over activities occurring within Lincoln County. Damages shall be measured by the cost of restoring the ecosystem or natural community to its state before the injury, and shall be paid to Lincoln County to be used exclusively for the full and complete restoration of the ecosystem or natural community.[63]

The citizen group Lincoln County Community Rights formed a political committee to circulate petitions for voter signatures in order to qualify the proposed ordinance to appear on the ballot as an initiative by the people. The initiative appeared on the ballot for the May 16, 2017, election, and won by a margin of 61 votes, making Lincoln County the first county in the United States to ban aerial pesticide spraying through the vote of the people.[64]

But corporate interests were not so pleased. A large corporate law firm found an individual in Lincoln County willing to sue the county, and they brought a lawsuit in the county court to attempt to invalidate the new ordinance, claiming it is pre-empted by the Forest Practices Act, and violates the Oregon Constitution's provisions on legal remedies, taking of property and equal privileges and immunities.[65] They later added a limited liability company, Wakefield Farms, LLC, that owns property in the county, noting that 'Wakefield is a corporation within the meaning of the Measure [the ordinance] and, therefore, subject to the prohibitions and restrictions the Measure imposes on corporations but not on natural persons'.[66]

Lincoln County Community Rights filed an unopposed motion to intervene, which the court granted. The next month, the Siletz River Ecosystem filed an unopposed motion to intervene. 'Unopposed' means that the other parties were not fighting the motion. So here, a corporate law firm and its clients did not oppose the Siletz River Ecosystem becoming a party in the case. That itself suggests that ecosystem rights is no longer 'unthinkable'. But the judge wasn't so sure the legal system was ready. She ordered a hearing.

On Monday, September 11, 2017, the attorneys and parties gathered at the Lincoln County Courthouse in Newport, Oregon for the oral argument on whether the River could be a party in court. Ann Kneeland, the attorney representing the Siletz River, argued that the River had substantial legal interests at stake, and that it had rights and standing recognised by the law at issue in the case. She analogised to cases where the courts recognised an association's standing because its members would have standing:

> I would suggest that an ecosystem is sort of a quintessential collective interest. It operates collectively among its members, if we think about its members as the subcomponents of an ecosystem. And that, I think, is flushed out helpfully by the definition of ecosystem in the ordinance itself in section 2(b), which defines an ecosystem as 'wetlands, streams, rivers, aquifers, and other systems, as well as all naturally occurring habitats that sustain wildlife, people, flora and fauna, soil dwelling, and aquatic organisms.' And so to the extent that those can be seen as the ecosystem's members, they certainly share a common interest in their ability to exist, persist, and flourish. And in that sense, share a collective interest.

The attorney and judge also discussed what it means for an ecosystem to be a person. After the attorney introduced the issue the judge raised her concerns:

> Don't you agree, and have to concede, that the definition of 'person' that includes organizations and includes the corporations, there are statutes that allows a court to enter judgments against the corporation. And there is a mechanism for those organizations to be held liable in litigation. For judgments to be obtained in favor or against them. …

Attorney Kneeland replied, 'I think you're correct. There would be no way to enter any kind of monetary or damage judgment against the ecosystem'.

'Nor in favor', added the Judge.

Kneeland replied, 'I guess I would argue in favor – if the favorable ruling would retain the law then the ecosystem would be retaining its rights. And that is a judgment, essentially in favor, in a meaningful way for the Ecosystem. In this particular case, it would mean that the Ecosystem would function without contamination by aerial pesticide sprays and that is a significant gain to the Ecosystem'.

The judge wasn't convinced, and contrasted ecosystem parties with corporations, where there is a robust body of legal 'mechanisms to hold them accountable, and how that would work, and who actually is responsible for a judgment in the corporation'.

In addition, they discussed the issue of whether the Siletz River ecosystem had legal interests separate from the existing parities in the case, namely the county or community group that advocated for the initiative. Attorney Kneeland's response recognised that this is one of the central issues with reconceptualising ecosystems in the law:

> Thank you for the question. I think this is one of the pieces that is inherently challenging about the Rights of Nature concept, in that we as lawyers and as the society have to grapple with, and to stretch ourselves to understand, that the ecosystems and natural communities and nature at large has interests that are separate from ours. That we aren't forever stewards of nature, in the sense that nature serves our interests. But that it stands as an independent entity. … And so by being allowed to intervene, it gives an acknowledgment to those separate rights that the Ecosystem has to defend itself and recognize and protect its own rights. And that that, while substantially overlapping with the human interest, is not the same interest, necessarily. And as I said it is a stretch, I think, for people in general because we are so accustomed to thinking of nature as existing, sort of, in our service.

The judge ultimately denied the Siletz River's motion to intervene, although she encouraged developing the issue and recognised that 'it's one that is gaining some traction, and I think will become an issue and raised in other litigation, whether it's here or somewhere else'.

Just because the judge was not ready to accept the procedural rights of the Siletz River Ecosystem to appear as a party in court, it does not necessarily mean that the substantive rights the Ecosystem has through Lincoln County's Freedom from Aerially Sprayed Pesticides Ordinance will not be recognised. That is a separate question, but as described above, it is one that runs up against doctrines in our legal system – corporate constitutional rights, ceiling pre-emption, Dillon's Rule – that are designed to prevent local democratic expansion of rights and protections for health, safety and welfare. 'Ceiling preemption' is a legal term for the doctrine that prohibits local levels of government, such as a city or county, from expanding health, safety and welfare protections beyond the level provided by state law. It says that the state (or federal) government can set a 'ceiling' on health, safety and welfare protections that local governments cannot go above.[67]

The fate of the entire Freedom from Aerially Sprayed Pesticides Ordinance is yet to be determined. As of February 2019, the parties are still waiting for the judge's opinion on summary judgment, which will likely be appealed along with the order denying the River's intervention.

### Rights for the Salish Sea, Washington State and British Columbia

People in the Salish Sea, the inland sea composed of Puget Sound, the Strait of Juan de Fuca and Georgia Strait, in Washington state and British Columbia, are starting to look to ecosystem rights to protect this inland marine ecosystem.

Most of the human population of Washington and British Columbia live in watersheds that drain to the Salish Sea. The major metropolitan areas of Vancouver, BC, and Seattle,

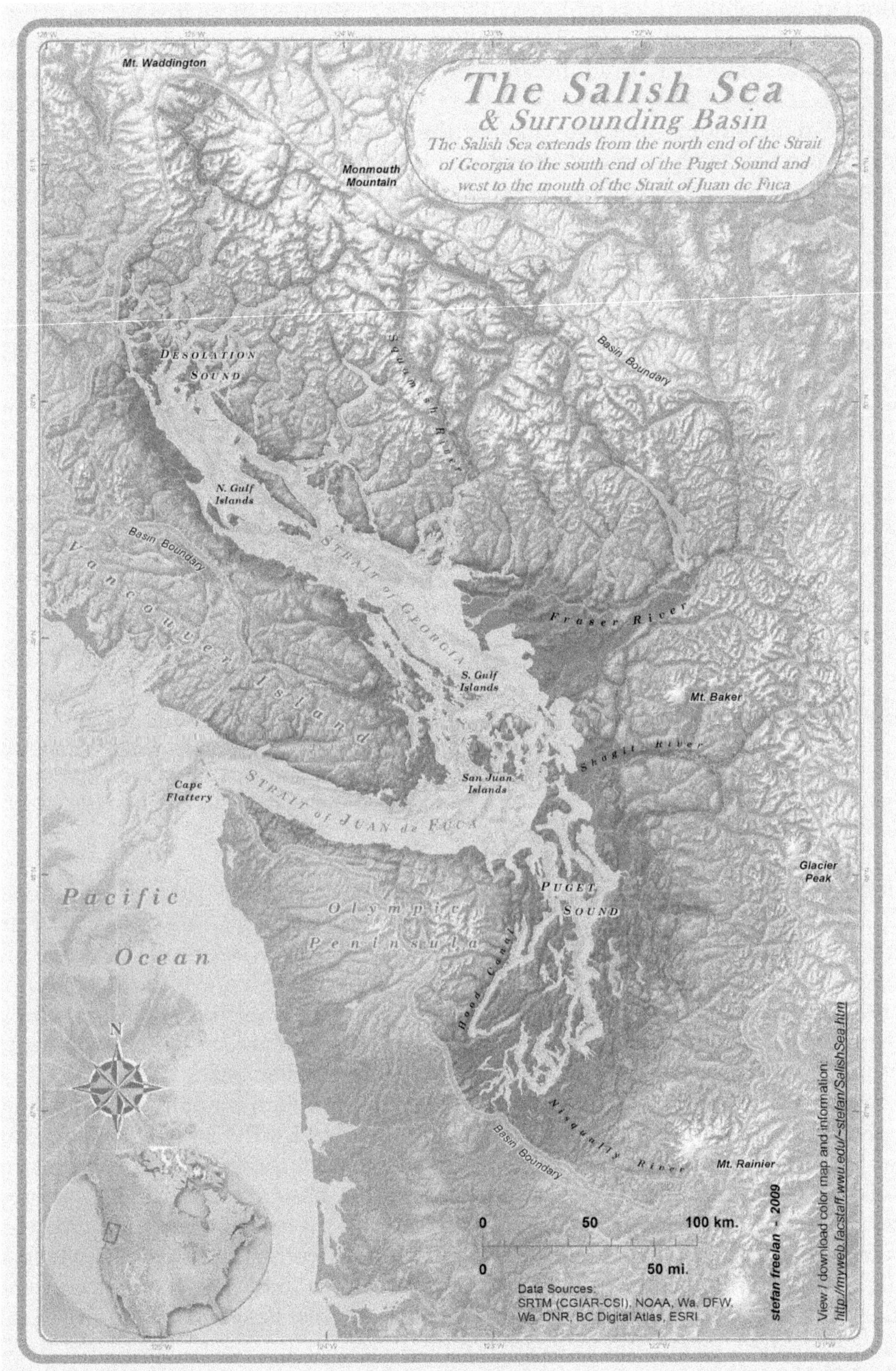


Map of the Salish Sea and surrounding basin. (Map by Stefan Freelan, Western Washington University. 2009. http://staff.wwu.edu/stefan/salish_sea.shtml)

Washington, contribute runoff pollution into the Sea. Many of these toxins stay in the fatty tissues of animals and get concentrated further up the food chain, from phytoplankton, to zooplankton, to forage fish, to salmon and eventually to the resident killer whales (orcas). This process is called bioaccumulation. As in Lake Erie, runoff containing other pollutants has caused algae blooms that consume all the dissolved oxygen and cause dead zones in areas of the Sea, such as the southern end of the natural fjord called Hood Canal. In addition, due to a number of factors, salmon populations in the Sea are in decline, with several populations listed as endangered under the Endangered Species Act.[68]

While these trends are not new, they hit a critical threshold in 2018 when scientists reported that the southern resident killer whales, the orcas that spend most of their time in the Salish Sea and eat mostly Chinook salmon (rather than also eating other marine mammals as the transient orcas do), are a few individuals away from losing their ability to successfully reproduce in the wild as a population.

Washington State's governor issued an Executive Order on March 14, 2018, which summarised the threats to the Southern Resident Killer Whales:

> WHEREAS, three primary factors threaten Southern Resident populations: (1) prey availability, (2) legacy and new toxic contaminants, and (3) disturbance from noise and vessel traffic. The health of Southern Residents and Chinook salmon are tightly linked. Recent scientific studies indicate that reduced Chinook salmon runs undermine the potential for the Southern Resident population to successfully reproduce and recover. Both Southern Residents and Chinook salmon populations are adversely impacted by warming oceans and ocean acidification due to climate change. Presence of contaminants and accumulation of pollutants in Washington's waters are also linked to the decline of Southern Residents. Key sources of contamination in storm water runoff remain to be addressed and the potential for a catastrophic oil spill continues to threaten Southern Residents and the entire ecosystem of Puget Sound. In addition, increased boat and ship traffic has caused greater underwater noise that interferes with Southern Resident critical feeding and communication.[69]

Unfortunately, despite the urgency of losing one of the iconic marine animals of the Pacific Northwest, the Governor's Executive Order called for measures that will merely slow the rate of destruction. For example, while the Executive Order calls for 'strategies for quieting state ferries in areas most important to Southern Residents', oil and gas companies (and now even the Canadian government) are building pipelines from the Alberta tar sands to the Salish Sea for vessel export, where hydrocarbons will get loaded on tankers, which means many more large vessels navigating through the Salish Sea.

Recognising that the same mindset that produced a problem is not likely to fix it, people throughout the region are organising to recognise rights for the Salish Sea. For example, in Gig Harbor, WA, the community group Legal Rights for the Salish Sea 'proposes a new law defining the Salish Sea as a "legal entity subject to basic rights" – no longer property and owned by the State, but instead acknowledged as an indivisible and living whole with inherent rights including the right to life, health and well-being, diversity, clean water and air, and full and prompt restoration and representation'.[70] The group spearheaded a coalition to draft a Declaration of the Rights of the Southern Resident Orcas, which calls on 'national, state, provincial, and local governments and their citizens [to] follow the Tribal and First Nations in recognizing the inherent rights of the Southern Resident Orcas and the ecosystems upon which they depend'.[71]

In other parts of the Salish Sea watershed, people are following the example of Lincoln County Community Rights and working to prohibit aerial pesticide spraying, which contributes to water pollution.[72] Ultimately, the root causes of the destruction of the Salish Sea – epitomised by the decline of the orca whale community – stem from activities as vast and complex as oil and gas, urban transportation, agriculture, forestry and international trade. By changing the legal system to recognise and enforce Rights of Nature and the local government's ability to protect those rights, we may actually be able to stop and reverse the slow destruction of places like the Salish Sea, or at a larger scale, Mother Earth.

## Conclusion

In October 2017, people involved in the movement for Rights of Nature met in New Orleans, Louisiana, and crafted principles for Rights of Nature lawmaking in order to ensure that this emerging structural change to our legal system does not get watered down into a subset of regulatory environmental law. Those New Orleans principles outline the plight of nature and the existing law's failure to recognise Nature's rights. They concluded with several standards for what Rights of Nature laws should include:

1. Recognise inherent Rights of Nature to exist, flourish, evolve and regenerate, and to restoration, recovery and preservation;
2. Guarantee the right of Nature to appear as the real party in interest in administrative proceedings and legal actions affecting these rights;
3. Provide that those committed to protecting these rights be authorised to represent them in these proceedings; and
4. Require that damages derived from these proceedings be used solely to protect Nature and restore Nature to its prior natural state.

These principles can guide communities seeking to protect themselves and their homes from ecological destruction.

No one else is going to do this for us. It is up to each of us to work with our local community to do what people in Highland Township, Toledo, Lincoln County and around the Salish Sea are doing: demand that the law accord ecosystems the right to exist, flourish, evolve and regenerate, and elevate those rights over now-dominant corporate interests.

## Notes

1. Michelle Newman contributed to this article in her personal capacity. The views expressed are her own and do not necessarily represent the views of the NRDC.
2. Moyer, B. et al. *Doing Democracy: The MAP Model for Organizing Social Movements.* 2001. New Society Publishers.
3. For example, in an 1829 case from the North Carolina Supreme Court, the court opined on whether the state could interfere with a master's treatment of his own slave: 'I repeat that I would gladly have avoided this ungrateful question. But being brought to it, the Court is compelled to declare, that while slavery exists amongst us in its present state, or until it shall seen fit to the Legislature to interpose express enactments to the contrary, it will be imperative

duty of the Judges to recognize the full dominion of the owner over the slave, except where the exercise of it is forbidden by statute. And this we do upon the ground, that this dominion is essential to the value of slaves as property, to the security of the master, and the public tranquillity, greatly dependent upon their subordination; and in fine, as most effectually securing the general protection and comfort of the slaves themselves'. *State v. Mann*, 2 Dev. 263 (N.C. 1829).

4. See generally Winkler, A. 2018. *We the Corporations: How American Businesses Won Their Civil Rights*. W. W. Norton & Co.
5. The history of legal rights for animals, as well as ecosystems, is detailed in Nash, R. F. 1989. *The Rights of Nature: A History of Environmental Ethics*. University of Wisconsin Press.
6. Leopold, A. 1966. *A Sand County Almanac*. Oxford University Press.
7. Stone, C. D. 1972. 'Should Trees Have Standing? – Toward Legal Rights for Natural Objects'. 45 *Southern California Law Review* 450.
8. Justice W. O. Douglas. United States Supreme Court Justice Douglas's dissenting opinion (citations and footnotes omitted). United States Supreme Court *Sierra Club v. Morton*, (1972) No. 70–34. Argued: November 17, 1971 Decided: April 19, 1972. https://supreme.justia.com/cases/federal/us/405/727/ (Accessed 10 July 2018); see also Stone, C. 1985. 'Should Trees Have Standing Revisited', 59 *Southern California Law Review* 1 (for history of how the law review article got to Justice Douglas).
9. Stone, C. 1985. 'Should Trees Have Standing Revisited', 59 *Southern California Law Review* 1, 4–5.
10. Justice Antonin Scalia delivered opinions that all limited access to the courts for environmental plaintiffs in: (1) *Lujan, Secretary of the Interior v. Defenders of Wildlife, et al.* Certiorari to the United States Court of Appeals for the Eighth Circuit. No. 90–1424. Argued December 3, 1991 Decided June 12, 1992. https://supreme.justia.com/cases/federal/us/504/555/case.pdf (Accessed 10 July 2018); (2) *Steel Co., a.k.a. Chicago Steel & Pickling Co. V. Citizens for a Better Environment* Certiorari to the United States Court of Appeals for the Seventh Circuit. No. 96–643. Argued October 6, 1997 Decided March 4, 1998. https://caselaw.findlaw.com/us-supreme-court/523/83.html (Accessed 10 July 2018); and (3) *Summers et al. v. Earth Island Institute et al.* Certiorari to the United States Court of Appeals for the Ninth Circuit No. 07–463. Argued October 8, 2008 Decided March 3, 2009. https://www.supremecourt.gov/opinions/08pdf/07-463.pdf (Accessed 10 July 2018).
11. In Hall, K. L., Findkleman, P. and Ely, J. W. Jr. 2004. *American Legal History: Cases and Materials*. Oxford University Press, Inc., New York, NY., 354 p.
12. *U.S. Supreme Court. Scott v. Sandford*, 60 U.S. 19 How. 393 393 (1856). https://supreme.justia.com/cases/federal/us/60/393/ (Accessed 10 July 2018).
13. Douglass, F. 'West India Emancipation' speech. Canandaigua, NY. August 3, 1857. http://www.blackpast.org/1857-frederick-douglass-if-there-no-struggle-there-no-progress (Accessed 28 August 2018).
14. Abraham, L. June 16, 1858. 'House Divided' speech. Springfield, IL. http://www.abrahamlincolnonline.org/lincoln/speeches/house.htm (Accessed 10 July 2018).
15. U.S. Supreme Court. *Lochner v. New York*, 198 U.S. 45 (1905) No. 292. Argued February 23, 24, 1905. Decided April 17, 1906. https://supreme.justia.com/cases/federal/us/198/45/ (Accessed 10 July 2018).
16. Fourteenth Amendment to the United States Constitution. https://www.archives.gov/founding-docs/constitution (Accessed 31 August 2018).
17. The Fireside Chats of Franklin Delano Roosevelt. March 9, 1937, page 35. http://www.public-library.uk/ebooks/34/80.pdf (Accessed 10 July 2018).
18. *West Coast Hotel Co. v. Parrish*, 300 U.S. 379 (1937). Argued December 16, 17, 1936. Decided March 29, 1937. https://supreme.justia.com/cases/federal/us/300/379/ (Accessed 10 July 2018).
19. Wright, G. 2006. *Slavery and American Economic Development*. Louisiana State University Press, Baton Rouge, LA, 152 pp.
20. Orr, D. 2000. '2020: A Proposal'. *Conservation Biology*, 14:338–341.
21. The District of Columbia Compensated Emancipation Act. https://www.senate.gov/artandhistory/history/common/generic/DCEmancipationAct.htm (Accessed 11 July 2018).

22. Fourteenth Amendment, Section 4: 'The validity of the public debt of the United States, authorized by law, including debts incurred for payment of pensions and bounties for services in suppressing insurrection or rebellion, shall not be questioned. But neither the United States nor any State shall assume or pay any debt or obligation incurred in aid of insurrection or rebellion against the United States, or *any claim for the loss or emancipation of any slave*; but al such debts, obligations and claims shall be held illegal and void' (emphasis added). Available at Legal Information Institute. 14th Amendment [Section 4]. https://www.law.cornell.edu/constitution/amendmentxiv (Accessed 11 July 2018).
23. McKibben, B. August 2, 2012. 'Global Warming's Terrifying New Math'. *Rolling Stone.*
24. Ibid.; see also Credit Suisse Estimates that Global Wealth in 2017 Is $280 Trillion USD. Global Wealth Report 2017. https://www.credit-suisse.com/corporate/en/research/research-institute/global-wealth-report.html. (Accessed 11 July 2018).
25. Fredrick Douglass' 1857 speech. http://www.blackpast.org/1857-frederick-douglass-if-there-no-struggle-there-no-progress (Accessed 11 July 2018).
26. Fitzsimmons, E. G. August 3, 2014. 'Tap Water Ban for Toledo Residents'. *New York Times.* https://www.nytimes.com/2014/08/04/us/toledo-faces-second-day-of-water-ban.html (Accessed 20 June 2018).
27. (a) Gurian-Sherman, D. 2014. 'What Toledo's Water Crisis Reveals about Industrial Farming'. https://civileats.com/2014/08/05/what-toledos-water-crisis-reveals-about-industrial-farming/ (Accessed 20 June 2018); and (b) University of Tennessee at Knoxville. (b) 'Virus Infection May Be Linked to Toledo Water Crisis, Study Shows'. *ScienceDaily.* www.sciencedaily.com/releases/2017/05/170531110745.htm (Accessed 20 June 2018).
28. *Ibid.*
29. Horan, J. November 21, 2017. 'Big Ag vs. Lake Erie: How Ohio's Biggest Industry Threatens Its Greatest Resource'. *Belt Magazine.* http://beltmag.com/big-ag-vs-lake-erie/ (Accessed 20 June 2018). 'According to the US Department of Agriculture's Natural Resource Conservation Service, of the 4.9 million acres in the Western Lake Erie Basin, 4.2 million acres are in the Maumee River Watershed. Of those 4.2 million acres, 71% are used for agricultural purposes. Eighty-five percent of total phosphorus delivered to the lake by the Maumee River comes from farm fertilizer and manure'. *Ibid.*
30. In the first section, the proposed Bill of Rights declares:

    (a) Rights of Lake Erie Ecosystem. Lake Erie, and the Lake Erie watershed, possess the right to exist, flourish, and naturally evolve. The Lake Erie Ecosystem shall include all natural water features, communities of organisms, soil as well as terrestrial and aquatic sub ecosystems that are part of Lake Erie and its watershed.
    (b) Right to a Clean and Healthy Environment. The people of the City of Toledo possess the right to a clean and healthy environment, which shall include the right to a clean and healthy Lake Erie and Lake Erie ecosystem.
    (c) Right of Local Community Self-Government. The people of the City of Toledo possess both a collective and individual right to self-government in their local community, a right to a system of government that embodies that right, and the right to a system of government that protects and secures their human, civil, and collective rights.

    The Lake Erie Bill of Rights: Citizens['] Initiative. http://www.lakeerieaction.org (Accessed 11 July 2018).
31. Section 3, on enforcement, states:

    (b) The City of Toledo, or any resident of the City, may enforce the rights and prohibitions of this law through an action brought in the Lucas County Court of Common Pleas, General Division. …
    (c) Governments and corporations engaged in activities that violate the rights of the Lake Erie Ecosystem, in or from any jurisdiction, shall be strictly liable for all harms and rights violations resulting from those activities.

(d) The Lake Erie Ecosystem may enforce its rights, and this law's prohibitions, through an action prosecuted either by the City of Toledo or a resident or residents of the City in the Lucas County Court of Common Pleas, General Division. Such court action shall be brought in the name of the Lake Erie Ecosystem as the real party in interest. Damages shall be measured by the cost of restoring the Lake Erie Ecosystem and its constituent parts at least to their status immediately before the commencement of the acts resulting in injury, and shall be paid to the City of Toledo to be used exclusively for the full and complete restoration of the Lake Erie Ecosystem and its constituent parts to that status.

32. Ibid.
33. (a) Ohio Const. art. II, § 1a; (b) See also https://www.supremecourt.ohio.gov/Clerk/ecms/#/caseinfo/2018/1824.
34. Davis-Cohen, S. February 27, 2019. 'Toledo Residents Vote to Recognize Personhood for Lake Erie', *The Progressive*. https://progressive.org/dispatches/toledo-residents-vote-to-recognize-personhood-for-lake-erie-davis-cohen-190227/ (Accessed 27 February 2019).
35. 'The Marcellus Shale Gas Boom in Pennsylvania: Employment and Wage Trends'. Bureau of Labor Statistics. Monthly Labor Review. 2014. https://www.bls.gov/opub/mlr/2014/article/the-marcellus-shale-gas-boom-in-pennsylvania.htm (Accessed 21 June 2018).
36. Warner, B. and Shapiroy, J. 2013. 'Fractured, Fragmented Federalism: A Study in Fracking Regulatory Policy'. *Journal of Federalism*, 43:474–496. http://citeseerx.ist.psu.edu/viewdoc/download?doi=10.1.1.860.7039&rep=rep1&type=pdf (Accessed 21 June 2018).
37. Legere, L. 'Sunday Times Review of DEP Drilling Records Reveals Water Damage, Murky Testing Methods'. *The Times-Tribune*, May 19, 2013. https://www.thetimes-tribune.com/news/sunday-times-review-of-dep-drilling-records-reveals-water-damage-murky-testing-methods-1.1491547 (Accessed 21 June 2018).
38. Drajem, M. 2013. 'EPA Official Links Fracking and Drinking Water Issues in Dimock, Pa'. *The Washington Post*, July 29, 2013. http://www.washingtonpost.com/politics/epa-official-links-fracking-and-drinking-water-issues-in-dimock-pa/2013/07/29/7d8b34b2-f8a1-11e2-afc1-c850c6ee5af8_story.html (Accessed 21 June 2018) ('The internal EPA report concludes that the causes of gas migration could be drilling, spills or fracking. "In some cases the aquifers recover [in less than a year] but, in other cases the damage is long term [longer than three years]", the report says'.).
39. (a) Won-Young Kim, W. -Y. 2013. 'Induced Seismicity Associated with Fluid Injection into a Deep Well in Youngstown, Ohio'. *Journal of Geophysical Research: Solid Earth*, 118:3506–3518; (b) Choi, C. Q. 2013. Fracking Practice to Blame for Ohio Earthquakes. *LiveScience*. https://www.livescience.com/39406-fracking-wasterwater-injection-caused-ohio-earthquakes.html (Accessed 21 June 2018); (c) Frohlich, C. 2012. 'Two-Year Survey Comparing Earthquake Activity and Injection Well Locations in the Barnett Shale, Texas'. *Proceedings of the National Academy of Sciences*, 109:13934–13938; and (d) Begley, S. June 11, 2013. 'Distant Seismic Activity Can Trigger Quakes at "Fracking" Sites'. *Reuters*. https://www.reuters.com/article/science-fracking-earthquakes/distant-seismic-activity-can-trigger-quakes-at-fracking-sites-idUSL1N0FG0VT20130711 (Accessed 21 June 2018).
40. FrackTrack.org. Operator: Seneca Resources Corp. http://fracktrack.org/searchexp.php?p=op_name&val=OGO-15547&pg=6&cnt=2&ord=muni (Accessed 21 June 2018).
41. The full text of the Ordinance is available at: https://archive.org/stream/554935-highland-township-community-rights-and/554935-highland-township-community-rights-and_djvu.txt (Accessed 21 June 2018). The Township amended the Ordinance in March 2015 but maintained the rights proclaimed therein.
42. Ibid. Section 3(a)-(e).
43. Ibid. Section 4(a).
44. Ibid. Section 5(b)-(c).
45. *See Seneca Res. Corp. v. Twp. of Highland*, 863 F.3d 245, 249 (3d Cir. 2017).
46. (a) Appeal Request/Underground Injection Control Permit PAS2D025BELK/Seneca Resources Corporation. https://yosemite.epa.gov/oa/EAB_Web_Docket.nsf/(Filings)/D43EF5D5A5078

01085257CA0004C5A53/$File/Swanson...UIC%2014-03.pdf (Accessed 12 July 2018); (b) *Seneca Resources Corporation v. Township of Highland, Elk County, Pennsylvania; Highland Township Board of Supervisors, Elk County, Pennsylvania.* (Pursuant to Rule 12(a) Fed. R. App. P.) http://www2.ca3.uscourts.gov/opinarch/163592p.pdf (Accessed 12 July 2018). and (c) *Seneca Resources Corp v. Township of Highland,* No. 16–3592 (3d Cir. 2017) https://law.justia.com/cases/federal/appellate-courts/ca3/16-3592/16-3592-2017-07-17.html (Accessed 12 July 2018).

47. The first ecosystem to attempt to intervene in a lawsuit to defend its rights was the Little Mahoning Watershed, which filed for intervention in *Pennsylvania General Energy Co., LLC v. Grant Township,* 1:14-cv-209 (W.D. Pa.), a 2014 lawsuit that sought to protect Grant Township from a proposed fracking waste injection well. See also Nicholson, C. 2015. Press Release: Ecosystem, Community Group, and Municipal Authority File for Intervention in Lawsuit to Defend Community from Injection Well. Community Environmental Legal Defense Fund. https://celdf.org/2015/08/press-release-ecosystem-community-group-and-municipal-authority-file-for-intervention-in-lawsuit-to-defend-community-from-injection-well/ (Accessed 20 June 2018).
48. 2015. Mem. in Support of Mot. to Intervene. *Seneca Resources Corp. v. Highland Twp.,* C.A. No. 15–60, (ECF No. 33:9).
49. *Seneca Resources Corporation v. Highland Township, Elk County, Pennsylvania* et al., No. 1:2015cv00060 – Document 44 (W.D. Pa. 2016). https://law.justia.com/cases/federal/district-courts/pennsylvania/pawdce/1:2015cv00060/222023/44/ (Accessed 12 July 2018).
50. Abraham, C. August 11, 2916. 'Highland Twp. Rescinds 'Bill of Rights' Ordinance'. *The Bradford Era.* http://www.bradfordera.com/news/highland-twp-rescinds-bill-of-rights-ordinance/article_bc658a3a-5f6d-11e6-bb86-672f0a970d3f.html (Accessed 21 June 2018).
51. Seneca Res., 863 F.3d at 253.
52. *See Ibid.* at 248 n.4.
53. (a) 53 Pa. C.S.A. §290 and (b) Weidenboerner, K. November 9, 2016. 'Highland Township Votes in Home Rule Charter'. *Courier Express.* (DuBois, Pa.). http://www.thecourierexpress.com/news/local/highland-township-votes-in-home-rule-charter/article_833142ae-b155-55fa-8477-5864bde37281.html (Accessed 21 June 2018).
54. Nicholson, C. August 15, 2016. 'Press Release: Highland Township Bans Injection Wells, Rebukes Corporate Rule; Voters Make History, Adopting Home Rule Charter Despite Industry Allies' Threats and Intimidation'. *Community Environmental Legal Defense Fund.* https://celdf.org/2016/11/pr-highland-township-bans-injection-wells-rebukes-corporate-rule/ (Accessed 20 June 2018).
55. *Seneca Res. Corp. v. Highland Twp.,* No. 16–289 (W.D. Pa. Nov. 30, 2016).
56. (a) *Seneca Res. Corp. v. Highland Twp.,* No. 16-cv-289, 2017 WL 4354710 (W.D. Pa. Sept. 29, 2017) (b) Schellhammer, M. December 21, 2017. 'Case Closed: Highland Twp.'s Home Rule Charter Stripped'. *The Bradford Era.* http://www.bradfordera.com/news/case-closed-highland-twp-s-home-rule-charter-stripped/article_dbd00682-e600-11e7-bdb8-a73cd5794a0c.html (Accessed 21 June 2018); and (c) Trego and Garrett, D. 2017. 'Pennsylvania and Federal Law Preempt Municipal Home Rule Charter'. MGKF Litigation Blog. https://www.lexology.com/library/detail.aspx?g=24fa547b-0d64-4929-b6aa-d6c23d3e03af (Accessed 21 June 2018).
57. *Seneca Res. Corp. v. Highland Twp.,* No. 16-cv-289, 2017 WL 4171703, at *5–6 (W.D. Pa. Sept. 20, 2017).
58. Talberth, J. and Fernandez, E. 2015. *Deforestation, Oregon Style.* World Resources Institute, Global Forest Watch Report, Appendix 2. September 2015.
59. Forest Practices Act. https://www.oregon.gov/ODF/Working/Pages/FPA.aspx (Accessed 13 July 2018).; Bernstein, L. et al. December 2013. *Oregon's Industrial Forests and Herbicide Use: A Case Study of Risk to People, Drinking Water and Salmon.* Introduction, page 70. Beyond Toxics. (Accessed 28 August 2018)
60. See 'Agency Oversight and Public Input' section in Bernstein et al., page 70.
61. Freedom from Aerially Sprayed Pesticides of Lincoln County. http://www.lincolncountycommunityrights.org/wp-content/uploads/2015/08/Filed-Ordiance_Freedom-from-Aerially-Sprayed-Pesticides-Ordinance-of-Lincoln-County.pdf (Accessed 23 June 2018).

62. Ibid.
63. Ibid.
64. Lincoln County Community Rights. http://www.lincolncountycommunityrights.org/about/ (Accessed 23 June 2018).
65. *Rex Capri, Wakefield Farms, LLC, Plaintiffs, v. Dana W. Jenkins, Lincoln County, Defendants, and Lincoln County Community Rights, Intervenor-Defendant. Case No. 17CV23360.* [Complaint filed June 6, 2017] http://www.co.lincoln.or.us/sites/default/files/fileattachments/county_counsel/page/4891/plaintiffs_memo_of_law_opposing_countys_and_intervenors_cross_-msj_9.18.17.pdf (Accessed 13 July 2018).
66. Amended Complaint, *Capri v. Jenkins,* filed June 15, 2017 (Case No. 17CV23360, Circuit Court, Lincoln County, OR).
67. Centers for Disease Control, Public Health Law Program. 'Preemption of Local Public Health Laws'. (Accessed 28 August 2018). Hodge, J. G. Jr. and Corbett, A. June 30, 2016. 'Legal Preemption and the Prevention of Chronic Conditions'. *Prev. Chronic. Dis.* (Accessed 28 August 2016).
68. These causes are often described as the 'four Hs:' habitat loss, harvest, hatcheries, and hydro. See Montgomery, David, R. 2003. *King of Fish: The Thousand-Year Run of Salmon.* Westview Press, Cambridge, Massachusetts. 290 pp.
69. Inslee, Jay. Executive Order 18-02: Southern Resident Killer Whale Recovery and Task Force https://www.governor.wa.gov/sites/default/files/exe_order/eo_18-02_1.pdf (Accessed 25 June 2018).
70. Legal Rights for the Salish Sea – Frequently Asked Questions. http://legalrightsforthesalishsea.org/faqs/ (Accessed 25 June 2018).
71. http://legalrightsforthesalishsea.org/petition/ (Accessed 4 February 2019).
72. Snohomish County Community Rights. https://www.facebook.com/sccr2017/ (Accessed June 2018).

## Appendix: Nature's Rights through Lawmaking in the United States

### 1. Rex Capri, Wakefield Farms, LLC, Plaintiffs vs. Dana W. Jenkins, Lincoln County, Defendants, and Lincoln County Community Rights, Intervenor-Defendant

**Transcript of Intervenor's motion for intervention of the Siletz River Ecosystem.**

On Monday, September 11, 2017, the attorneys and parties gathered at the Lincoln County Courthouse in Newport, Oregon. The judge started things off. 'Okay, I'm going to call the case of Rex Capri, Wakefield Farms versus Dana Jenkins, and then we have the intervenors, that are represented by Lincoln County Community Rights. This is a time set for hearing on the intervenor's motion to include or intervene yet another party, the Siletz River Ecosystem.

'It was presented to the Court as an unopposed Motion to Intervene. I think the Declarations set forth correctly that the County had indicated and the other parties to the litigation indicated they were not taking a position on the motion. I read the motion, the declaration. I think I contacted Ms. Kneeland [attorney for the Siletz River Ecosystem and Lincoln County Community Rights] after I read that. I wasn't convinced the court had the legal authority, but was in agreement with the authority that was presented – that the Court would have to include them as a party. I invited you to submit further briefing or argument on the issue, which is why we scheduled this hearing. Are the parties prepared to proceed?'

Attorney Kneeland replied, 'Yes, Your Honor.'

'Okay, Ms. Kneeland, go ahead, this is your Motion.'

'Thank you, your honor, Ann Kneeland for the Siletz River Ecosystem. I'm here with Carol Von Strum, who had signed the declaration, herself a part of the Siletz River Ecosystem, and in support of the Motion to Intervene. And I did want to thank the Court for the opportunity to appear today and supplement our argument. I would just correct the Court that I am representing the Siletz River Ecosystem as an independent party from the Lincoln County Community Rights, who has already intervened. So we are appearing independently in our request to appear as a party.

'As this Motion was filed and now is before the Court I feel safe to say that it probably was unexpected. And I think with that both in the minds of the Court and the society and community at large brings up questions about can an ecosystem be a party to litigation, can an ecosystem have cognizable interests to appear and be heard in a lawsuit, and can an ecosystem have rights, let alone enforceable rights. And I think these are certainly logical questions that come to mind when we think about nature, or part of nature, appearing in court.

'But I would like to suggest that these are not the first time that we have considered these questions, and courts, and society in general, has considered these questions when thinking about things that we think of historically as property, whether they have rights in court. And I'm thinking here of African-Americans, Native Americans, women, children, certainly other people that at times we thought of as property. And corporations too, at times, were not recognized as persons legally. And that at that time, we thought that those persons were sort of equally unqualified, perhaps, to be rights-bearing entities in court, just as the Ecosystem is today. And so I just encourage the Court to be mindful of the request for nature, in this case the Siletz River Ecosystem, to appear and defend its rights that have been codified in the law also before this Court.

'And suggest also that just like the rights that were extended to the parties that I've knowledgeable [sic] in history, that we have a moral imperative at this time also to recognize ecosystems and natural communities as having rights to exist, persist, and flourish, and that they should be entitled to appear in Court to defend those rights.

'I would like to just give a little bit of context for this request, because it is novel and surely the first time this Court has given consideration to it. That on an international scale, the Rights of Nature is a concept that is gaining traction not only in government and in courts, but in these countries worldwide.

'In Ecuador in 2008, the people voted to adopt a constitutional amendment that recognized the Rights of Nature. In 2011, the first enforcement action was filed under that constitutional amendment by the Vilcabamba River, that filed to stop a government highway project that was interfering with its rights to exist and maintain itself. And in that proceeding the river won, it prevailed under that constitutional amendment. So that is certainly a significant hallmark in the journey for nature to be seen as a legal entity with rights.

'In Bolivia in 2010, that country also adopted a law of the rights of mother earth. Also later recognizing statutory Rights of Nature in their country's laws.

'New Zealand also, in 2012, has passed two laws recognizing both a former park that had been indigenous lands and the Whanganui River to be legal persons. In 2015, New Zealand subsequently passed laws recognizing animals as sentient beings.

'In India, also in 2012, the high court of northern India recognized the Ganga and Yamuna Rivers as legal persons, living persons, and the jurisprudence in that country will continue in that vein.

'And similarly in this country, although we haven't taken as significant steps, parts of nature have appeared in federal litigation as parties: a river, a marsh, the Marbled murrelet, a sea turtle, a brook has. And also, other laws around this country, very similar to the law that was passed in this community, recognizing the Rights of Nature and the rights of those natural communities and ecosystems to appear in court to defend and enforce those rights.

'Also in this country in 2016, the general council of the HoChunk Nation adopted an amendment to their tribal constitution to recognize Rights of Nature.

'So these are important hallmarks in what is the evolution of the Rights of Nature. And again indicative of the climate around the world to step up to this moral call to give and recognize nature's rights in our legal system and in our cultures.

'I do understand from the Court's email and from your statement here that there are reservations about the legal authority to support this motion, specifically under ORCP 33(c) [Oregon Rule of Civil Procedure concerning a party intervening into a case]. I believe that consistent with an evolution of legal rights, that the law is sufficient to justify it. Although I recognize that it certainly is a novel application for intervention.

'I would first say that with respect to the general procedures for applying for a motion to intervene, the Siletz River Ecosystem has complied at least in two important respects. And that is we have filed a declaration by a human person who is a part of that ecosystem just as she would be a member of an organization or association, similarly situated to Barbara Souse and Maria Davis who had filed declarations in support of Lincoln County Community Rights' Motion to Intervene that was granted by this Court. And that declaration, very much in the same way, lays out the requirements for the Ecosystem to intervene.

'Again in accordance with the general procedures for ORCP 33(c), the Motion and the Declarations have substantiated that the Ecosystem has a substantial legal interest at stake in this proceeding. The law that has been adopted recognizes rights, in section 3(c), for natural communities and ecosystems to be free from the aerial spray of pesticides in Lincoln County, and also in sections (b) and (c) have given standing to the ecosystems and communities themselves, and the human beings within those ecosystems, to file legal action to defend and enforce those rights. So between the rights that have been recognized and the right to defend them the Ecosystem has significant rights at stake in this case that will either be advanced or dismissed by operation of the judgment of this Court. And

I think that complies with the requirements for having interest at stake in the case law supporting motions to intervene.

'It is necessary to acknowledge that there are no cases in Lincoln County, in the State of Oregon, or really any other state with similar intervention laws, that I can point to to say that this is right, that this has been done before. But with that said, I would encourage the Court to look to *Renler v. Lincoln County*, a 1986 case from the [Oregon] Supreme Court, that looks not to a natural community or ecosystem, but to an association, in terms of what their collective interest was to appear and intervene into a case under ORCP 33(c). And the Court looked at that membership organization and assessed its collective interests, and in that vein the Court said the following: "If an association has standing to pursue a collective interest of its members, it also makes little difference whether the association chooses to organize in the form of a nonprofit membership corporation as the intervenor has done. In either event, the underlying questions are first, whether the asserted collective interest is one that the members themselves could pursue, and second, whether the organization is representing the position of its members on the disputed issues".

'And I think looking at this motion in that frame, I would suggest that an ecosystem is sort of a quintessential collective interest. It operates collectively among its members, if we think about its members as the subcomponents of an ecosystem. And that, I think, is fleshed out helpfully by the definition of ecosystem in the ordinance itself in section 2(b), which defines an ecosystem as 'wetlands, streams, rivers, aquifers, and other systems, as well as all naturally occurring habitats that sustain wildlife, people, flora and fauna, soil dwelling, and aquatic organisms'. And so to the extent that those can be seen as the ecosystem's members, they certainly share a common interest in their ability to exist, persist, and flourish. And in that sense, share a collective interest. They not only share that collective interest, but because of the law itself, and section 5(c), they are authorized and do have standing to pursue those interests. And to pursue them collectively as an ecosystem, or arguably pursue them individually as a river or as a wetland. And so in the same way that an association reflects the values of its collective membership, I would argue that the ecosystem, likewise, reflects the interests of its collective membership, so to speak, and has the right to pursue those claims individually.

'And so as to both of those questions that the court identified, I would argue that the ecosystem has the very same collective rights and interests as are explained in the court pertaining to the association and should thereby, based on that case, be authorized to intervene.

'I recognize that the rule itself says 'person' in terms of who the rule is authorizing to appear. And if we think of that term as a legal person, then clearly I cannot say to the Court that the Ecosystem is a legal person. But I can say that there are no cases in which the courts have looked at, or the Oregon Courts have looked at, the meaning of "person" within ORCP 33 that I could find. There was no discussion of an applicant applying to be an intervenor and the court deciding that it was not a "person" as to the meaning of that word.

'So in that light I would also suggest, as far as I've been able to determine, the Court has not taken a position on the meaning of the word "person". Does it mean "legal person" or otherwise – I would suggest that there is some latitude here. In the land use context, and quasi-judicial hearings in that regard, this Oregon statute provides a broader definition of "person". And now I'm speaking of ORS 197.015 subsection 14 that defines "person" to be "any individual, partnership, corporation, association, governmental subdivision, or agency, or public or private organization of any kind". And I would suggest that this Court would have the latitude to interpret the word "person" in accordance with another Oregon statute pertaining specifically to organizational and association standing to appear in a judicial proceeding. And so that is what that statute pertains to, to establish organizational standing for the purpose of appearing in those hearings. And that would be a definition that would support the ecosystem as a public organization of any kind, which it certainly is.'

The judge raised her concerns, 'Don't you agree, and have to concede, that the definition of "person" that includes organizations and includes the corporations, there are statutes that allows a court to enter judgments against the corporation. And there is a mechanism for those organizations to be held liable in litigation. For judgments to be obtained in favor or against them. And there is not a similar, which raised the conundrum here, which is that without a person as defined, if a judgment were to be entered either for or against such an entity as you are describing in the ecosystem, what is the – how do you imagine that – or what would you, or under the law, is there really a mechanism for that if your party were to prevail, or not, in litigation, doesn't that create the – the Court would be without legal authority to enter a judgment?'

Attorney Kneeland replies, 'I think you're correct. There would be no way to enter any kind of monetary or damage judgment against the ecosystem.'

'Nor in favor,' added the judge.

'I guess I would argue in favor – if the favorable ruling would retain the law then the ecosystem would be retaining its rights. And that is a judgment, essentially in favor, in a meaningful way for the Ecosystem. In this particular case, it would mean that the Ecosystem would function without contamination by aerial pesticide sprays and that is a significant gain to the Ecosystem. In terms of damages, it's a curious question, whether a mechanism exists[1] – in other contexts, like say people who are judgment proof who are ruled against in a monetary or damages capacity, they can't pay. So it can't be just that it can't pay.'

Judge, 'It's not that it can't pay. It's that there is a mechanism to enforce a judgment against, and that there are laws that support that and statutes have been enacted that specifically address how, say, a corporation – and I know people struggle with a corporation being a person – but part of that is that you can enter a judgment against them and they can be held liable. They are a person under the law for that purpose. And there are mechanisms to hold them accountable, and how that would work, and who actually is responsible for a judgment in the corporation. And sort of, that, for the Court, that's what really guides its rational in deciding what would be a permissible party in a case: is the mechanism for a lawful judgment to be entered upon resolution of that case. And that's where there is some disconnect here if the Court were to allow the Ecosystem to be a party in this particular litigation.'

Attorney Kneeland, 'I understand your concern and I guess all I can say in that respect is that other countries have passed these laws, have empowered ecosystems to be legal persons, and as I explained, in Ecuador have actually appeared and prevailed. They did win, so it wasn't a question of how to recover in the case of damages or an adverse ruling. However, I think what we have to think about in terms of the moral imperative to recognize the Rights of Nature, to understand nature as a living, breathing organism on which we all depend, and that thereby needs to be a rights-bearing entity. As previous property interests of people had that same compulsion, that necessarily prevails over considerations of monetary claims.

'I think maybe the other complicated piece about ecosystems is that they don't exact harm in the same way as a corporation does or a human person does. And so in the sense of what would that judgment look like, it's a harder comparison to make. And I guess I would suggest that in these times where the courts and society are shifting their cultural and judicial understanding of rights-bearing entities, that these are some of the, essentially, kinks that will be worked out. But that it's not sufficient to stop the movement to recognize the rights of these natural communities that are essential to the functioning of the planet, and thereby necessitate the rights associated with them, for their survival and our survival. But I can't provide a clear answer to that question.'

Judge, 'My last question, Ms. Kneeland, has to do with the Court did exercise is discretion under ORCP 33 in allowing the joinder of Lincoln County Community Rights organization in this litigation. And when I'm looking at the motion and the declaration in support of that, really the premise is that this organization aligns itself with the rights of the Ecosystem. So how is it that joining – one of the considerations of the Court in permissive joinder: is there a separate interest? And if I understand correctly, if I'm reading parts of this, "that the advancement of the rights are essential to the purpose and the mission of Lincoln County Community Rights", is that how are the rights of the Ecosystem different than that which Lincoln County Community Rights is representing in this litigation, because I've already granted that permissive joinder and I understood the organization to be asserting the rights of the Ecosystem?'

Attorney Kneeland, 'Thank you for the question. I think this is one of the pieces that is inherently challenging about the Rights of Nature concept, in that we as lawyers and as the society have to grapple with, and to stretch ourselves to understand, that the ecosystems and natural communities and nature at large has interests that are separate from ours. That we aren't forever stewards of nature, in the sense that nature serves our interests. But that it stands as an independent entity.

'And that because I represent both groups it is clear that their interests are aligned. But I wouldn't say that they are entirely the same. And that the Ecosystem necessarily has its own interests in existing, persisting, and flourishing in a way that cannot be, or is distinct from, what a human person's interest is in terms of advocating for those rights. And so I acknowledge that it is certainly Lincoln County Community Rights' mission and objective to see this law upheld, and also they crafted and campaigned for a law that recognized the Rights of Nature, and they value those rights. I just would urge the Court, and everybody, as we grapple with Rights of Nature to understand that we need to see nature as having independent rights.

'And so by being allowed to intervene, it gives an acknowledgment to those separate rights that the Ecosystem has to defend itself and recognize and protect its own rights. And that that, while substantially overlapping with the human interest, is not the same interest, necessarily. And as I said it is a stretch, I think, for people in general because we are so accustomed to thinking of nature as existing, sort of, in our service. And therefore our protection of it, in sort of a paternalistic sense, is what we see as meritorious. But I would argue that even where you have another paternalistic interest, even though it is in the aid of the subjugated interest, that interest nevertheless has its own freestanding interest, and that is the independent interests of the Rights of Nature, and for an ecosystem to appear and be acknowledged as an independent defendant, intervenor-defendant in this case.

'And I would encourage the Court to recognize that separate interest and grant the Motion on that basis.'

Judge, 'Okay, thank you.'

Attorney Kneeland, 'I have nothing further Your Honor, thank you very much for the time.'

Judge, 'Thank you.'

Papers shuffle as the attorneys trade places; then the judge continues, 'Mr. Belmont?'

Attorney Belmont, 'Thank you, Your Honor. Wayne Belmont for Lincoln County. And first and foremost I want to say that the County did not, and does not, oppose this Motion to Intervene.

'But we did reserve the right to address issues of justiciability and standing in our responsive pleadings, and would do so depending on what the Court does rule in this case. Our concern is a little bit with sort of the posture of where we're at with this. What we do not want to see is an interlocutory order that might be appealable to the Court of Appeals if the Court, for instance, denied standing and that were to be appealed and reduced to a limited judgment for appeal. We don't want to stop the rest of the proceedings.

'So what we would ask – I mean the Court's going to make a ruling whatever direction you want to go, but we would like to see the actual ruling as part of the whole decision-making process here for what you're doing in the entire case. And I think that's in part because of what you said there: that the Answer that was filed in this particular case really offers no substantive legal issues different than that which was already filed by Lincoln County Community Justice [sic]. So there aren't going to be any issues separate from that, that would be decided or not decided depending on your ruling in this particular case. And we just wanted to make sure that we got direction from the Court on all the issues that have been raised – this is only one part of a greater number of issues that are being raised as part of this case, and we'd like to see those all decided at once, so that if any side decides to appeal it, we can take everything up on one appeal, and not have this thing, for instance, stayed for a period of time while we're deciding some parts of it at the front end of this.

'So that's really, I think, all I'd like to say. We did reserve the right to, again, address the issues of justiciability and standing and I think the Court has some concerns with that, and we would do that in our responsive pleadings if this issue is not decided.'

Judge, 'Ms. Kneeland, with respect to the procedural – I guess – request, on behalf of Mr. Belmont, and it would, again, if the Court's ruling were to be appealed, necessarily, it could stay the litigation, and it sounds like all sides want the merits of the case resolved. With respect to that?'

Attorney Kneeland, 'So, I guess with respect to that Your Honor, I would ask that the Court make – I would ask – a prompt ruling on the motion. If the Ecosystem were allowed to intervene, our responses in the case, for all parties, are due on Monday. And while the Ecosystem hasn't filed to make a counter-claim or other different affirmative defenses, if the Ecosystem would still have the right to make different arguments in its responsive pleading, and would have the opportunity to submit those by the Monday deadline if a prompt ruling were forthcoming. I am comfortable saying that it would not be the intention of the Ecosystem to file for an interlocutory appeal. I think it is in all parties' interests that this case move forward. And would not want to stop the Court reaching the underlying merits for the purpose of reaching that appeal in the meantime. So in that sense I'd ask the Court to make a prompter ruling on the motion, if possible. Thank you.'

Judge, 'Ms. Kneeland, I was intrigued by the motion, which is why I wanted to invite you to submit further authority in the light of the opportunity to make the record that you have before the Court regarding this issue. I agree it's a novel concept. But as you indicated in your briefings with the Court, it's one that is gaining some traction, and I think will become an issue and raised in other litigation, whether it's here or somewhere else.

'It is a narrow issue for this Court in this particular case. And honestly, as a concept, I don't struggle with it as much as I do in this particular case when I look at the interests as they are with the Lincoln County Community Rights aligning themselves with the interests of the Ecosystem and allowing their joinder. I don't view those interests as being separate and apart from the Ecosystem significant enough that it would – and I know there is a separate philosophical discussion to take place with respect to that – but again, within the context of this particular litigation, what I don't find are those rights of the Ecosystem to be so different that a separate interest would further this litigation, and representations of that party in this litigation.

'And I accept that, again, the purpose and the mission of Lincoln County Community Rights is to advance the interests of the Ecosystem. And that's what I understood it to represent when I allowed the joinder of Lincoln County Community Rights in this case. And that again, ultimately the Court will have to rule on the Ordinance, and again on the merits of the case it's a very narrow issue for the court: it's not one under ORCP 33(c).

'Ms. Kneeland, as you point out, the Court has little guidance nationally, locally, statewide, on allowing, recognizing, an ecosystem as a 'person' under the statute. It is significant to the Court, the other persons that are defined under statute are organizations and entities and corporations for which the Court has the ability and the authority to enter some type of judgment when ultimately the case is resolved, either in favor or against. And which creates an issue for the Court in allowing the Ecosystem to be joined as a party in this particular lawsuit.

'And I do not, in the evaluation of the merits of this case, and that which Lincoln County Community Rights group has advanced, that there is an interest separate and apart significant enough to allow intervenor status. I found that with respect to Lincoln County Community Rights that their position is that they would not be adequately represented by the parties in this case, and I agreed. That's why I allowed intervention, so those rights are adequately represented. I can't make the same finding with respect to the Ecosystem itself.

'It is a novel issue and I wanted you to have the opportunity to develop this record, so that it could be reviewed and somebody else could be of a different opinion. I think these questions should be resolved at some other level. Somebody may find additional authority, and it could be a compelling argument at some other level than the trial court. And I welcome that opportunity for other persons to review my decision here. Which is why I wanted you to develop the record a bit more, and you've done that and I appreciate the time you've spent on it.

'But I am going to deny the motion for intervenor status on behalf of the Siletz River Ecosystem.'

The Court wrapped up the hearing with scheduling matters related to the Motions for Summary Judgment that would likely resolve the underlying issues of the validity of the Ordinance.

At the end, the judge emphasized that she invites the Siletz River Ecosystem to appeal, 'I think this is an issue that will be gaining more interest in the future, for the reasons that [Attorney Kneeland] just stated.'

## 2. Lake Erie Bill of Rights

### Establishing A Bill of Rights for Lake Erie, Which Prohibits Activities and Projects That Would Violate the Bill of Rights.

*We the people of the City of Toledo* declare that Lake Erie and the Lake Erie watershed comprise an ecosystem upon which millions of people and countless species depend for health, drinking water and survival. We further declare that this ecosystem, which has suffered for more than a century under continuous assault and ruin due to industrialization, is in imminent danger of irreversible devastation due to continued abuse by people and corporations enabled by reckless government policies, permitting and licensing of activities that unremittingly create cumulative harm, and lack of protective intervention. Continued abuse consisting of direct dumping of industrial wastes, runoff of noxious substances from large scale agricultural practices, including factory hog and chicken farms, combined with the effects of global climate change, constitute an immediate emergency.

*We the people of the City of Toledo* find that this emergency requires shifting public governance from policies that urge voluntary action, or that merely regulate the amount of harm allowed by law over a given period of time, to adopting laws which prohibit activities that violate fundamental rights which, to date, have gone unprotected by government and suffered the indifference of state-chartered for-profit corporations.

*We the people of the City of Toledo* find that laws ostensibly enacted to protect us, and to foster our health, prosperity, and fundamental rights do neither; and that the very air, land, and water – on which our lives and happiness depend – are threatened. Thus it has become necessary that we reclaim, reaffirm, and assert our inherent and inalienable rights, and to extend legal rights to our natural environment in order to ensure that the natural world, along with our values, our interests, and our rights, are no longer subordinated to the accumulation of surplus wealth and unaccountable political power.

*We the people of the City of Toledo* affirm Article 1, Section 1, of the Ohio State Constitution, which states: 'All men are, by nature, free and independent, and have certain inalienable rights, among which are those of enjoying and defending life and liberty, acquiring, possessing, and protecting property, and seeking and obtaining happiness and safety'.

*We the people of the City of Toledo* affirm Article 1, Section 2, of the Ohio State Constitution, which states: 'All political power is inherent in the people. Government is instituted for their equal protection and benefit, and they have the right to alter, reform, or abolish the same, whenever they may deem it necessary; and no special privileges or immunities shall ever be granted, that may not be altered, revoked, or repealed by the general assembly'.

*And since all power of governance is inherent in the people,* we, the people of the City of Toledo, declare and enact this Lake Erie Bill of Rights, which establishes irrevocable rights for the Lake Erie Ecosystem to exist, flourish and naturally evolve, a right to a healthy environment for the residents of Toledo, and which elevates the rights of the community and its natural environment over powers claimed by certain corporations.

**Section 1: Statements of Law – A Community Bill of Rights**

a. Rights of Lake Erie Ecosystem. Lake Erie, and the Lake Erie watershed, possess the right to exist, flourish, and naturally evolve. The Lake Erie Ecosystem shall include all natural water features, communities of organisms, soil as well as terrestrial and aquatic sub ecosystems that are part of Lake Erie and its watershed.
b. Right to a Clean and Healthy Environment. The people of the City of Toledo possess the right to a clean and healthy environment, which shall include the right to a clean and healthy Lake Erie and Lake Erie ecosystem.
c. Right of Local Community Self-Government. The people of the City of Toledo possess both a collective and individual right to self-government in their local community, a right to a system of government that embodies that right, and the right to a system of government that protects and secures their human, civil, and collective rights.
d. Rights as Self-Executing. All rights secured by this law are inherent, fundamental, and unalienable, and shall be self-executing and enforceable against both private

and public actors. Further implementing legislation shall not be required for the City of Toledo, the residents of the City, or the ecosystems and natural communities protected by this law, to enforce all of the provisions of this law.

**Section 2: Statements of Law – Prohibitions Necessary to Secure the Bill of Rights**

a. It shall be unlawful for any corporation or government to violate the rights recognized and secured by this law. 'Corporation' shall include any business entity.
b. No permit, license, privilege, charter, or other authorization issued to a corporation, by any state or federal entity, that would violate the prohibitions of this law or any rights secured by this law, shall be deemed valid within the City of Toledo.

**Section 3: Enforcement**

a. Any corporation or government that violates any provision of this law shall be guilty of an offense and, upon conviction thereof, shall be sentenced to pay the maximum fine allowable under State law for that violation. Each day or portion thereof, and violation of each section of this law, shall count as a separate violation.
b. The City of Toledo, or any resident of the City, may enforce the rights and prohibitions of this law through an action brought in the Lucas County Court of Common Pleas, General Division. In such an action, the City of Toledo or the resident shall be entitled to recover all costs of litigation, including, without limitation, witness and attorney fees.
c. Governments and corporations engaged in activities that violate the rights of the Lake Erie Ecosystem, in or from any jurisdiction, shall be strictly liable for all harms and rights violations resulting from those activities.
d. The Lake Erie Ecosystem may enforce its rights, and this law's prohibitions, through an action prosecuted either by the City of Toledo or a resident or residents of the City in the Lucas County Court of Common Pleas, General Division. Such court action shall be brought in the name of the Lake Erie Ecosystem as the real party in interest. Damages shall be measured by the cost of restoring the Lake Erie Ecosystem and its constituent parts at least to their status immediately before the commencement of the acts resulting in injury, and shall be paid to the City of Toledo to be used exclusively for the full and complete restoration of the Lake Erie Ecosystem and its constituent parts to that status.

**Section 4: Enforcement – Corporate Powers**

a. Corporations that violate this law, or that seek to violate this law, shall not be deemed to be 'persons' to the extent that such treatment would interfere with the rights or prohibitions enumerated by this law, nor shall they possess any other legal rights, powers, privileges, immunities, or duties that would interfere with the rights or prohibitions enumerated by this law, including the power to assert state or federal preemptive laws in an attempt to overturn this law, or the power to assert that the people of the City of Toledo lack the authority to adopt this law.
b. All laws adopted by the legislature of the State of Ohio, and rules adopted by any State agency, shall be the law of the City of Toledo only to the extent that they do not violate the rights or prohibitions of this law.

### Section 5: Effective Date and Existing Permit Holders

This law shall be effective immediately on the date of its enactment, at which point the law shall apply to any and all actions that would violate this law regardless of the date of any applicable local, state, or federal permit.

### Section 6: Severability

The provisions of this law are severable. If any court decides that any section, clause, sentence, part, or provision of this law is illegal, invalid, or unconstitutional, such decision shall not affect, impair, or invalidate any of the remaining sections, clauses, sentences, parts, or provisions of the law. This law would have been enacted without the invalid sections.

### Section 7: Repealer

All inconsistent provisions of prior laws adopted by the City of Toledo are hereby repealed, but only to the extent necessary to remedy the inconsistency.

## Note

1. The Ordinance does provide that, in the context of enforcing or defending the rights of ecosystems or natural communities, 'Damages shall be measured by the cost of restoring the ecosystem or natural community to its state before the injury, and shall be paid to Lincoln County to be used exclusively for the full and complete restoration of the ecosystem or natural community'. Section 5(c).

# 18

# The Experiment with Rights of Nature in India

**Kelly D. Alley and Tarini Mehta**

**CONTENTS**

Summary ... 365
Introduction ... 366
Precursors of Nature's Rights in India: Public Interest Litigation ... 367
Standing for Nature: The National Green Tribunal ... 369
Provisions Close to Rights of Nature ... 370
Standing and Other Procedural Matters ... 371
Ganga and Yamuna Rivers as Persons ... 372
The Problem of Guardianship ... 374
Other Considerations: Religious Doctrines of Personhood ... 375
Overlapping Institutions and Politics in 'Guardianship' ... 377
Conclusion ... 379
Notes ... 380

## Summary

On March 20, 2017, the High Court of Uttarakhand, India, ruled that, 'The Rivers Ganga and Yamuna, all their tributaries … are declared as juristic/legal persons/living entities having the status of a legal person with all corresponding rights, duties and liabilities of a living person in order to preserve and conserve river Ganga and Yamuna' (*Mohd Salim v. State of Uttarakhand and others*). In the ruling, the judges appointed three guardians as 'persons in loco parentis' [meaning 'in place of a parent'] as the human face to protect, conserve and preserve the Rivers Ganga and Yamuna and their tributaries. The public celebration of this judgment drew legitimacy from the transnational discourses and legal initiatives giving 'rights to Nature', but it omitted acknowledgment of other human designs for water, most notably the intention to centralise control of all basin extractions and uses into a single authority.

The river Ganga and its tributaries are known worldwide for producing food and water for large populations. Surface water is burdened with pollution. The judgment presents an interesting example of a negotiation of categories of personhood, goddess and natural resource. However, the Supreme Court stayed this High Court judgment so that government departments could avoid the accountability that a rights-based ruling would require. This paper argues that the continuation of multiple institutions of decision-making and judgment, including the National Green Tribunal, is a better path to protecting these sacred rivers than the path proposed in the landmark ruling.

DOI: 10.1201/9780429505959-18

Ritual bathing in the Ganga (Ganges) River in 2001. (Photograph by Kelly D. Alley.)

## Introduction

In India and worldwide, the river Ganga (Ganges) has been a Mother, Goddess, purifier and sustainer of all life for Hindus for millennia.[1] The cleaning of Mother Ganga, on the other hand, is a more recent invention.[2] This invention now comprises a number of complicated, and largely unsuccessful, plans to arrest the mounting pollution and deteriorating water quality of this sacred river.[3] This deterioration directly feeds the growing use, overuse and depletion of groundwater in this basin and others across the country.[4]

Today around one billion people reside in the Ganga Basin. The Ganga passes by more than 30 major cities and towns on the traverse from the Himalayas to the Bay of Bengal. The Upper Ganga Plain in the state of Uttar Pradesh is the most industrialised part of the river basin, home to sugar factories; leather tanneries; textile industries of cotton, wool, jute and silk; food-processing industries related to rice, dal and edible oils; paper and pulp industries; heavy chemical factories and fertiliser and rubber manufacturing units. A number of these industries discharge wastewater that contains hazardous chemicals and pathogens. Four major thermal power plants also depend upon water from the Ganga and the heated return flow affects river ecology. The discharge of untreated municipal sewage draining from the urban centres makes up three-fourths of the pollution of the river. Unfortunately, the building of sewage treatment plants to break down this sewage and the enforcement of compliance to discharge regulations for municipalities and industries have been hampered by Centre-State politics, funding and infrastructure problems and other socioeconomic complications related to revenue flows at the city and state levels. Over the past few decades, concerned citizens have been using the courts and the main environmental tribunal in the country, the National Green Tribunal, to push agencies to do the work they are responsible for doing to remediate water quality and prevent further pollution.

In this context, a Rights of Nature approach was recently attempted by High Court Justices in the case titled *Mohd Salim v. State of Uttarakhand and others*. This was an earnest attempt by judicial authorities to enforce water conservation, but the Supreme Court squarely quashed it a few months later. The case helps us to understand what a Rights of Nature framework would look like in India and what its problems may be in terms of religious values, government structure, politics and citizenship. A discussion of these issues can help to shed light on the feasibility of a Rights of Nature framework when considered in terms of specific domestic constraints and opportunities.

## Precursors of Nature's Rights in India: Public Interest Litigation

Many activists around the world have argued that a Rights of Nature framework should replace the current regulatory or market-based legal frameworks to guarantee nature's integrity and freedom to flourish.[5] Along with this, some have added that the framework should be designed, created, supported and enforced at all levels of government. While the United States has not made much headway in setting up a legal framework for the Rights of Nature, other countries offer interesting, experimental examples. In this regard, the initiatives recently executed in India provide another window into deliberating on the utility of a Rights of Nature framework. This case, like the others in Ecuador, Bolivia and New Zealand, allows us to ask whether this framework can assist in conserving riverine ecology and reducing ongoing water pollution. In this chapter, we will present the unique case of Rights of Nature in India by focusing specifically on efforts to preserve river basins and water supplies.

India's legal framework is already generous to petitioners and to notions of fundamental rights. These liberties and their historical development are precursors to a nature's rights framework. The advocacy aspect of a Rights of Nature approach can be found in India's tradition of public interest or social-action litigation which emerged along with investigative journalism, human rights and environmental activism over the last three decades. A unique kind of judicial populism has evolved through public interest litigation, wherein charismatic justices and petitioners have become central to the power of the law.[6] The Hon'ble Mr. Justice S. P. Bharucha explains the opening of legal debate in the early years of the Indian Union in this way: 'There was then the age of expanding the scope and ambit ["extent"] of fundamental rights, using, particularly, the wide language of Article 21. "Due process" deliberately eschewed by the founding fathers was judicially brought into play by holding that a "procedure established by law" had to be just and fair'.[7]

Justice Krishna Iyer was the first to lay the conceptual foundation for what is now called public interest litigation. In 1976, while ruling on an industrial dispute, Justice Iyer made the following comments:

> Our adjectival [relating to] branch of jurisprudence [knowledge of law], by and large, deals not with sophisticated litigants [people involved in a law suit] but the rural poor, the urban lay and the weaker societal segments for whom law will be an added terror if technical mis-descriptions and deficiencies in drafting pleadings and setting out the cause-title create a secret weapon to non-suit a part. Where foul play is absent, and fairness is not faulted, latitude is a grace of processual justice. Test litigations, representative actions, pro bono publico [for the public good] and like broadened forms of legal proceedings are in keeping with the current accent on justice to the common man and a necessary disincentive to those who wish to bypass the real issues on the merits by suspect reliance on peripheral procedural shortcomings. Even Article 226, viewed on wider perspective, may be amenable to ventilation of collective or common grievances, as distinguished from assertion of individual rights, although the traditional view, backed by precedents has opted for the narrower alternative. Public interest is

> promoted by a spacious construction of locus standi [to be heard in court] in our socio-economic circumstances and conceptual latitudinarianism permits taking liberties with individualization of the right to invoke the higher courts where the remedy is shared by a considerable number, particularly when they are weaker. Less litigation, consistent with fair process, is the aim of adjectival law.[8]

In the early 1970s, Justice Bhagwati and Justice Iyer were involved with the establishment of legal-aid schemes for 'poor persons and members of backward classes'.[9] Later, Justice Bhagwati articulated in a more decisive way the nature of public interest litigation. In 1988, he summarised the court's position on judicial procedure as follows:

> This Court will readily respond even to a letter addressed by such individual acting pro bono publico. It is true that there are rules made by this Court prescribing the procedure for moving this Court for relief under Article 32 and they require various formalities to be gone through by a person seeking to approach this Court. But it must not be forgotten that procedure is but a handmaiden of justice and the cause of justice can never be allowed to be thwarted by any procedural technicalities. The court would therefore unhesitatingly and without the slightest qualms of conscience cast aside the technical rules of procedure in the exercise of its dispensing power and treat the letter of public-minded individual as a writ petition and act upon it ....[10]

In this pronouncement and others, Justice Bhagwati contended that social conditions require a relaxation of procedure. The social conditions are of two dimensions. On the one hand, inequality and differential access to justice and power prevailed; on the other, particular kinds of executive practises were at work. These executive practises were carried out through inaction, as a breach of the law or dereliction of duty. Along with the need for procedural relaxation, Justice Bhagwati stressed the judicial remedy of mandamus ['We command' is a judicial remedy in the form of an order from a superior court]. Any member of the public with sufficient interest can call for judicial redress for any public injury arising from breach of public duty or from the violation of some provision of the Constitution or the law. A citizen can seek enforcement of public duty or observance of such constitutional or legal provisions by standing for the public interest.

Justice Bharucha also commented on the central role that the remedy of mandamus plays in public interest litigation, stating that the remedy has 'canvassed the mandate of Article 14 that no one is above the law'.[11] He added that the prosecutions that were launched pursuant to the continuing mandamus in Vineet Narain [the Hawala case] 'produced reassurance in the people that the Courts would not turn a blind eye only because powerful men in public places were involved'.[12] He continued:

> Public interest litigations have also led to the Courts' pronouncements in pollution and environment matters. I would be the first to concede that these pronouncements have sometimes been a mixed blessing, but we must remember that in these matters, as in so many others, the Court has had to step in because the legislature and the executive had not acted upon their obligation to protect the quality of life.[13]

These discussions have also raised issue with the particular kind of lawyering required. When introducing a publication series on public interest litigation, a contributing author wrote: 'It compels the lawyer to unshackle herself from legalese, reach out to and inform herself of other disciplines. It teaches her to assimilate the legal principles into problem situations and present them before the court in a "judicially manageable" form. PIL throws up several opportunities for innovation and creativity'.[14]

## Standing for Nature: The National Green Tribunal

The Supreme Court, once the primary court for environmental litigation through a legally unique 'public interest cell', has now ceded most of its environmental decision-making to the National Green Tribunal. The National Green Tribunal was established in 2010 to oversee cases concerned with environmental degradation, combining legal deliberation with scientific assessment and monitoring.[15] The National Green Tribunal was created after many years of discussion on the best ways to deal with environmental issues and dispense environmental justice in India. It is composed of judges and expert members with knowledge of science and environment, and the expectation is that they decide matters on the basis of environmental merits and citizen concerns. The majority of the cases taken up by the Tribunal are related to pollution (31%) and environmental clearances (35%). Environmental clearances are required for 39 types of projects in order to assess and thereby avoid or minimise their environmental impacts. This involves screening, scoping and evaluation of a proposed project. At the same time, several judgments passed by the National Green Tribunal have brought public scrutiny, raising questions about the soundness of its decisions.

As part of this wave of transferring cases, in 2016, the Supreme Court passed one of the most important environmental cases, commonly known as the Ganga Pollution Case, to the National Green Tribunal. In 1985, M.C. Mehta filed a writ petition in the Supreme Court charging that, despite the strides made in the legal code, government authorities had not taken effective steps to prevent environmental pollution. Mehta issued a writ of mandamus to restrain leather tanneries and the municipal corporation of Kanpur from disposing of industrial and domestic effluent in the river Ganga. The court subsequently bifurcated the petition into two parts. The first dealt with the tanneries of Kanpur and the second with the Municipal Corporation. Called Mehta I and Mehta II in legislative digests, they became the 'Ganga Pollution Cases' and the most significant water pollution litigation in the short history of Indian environmental law.

In the original petition of 1985, Mehta requested the court to order the leather tanneries of the Jajmau district of Kanpur to stop discharging their untreated effluent into the river. He also claimed that the Municipal Corporation of Kanpur was not undertaking treatment of domestic sewage. The petition named 89 respondents; among them were 75 tanneries of the Jajmau district of the city, the Union of India, the Chair of the Central Pollution Control Board, the Chair of the Uttar Pradesh Pollution Control Board and the Indian Standards Institute.

At the time the Ganga Pollution cases were being heard, the Government of India was initiating its first environmental scheme to combat river pollution in northern India. The Ganga Action Plan (hereafter GAP) was created to address problems of waste management by diverting and treating industrial and municipal effluent before it reached the Ganga. Executed through the Ministry of Environment and Forests, this central government scheme provided grants to create this infrastructure in the largest cities bordering the river Ganga in three states. The original plan was to construct these facilities and then turn them over to city municipalities for long-term operation and maintenance.

Entering the mid-1990s, the Supreme Court stepped up its efforts to check the centralising efforts of the government by passing orders that reprimanded government agencies for dereliction of duty. By 1995, the Supreme Court had fined over 200 industries in the Ganga basin, penalised the State Pollution Control Boards for false reporting and pressed the Ministry of Environment to streamline its proposals for treatment plants through a less wieldy set of supervisory committees. It appeared that the struggle for power was certainly intensifying on paper, as the judiciary, through courtroom dramas, fines and punishments, sought to check the power of the executive branch and industries throughout the country. Yet this judicial activism proved profoundly limited by the very system it sought to check: by calling on the same agencies it reprimanded to implement its orders, the Supreme Court was rendered profoundly ineffective.

The recent shift of environmental cases to the National Green Tribunal marks a major shift from pressure through 'continuing mandamus' in the Supreme Court to pressure from the persistent complaints of citizens through the National Green Tribunal. In the Tribunal, the Judges call forward stakeholders and direct collaborative solutions. In prior Supreme Court cases, continuing mandamus occurred in cases when monitoring of compliance was directed by the court and considered necessary to the functioning of government and the effective use of its assets. When the Supreme Court agreed to cede responsibility to the tribunal, the media accused it of 'passing the buck'. The Court responded, 'For us, to monitor week after week, months after month, it is difficult ... it [you] can go there [National Green Tribunal] and if you have a problem you can come back to us'. At the end of April 2017, the Supreme Court passed another key case on the degradation of the Yamuna river and floodplains to the Tribunal.

Originally in 1994, the Supreme Court had taken *suo motu* cognisance of the pollution in the Yamuna on the basis of a news article reporting this. In 2017, a bench consisting of Chief Justice J.S. Kehar and Justice D.Y. Chandrachud decided that the monitoring of the various projects and actions to clean the river should be done by the National Green Tribunal and transferred the more than two decades–long public interest litigation to it. The Supreme Court held that there should not be 'parallel proceedings' on the same issue.[16] As one informant explained to Alley, 'When cases are sent by the Supreme Court to the [National Green Tribunal], the [National Green Tribunal] is more motivated to hear them'.[17] This was the general sentiment during the tenure of Tribunal Chairperson Swatantra Kumar; more recently, under new leadership in the Tribunal, advocates are submitting their petitions back in the Supreme Court and High Courts, hoping for a better hearing.

## Provisions Close to Rights of Nature

The Indian Constitution is not unique in incorporating articles that declare the responsibilities of the state and its citizens to protect and improve the environment and safeguard forests and wildlife. Like those used in other countries, India's constitutional articles were adopted in response to international conferences and conventions. The first provisions in law were made through the Forty-Second Amendment to the Indian Constitution.[18] Passed in 1976, the Forty-Second Amendment responded to the Stockholm Declaration adopted by the International Conference on Human Environment in 1972. That Declaration confirmed the responsibility of each member of society to protect and improve the environment. In conformity with these objectives, the Forty-Second Amendment inserted Article 48A into the Directive Principles of State Policy in Chapter 4 of the Constitution. This declared the State's responsibility to protect and improve the environment and safeguard the forests and wildlife of the country. Another provision, inserted in Article 51A(g), stipulates the duty of every citizen to 'protect and improve the natural environment including forests, lakes, rivers and wildlife and to have compassion for living creatures'. Both Article 48A and Article 51A impose an obligation on the government and the courts to protect the environment for the people and the nation.

Along with these provisions, the Indian legal system provides a few other sources of law for addressing environmental and especially water pollution problems. The Water

(Prevention and Control of Pollution) Act of 1974[19] ('the Water Act') was the first to specifically set out a concern for environmental protection. Parliament adopted minor amendments to the Water Act in 1978 and revised it in 1988 to conform to the provisions of the Environment (Protection) Act of 1986.[20] Although the Constitution determined that water would be a subject in the State List and would therefore fall under the purview of the state, the Water Act empowered the Union Government to legislate in the field of state control. That power was affirmed when all states in the Union approved the Act.

The administrative regulation under the Water Act provided for the establishment of a Central Pollution Control Board and, under this, a Board in each state of the Union. These Boards develop plans for the control and prevention of pollution. The Central Board plans and executes a national programme for preventing pollution, carries out research, compiles data and advises the Government on water and air pollution matters. The State Boards implement the Water Act by inspecting industrial and wastewater treatment plants and conducting research on water quality and sewage treatment methods. Both Boards have been awarded the authority to set standards for water quality, air quality and emissions and effluents from industry and other sources.

Although environmental protection is included in the Directive Principles of State Policy and the Fundamental Duties of the Indian Constitution, environmental rights are not listed under the justiciable Fundamental Rights of the Indian Constitution. Hence, they are not directly enforceable. However, by utilising these provisions on environmental protection to flesh out the constitutional right to life, the Supreme Court has held that 'a citizen has a right to have recourse to [the remedies provided by] Article 32 of the Constitution for removing the pollution of water or air which may be detrimental to the quality of life'.[21] In other words, the courts have been willing to read the Directive Principles into the Fundamental Rights. Environmental cases have also benefitted from the procedural advantages attached to the enforcement of constitutional rights. The Indian courts' willingness to apply an ecological norm has advanced a broader judicial commitment to the rectification of the perceived failures of other branches of government.

## Standing and Other Procedural Matters

As the preceding quotes reveal, justices in the postindependence period were concerned with opening up uses of the law to citizens of all socioeconomic classes, to act upon their ideals of equality and justice. In the process of establishing these ideals, justices defined procedural matters in unique ways and set these procedures apart from those used in other countries. As litigants began to approach the court with concerns of a public interest nature, justices began a vigorous debate over standing and procedural matters. Concerned citizens discussed what was meant by the public interest and who could represent or argue for the constitutional rights of large numbers of citizens. The short history reveals a unanimous interest in relaxing the rule of standing. In *Akhil Bhartiya Soshit Karamchari Sangh v. Union of India*,[22] Justice Iyer commented on the issue of standing:

> Our current processual jurisprudence is not of individualistic Anglo Indian mould. It is broad based and people oriented, and envisions access to justice through 'class actions' 'public interest litigation' and representative proceedings. Indeed little Indians in large numbers seeking remedies in courts through collective proceedings, instead of being driven to an expensive plurality of litigations, is an affirmation of participative justice in our democracy. We have no hesitation in holding that the narrow concept of 'cause of action' and 'person aggrieved' and individual litigation is becoming obsolescent in some jurisdictions.[23]

In many cases, justices have grappled with a determination over the 'person aggrieved'. Generally in public interest cases, the issue of standing has been linked to other procedural matters. For instance, justices broadened the guidelines for writing an acceptable writ, building upon the general sense since early postindependence jurisprudence that the general format should be flexible to various forms of appeal. In 1956, Justice Mukherjee claimed that, 'Under Article 32, the court enjoys a broad discretion in the matter of framing the writs to suit the exigencies of the particular case and it would not throw out the application of the petitioner simply on the ground that the proper writ or direction has not been prayed for'.[24]

Looking across the many environmental cases of the past 20 years, we find that High Court and Supreme Court Justices have accepted letters, appeals and newspaper editorials as writ petitions for the public interest. In these letters and appeals, petitioners have written to protect fresh water bodies and coastal zones, forests, national monuments, planning provisions and urban heritage sites, among other national goods. Petitioners have used Articles 32 and 226 to put pressure on state offices and on private agents of industrial and technological development to manage and adequately dispose of waste by-products. Justices have appointed amicus curiae (an impartial adviser) to assist petitioners in approaching the court for the public interest.

While accepting a relaxed notion of standing and a broad range of petition types, the Indian courts have also exercised *suo motu* powers (Latin: 'of his or its own accord', an action initiated by an authority on its own). Justices have exercised the *suo motu* power to intervene directly in the administration of a State or private project.

## Ganga and Yamuna Rivers as Persons

We now turn to the recent high-profile ruling designating the Ganga and Yamuna rivers as juristic persons. This case was heard in the High Court of the state of Uttarakhand and not by the National Green Tribunal. On March 20, 2017, the High Court ruled in *Mohd Salim v. State of Uttarakhand and Others* that, 'The Rivers Ganga and Yamuna, all their tributaries, streams, every natural water flowing with flow continuously or intermittently of these rivers, are declared as juristic/legal persons/living entities having the status of a legal person with all corresponding rights, duties and liabilities of a living person in order to preserve and conserve river Ganga and Yamuna'.[25]

The immediate motivation for the landmark ruling, as Lokgariwar and others have noted,[26] comes from the case of the Whanganui river in New Zealand. In that case, the river was declared a living entity with full legal rights by the country's parliament after a long push by the Maori, an indigenous group.[27] In India, a discussion on the need to grant personhood to these rivers started in 2014 when members from the Community Environmental Legal Defense Fund met with members of the Global WASH Alliance-India and Ganga Action Parivar. Together they developed the National Ganga River Rights Act, which recognised the rights of the Ganga River basin and the people of India to a healthy river ecosystem.[28]

Led by the head of a popular ashram in Rishikesh, the Ganga Action Parivar and the WASH Alliance gathered a group of 25 religious leaders and called upon the government to declare Ganga a person, as a way to enforce stronger steps to protect her. They proposed the new law to 'grant legal rights to the Ganga' and submitted their letter of request to the Union Science and Technology Minister at an event organised by Global Interfaith WASH Alliance and UNICEF-India.

The leader of this group noted to the media that: 'We feel that by enabling the Ganga to secure its own legal rights to survive and thrive, just like a company or a person, the same can later be done for all other rivers such as the Indus'.[29] On the need for a new law despite one already being in place, he said the Water (Prevention and Control of Pollution) Act, 1974, had been 'rendered dry' over the last four decades. 'In 1974, a Water Act was passed that called for up to seven years [in] jail for repeat polluters. Yet 42 years later, our nation is in trouble. … 80% of our drinking water is polluted, mostly with sewage. The Water Act is a dry piece of legislation', he said.[30]

This meeting did not result in a more serious consideration of an Act or law on the part of the government until the High Court ruling was promulgated. In that case, the advocate brought the New Zealand Rights of Nature example to the attention of the High Court Judges when arguing for river protection and pollution prevention. The lead justice took it upon himself to research this case and other Rights of Nature initiatives as he formed his judgement. In the ruling, the lead justice, and the supporting justice on the bench, appointed three guardians—the director of Namami Gange (a central government official), the chief secretary of the state of Uttarakhand (a state government official) and the advocate general of the state (a state government official) – as 'persons in *loco parentis* [Latin: 'in place of a parent'] as the human face to protect, conserve and preserve the Rivers Ganga and Yamuna and their tributaries'. By this order, the officers were bound 'to uphold the status of [the rivers] and also to promote the health and well-being of these rivers'.[31] The court confirmed that any harm done to the river would be a cognisable offence and the state would initiate criminal proceedings without waiting for a petitioner. Ten days later, as people were still trying to understand the implications of this order, the same Justices went on to designate the glaciers, lakes and wetlands of these basins as 'legal persons' in another case related to resource uses in the state.[32] They were intentional in their interest in strengthening the Rights of Nature approach even further when they stated,

> '… the Glaciers including Gangotri & Yamunotri, rivers, streams, rivulets, lakes, air, meadows, dales, jungles, forests wetlands, grasslands, springs and waterfalls, legal entity/legal person/juristic person/juridicial person/ moral person/artificial person having the status of a legal person, with all corresponding rights, duties and liabilities of a living person, in order to preserve and conserve them. They are also accorded the rights akin to fundamental rights/legal rights'.[33]

This second ruling was the result of a petition filed by Lalit Miglani and argued by his brother, also an advocate in the same High Court. The previous orders in this second case aimed to regulate plastic waste and the discharge of sewage into the Ganga. This second case, the Lalit Miglani PIL, revealed the gross negligence of governmental authorities (both central and state) in discharging their statutory duties to prevent pollution of the Ganga. In an effort to ensure the protection of the natural resources of the region, the justices reaffirmed and, in fact, expanded their first judgment on the juristic personhood of the Ganga and Yamuna by including other natural resources within this notion of juristic rights. The justices once again appointed five legal representatives for the management of the glaciers, rivers, streams and other resources 'in loco parentis'.[34] They declared that designating ecosystems as juristic persons was an imperative measure for their preservation. The judges also directed the establishment of sewage treatment plants and directed action against and closure of polluting industries and institutions such as ashrams (religious institutions).[35]

While the ruling on personhood in *Mohd. Salim* is ostensibly for environmental conservation and protection, its roots in religious sentiments and philosophy cannot be set aside. Justice Sharma began his presentation on these aspects by stating that, 'Rivers Ganges and Yamuna are worshipped by Hindus. These rivers are very sacred and revered'. He continued, 'The Ganga is also called "Ganga Maa". It finds mentioned [sic] in ancient Hindu scriptures including "Rigveda"'.[36] With this statement, he mentioned other cases in which a Hindu idol had been seen as a juristic entity capable of holding property.

## The Problem of Guardianship

Both these cases prompt the question: In the Indian context, is it possible for a natural entity to have legal standing on the basis of notions of standing already in existence? As La Follette and Maser argue:

> It is of course no answer to say that Nature, or any being of Nature, should be denied rights because it cannot speak. The American legal system gives rights to many entities that cannot speak: cities, states, and the federal government are the most common examples. This could be multiplied to include water districts; religious organizations; social service organizations, such as the United Way; trusts and estates; joint venture partnerships; and many others. All these entities hire attorneys to speak for them and argue their cases in court.[37]

Western industrialised society, La Follette and Maser note, seems to find little or no intrinsic value in Nature unless it is demonstrably 'good for something' or can be converted into something of material value.[38] By contrast, the Ganga is a Goddess with her own sacred power and history. Ganga is a Goddess and also a Mother, and some also consider her a person or personified Goddess. So in these Hindu interpretations, it is feasible to some that she should be able to avail herself of the right-to-life provisions in the constitution.[39] Before looking into this religious interpretation of Goddess and personhood, however, we need to consider what guardianship means in the transnational legal discourse.

As La Follette and Maser note, the questions to be answered in the case of a surrogate who advocates for Nature include the following: Who could be qualified? How would they be selected? What obligations would they have? And what powers would be granted and by whom? Most importantly, to whom would a Nature's Rights guardian be accountable, and according to what standards? They write:

> A guardian (or ombudsman, to use a term for a governmentally appointed advocate, usually with limited powers) could be a government agency, but might prudently be expanded to include qualified environmental organizations that could act as additional, independent, guardians with rights to advocate for species or ecosystems in planning, and also through litigation. This is a system already in use in Germany. Guardians could be given additional roles in protecting resources of the global commons, such as oceans or the air: monitoring the health of the commons, and having powers to enforce treaties and national laws affecting the commons. The guardian would have the power to appear before international rule-making and legislative bodies on behalf of the aspect of Nature the guardian represents, and also have powers to bring suit if the commons is endangered by human activity.[40]

In order for any guardianship or ombudsman to enforce Nature's Rights, the legal system must give Nature, or its surrogate, 'standing' to fight for rights and defend against harms done to them. As the history of Indian litigation shows, justices have sufficiently liberalised the notion and procedures for standing to allow a range of actors to argue for environmental rights. Likewise, in the United States, groups have been able to sue on the basis of recreational, aesthetic or ecological losses, essentially 'for' the place or being that will be harmed.[41]

The notion of standing in the High Court ruling in India hinges on the role of the guardians who are designated as three government officials. However, this judgment would not alter the ability of a public-spirited person to file public interest litigation in case of pollution or environmental harm to the Ganga or Yamuna. In such a case they could implead the three guardians as respondents. Nonetheless, there are other complications that arise as a result of such a framework. Before considering these problems, we need to understand the religious interpretations of these sacred rivers and the ways devotees understand the goddess as a divine being, a Mother and a person.

## Other Considerations: Religious Doctrines of Personhood

The Hindu religious view pertaining to deities is foundational to this judgment and others, particularly in terms of the personification of different elements of nature as gods and goddesses. Of the six Astika schools of Hindu philosophy, Vedanta is one of the most popular and influential. It has three schools of thought – Dvaita, Vishishta Advaita and Advaita – and all invoke different forms of personhood for divinity. While Dvaita is the philosophy of dualism, Vishishta Advaita is qualified monism and Advaita is monism.

In terms of the spiritual journey, these three epitomise the progression of the consciousness toward the realisation of ultimate reality, exemplified by Advaita philosophy. While followers of Dvaita and Vishishta Advaita believe in a personal god and explain devotion as the path to liberation, Advaita metaphysics holds that the world has no existence separate from Brahman, the ultimate reality. The self that is experiencing, or the jiva, and the universal self, or atman, are the same; they are both Brahman, despite the fact that the individual self seems distinct. The Brahman of Advaita Vedanta advocated by Sankara is impersonal. It transcends all attributes and thoughts. As described by Sankara, 'Brahman [ultimate reality] is without parts or attributes ... one without a second'. This Nirguna Brahman is the ultimate reality as it truly is. It, however, becomes a personal god because of its relationship with the principle of Maya. The Advaita Vedanta school distinguishes between Nirguna Brahman (attributeless) and Saguna Brahman (with attributes); Saguna Brahman is illusory.

The abstract philosophy of Advaita is difficult to grasp and comprehend; thus, the value of the concept of a personal god is acknowledged. Depending on the tradition one adheres to, deities are either concepts beneficial for spiritual progress or real beings toward whom one surrenders with devotion. Either way, they are a key element of Hindu philosophy and religious practise; additionally, the expansive and multifarious Hindu traditions create manifold deifications to natural elements and qualities recognised as valuable for human life and society, with some ritual practises traceable to the Vedas.

Personhood of deities is therefore a key aspect of the living spiritual traditions of India. In relation to this personification, broadly speaking, two different approaches can be identified, one that considers deities personifications of abstract energies and qualities and another that sees deities as real beings embodying divine energies and qualities. Both views result in worship, with the difference that, in the former case, the spiritual aspirant's worship is directed toward the energy symbolised by the deity. In the latter case, the deity epitomises the energy or quality that is the object of devotion. In both cases, the representation of the deity is of importance, as is the emotional connection developed with that form and the narrative around it. To give an example, during the Hindu festival of Diwali, millions of Hindus pray to Goddess Lakshmi and light lamps to urge her to enter and bless their homes, firm in the belief that through proper worship the goddess will be propitiated and pleased. Ganga is another such deity. She is worshipped as a goddess, and places of worship have developed along her banks, including the ancient towns of Varanasi and Haridwar. For the devout Hindu, she is a mother, protector, remover of sin and purifier of matter and consciousness.

In the worship of Ganga, one finds the two approaches commingling, for Ganga is at once 'the Supreme Shakti of the Eternal Shiva' and also a real being.[42] Several mythological stories about Ganga as the personified goddess exist in scriptures. Devotees bathe in her waters to be cleansed of their sins; the ashes of the dead are immersed in her waters, which leads the departed soul to a higher birth; her name is chanted with the belief that it will bestow freedom from poverty and protection, even lead to liberation. For the Hindu mind, Ganga is supreme among rivers, an 'archetype of sacred water'. Other rivers are said to be like the Ganga; others are said even to be the Ganga. Such is the strength of the belief in the personhood of the Ganga and in her divinity and powers. Yamuna, Godavari, Saraswati, Sindhu and Kaveri are the other sacred rivers of India. Two of the most sacred Hindu pilgrimage places, Gangotri and Yamunotri, are the sources of these two sacred rivers. The river Yamuna is a tributary to the Ganga and, like Ganga, Goddess Yamuna has been personified. In Hindu mythology, Yamuna is also a mother who sustains and provides, but she is more concerned with the blessings of this life than Ganga, whose role is purification and preparation for death. Yamuna is the daughter of Surya (the sun god) and Saranyu. The lord of death, Yama, is her brother.[43]

Uma Bharti, when she was Minister of Water Resources, put her own sentiments in this way at an opening statement for a climate conference in 2016:

> I never looked at Ganga from the religious point of view. Because the Hindu point of view is such that we look at everything with a religious point of view. So the religious point of view is always hidden there. It is always there. It cannot be without it. A Hindu vision will never be without a spiritual vision. Trees, stones, rivers, animals, insects, stars, sky, water, air. God is existing in everything, everywhere. Ganga is the very cool economic flow of this country. Because Bihar and UP (Uttar Pradesh), the biggest populated states, are completely dependent totally on Ganga. And Ganga is a story of how we destroy rivers. So Ganga becomes a model of how we save rivers. In that ecological flow is necessary, cleaning of Ganga through various methods is very necessary. And saving the rivers of this country because I always say that rivers and women have to fight for their own existence. Nobody helps them. They create their own existence, they save their own existence. It is very difficult for them. They struggle a lot. So the flow of women's growth and the flow of the river also.[44]

However, the religious and cultural basis for the High Court decision begs the question of whether the river's value for conservation arises solely from this connection to values or from considerations of it as a highly polluted, transboundary river vital for the livelihoods of millions. It appears, in this case, that environmental conservation of natural resources does not have importance as a value standing alone, but is based on the cultural, social or religious understandings of these resources. The leading Justice Sharma of the Uttarakhand High Court justified his position by stating, 'The extraordinary situation has arisen since Rivers Ganga and Yamuna are losing their very existence. This situation requires extraordinary measures to be taken to preserve and conserve

Rivers Ganga and Yamuna'.[45] Yet, in the next paragraph, the judgment launched into an analysis of the religious and cultural importance of these rivers. Rooting the judgment in the Hindu faith has usually been an implicit rather than explicit justification in cases related to the Ganga; the justices in this case are therefore going beyond what has been customary in the legal procedures of the Supreme Court and National Green Tribunal. It appears that they did so to argue for the concept of personhood, using precedents of personhood and property rights pertaining to Hindu deities. The personhood attributed to these rivers in these rulings was not just a metaphor for asserting the Rights of Nature to serve the purpose of environmental protection. It was also an expression of the divine rights these rivers have held in colonial and postindependence law as well as Hindu faith, practise and mythology.

## Overlapping Institutions and Politics in 'Guardianship'

Moving on to institutional and bureaucratic considerations, it is not clear how personhood based on religious beliefs and prior rulings on deities' rights will help in better management and conservation. There are many agencies involved in controlling water uses, monitoring these rivers and guiding decision-making on water allocation. How would they be included in decisions on rivers' rights? Since the rulings did not draw in and expand the roles of religious leaders or communities in guardianship, it is not clear where the innovation lies in terms of on the ground enforcement of protections and regulations. It is doubtful that there would be any additional benefit by making the director of Namami Gange, the Chief Secretary of the State of Uttarakhand, and the Advocate General of the State 'persons in loco parentis as the human face to protect, conserve and preserve the Rivers Ganga and Yamuna and their tributaries'.[46] These and other agencies responsible for conserving these rivers have already been failing miserably; adding the additional title of guardian will in itself not bring much change and will only narrow the range of decision-makers, potentially leading to fewer checks and balances on power.

Apart from its religious foundations, the judgment delivered by Justices Rajiv Sharma and Alok Singh in *Mohd Salim v. State of Uttarakhand and Others* was concerned with issues of federalism and tested whether a state – through its judiciary – could order the central government to take steps to protect the river.[47] States have significant power over water in the state list of the Constitution (i.e. the list of subjects on which state legislatures are empowered to enact laws). In earlier orders in the Mohd Salim case, the High Court had directed that encroachers along the Ganga canals and riverbanks be evicted, and the central government was ordered to clarify the division of canal land and authority between Uttarakhand and Uttar Pradesh (Uttarakhand was carved out of Uttar Pradesh in 2000). But there were other institutional tangles just under the surface.

In the December 2016 hearing, the Court ordered that there should be no more delay in constituting a Ganga Management Board under the Irrigation Department. This had been mandated under Section 80(2)(b) of The Uttar Pradesh Reorganization Act 2000.[48] After the reorganisation into the two states of Uttar Pradesh and Uttarakhand, there remained faulty allotments of control and maintenance along the many canals that distribute river water to agricultural fields in the region. The justices saw that the Ganga Management Board was needed to solve canal and related property disputes and coordinate river uses between central and state governments in this river basin, but the actual ruling on personhood put jurisdiction of Ganga matters under the Namami Gange project at the central government level. Even though the justices appeared to advocate for coordination between centre and states, by involving one representative from each of the states of Uttar Pradesh and

A cow drinking from the polluted Yamuna River in 2018. The Yamuna is a tributary of the Ganga and a major river of its own as well. (Photograph by Kelly D. Alley.)

Uttarakhand and a chairman of the board nominated by the central government, the configuration appeared messy.[49] It was potentially problematic, as the river Ganga passes through five states.

During the hearing in March 2016, an official in the Ministry of Water Resources told the court that, despite the long correspondence, neither the states of Uttar Pradesh or Uttarakhand had cooperated with the central government in constituting the Ganga Management Board. This was a problem of noncooperating state governments that were ruled by opposition parties. In the March hearing, the judges asked for finalisation of committee membership within 60 days.

As already mentioned, the invocation of fundamental rights in the protection of the environment has been extensive. In the current case, however, the immediate and more powerful administrative interest of the central government was not to grant Ganga and Yamuna rivers fundamental rights, but to constitute the Ganga Management Board to rein in noncooperating states. By doing so, the central government could use a centralised method of planning for all water uses in these basins, including hydropower, irrigation, potable supply and wastewater and water treatment. The justices agreed with this need for a Board. In the March 2017 hearing, while fussing at all parties for delaying the constitution of the Board, the judges issued the order on personhood for the rivers, in effect joining the interest in centre-state coordination with a Rights of Nature approach. They stated, 'The Constitution of the Ganga Management Board is necessary for the purpose of irrigation, rural and urban water supply, hydropower generation, navigation, industries. There is utmost expediency to give legal status as a living person/legal entity to rivers Ganga and Yamuna r/w Articles 48-A and 51A(g) of the Constitution of India'.[50]

This personhood ruling became news across environmental policy communities as an achievement. But it supported another agenda to centralise authority. The problem with centralising authority means that vetting, monitoring and investment in projects would be narrowed to the choices of the central government agencies involved. We will say more about this in the concluding section.

Quite surprisingly, in July 2017, the State Government of Uttarakhand hired a well-known and powerful barrister to represent them in a special leave petition or SLP to the Supreme Court to stay this High Court order on personhood.[51] The Supreme Court stayed the ruling pending a final hearing, which has not yet occurred. At the time, the media reported the motivations as follows:

> The order had put the state government in a quandary. Since the rivers flow through several states, only the Centre could frame rules for their management. The ruling also raised questions like whether the victim of a flood in the rivers can sue the state for damages and also about whether the state and its officers will be liable in case of pollution in the rivers in another state through which it flows.[52]

In the petition, it appears that the state government does not want to assume the liability and responsibility for the grievances that people could bring to the court in the name of Ganga's rights or when suing the rivers as primary agents of floods and other disasters. In addition, the Supreme Court appeared to argue that, even though water is a state right, the central government needed to be in control since the rivers are transboundary and flow through many states. State rights mean the state governments are the primary decision-makers. Again, the interest in gaining more centralised control is at the root of the legal manoeuvres, and the Supreme Court appeared to support that interest. But the ruling was stayed because it brought on a host of other responsibilities and liabilities that both state and central governments did not want to bear. A few months later, the central government filed a stay against the other High Court ruling on personhood for glaciers, streams and others, registering their continued disinterest in upholding the responsibilities of a Rights of Nature approach. The final hearings have been pending for almost 2 years, so the conclusion on personhood remains unclear. Yet the general view among advocates working in the Supreme Court is that the personhood ruling will be overturned, bringing the problems back to the status quo.

## Conclusion

As La Follette and Maser note: 'The best way to achieve honest, factual ecosystem monitoring may be through a combination of a Nature's advocate or prosecutor with enforcement power, operating in an independent governmental capacity, and citizen enforcement at every level of the planning and decision-making process'.[53] In this sense, regulation and enforcement appear to work best when there are multiple players checking each other's powers and vetting each other's behaviours. This has been the state of affairs in India, where the High Courts, Supreme Court and National Green Tribunal have permitted citizens to check the powers and behaviours of government ministries, departments and projects through their writ petitions.

The National Green Tribunal, the High Courts, and Supreme Court are important counterbalancing forces in a political and institutional environment where noncompliance is rampant. This is why the ruling on personhood was not in fact a viable solution to the ongoing pollution and ecological problems in these river basins. The appointment of three guardians would have narrowed the checks and balances on the powers and behaviours of those implementing policies and projects for these rivers. It would have eliminated or sidelined the powers of the National Green Tribunal to allow citizen monitoring and policy input. It would have centralised authority to decide water allocations to the Namami Gange department, a central governmental body and two state officials, and it would have mandated a Ganga Management Board with decision-making powers to decide all the uses of these river waters. Such centralised decision-making rights would undermine judicial oversight and citizen input.

The ongoing problems of compliance with court orders involving pollution and overuse of these river resources have frustrated the Courts and the National Green Tribunal. In the cases described in this paper, these frustrations have pushed conservation-minded judges to look for new tools to regulate water users. The personhood ruling is one such tool that the judges reached for. They overshot, because the proposed guardianship would have set a new precedent for responsibilities and liabilities that government agencies did not want to uphold.

The Personhood decision of the High Court, currently stayed by the Supreme Court pending final judgment, also raises issues related to the role of guardians and what would amount to dereliction of their duties. In this regard, an interesting comparison can be made with the law related to guardians of vulnerable persons such as children, victims, detained persons, mentally challenged persons and so forth. Indeed, the guardianship law paradigm fits better than, for instance, the public trust doctrine, when one adopts the personhood model. However, the complications one sees in guardianship cases could very well arise here too, along with additional complications, because the rivers would not be able to express their interests and wishes in the way other vulnerable persons may be able to do. What then is in the best interests of these rivers? Because these interests would be decided by state and society, such an approach remains anthropocentric and in effect undermines the very purpose of endowing such natural resources with rights that do not cater directly to human intention. These are some of the challenges that a Rights of Nature approach would need to deal with and address, not only in India but in other contexts where specific individuals or entities are granted guardianship roles.

This does not mean that a Rights of Nature framework is not possible in India or elsewhere. The concerns for thriving ecosystems are critical to the sustainability of human life, and a more eco-centric way of understanding conservation and ensuring compliance would be beneficial. But to limit the potential human misuses of such an approach, a critical understanding of the values, institutions and political intentions of movements and countermovements involving Nature's Rights, including the specific definitions and parameters of guardianship, needs to be applied.

## Notes

1. (a) Alley, K.D. and Drew, G. 2012. 'Ganga' in *Hinduism, Oxford Bibliographies*. Oxford, UK: Oxford University Press; (b) Eck, D.L. 1982. 'The Goddess in Hindu Sacred Geography' in *The Divine Consort: Radha and the Goddesses of India*. Hawley, J.S. and Wulff, D.M. (eds). Boston, MA: Beacon Press.
2. (a) Alley, K.D. 2002. *On the Banks of the Ganga: When Wastewater Meets a Sacred River*. Ann Arbor, MI: University of Michigan Press; (b) Alley, K.D. 2014. "Ganga and Varanasi's Waste-Water Management: Why Has It Remained Such an Intractable Problem? SANDRP South Asia Network on Dams, Rivers and People (blog); (c) Alley, K.D. 2015. "Killing a River: Failure of Regulation" in *Living Rivers, Dying Rivers: A Quest through India*. Iyer, R. (ed). Oxford, UK: Oxford University

Press; (d) Alley, K.D. and Drew, G. 2012. "Ganga" in *Hinduism, Oxford Bibliographies*. Oxford, UK: Oxford University Press; (e) Haberman, D.L. 2006. *River of Love in an Age of Pollution: The Yamuna River of Northern India*. Berkeley, CA: University of California Press; (f) Markandaya, A. and Murty, M.N. 2000. *Cleaning Up the Ganges: A Cost-Benefit Analysis of the Ganga Action Plan*. New Delhi: Oxford University Press.

3. (a) Rauta, R. 2015. "The Ganga: A Lament and a Plea" in *Living Rivers, Dying Rivers: A Quest through India*. Iyer, R. (ed). Oxford, UK: Oxford University Press; (b) Sanghi, R. (ed). 2014. *Our National River Ganga: Lifeline of the Millions*. New York, NY: Springer; (c) Tare, V. and Roy, G. 2015. "The Ganga: A Trickle of Hope" in Iyer, R. (ed). *Living Rivers, Dying Rivers: A Quest through India*. Oxford, UK: Oxford University Press.
4. Scott, C.A. and Sharma, B. 2009. 'Energy Supply and the Expansion of Groundwater Irrigation in the Indus-Ganges Basin'. *International Journal of River Basin Management* 7(2): 119–124.
5. (a) La Follette, C. and Chris, M. 2017. *Sustainability and the Rights of Nature: An Introduction*. Boca Raton, FL: CRC Press; (b) O'Donnell, E.L. 'At the Intersection of the Sacred and the Legal: Rights for Nature in Uttarakhand, India'. *Journal of Environmental Law*, 30(2018): 135–144; (c) O'Donnell, E.L. and Talbot-Jones, J. 'Creating Legal Rights For Rivers: Lessons from Australia, New Zealand, and India'. *Ecology and Society*, 23(2018): 7; (d) Studley, J. "Why Shouldn't Sacred Natural Sites Be Declared as Juristic Persons Predicated on Spiritual Governance? http://sacrednaturalsites.org/2017/06/declaring-sacred-natural-sites-as-juristic-persons/
6. Bhuwania, A. 2017. *Courting the People: Public Interest Litigation in Post-Emergency India*. Cambridge, UK: Cambridge University Press 157 pp.
7. (a) Alley, K.D. 2009. 'Legal Activism and River Pollution in India'. *Georgetown International Environmental Law Review*, 21(2009): 793; (b) Bharucha, S.P. 1998. 'Golden Jubilee Year of the Constitution of India and Fundamental Rights' in *Supreme Court on Public Interest Litigation VII*. Kapur, J. (ed)., p. 797.
8. (a) Alley, K.D. 2009. 'Legal Activism and River Pollution in India'. *Op. cit.* pp. 797–798; (b) Kapur, J. 1998. 'Preface' in *Supreme Court on Public Interest Litigation X-XV*. Kapur, J. (ed)., p. 798.
9. (a) Alley, K.D. 2009. 'Legal Activism and River Pollution in India'. *Op. cit.*, p. 798; (b) Ganguli, A.K. 1998. 'In Public Interest: A Review of PIL in the Supreme Court'. in *Supreme Court on Public Interest Litigation A-1*. Kapur, J. (ed)., p. 798.
10. (a) Alley, K.D. 2009. 'Legal Activism and River Pollution in India'. *Op. cit.*, p. 798; (b) Kapur, J. (ed). 1998. *Supreme Court on Public Interest Litigation*. p. 798.
11. (a) Alley, K.D. 2009. 'Legal Activism and River Pollution in India'. *Op. cit.*, p. 799; (b) Bharucha, S.P. Golden Jubilee Year of the Constitution of India and Fundamental Rights. *Op. cit.*
12. Bharucha, S.P. Golden Jubilee Year of the Constitution of India and Fundamental Rights. *Op. cit.*
13. (a) Alley, K.D. 2009. 'Legal Activism and River Pollution in India'. *Op. cit.*, p. 799; (b) Bharucha, S.P. Golden Jubilee Year of the Constitution of India and Fundamental Rights. *Op. cit.*
14. (a) Alley, K.D. 2009. 'Legal Activism and River Pollution in India'. *Op. cit.*, p. 799; (b) Muralidhar, S. 1999. 'PIL in 1999 in the Supreme Court' in *Supreme Court on Public Interest Litigation V*. Kapur, J. (ed)., pp. V–XII.
15. (a) Amirante, D. 2012. 'Environmental Courts in Comparative Perspective: Preliminary Reflections on the National Green Tribunal of India'. *Pace Environmental Law Review*, 29(2012): 441–468; (b) Kumar, A. 2016. 'National Green Tribunal: A New Mandate towards Protection of Environment'. http://www.legaldesire.com/national-green-tribunal-a-new-mandate-towards-protection-of-environment/ (accessed 30 September 2018); (c) Shrotria, S. 2015. 'Environmental Justice: Is the National Green Tribunal of India effective?' *Environmental Law Review*, 17(2015): 169–88.
16. PTI. 'Supreme Court transfers to NGT Plea over Yamuna River Pollution'. *Economic Times*, April 24, 2017. http://economictimes.indiatimes.com/articleshow/58341955.cms?utm_source=contentofinterest&utm_medium=text&utm_campaign=cppst (accessed 20 January 2019).
17. Comment by advocate practicing in the Tribunal, September 2017.

18. The Constitution (Forty-second Amendment) Act, 1976. https://www.india.gov.in/my-government/constitution-india/amendments/constitution-india-forty-second-amendment-act-1976 (accessed 11 June 2018); (b) Forty-second Amendment of the Constitution of India. https://en.wikipedia.org/wiki/Forty-second_Amendment_of_the_Constitution_of_India (accessed 11 June 2018)
19. Water (Prevention and Control of Pollution) Act of 1974. http://www.indiawaterportal.org/sites/indiawaterportal.org/files/e7402_1.pdf (accessed 11 June 2018); (b) Hewameealla, S. 'Control of Water Pollution: Constitutional Aspect in India'. http://dl.lib.mrt.ac.lk/handle/123/12556 (accessed 14 August 2019)
20. The Environment (Protection) Act of 1986. http://toxicslink.org/docs/rulesansregulation/THE-ENVIRONMENT-(PROTECTION)-ACT-1986.pdf (accessed 14 August 2019)
21. Cullet, P. and Koonan, S. 2011. *Water Law in India: An Introduction to Legal Instruments* (2nd edition). New Delhi: Oxford University Press India, p. 124.
22. Akhil Bhartiya Soshit Karamchari Sangh. 'Through Its Secretary and Another vs. Union of India through Its Secretary Ministry of Railway and Others'. http://www.youthforequality.com/supreme-court-cases/34A.pdf (accessed 11 June 2018).
23. 'Narmada Bachao Andolan v. The State of Madhya Pradesh on 30 September, 2014'. https://indiankanoon.org/doc/118282924/ (accessed 11 June 2018).
24. *Charanjit Lal Chowdhury v. Union of India*. http://www.nja.nic.in/P-950_Reading_Material_5-NOV-15/6&7%20Charanjit%20Lal%20&%20%20Namit%20Sharma.pdf (accessed 11 June 2018).
25. (a) *Mohd. Salim v State of Uttarakhand* Writ Petition (PIL) No.126 of 2014, Order dated 20 March, 2017; (b) Dasm M. 2017. 'Ganga and Yamuna Are Now Legal Entities: What Does This Mean and Is It a Good Move?' *The News Minute*, March 21, 2017. https://www.thenewsminute.com/article/ganga-and-yamuna-are-now-legal-entities-what-does-mean-and-it-good-move-58999 (accessed 11 June 2018).
26. (a) Gopalan, R. 2017. 'Why the Court Ruling to Humanise the Ganga and Yamuna Rivers Rings Hollow'. *The Wire*, March 27, 2017; (b) Lokgariwar, C. 2017. 'The Sad State of These Persons Called Ganga & Yamuna – Can State Protect Them? Blog of the South Asian Network for Dams, Rivers and People', April 11, 2017. https://sandrp.wordpress.com/2017/04/11/the-sad-state-of-these-persons-called-ganga-yamuna-can-state-protect-them/ (accessed 30 September 2017).
27. O'Bryan, K. 2017. 'Giving a Voice to the River and the Role of Indigenous People: The Whanganui River Settlement and River Management in Victoria'. *Australian Indigenous Law Review*, 20: 48–77.
28. 'A Campaign to Establish Rights of the Ganges River'. https://celdf.org/2016/09/blog-gangas-rights-rights/ (accessed 11 June 2018).
29. 'Religious Leaders Submit Proposal to Government to Protect Ganga'. https://www.business-standard.com/article/pti-stories/religious-leaders-submit-proposal-to-govt-to-protect-ganga-116090200871_1.html (accessed 11 June 2018).
30. Ibid.
31. Gopalan, R. 2017. 'Why the Court Ruling to Humanise the Ganga and Yamuna Rivers Rings Hollow'. *The Wire*, March 27, 2017
32. W.P.PIL No. 140 of 2015, *Lalit Miglani v. State of Uttarakhand*.
33. Ibid.
34. Ibid.
35. Ibid.
36. *Mohd. Salim v. State of Uttarakhand* Writ Petition (PIL) No.126 of 2014, Order dated 20 March, 2017, pp. 4–5
37. La Follette, C. and Maser, C. 2017. *Sustainability and the Rights of Nature: An Introduction. Op. cit.*, p. 89.
38. Ibid.
39. A sample of residents of Varanasi gathering on the Ganga banks were surveyed in October 2018 on this point. About half agreed that Ganga could be considered a 'person'. All agreed that she was a Mother and should have rights.
40. Ibid. p. 90.

41. La Follette and Maser, *op. cit.*
42. Eck, D.L. 1982. 'Ganga: the Goddess in Hindu Sacred Geography' in *The Divine Consort: Radha and the Goddesses of India*. Hawley, J.S. and Wulff, D.M. (eds). Boston, MA: Beacon Press, pp. 166–83.
43. Haberman, D L. 2006. *River of Love in an Age of Pollution: The Yamuna River of Northern India.* Berkeley, CA: University of California Press..
44. Lecture presented at 3rd India Water Forum (IWF) 2016, 'International Water Convention', 'Water for Sustainability: towards Development And Prosperity'. 4.20.2016, India Habitat Centre in Delhi.
45. Krishna, G. 2017. 'A Verdict That Could Stall River Linking Project'. http://www.rediff.com/news/column/a-verdict-that-could-stall-river-linking-project/20170526.htm. May 26, 2017. (accessed 11 June 2018).
46. Gopalan, R. 'Why the Court Ruling to Humanise the Ganga and Yamuna Rivers Rings Hollow'" *Op. cit.*
47. *Mohd. Salim v. State of Uttarakhand & others*, High Court of Uttarakhand at Nainital Writ Petition (PIL) No.126 of 2014 http://hindi.indiawaterportal.org/sites/hindi.indiawaterportal.org/files/WPPIL-126-14%20HC-UTTARAKHAND%20ORDER%20ON%20GANGA%20AND%20YAMUNA%20RIVER%20RIGHTS-1.pdf (accessed 11 June 2018).
48. The Uttar Pradesh Reorganization Act 2000. http://legislative.gov.in/sites/default/files/A2000-29.pdf (accessed 11 June 2018).
49. Ibid.
50. *Mohd. Salim vs. State of Uttarakhand* Writ Petition (PIL) No.126 of 2014, Order dated 20 March, 2017, p. 11
51. *State of Uttarakhand & Others v. Mohd Salim & others*, Petition for Special Leave to Appeal 016879/2017 (Supreme Court of India) 7 July 2017.
52. 'SC Stays Uttarakhand HC Order on Ganga, Yamuna Living Entity Status'. *Indian Express News Service*, July 8, 2017, http://indianexpress.com/article/india/sc-stays-uttarakhand-hc-order-on-ganga-yamuna-living-entity-status-4740884/
53. La Follette, C. and Maser, C. *Sustainability and the Rights of Nature: An Introduction. Op. Cit.*, p. 100.

# 19

# *Caring for Country and Rights of Nature in Australia: A Conversation between Earth Jurisprudence and Aboriginal Law and Ethics*

**Mary Graham and Michelle Maloney**

**CONTENTS**

Introduction ..... 385
Part I: Earth Jurisprudence and the Rights of Nature ..... 386
Part II: Aboriginal Law and Ethics ..... 388
  'The Land is the Source of the Law' ..... 389
  'You Are Not Alone in the World' ..... 390
    Custodial Ethic ..... 391
    Locality ..... 391
    Autonomy ..... 391
    Balance ..... 392
Part III: Earth Jurisprudence from the Perspective of Aboriginal Law ..... 392
Conclusions ..... 396
Notes ..... 398

Western: What is the meaning of life?
Aboriginal: What is it that wants to know?

## Introduction

The global ecological crisis is now well documented.[1] Throughout the twentieth century, we have seen unprecedented deforestation, biodiversity loss, air and water pollution and the escalating disruption of entire components of the Earth System, such as anthropogenic climate change. Human impacts on the planet have become so significant that it was proposed in 2000 that we have moved into a new geological epoch – the 'Anthropocene'.[2]

In Australia, the ecological crisis is evidenced by the bleaching of the exquisite Great Barrier Reef; devastating biodiversity loss due to industrial-scale agriculture and land clearing; the death of rivers and waterways in the Murray-Darling Basin due to unsustainable, industrial-scale irrigation and hottest-on-record summers that have killed entire colonies of heat sensitive animals such as the mega-bats known in Australia as 'flying foxes'.

The failure of modern human societies to effectively govern our relationship with the natural world requires a *systemic analysis*, an analysis that enables us to see the underlying causes behind our insatiable consumption of nature and to rethink the very foundations of our legal, political, economic and cultural worldviews. This systemic analysis requires us to

DOI: 10.1201/9780429505959-19

focus on the structures we create in human societies and the logic and power imbalances that have brought us to where we are.

In this chapter, we explore the capacity of Earth jurisprudence to offer a framework for conducting this type of systemic analysis, to help us understand and respond to the ecological crisis. And we focus particularly on its relevance in the Australian context.

We ask: What does the theory and practise of Earth jurisprudence and Rights of Nature offer to the peoples, landscapes and biodiversity of Australia? And even if these concepts are useful within the Western culture and legal system in Australia, do they offer any insights or value to First Nations peoples and First Nations laws?

We write from the perspective of two very different cultural traditions: a Kombumerri woman, whose people have lived on their land and sea country, their 'traditional estate' on the East coast of Australia, for millennia, and a woman who is a descendant of the Irish convicts who were sent, unwillingly, to Australia less than 200 years ago, as part of the British imperial and colonial project.

We come from extremely different cultural backgrounds, but we are passionately interested in similar questions: How do we respond to the ecological crisis in Australia? How do we create a culture where all Australians care for the environment and for one other, and build a sustainable future? And how do we build this future in a nation-state that has not yet dealt with the horrors of its colonial past, or the ongoing violence of its colonial present? A sustainable future for Australia must be built on justice for the First Nations Peoples who have endured colonisation since 1788, when the British Empire claimed the continent as its newest penal colony.

In Australia, the experience of the Australian Earth Laws Alliance is that the key concepts within Earth jurisprudence resonate very powerfully with people from the Western cultural tradition, as it helps people from this culture reveal and 'unpack' the deeply flawed basis of Western industrial society's approach to commodifying and using up the living world.[3] What is important for the future of Earth jurisprudence in Australia is whether the theory and practise is of any use or relevance to First Nations Peoples.

In Part I of this paper, we introduce Earth jurisprudence and the Rights of Nature and discuss how these concepts are inviting people in industrialised societies to rethink their relationship with the Earth community. In Part II, we introduce some of the foundational concepts of Aboriginal law and ethics, including the key concept that the land is the source of the law. In Part III, we discuss four key elements of Earth jurisprudence and examine them from the perspective of Aboriginal law.

We argue that Earth jurisprudence and the Rights of Nature offer a valuable framework for Western peoples to critique and respond to the impacts of their own culture on the living world. We also suggest that there are many elements of Earth jurisprudence, as articulated by Thomas Berry, that resonate with – indeed, that draw inspiration from – First Nations Peoples laws, ethics and culture. But we also suggest that there are several elements – including the concept of the Rights of Nature – that are not immediately compatible with Aboriginal law, and which will need further analysis and conceptual refinement if they are to be made more relevant to Australian society.

## Part I: Earth Jurisprudence and the Rights of Nature

Earth jurisprudence is a useful theory to critique contemporary industrial society and embark on the process of rethinking and recreating healthy relationships between humans and the rest of the living world. Earth jurisprudence has emerged from the logic of Western thinking to challenge Western society's cultural values and norms.

The term 'Earth jurisprudence' was coined by deep ecologist, 'geologian' and Earth scholar Thomas Berry[4] (1914–2009). He proposed, in his 1999 book *'The Great Work: Our Way into the Future'*, that the challenge for humanity is to understand the underlying, systemic reasons for the ecological crisis, and to transform our relationship with the natural world from one of destruction to one of mutually beneficial support.[5] He suggests that acting ethically and living within Earth's natural capacities requires that we look to a new jurisprudence, a new way of governing ourselves for the challenges and possibilities of the twenty-first century so as to protect the integrity of Earth systems.[6]

Berry proposes that the primary cause of the ecological crisis is anthropocentrism – a belief by people in the Western industrialised world that we are somehow separate from, and more important than, the rest of the natural world.[7] Berry argues that this anthropocentric world view underpins all the governance structures of contemporary industrial society – economics, education, religion, law – and has fostered the belief that the natural world is merely a collection of objects for human use.[8]

Berry suggested that the time had come for an ecocentric, or Earth-centred foundation for human societies. He stated that: '[E]very component of the Earth Community has three rights: the right to be, the right to habitat, and the right to fulfil its role in the ever-renewing processes of the Earth Community'.[9] He stated that these rights 'originate where existence originates. That which determines existence determines rights'.[10] Thus, existence and the laws of the emerging Universe, of Earth's functions, are the highest laws, and human-made laws need to be in alignment with them.

Earth jurisprudence, then, is an emerging theory of law and governance that requires a radical rethinking of humanity's place in the world, to acknowledge the history and origins of the Universe as a guide and inspiration to humanity and to see our place as one of many interconnected members of the Earth community.[11] Berry and the broader Earth jurisprudence movement acknowledge the inspiration and guidance that indigenous cultures and wisdom can provide to industrialised societies and the development of Earth jurisprudence.

Many of the key elements of Earth jurisprudence and ecocentrism have long been debated in environmental philosophy and human ecology, and ecocentrism in the law has been explored by many writers.[12] In addition, many of the critiques directed at contemporary or 'traditional' environmental law by Earth jurisprudence have also been articulated by other writers. Traditional environmental law has been criticised as being embedded in industrial society's pro-growth governance culture, and simply 'legalising' severe environmental harm, rather than effectively protecting the Earth community.[13] It has also been criticised for being incapable of calculating or managing the cumulative impacts of human activities and ignoring the reality of ecological limits.[14]

The work of Berry, and the many Earth laws and Rights of Nature advocates who have followed him, builds on this body of work, but it can be argued that it also offers something new. In addition to being a critical theory stimulating a growing body of literature,[15] Earth jurisprudence and Rights of Nature are increasingly becoming practical and constructive tools that are directly inspiring and informing innovative practise by lawyers and activists around the world. For example, organisations have been created in many countries around the world to explore and implement Earth jurisprudence.[16] Rights of Nature laws are emerging around the world, and now exist in the United States, Ecuador, Bolivia and Mexico, and laws recognising nature as having 'legal personhood' exist in New Zealand, India, Colombia and Bangladesh.[17]

Earth jurisprudence is also incorporated into a broader movement advocating that Western law evolve beyond traditional environmental law, towards 'ecological law and governance', which is articulated by a range of academics, networks and organisations.[18]

## Part II: Aboriginal Law and Ethics

Australian Aboriginal peoples' culture is ancient[19] and has made certain observations over many millennia about the nature of Nature, spirit and being human. This in turn has developed a unique Aboriginal logic and culture. The most basic questions for any human group, despite advances in technology, have not changed much over time; they include: How do we live together (in a particular area, nation or on Earth), without killing one another off? How do we live without substantially damaging the environment? Why do we live? We need to find the answer to these questions in a way that does not make people feel alienated, lonely or murderous.

Aboriginal law is very different from Western notions of law. As stated by eminent First Nations Law Professor, Irene Watson:

> First Nations Peoples' law is of the beginning: of the first songs, sung by the ancestors. When the first steps were walked across the ruwe, country was sung into creation. Law conceived as a way of living is difficult to write about and cannot simply be described or easily translated into a foreign language that is empty of the ideas that our law ways carry. Our law was not written in the way in which the West conceives of writing. Law was painted in ceremonial design and symbols were marked on boundary markers, identifying traditional owners and their ngaitjis. *The differences between Nunga and non-Aboriginal legal systems are so extensive that there is no basis upon which comparison can be drawn* ... (Aboriginal) law is the essential basis of social conduct: respect, reciprocity and caring for country to name a few. These ethical principles convey the essential nature of the law.[20] (emphasis added)

A chapter as short as this cannot hope to do justice to summarising the complexity of Aboriginal law, but we will try to provide a brief overview of key concepts. There are two basic precepts of Aboriginal Law: *the land is the source of the law* and *you are not alone.*

Mountain ash (*Eucalyptus regnans*), the world's tallest flowering tree species, in the Tahune Forest Reserve, Tasmania, Australia. (Photograph by James K. Lee.)

## 'The Land is the Source of the Law'

In Aboriginal law, the land is a sacred entity, not property or real estate; it is the great mother of all humanity and is the source of life and law.

In Aboriginal law, *the two most important kinds of relationship in life are, first, those between land and people and, second, those amongst people themselves, the second always being contingent upon the first.* The land, and how we treat it, is what determines our human-ness. Because land is sacred and must be looked after, the relationship between people and land becomes the template for society and social relations. Therefore, all meaning comes from land, or 'country'.

Nonindigenous people often know about our connection to country through a concept known as 'the Dreaming'. This English term – 'the Dreaming' or 'Dreamtime' – was created by Westerners studying Aboriginal people in Australia in the late 1800s.[21] Using this term to encapsulate the complexity of Aboriginal law, religion and culture is simplistic and insufficient, but the concept entered the English language through the work of English speaking anthropologists and has 'stuck'. 'The Dreaming' is often described as the time of creation, but as Nicholls explains:

> (it) isn't something that has been consigned to the past but is a lived daily reality … it incorporates creation and other land based narratives, social process including kinship regulations, morality and ethics. This complex concept informs people's economic, cognitive, affective and spiritual lives.[22]

Nicholls goes on to explain that there is in fact, no universal, 'pan-Aboriginal' word to represent the complex beliefs and legal systems of the nations and peoples in the continent now known as Australia. Before 1788, there were more than 500 autonomous nations across the continent, with approximately 250 separate Aboriginal languages and between 600 and 800 dialects.[23]

> The Warlpiri people of the Tanami Desert describe their complex of religious beliefs as the Jukurrpa. …. The Kija people of the East Kimberley use the term Ngarrankarni (sometimes spelled Ngarrarngkarni); while the Ngarinyin people (previously spelled Ungarinjin, inter alia) people speak of the Ungud (or Wungud). 'Dreaming' is called Manguny in Martu Wangka, a Western Desert language spoken in the Pilbara region of Western Australia; and some North-East Arnhem Landers refer to the same core concept as Wongar – to name but a handful.[24]

We acknowledge and respect the complexity of legal systems that have evolved across hundreds of nations over thousands of years, and rather than analyse the flawed concept of 'the Dreaming', we offer some key concepts that help us explore our 'conversation' between Aboriginal law and ethics, and Earth jurisprudence.

A fundamental element of Aboriginal first laws, the laws between people and land, is the law of obligation. As the land created us, so we are always going to be obligated to it. Not just our life but our *existence*, the whole of our existence and all meaning that underpins and surrounds it, that lives through us. All the flora and fauna, every living thing, all the landforms and features of the land, they are all our ancestors, because they all came before us. They helped us emerge and become and stay human, to develop us further as human beings and to create culture. Literally, the grass we walk on, the soil we walk on, the plants and animals we eat – these all made us human and gave us meaning and identity. And it all came about through our relationship with the land.

This deep relationship with the land created the fundamental nature of our law and culture, and our logic and way of thinking: *our relationist ethos.* Because we are always obliged to the land, always thankful for it, we are in turn obliged to look after it. This produces reciprocity – the land looks after us, we look after it, it looks after us, we look after it, and right across the country, North, South, East and West, all the different groups have their own particular relationship with their particular part of the land, their country.

Although Aboriginal people across Australia are Westernised to different degrees, our peoples' identity is essentially always embedded in land and defined by their relationships to it and to other people. The sacred web of connections includes not only kinship relations and relations to the land, but also relations to all living things.

## 'You Are Not Alone in the World'

As noted above, the two most important kinds of relationship in life are first those between land and people and second those amongst people themselves, the second always being contingent upon the first.

These 'second laws', the laws between people, are also built on the relationist ethos. Aboriginal Law is grounded in the perception of a psychic level of natural behaviour, the behaviour of natural entities. Aboriginal people maintain that humans are not alone. They are *connected and made* by way of relationships with a wide range of beings, and it is thus of prime importance to maintain and strengthen these relationships.

Aboriginal people have a kinship system that was and still is organised into clans. One's first loyalty is to one's own clan group. It does not matter how Western and urbanised Aboriginal people have become, this kinship system never changes. It has been damaged by colonisation, which led to cultural genocide, Stolen Children and westernisation, but has not been altered substantially. Every clan group has its own explanation of existence – as mentioned earlier, this is sometimes referred to, simplistically, as the clan's 'Dreaming'. *In Aboriginal law and culture, we believe that a person finds their individuality within the group.* To behave as if you are a discrete entity or a conscious isolate is to limit yourself to being an observer in an observed world.

Over vast periods of time, Aboriginal people invested most of their creative energy in trying to understand what makes it possible for people to act purposively, or to put it another way: What is it exactly that makes us human? What Aboriginal people have done is map the great repertoire of human feeling to such an extent that its continuities with the psychic life of the wider world become apparent.

There are four essential attributes to the relationist ethos – a custodial ethic, locality, autonomy and balance.

Blue-winged kookaburras (*Dacelo leachii*) on Magnetic Island, Queensland, Australia. (Photograph by James K. Lee.)

### Custodial Ethic

Aboriginal law and ethics involve daily practise. They are not taught in abstract; they have to be applied to all aspects of life, every day. First, we look after the land – this is the template for society and how the law of obligation comes into being, and this is how we know how to look after each other. *Ethics only come from having empathy and from looking after something outside ourselves.*

The understanding that the land is the source of the law is so ingrained in Aboriginal law that much of the law known from stories, paintings and dance was actually devoted to 'second laws'; teaching us how to live by the law and how to live together. Aboriginal law doesn't try to change human behaviour; it accepts the full range of human emotions, flaws and actions. But it does set out how it should all be handled in order to build peace and stability.

### Locality

The second element of the relationist ethos – locality – is interwoven with the first. Locality is everything for Aboriginal people. In European culture, Descartes said 'I think, therefore I am'. If there were an Aboriginal equivalent, it would be: 'I am located, therefore I am'.

Locality refers to peoples' connection not just to country or nature generally, but to the region they come from, the particularity of their land, the 'traditional estate' of a clan or language group. Identity and character come from the land itself, the shape and the form of it; whether it is desert, rainforest, saltwater, freshwater, mountains or plains: every part of the land has its own character. So the character of the land is the basis of the character of the people, not just in terms of our relationist ethos, but in the actual character of the people.

Western science has now demonstrated what Aboriginal people have always known: that our people have an ancient connection to particular parts of the continent. Human genome testing has shown Aboriginal people have been continuously present in their home regions for at least 50,000 years. This is some of the longest continual connection to country known to exist anywhere in the world and has resulted in people developing specific characteristics connected to their regions – such as desert dwellers having a greater physical capacity to cope with freezing night-time temperatures than other peoples.[25]

This ancient, stable connection to country has many implications for our culture. An example that many European descendants find especially intriguing is that in Aboriginal law, there is no concept of long-term warfare, and there is no concept of taking another group's land. There were rules of engagement for conflict between families, between clans, between nations – there were even rules about how to manage your traditional enemies. But the idea of someone invading or taking control of someone else's country is totally unheard of. It is not even found in Aboriginal creation or 'Dreamtime' stories.

### Autonomy

A third essential element of the relationist ethos is autonomy. The concept is connected to land, locality and other aspects of Aboriginal law. In Aboriginal law, every person is an autonomous being; no one can be enslaved or treated as a lesser being. And this extends into all aspects of society. So, for example, the argument goes something like: 'My family, my community, is an autonomous one. The one next door is also autonomous, and our right to autonomy must not transgress the other one's autonomy'. And while this didn't prevent conflict, the law contained the rules for handling conflict, without wiping each other out and without taking each other's land.

This strong foundation of autonomy is responsible for an important element of Aboriginal governance and society – the absence of hierarchies. In clans and language groups, everyone had their place in the collective and no one was more important than anyone else. Governance structures included elders who had authority that was earned through proper conduct – by following the law. Authority and power were separated so that the cohesion and strength of the group were always maintained.

### *Balance*

Like many other ancient cultures, Aboriginal societies focussed on balance and harmony. People sought balance in everything: between people and the land; between men and women, with men's law and women's law; between different families and clan groups. In a relationist ethos, the task is to nurture, practise and grow this balance. In Aboriginal law, it is understood that you build, or grow, your knowledge and proper conduct every day; and you grow your attachment to place and personhood, your identity and your autonomy.

To work toward restoring and maintaining balance takes many complex rules of interaction and engagement. One rule that, in many ways, is in stark contrast to today's 'celebrity' or 'social media' culture, is the rule that ego must be managed. In Aboriginal law, ego is considered a volatile substance; one that should be respected and acknowledged (not suppressed as such), but which needs active management and containment. To allow an individual's ego to override the well-being of the family, clan or community is to allow volatility, instability and unfairness.

This section has provided just a brief introduction to some of the foundational concepts in Aboriginal law, but we hope it has provided enough to now allow an analysis of Earth jurisprudence from the perspective of Aboriginal law.

## Part III: Earth Jurisprudence from the Perspective of Aboriginal Law

Earth jurisprudence is a theory that has emerged from within Western industrial society, to 'talk back' to and challenge the devastating environmental impacts of Western industrial society. In this section we briefly review some of its key concepts from the perspective of Aboriginal law.

In Australia, the experience of the Australian Earth Laws Alliance is that the key concepts within Earth jurisprudence resonate very powerfully with people from the Western cultural tradition, as they help people from this culture reveal and 'unpack' the deeply flawed basis of Western industrial society's approach to objectifying, commodifying and using up the living world.[26] What is critically important for the future of Earth jurisprudence in Australia is whether the theory and practise can 'go deeper', and engage in a dialogue with First Nations Peoples.

As already noted, in *The Great Work*, Thomas Berry suggests that industrial society's laws, religious traditions, educational institutions and economic systems need to reflect and protect the biophysical realities of the Earth community. Earth jurisprudence advocates for a legal system with a number of critical elements; in this section we briefly examine four of the foundational concepts.

First, Earth jurisprudence acknowledges that the Universe is the *primary lawgiver*. In contrast to the current Western legal system that sees human laws as the highest authority for human society (and, implicitly, for all other life forms and ecological systems), Earth jurisprudence sees the laws of the Universe, the 'Great Jurisprudence' or 'Great Law' as providing the fundamental parameters of the Earth Community, including human societies.[27] The concept of 'the Great Law' is seen as problematic even by many Western scholars. It is argued that it's very difficult to clarify what the Great Law is, or to specify how people can live within those 'laws'.[28] Nonetheless, the idea that nature or the Universe is the primary lawgiver sees Earth jurisprudence as explicitly advocating for an idea of human societies living within the rules or limits of the natural world. In other words, it places human relationships with the nonhuman world as primary.

This concept of a Great Law being the primary law, and human laws needing to fit within the Great Law, is in some ways comparable to First Nations Peoples' ancient first law: that the land is the source of the law. However, as First Nations Peoples have lived by these first laws since time immemorial, and these laws are connected to *specific territories and regions* around Australia, it is difficult to compare them to the more 'general' rules articulated by Earth jurisprudence. Nonetheless, the idea of the Great Law is an important one to reintroduce to Western culture, as it begins the process of bringing Western people back to a sacred appreciation and respect for Mother Earth and can assist in the process of transforming the Western legal system.

A second foundational concept in Earth jurisprudence is the view that the Earth is an interconnected community and there is a *relationship-based existence* between humanity and the rest of the Earth Community. This contrasts with the current Western legal view that creates legal relationships among people, and between people and corporations, through constructs like property law, but commodifies and exploits all other aspects of the natural world.[29] By framing the natural world as a community, Earth jurisprudence imposes greater constraints on humanity's actions than our current legal system does. By claiming that 'the primary concern of the human community must be the preservation of the comprehensive community', Berry argued for a human world that works to ensure that all members of the Earth Community can thrive and continue their evolutionary journey.[30]

This concept of an 'Earth community' is in some ways comparable to the Aboriginal legal system, which is based on a relationist ethos. As outlined in Part II of this paper, First Nations people understand the deep connection and relationist basis of human existence. However, as noted above, it is difficult to compare general statements in Earth jurisprudence to the ancient, specific, deep connection that First Nations Peoples have in practise. However, inviting Westerners to see the world as being made up of relationships is an important way to challenge them to rethink the deepest beliefs in their culture.

Third, many advocates of Earth jurisprudence have argued that the Earth Community and all the beings that constitute it have rights, including the right to exist, to habitat or a place to be and to participate in the evolution of the Earth Community.[31] Berry argued that 'nature's rights should be the central issue in any … discussion of the legal context of our society'.[32] This view contrasts with the dominant approach in the Western legal system, which privileges and grants rights only to humans and selected human constructs, such as corporations or government entities. Granting rights to Nature is a radical rethinking of the role of our anthropocentric legal system, yet the idea appears to be taking hold in many jurisdictions.

Berry distinguishes the rights of Nature from other legal rights by saying they are 'analogous':[33] that is, these rights are already existent; they are not created by human law but rather are created by the very act of the Universe bringing forth its evolutionary processes. These rights of Nature come from the same source as human rights: the Universe itself.[34] Therefore, it is the work of Earth jurisprudence to develop and advocate for cultural, legal – and even spiritual – change that recognises these already existing 'rights' and provides legal consideration and protection of those rights.

This articulation of rights is one of the areas in which Earth jurisprudence diverges from any comparable concepts in Aboriginal law. As outlined above, Aboriginal laws are built upon a custodial ethic, the concept of reciprocity and obligation. While certain people within Aboriginal communities have a right to certain aspects of knowledge and law, rights-based laws and the notion of any living thing having a 'right' is not a conceptual element in the Aboriginal legal system. As Aboriginal law and ethics are based on relationships, and based on a perpetual obligation to look after land and each other, there is no way – and no need – to bring a Rights of Nature framework into Aboriginal law. Rights-based law is seen as a conflict-oriented approach to resolving different land use demands. This contrasts with the fundamental goal of Aboriginal societies, which is to create stability, harmony and balance.

A further, fundamental problem with 'rights' is that they are seen as a distinctly Western legal concept, which has emerged from the colonial project and is strengthened by the international legal system created by – and used against First Nations Peoples by – colonial powers.[35] However despite these conceptual clashes, around the world there are many indigenous peoples who are embracing the Rights of Nature concept and implementing or adopting Rights of Nature laws.[36] These groups see the Rights of Nature as a conceptual framework that can be used *within* Western law to challenge fundamental principles *of* Western law such as private property rights that permit environmental destruction.

The recent legal developments in New Zealand, India, Colombia and Bangladesh that have seen legislation and court cases recognise rivers, forests and other living entities as having 'legal personhood' can be differentiated from jurisdiction-wide Rights of Nature laws and are, in some cases, much closer to some of the concepts of interrelationships in Aboriginal law. For example, New Zealand's recognition of legal personhood status for the Whanganui River, Urewera Forest and Mount Taranaki, which has emerged from Settlement Agreements under the Treaty of Waitangi, is based on Maori Peoples' spiritual connection to these living ancestors and offers interesting and important connections between First Nations Peoples' laws and Western legal constructs.[37]

We see Rights of Nature laws and legal personhood laws not as ends in themselves, but as important building blocks between cultural and legal systems that can support a transformation of Western law, from human-centred to Earth-centred governance that is appropriate to the culture and the specific ecology of a place. Rights of Nature and legal personhood laws offer a way for Westerners to see their own legal system differently and to reconfigure Western laws in order to start building a new jurisprudence of Earth-centred, relationist law. Where appropriate, and with the permission and leadership of First Nations Peoples, such laws can also offer a legal and conceptual 'bridge' between First Nations and Western laws.

In Australia, we are seeing some First Nations Peoples beginning to use the 'framing' of Rights of Nature in documents, statements and initiatives. In 2016, First Nations Peoples in the region now known as the Kimberley created an historic declaration – the Fitzroy River Declaration – which sets out their intention to protect and manage the River, and it recognises that 'the River is a living ancestral being and has a right to life'.[38]

Bennett's wallaby (*Macropius rufogriseus*) on kunyani/Mount Wellington, Hobart, Australia. (Photograph by Mat Palmer. Adobe Stock, licensed by the Australian Earth Laws Alliance.)

In 2017, sixteen Aboriginal nations from across the northern Murray-Darling Basin signed a Treaty to work together and have a united voice on issues of importance to them.[39] The Treaty, known as the Union of Sovereign First Nations of the Northern Murray-Darling Basin, includes a pledge to uphold the 'rights of Mother Earth', as follows:

> [t]he rights of Mother Earth are upheld by all Nations …. And we pledge our commitment to ensuring 'respect' and preservation of her inalienable rights and all things natural. We acknowledge that these guarantees are the absolute inherent rights to the human condition.[40]

Also in 2017, the Victorian government passed the Yarra River Protection (Willip-gin Birrarung Murron) Act, which enables the identification of the Yarra River and the many hundreds of parcels of public land it flows through as one living, integrated natural entity for protection and improvement. While this legislation does not change the legal status of the river or explicitly refer to the Rights of Nature, its acknowledgment of the river as a living entity is important in Australian law, and it is the first time in Australia that Aboriginal language and custodial responsibilities are recognised in connection with and responsibility for this important waterway.[41]

A fourth, and important, element of Earth jurisprudence is the idea of Earth Democracy. Many advocates for the Rights of Nature embed Nature rights within a framework of 'Earth Democracy'. Earth Democracy has been defined as an attempt to fuse ecocentric ethics with deeper forms of human democracy and public participation.[42] It promotes the idea that all human and nonhuman life forms are borne of Earth, and as evolutionary companions, we all have a right to exist, thrive and evolve. In terms of *human* relationships, Earth Democracy is a concept that examines power, privilege and inequity, and rejects them in favour of the idea that all people have the right to their own self-determination, particularly when it comes to Earth stewardship within their local communities. It is important to recognise that under an Earth jurisprudence approach, human rights and 'democracy' are an interdependent and correlative subset of Earth rights; humanity cannot be healthy and secure if Earth is veering towards depletion and overextraction.

From a First Nations legal point of view, the idea of Earth Democracy is an important one for Westerners to embrace, and some elements of the concept resonates with Aboriginal law. As outlined above, Aboriginal law is built on deep respect for each other's autonomy; it is nonhierarchical, and governance structures exist to separate power and authority so that power is diffused throughout the community. Consequently, notions of Earth Democracy – particularly when articulated as including an obligation to care for the local environment and local place – can be seen as important for Westerners, so that they take responsibility for their impact on the country.

The idea that a small elite could control the rest of a community was unheard of in Aboriginal society. Everyone has an equal value in society, and everyone has an obligation to care for country and for one other. When a controlling ethic or spiritual basis emerges in a society that prioritises ego over community, or the individual over the collective, then the journey to societal and spiritual hierarchies begins, and unfairness and injustice follow.

It is the notion of Earth Democracy, and local communities' claims for ecological and social justice, that are driving some initial Rights of Nature campaigns in Australia. The Rights of Nature concept is receiving increasing attention from non-Indigenous communities in Australia, and to date, all non-Indigenous communities interested in the Rights of Nature are engaging with First Nations Peoples to explore how such concepts might (or might not) be used by people working together in a local region.

2018 saw a number of community-driven initiatives that reflect how Rights of Nature can capture peoples' imagination in a way that traditional environmental law does not, and how Rights of Nature framing and strategies can be embraced by people seeking to support Earth Democracy.

In March 2018, more than 100 local people rallied in support of the Margaret River in Western Australia.[43] Their signs, banners and strategy focused on giving the river its own voice and its own legal rights. One of the local advocates for rights of the river said that people 'understood the idea of recognising the river as a living entity, because we all know it's more than just a resource, it's alive and it has a right to exist'. (Personal communication, March 24, 2018.). In the Blue Mountains, community members concerned about threats to the Blue Mountains World Heritage Area have worked with the Australian Earth Laws Alliance (AELA) to draft a Local Council statement which they hope to gain support for, and to advocate for local recognition of the Rights of Nature in the Blue Mountains local council area.

AELA was also invited to develop model Rights of Nature laws for the world's largest coral reef community, the Great Barrier Reef. AELA drafted laws for all three levels of Australian government – a model law for Local Councils in the Great Barrier Reef Catchment, a State law recognising the Rights of the Reef and a proposed amendment for the Federal Constitution. These model laws are to demonstrate what is possible in Australia, and how Rights of Nature laws might be crafted to embrace Earth Democracy and recognition of First Nations Peoples' rights and obligations to care for land and sea country.[44]

## Conclusions

To conclude, we return to our questions posed at the beginning of this paper: Does the theory and practise of Earth jurisprudence have any relevance to First Nations

Peoples in Australia? Can it provide any assistance in building a sustainable future for all Australians?

From the analysis provided in this paper, it can be seen that Earth jurisprudence draws on and has some comparable theoretical concepts to those in First Nations Peoples' laws. The idea of a Great Law, the idea that we are interconnected with all life on Earth and some of the concepts connected to the ideas of 'Earth democracy' resonate to some extent with First Nations Peoples' laws.

However there are also significant differences. The notion of Rights of Nature does not have a comparable conceptual basis in First Nations laws, but it can be interpreted as a positive development in Western law, because it can bring Westerners into a different frame of mind when considering their place in the world and the impact of their current legal system. The growing use of 'legal personhood' constructs is also of value, as they try to represent some aspects of First Nations Peoples' relationist ethos and the laws of obligations within Western laws.

We conclude with the following three suggestions:

First, Earth jurisprudence offers a useful critique of Western law and industrial society's underpinning, anthropocentric belief system that has been a key driver of the current ecological crisis. It offers Westerners a way to understand how the current ecological crisis has emerged, how their culture has become disconnected from Nature and how they can start to respond personally and collectively to challenge the destruction that their culture has unleashed on the world and rebuild a relationist ethos.

Second, in Australia, Earth jurisprudence may offer a bridge between Western and First Nations Peoples' culture and law. If they so wish, First Nations Peoples can use the language of Earth jurisprudence and the Rights of Nature to facilitate conversations with Westerners about a relationist ethos, conversations that might otherwise be difficult to have in Australia's human-centred, colonial society. Earth jurisprudence can enable Westerners to see their own world differently, and in turn may help them be more open to understanding the distinct, ancient culture and law that have always existed and continue to exist in Australia, and the way that colonisation has ravaged this ancient culture since 1788.

Third, for Earth jurisprudence and Rights of Nature to be truly relevant to the Australian context, we need to work together – Aboriginal and non-Aboriginal people – to explore what these ideas mean in *this* place, in the various regions of *this* continent. Only together can we explore whether these ideas might, or might not, be useful in protecting our precious living world, and only together can we see if these approaches might play a part in disrupting the colonial project in Australia sufficiently to, as Irene Watson describes it, 're-affirm' the first laws of the first Peoples in Australia.

Old Aboriginal people have often stated that non-Aboriginal people in Australia 'have no Dreaming', that is, they have no collective spiritual identity, and no true understanding of having a correct or 'proper' relationship with land. Many non-Aboriginal Australians recognise this themselves and are working, planning and creating, quite often in dialogue with Aboriginal people, to change this situation. This paper is one of the many efforts to work together to discuss and build a truly sustainable future for all life on this amazing, ancient continent.

## Notes

1. Millennium Ecosystem Assessment. 2005. *Ecosystems And Well-Being: Synthesis.* Island Press.
2. Crutzen, P. and Stoermer, E. 2000. The 'Anthropocene' *Global Change Newsletter,* 41, p. 17.
3. Australian Earth Laws Alliance has given over 100 workshops, presentations and community engagement discussions since its creation in 2012, and every outreach activity has brought more people who are interested in and inspired by Earth-centred governance into the wider movement and local network. www.earthlaws.org.au
4. Berry often described himself as a 'geologian', as he studied the Earth rather than theology. See Cullinan, C. 2002, 2011. *Wild Law: A Manifesto for Earth Justice.* White River Junction, VT: Chelsea Green Publishing.
5. Berry, T. 1999. *The Great Work: Our Way into the Future.* New York: Bell Tower/Random House.
6. Ibid, p. 161.
7. Ibid, p. 182.
8. Ibid, p. 4.
9. Cullinan, C. 2002, 2011. *Wild Law: A Manifesto for Earth Justice.* White River Junction, VT: Chelsea Green Publishing, p. 103.
10. Ibid.
11. Swimme, B. and Berry, T. 1992. *The Universe Story: From the Primordial Flaring Forth to the Ecozoic Era – A Celebration of the Unfolding Cosmos.* New York: Harper Collins.
12. See, for example (a) Stone, C. 2010. *Should Trees Have Standing?: Law, Morality, and the Environment.* Third Edition. Oxford University Press; (b) Nash, R.F. 1989. *The Rights of Nature: A History of Environmental Ethics.* Madison, WI: University of Wisconsin Press; (c) Bosselmann, K. 1994. 'Governing the Global Commons: The Ecocentric Approach to International Environmental Law'. In: Prieur, M. and Doumbe-Bille, S. (eds.) *Droit De L'environnement et Developpement Durable.* Limoges.
13. Linzey, T., with Campbell, A. 2009. *Be the Change: How to Get What You Want in Your Community.* Layton, UT: Gibbs Smith.
14. Guth, J. 2008. 'Law For The Ecological Age'. *Vermont Journal of Environmental Law,* 431.
15. See, for example (a) Cullinan, C. 2002, 2011. *Wild Law: A Manifesto for Earth Justice.* Chelsea Green Publishing, White River Junction, VT; (b) Burdon, P. (ed.). 2011. *Exploring Wild Law: The Philosophy of Earth Jurisprudence,* Kent Town, South Australia: Wakefield Press; (c) Maloney, M. and Burdon, P. (eds.). 2014. *Wild Law – In Practice,* New York, NY and Oxford, UK: Routledge; (d) Rogers, N. and Maloney, M. 2017. *Law as if the Earth Really Mattered: The Wild Law Judgment Project,* Abingdon, Oxfordhsire, UK: Routledge; (e) La Follette, C. and Maser, C. 2017. *Sustainability and the Rights of Nature: An Introduction.* Boca Raton, FL: CRC Press; (f) Boyd, D. 2017. *The Rights of Nature, A Legal Revolution That Could Save the World.* Toronto, Canada: ECW Press.
16. For example, the Australian Earth Laws Alliance was created in early 2012 and was inspired by the work of Berry, T. and Cullinan, C. Other similar organisations include the Earth Law Center in the United States, Earth Advocates in Scotland and the Global Alliance for the Rights of Nature.
17. For a brief overview of Rights of Nature and legal personhood laws around the world, see Maloney, M. October 2018. 'Changing the Legal Status of Nature: Recent Developments and Future Possibilities'. *NSW Law Society Journal,* (49), pp. 78–79.
18. See for example, the Ecological Law and Governance Association (ELGA), https://www.elga.world/
19. Aboriginal people know they have lived in Australia since time immemorial. Current Western scientific knowledge says Aboriginal people have lived in Australia for at least 60,000 years – around 2,500 generations.
20. Watson, I. *Aboriginal Peoples, Colonialism and International Law: Raw Law.* Routledge, p. 22.
21. For an overview of how the term 'the Dreaming' entered the English language, see Nicholls, C. 23 January 2014. '"Dreamtime" and "The Dreaming": Who Dreamed Up

These terms?' *The Conversation*. https://theconversation.com/dreamtime-and-the-dreaming-an-introduction-20833 (Accessed 27 February 2019).
22. Ibid.
23. Ibid.
24. Ibid.
25. See, for example: https://www.abc.net.au/news/science/2017-03-09/dna-confirms-aboriginals-have-long-lasting-connection-to-country/8336284 (Accessed 27 February 2019).
26. Australian Earth Laws Alliance, http://www.earthlaws.org.au
27. Burdon, P. 2011. 'The Great Jurisprudence'. In: Burdon, P. (ed.) *Exploring Wild Law: The Philosophy of Earth Jurisprudence*. Kent Town, South Australia: Wakefield Press.
28. (a) Ibid; (b) Pelizzon, A. and Ricketts, A. 2015. 'Beyond Anthropocentrism and Back Again: From Ontological to Normative Anthropocentrism', *The Australasian Journal of Natural Resources Law and Policy*, 18(2), pp. 105–124.
29. (a) Cullinan, C. 2002, 2011. *Wild Law: A Manifesto for Earth Justice*. White River Junction, VT: Chelsea Green Publishing; (b) Graham, N. 2011. 'Lawscape: Property, Environment and Law'. *Journal of Environmental Law*, 23, p. 160.
30. Berry, T. 1999. *The Great Work: Our Way into the Future*. New York: Bell Tower/Random House, p. 80.
31. Berry, T. 2002. 'Rights of The Earth: We Need a New Legal Framework Which Recognises the Rights of All Living Beings.' In: Burdon, P. (ed.) *Exploring Wild Law: The Philosophy of Earth Jurisprudence*. Kent Town, South Australia: Wakefield Press.
32. Berry, T. 1999. *The Great Work: Our Way into the Future*. New York: Bell Tower/Random House, p. 80.
33. Tuff, N. and Berry, T. 2009. In: 'A Conversation with Thomas Berry', Berry says, 'Everything has rights. How could everything have rights? Well, it's an analogy. A tree needs tree rights. A bird needs bird rights …. To say that something exists is true, but not the same'. from *Minding Nature*, Winter 2009, 2(3). http://www.humansandnature.org/ (Accessed 27 February 2019).
34. Ibid.
35. Watson, I. *Aboriginal Peoples, Colonialism and International Law: Raw Law*. Abingdon, Oxfordhire, UK: Routledge, p. 7.
36. For example, see the decision by the Ho Chunk Tribe in the United States in 2016 to adopt Rights of Nature in their Constitution – https://celdf.org/2016/09/press-release-ho chunk-nation-general-council-approves-rights nature-constitutional-amendment/; also see the statement by the Ponca First Nations in 2018 – http://indigenousagain.com/first-tribe-u-s-recognizes-rights-nature-law/ (Accessed 27 February 2019).
37. For a brief overview of Rights of Nature and legal personhood laws around the world, see Maloney, M. 2018. 'Changing the Legal Status of Nature: Recent Developments and Future Possibilities', *NSW Law Society Journal*, (49), pp. 78–79.
38. See https://d3n8a8pro7vhmx.cloudfront.net/environskimberley/pages/303/attachments/original/1512653115/fitzroy-river-declaration.pdf?1512653115
39. See https://www.abc.net.au/news/2017–05-11/murray-darling-aboriginal-nations-sign-treaty/8518228
40. See http://nban.org.au/treaty/ For the relevant paragraph of the Treaty, see: http://nban.org.au/wp-content/uploads/2017/05/page-4.pdf (Accessed 15 March 2019).
41. For details, see: https://www.planning.vic.gov.au/policy-and-strategy/waterways-planning/yarra-river-protection (Accessed 19 March 2019).
42. Burdon, P. 2014. 'Wild Law and the Project of Earth Democracy'. In: Maloney, M. and Burdon, P. (eds.). *Wild Law – In Practice*. Abingdon, Oxfordhsire, UK: Routledge.
43. Gleeson-White, J. 2018. 'It's Only Natural: The Push to Give Rivers, Mountains and Forests Legal Rights'. *The Guardian*, March 31, 2018. https://www.theguardian.com/australia-news/2018/apr/01/its-only-natural-the-push-to-give-rivers-mountains-and-forests-legal-rights (Accessed 19 March 2019).
44. See https://rightsofnature.org.au/rightsofthereef/ (Accessed 15 March 2019).

# 20

## *Conclusion: Nature's Laws of Reciprocity*

**Chris Maser**

**CONTENTS**

Nature's Laws of Reciprocity ........ 402
Law 1: Everything Is a Relationship ........ 402
Law 2: All Relationships Are All Inclusive and Productive of an Outcome ........ 402
Law 3: The Only True Investment in Our Global Ecosystem Is Energy from Sunlight ........ 402
Law 4: All Systems Are Defined by Their Function ........ 402
Law 5: All Relationships Result in a Transfer of Energy ........ 402
Law 6: All Relationships Are Self-Reinforcing Feedback Loops ........ 403
Law 7: All Relationships Have One or More Trade-offs ........ 403
Law 8: Change Is a Process of Eternal Becoming ........ 403
Law 9: All Relationships Are Irreversible ........ 403
Law 10: All Systems Are Based on Composition, Structure and Function ........ 403
Law 11: All Systems Have Cumulative Effects, Lag Periods and Thresholds ........ 404
Law 12: All Systems Are Cyclical, but None Are Perfect Circles ........ 404
Law 13: Systemic Change Is Based on Self-Organized Criticality ........ 404
Law 14: Dynamic Disequilibrium Rules All Systems ........ 404
Law 15: This Present Moment, the Here and Now, Is All We Ever Have ........ 404

Restoration, as it is generally thought of, helps us to understand how a given ecosystem functions. As we strive to put it back together by reconstructing the knowledge of times past, we learn how to sustain the system's ecological processes and its ability to produce the products we valued it for in the first place and some generation might be able to value it for again in the future.

Similarly, restoration helps us understand the limitations of a given ecosystem or a portion thereof. As we slow down and take time to reconstruct a functional likeness of what was, we learn how fast we can push the system to produce products on a sustainable basis without impairing the system's ability to function sustainably.

Thus, the very process of restoring the land to biophysical well-being is the process through which we become attuned to the Rights of Nature and – through those Rights – to ourselves, as evidenced from the foregoing chapters. Restoration, in this sense, is both the means and the end, for as we learn how to repair the processes of the land, we heal its ecosystems, and as we heal the ecosystem, we heal the deep geography of ourselves. Simultaneously, we also restore both our options for products and amenities from the land and those of future generations. This act is crucial because our ethical obligation, as human trustees of Planet Earth, is to maintain the welfare of all life – present and future. To this end, maintaining healthy, viable ecosystems is an expression of the heart and the spirit of taking care of the Earth as a biological living trust. I use the word *spirit* on purpose, because it is derived from the Greek word for breath, which denotes life.

DOI: 10.1201/9780429505959-20

To achieve the level of consciousness and the balance of energy necessary to maintain the sustainability of ecosystems, we must focus our questions – social and scientific – toward understanding Nature's Laws of Reciprocity inherent in the governance of those systems – and our place within that governance. Then, with humility, we must develop the moral courage and political will to direct our personal and collective energy toward living within the constraints defined by Nature's Laws of Reciprocity – not by our economic/political ambitions – to honour the Rights of Nature.

## Nature's Laws of Reciprocity

Each law is an interactive strand in the multidimensional web of energy interchange that constitutes the universe, our world within it and how we affect our world.

### Law 1: Everything Is a Relationship

The universe is an all-inclusive relationship constituted of an ever-expanding web of biophysical feedback loops, each of which is novel and self-reinforcing. Each feedback loop is a conduit whereby energy is moved from one place, one dimension and one scale to another. And, all we humans do – ever – is practise relationships with the flow of energy within this web.

### Law 2: All Relationships Are All Inclusive and Productive of an Outcome

Every relationship is productive of a cause that has an *effect*, and the effect, which is the cause of another effect, *is* the product – but often contrary to human-desired, commodity-oriented, monetary outcomes.

### Law 3: The Only True Investment in Our Global Ecosystem Is Energy from Sunlight

The *only true investment* in the global ecosystem is energy from solar radiation (materialised sunlight). Everything else is merely the recycling of already-existing energy. In a business sense, for example, one makes money ('economic capital') and then takes a percentage of those earnings (after the fact) and recycles them into the infrastructure to facilitate making a profit by protecting the initial capital outlay. 'Biological capital', on the other hand, must be 'recycled' *before* the profits are earned, which means leaving enough of an ecosystem intact for it to function in a productively sustainable manner.

### Law 4: All Systems Are Defined by Their Function

The behaviour of every system depends on how its individual parts interact as functional components of the whole, not on what an isolated part is doing. The whole, in turn, can only be understood through the relationships, the interaction of its parts and its interdependent relationship to everything else.

### Law 5: All Relationships Result in a Transfer of Energy

Although a 'conduit' is technically a hollow tube of some sort, the term is used here to connote any system employed specifically for the transfer of energy from one place to another. Every living thing, from a virus to a bacterium, fungus, plant, insect, fish, amphibian, reptile, bird, mammal and every cell in our body, is a conduit for the collection, absorption, transformation, storage, transfer and expulsion of energy. In fact, the function of the entire biophysical system is tied up in the collection, absorption, transformation, storage, transfer and expulsion of energy – one gigantic, energy-balancing act, or perhaps more correctly, 'energy-juggling act'.

### Law 6: All Relationships Are Self-Reinforcing Feedback Loops

Everything in the universe is connected to everything else in a cosmic web of interactive feedback loops, all entrained in self-reinforcing relationships that continually create novel, never-ending stories of cause and effect. Everything, from a microbe to a galaxy, is defined by its ever-shifting relationship to every other component of the cosmos. Thus, 'freedom' (perceived as the lack of constraints) is merely a continuum of fluid relativity – precluding absolute freedom of anything, despite political/legal proclamations to the contrary.

### Law 7: All Relationships Have One or More Trade-offs

All relationships have trade-offs that may or may not be readily apparent or immediately understood. Each trade-off is couched in terms of a decision to change or not to change, largely on personal values. Thus, the consequences of today's decisions, be they wise or foolish, are passed forward, as an irreversible legacy of pending trade-offs, to all the generations of the future.

### Law 8: Change Is a Process of Eternal Becoming

Change is a universal constant – a *continual process that produces inexorable novelty.*

### Law 9: All Relationships Are Irreversible

Because change is a constant process orchestrated along the interactive web of universal relationships, it produces infinite novelty that precludes anything from ever being reversible.

### Law 10: All Systems Are Based on Composition, Structure and Function

We perceive objects by means of their obvious structures and/or functions. Structure is the configuration of elements or parts, be it simple or complex. The organisation or arrangement of a thing defines its structure. Function, on the other hand, is what a particular structure either can do or allows to be done to it, with it or through it. The characteristics governing social–environmental sustainability and the Rights of Nature that we must be concerned with are: (1) composition, (2) structure, and (3) function.

Composition determines the structure, and the structure determines the function. Thus, by negotiating the composition, we simultaneously negotiate both the structure and the function. On the one hand, once the composition is in place, the structure and the function are set – unless, of course, the composition is manipulated for a particular purpose, at which time both the structure and the function are altered accordingly. Nature's disturbance regimes, on the other hand, periodically alter an ecosystem's composition, structure, and function, despite our human desires or intentions.

### Law 11: All Systems Have Cumulative Effects, Lag Periods and Thresholds

Nature has intrinsic value only and so allows each component of an ecosystem to develop its prescribed structure, carry out its biophysical function, and interact with other components through their evolved, interdependent processes and self-reinforcing feedback loops. No component is more or less important than another – except in human valuation based on personal desire for a particular outcome. Though components may differ from one another in form, all are complementary in function.

Our intellectual challenge in decision-making (legal and otherwise) is to recognise that no given factor can be singled out as the sole cause of anything. In other words, cumulative effects, gathering themselves below the level of our conscious awareness (lag period), seem to suddenly cross a threshold and become visible. By then, it is too late to retract our decisions and actions, even if the outcome they cause is decidedly negative with respect to our intentions.

### Law 12: All Systems Are Cyclical, but None Are Perfect Circles

While all processes in Nature are cyclical, no cycle is a perfect circle. They are, instead, a coming together in time and space at a specific point, where one 'end' of a cycle approximates – but only approximates – its 'beginning' in a particular time and place. Between its beginning and its ending, a cycle can have any configuration of happenstance. Biophysical cycles can thus be likened to a coiled spring insofar as every coil approximates the curvature of its neighbour but always on a different spatial level (temporal level in Nature), thus never touching.

### Law 13: Systemic Change Is Based on Self-Organized Criticality

Large, complicated, interactive systems evolve naturally to a critical state in which even a minor event starts a chain reaction that can affect any number of internal elements and lead to a dramatic alteration in the system. Although such systems produce more minor events than catastrophic ones, chain reactions of all sizes are an integral part of system dynamics. Not understanding this, analysts have typically and erroneously blamed some rare set of circumstances (some exception to the rule) or some powerful combination of mechanisms when catastrophe strikes.

### Law 14: Dynamic Disequilibrium Rules All Systems

If change is a universal constant in which nothing is static, what, then, is a natural state? In answering this question, it becomes apparent that the 'balance of Nature' in the classical sense (disturb Nature and Nature will return to its former state after the disturbance is removed) is an illusion. In reality, Nature exists in a continual state of ever-shifting 'dis-equilibrium'. Ecosystems are always in an irreversible process of change and novelty, thereby altering their composition, structure, function and the resultant interactive feedback loops – irrespective of human desires and a false sense of control.

### Law 15: This Present Moment, the Here and Now, Is All We Ever Have

This eternal, present moment is all we ever have in which to act. The past is a memory, and the future an illusion that never comes. *Now* is the eternal moment. Nature's inviolable Laws of Reciprocity are enshrined in every moment of everyday life, whether we recognise them or not. As such, the person who honours them, while caring about the consequences of their decisions on future generations, will move forward honouring the Rights of Nature.

# *Index*

*Note*: Page numbers followed by *"n"* with numbers indicate notes.


**A**

Abandoned shrine in Egiri, 178
Aboriginal law and ethics, 386, 388
  Earth jurisprudence from perspective of, 392
  land, 389–390
Absolute property rights, 322
Act of the Input of Power, *see* *Stromeinspeisungsgesetz*
Adibe Lakes, 171–172
Advaita, 375
Advanced digital devices, 53
*Adventures of Pinocchio, The* (Collodi), 43
AELA, *see* Australian Earth Laws Alliance
African Convention on Conservation of Nature and Natural Resources, 182
African oil palm (*Elaeis guineensis*), 61
African teak, 173
Agenda for Growth and Jobs (2020), 314
Agenda for Sustainable Development Goals, 121
Agreement on Joint Regulation of Fauna and Flora in Lake Chad Basin, 182
Agricultural/agriculture, 60, 206
  genetics, 120
  salting of soil, 61
Air pollution, 285
Alder trees, 149
Alpha Centauri star system, 47
Alternative Nobel Prize, *see* International Right Livelihood Award
American legal system, 5
American Prairie Reserve Project, 9
*Amita Gautam Poudel vs. Prime Minister and Office of Council of Ministers and others*, 301
*Amour de soi-même*, 24
*Amour proper*, 24
Angelica, 153
Anglo-American traditions, 7–8
Animated Earth, 205–206
Animation, 41–43
Animism, 39
  impact on current collective mentality, 50–51
  in law, 51–52
Anthropocentrism, 387
Ants, 219
Apartheid South Africa, laws of, 30–31
Apex predators, 158
Arsenic, 4
Artificial inorganic fertilisers, 69
Ash, 153
Asia Development Bank, 133
Aspen hoverfly, 151
Assamese macaque, 126
Atlantic salmon, 158
Atlikimma cycle, 102
Australia
  Aboriginal Law and Ethics, 388–392
  Earth jurisprudence and Rights of Nature, 386–388
  Earth jurisprudence from perspective of Aboriginal Law, 392–396
Australian Earth Laws Alliance (AELA), 386, 396, 398*n*3
Auteur Cinema, 48–50
Autonomy, 391–392
*Aziza* spirit, 172

**B**

Bagmati River pollution, 300
Baltic Marine Environment Protection Commission (HELCOM), 240, 242
Baltic Sea, 239, 242, 247
  European Union Waters and, 239–240
*Bambi, a Life in the Woods* (Salten), 42
Bandeirantes, 216
Bark peelers, 103
BBC, 50
Bears, 103
Beetles, 149
Bennett's wallaby (*Macropius rufogriseus*), 395
Berry patches, 100
BFL, *see* Bhutan For Life
Bhutan, 113
  biodiversity, 114–115
  Gross National Happiness, 117
  yartsa goenbub, 115
Bhutan Biological Conservation Complex, 115
Bhutan For Life (BFL), 124–125

Bhutan King Jigme Singye Wangchuck, 117
Bhutan National Human Wildlife Conflicts Management Strategy, 126
Bill of rights for Lake Erie, 338–340
Bio-economy fosters techniques, 320
Bioaccumulation, 348
Biocentric approach, 283
Biocultural approach, 289
Biodigesters case, 285–286
Biodiversity, 313
Biological capital, 402
Biological diversity, 120
Biophilia, 23
Biosphere, 325–326
Biosphere Ethics Initiative, 15, 19
Biseni indigenes, 171
Black grouse, 149–150
Blue-winged kookaburras (*Dacelo leachii*), 390
Bog myrtle, 149
Bog woodland, 142
Borrow pits dug, 84
Boupere Lake, 172, 177–178
Bovines, 222
*Brilliant Green: the Surprising History and Science of Plant Intelligence* (Viola), 50
British Columbia, rights for, 346–349
*Bundesverfassungsgericht*, 193

C

Cadmium, 4
Cairngorms National Park in Scotland, 153
Calcium carbonate, 238
Caledonian Forest, 141–142
  deepening restoration process, 151–154
  founding of Trees for Life, 144–147
  historical background, 142–144
  initial steps in forest restoration, 144
  native pinewoods, 142
  reconnecting people with nature, 154–156
  reweaving web of life, 147–150
  Rights of Nature, 156–161
Calypso orchid, 158
Camelids, 214
CAP, *see* Common Agriculture Policy
Capercaillie, 150
Carbon neutrality, 120
Care, 29
*Carmina Gadelica* (Carmichael), 7
Cattle
  explosive dissemination of, 216–217
  in Llanos and Pantanal, 223–224
CBD, *see* Convention on Biological Diversity
CDU, *see* Christian Democratic Party
Cedar, 101–104
Celtic lands, 7
Central Pollution Control Board, 371
Champ (*Michelia champaca*), 122
Change as eternal becoming process, 403
Charity Moor Trees, 155
Charles Russell National Wildlife Refuge and tribal lands in Montana, 9
Chauvet Cave in France, 6
*Chocolat* (film), 49
Christian Democratic Party (CDU), 194
Christian Social Party (CSU), 194
*Chronicles of Narnia, The* (film), 41, 43
Cinema, Nature's intelligence portraying in, 40–41
CITES, *see* Convention on International Trade on Endangered Species of Wild Fauna and Flora
City dwellers, 51
Civilisation, 22
Civil rights, 284
Claims settlement process, 264–266
Clam gardens, 100
Clams, 98–101
Clean-up in Nigeria, 169–170
Clean Bhutan, 130
Clear-cutting forests, 61
Climate change, 79
  crisis of, 80–82
  Kiribati's response to, 82–85
  uncertainties for National sovereignty, 85–86
Climatic factors
  in Llanos, 208
  Pantanal, 211–212
Coal Mines Amendment Act (1903), 277*n*26
Commercial films
  designed for older children, 43–45
  intended for adults, 45–48
Commercial redress, 266
Common Agriculture Policy (CAP), 157
Common Fisheries Policy, 240, 248
Commonwealth Scientific and Industrial Research Organization (CSIRO), 62
Community, 30, 60
Complementarity, principle of, 282
Composition, 403
Comprehensiveness, principle of, 286
Comprehensive theory, 6
Conservation, 30
  awards received in Bhutan, 125
  challenges, 131–133
  history, 115–116

policies and laws, 118–120
protected areas of Bhutan, 116
strategy, 175–176
weeks, 154
Constitution, 282
Constitutional approach, 284
Constitutional Court of Ecuador, 180, 286, 288
Constitutional jurisprudence, 283
Constitutional principles, 283
Contracts clause, 336
Convention on Biological Diversity (CBD), 150, 182–183
Convention on International Trade on Endangered Species of Wild Fauna and Flora (CITES), 182
Cordyceps, *see* Yartsa goenbub (*Ophiocordyceps sinensis*)
Corporate constitutional rights, 335–336
Corporate state, 334–337
*Cosmopolitanism*, 21
Cost-shifting, 319
Country's education system, 160
Crabapple (*Malus fusca*), 103
Crabapple groves, 100
Creator, 95
Creole breeds, 223
Creole cattle, 226
Crystal Spring ecosystem, 341
CSIRO, *see* Commonwealth Scientific and Industrial Research Organization
CSU, *see* Christian Social Party
Cultural ecologists, 94
Cultural keystone' species, 100
Cumulative effects of system, 404
*Cupressus corneyana, see* Medicinal cypress
Custodial Ethic, 391
Cyanobacteria blooms, 240
Cyclone Pam, 82
*Cymbidium erythraeum, see* Edible orchid

D

*Dacelo leachii, see* Blue-winged kookaburras
Dam removal, 10–11
*Dances with Wolves* (film), 41, 46
DANTAK, 132
Dated Governance Paradigm, 250
Decision-making, rights of Nature affecting criteria for, 246–248
Declaration for the Rights of Mother Earth, 18
Deforestation problem, 61
DEP, *see* Department of Environmental Protection
Department of Environmental Protection (DEP), 341
Deregulatory phase in environmental law, 314
Desertification, 120
Desuung initiative, 131
Development model, 282
DHI, 123
Dillon's Rule, 336, 346
Directive Principles of State Policy, 370–371
'Disrespectful' engagement, 104
Diversity, 72
Domain of inquiry, 16
Dormant commerce clause, 336
Downy willow, 152
*Drafting a Code of Ethics for Biodiversity Conservation*, 20
'Dreaming' concept, 389
Dredging in Nigeria, 169
*Dred Scott*, 335
Druk Holding and Investment Corporation, 122
Dryads, 39–40
Due process, 367
Duty of Care, 326
Dvaita, 375
Dwarf birch, 152
Dynamite fishing, 176
Dzongs, 122

E

Earth, 326
community, 205–206
democracy, 395–396
ecosystem, 249
functions, 64
system, 385
Earth Jurisprudence, 326
in Australia, 386–388
from perspective of Aboriginal Law, 392
Earthworms, 219
EC, *see* European Commission
Ecclesiastical law, 333
ECNA, *see* Federal Ethics Committee on Non-Human Biotechnology
Ecological
balance, 153–154
crisis, 385, 387
diversity, 206
reconciliation, 25
relationships, 158
reserve, 286
trauma, 4
*Ecological Conscience, The* (Leopold), 30

*Ecological Governance:Toward a New Social Contract with the Earth* (Jennings), 28
Ecology, 19
Economic capital, 402
Ecosystems, 404
Ecotourism, 124
Ecuadoran Constitution, 19
Ecuadoran constitutional case law, 284
Ecuadoran legal system, 283
Ecuadoran mining authority, 287
Ecuador's legal system, 280
Edges, 73
Edible orchid (*Cymbidium erythraeum*), 121
EEG, *see* Erneuerbare-Energien-Gesetz
Ego, 392
Egophilia, 23
EIA, *see* Environment Impact Assessment
Ekman spiral, 70
*Elaeis guineensis, see* African oil palm
*Elements of the Philosophy of Right*, 18
Elk populations, 10
Elwha Dam, 11
Empathy, 29
Endangered Species (Control of International Trade and Traffic) Act, 182
Enforceability of substantive laws, 315
Enlightenment, 28
Environment(al)
  anthropologists, 94
  in court, 133
  degradation impact on indigenous sites, 176–178
  federalism in Nepal, 301–302
  impacts, 247, 280
  law, 317
  problems in Nepal, 297–298
  protection, 371
  quality, 176
  sector, 133
Environment Act (1986), 371
Environmental Defenders Office of Northern Queensland, 250
Environmental Performance Index, 398
Environmental sustainability, 176
  BFL, 124–125
  Bhutan and biodiversity, 114–115
  Bhutan's international commitments and national policies, 120–121
  conservation awards received in Bhutan, 125
  conservation challenges, 131
  conservation history, 115–116
  conservation milestones in Bhutan, 123–124
  conservation policies and laws, 118–120
  Gross National Happiness philosophy, 118
  guardians of peace and harmony, 131
  human-wildlife conflict, 126–128
  legacy of monarchs in sustainable environmental conservation, 116–118
  medical fungus problem, 128–130
  national planning and institutional frameworks, 121–123
  sustainable conservation, 126
  waste management and sustainable landscape, 130
Environment Impact Assessment (EIA), 299
Environment Protection Act (1997), 299, 302, 304
Environment Protection Council, 299
Environment Protection Fund, 299
Environment Protection Rule (1997), 299, 302
Erneuerbare-Energien-Gesetz (EEG), 194, 195
Esiribi Lakes, 171–172
Estuarine 'root gardens', 100
Estuary, 68
Ethical justification for Rights of Nature, 13, 15
  extending ethical roots of justice to roots, 16–18
  foundational principles, 24–25
  injustices, 29–31
  living ethics for living earth, 18–19
  motivations, 23–24
  rights of nature, 25–27
  rooted cosmopolitanism approach to law and life, 19–23
  *Sapere aude!*, 27–29
*Ethics of Identity, The*, 21
EU, *see* European Union
*Eucalyptus regnans, see* Mountain ash
Eulachon, 98–101
Eurasian lynx, 158
European aspen, 151
European beaver, 151
European brown bear, 158
European Commission (EC), 178
European conquest of new world, 215–216
  explosive dissemination of cattle, 216–217
  native grass *vs.* non-native grass, 217–219
European Parliament, 312
  Marine Strategy Framework Directive, 242
European traditions, 7–8
European Union (EU), 239, 311, 315
  environmental policy, 311
  Habitats and Birds Directive, 142, 150
  LIFE Fund, 150
  Sustainable Development Agenda, 314
  waters, 239–240
European Union Biodiversity Strategy, 313

European Union Environmental Law, 311
  Baltic Sea and surrounding countries, 239
  biosphere, 325–326
  challenges in implementation of, 314
  economics, 323–324
  green economy, 319–321
  Hierarchy of Rights, 325
  human right, 324–325
  law creating conditions, 321–323
  legal and institutional frameworks, 240–243
  Nature's rights legal framework, 327
  need for systemic approach, 313–314
  neoliberalism, economic growth and structure of law, 318
  and policy failing, 311–313
  sustainability models, 324
  systemic problems with environmental law, 315–318
Evolutionary processes, 73, 283
Evolving Biosphere Ethic, 20–23
Explosive dissemination of cattle, 216–217

F

FAO, *see* Food and Agriculture Organization
Fauna
  in Llanos, 210–211
  in Pantanal, 213
Faunus, 40
FDP, *see* Free Democratic Party
Feature Animation Studio, 42
Federal Ethics Committee on Non-Human Biotechnology (ECNA), 179
Federal Government of Nigeria, 181
Financialisation, 319
Findhorn Community, 145
Fire
  in Llanos, 209–210
  Pantanal, 212–213
Fisheries policies, 245–246
Fishing, 171
Fitzroy River Declaration, 394
Flies, 149
Flooded savannas, 219
Floods, 212
Flora, 40
Flow, 219
Flowering plants, 153
Food and Agriculture Organization (FAO), 249
Food forestry, 72
Foreshore and Seabed Act (2004), 277*n*32
Forest and Nature Conservation Act, 124
Forest bathing, 51
Forest Practices Act, 345
Forest restoration, 144
Forestry and National Park laws, 182
Forestry Commission Scotland, 144, 150, 155
Forestry Grant Schemes, 150
Formation water, 169
Forty-Second Amendment, 370
Foundational principles, 24–25
*Fox and the Child/Le renard et l'enfant, The* (film), 45
Free Democratic Party (FDP), 193
Friendliness, 30
Functional ecosystem, 149
Fundamental rights, 367

G

Gaea, 40
*Gaelic Otherworld, The* (Campbell), 7
Gaia, 205–206
Galapagos Marine Reserve of Ecuador, 250
Gallery forests, 208
Ganga Action Plan (GAP), 369
Ganga Management Board, 377–378
Ganga Pollution Case, 369
Ganga River, 366
GAP, *see* Ganga Action Plan
Gas flaring in Nigeria, 169
GDP, *see* Gross Domestic Product
GECC, *see* Gewog environmental conservation committee fund
Geriatric woodlands, 143
German constitution, 193
German *Energiewende*, 192, 195–201, 196
German energy system, 196
German Federal Constitutional Court, 193
German Parliament (*Deutscher Bundestag*), 192
Gewog environmental conservation committee fund (GECC), 128
Gewog of Lunana, 128
Gilbert Islands, *see* Republic of Kiribati
*Gilgalis*, 96–97, 104–105
Glacial lake outburst floods (GLOF), 131
Glines Dam, 11
Global capitalism, 322
Global ecological crisis, 385
Global Ocean Refuge System evaluation (GLORES evaluation), 246
GLORES evaluation, *see* Global Ocean Refuge System evaluation
Godavari Marble 1 case, 299–300
  legacy, 302–303

Godavari Marble 2 case, 303
facts of case, 304
holding, 305
issues, 304–305
Gold mining with mercury amalgamation, 220
Good Mind, 19
Grassroots system change
courts and system change, 334–335
courts prohibiting lawmaking, 335–336
recognising Rights of Nature, 337
Grazing, 157
in Llanos and Pantanal, 221–222
Great American Biotic Interchange, 213–214
Great Depression, 319
Great Forest Spirit, 48
Great Law, 393
Great Plains, 9
*Great Work, The* (Berry), 392
Green Bench, 133
Green economy, 319–321
Greener Ways and ReCiTi, 130
Green revolution, 61
Greens (*Die Grünen*), 194
Green socio-economic development, 120
Green tax, 124
Greenwashing, 8
Grey squirrels, 153
Grizzly Bear in Yellowstone National Park, USA, 48
Gross Domestic Product (GDP), 319
Gross National Happiness philosophy, 113, 115, 118
Grundgesetz, 193
Guardianship
overlapping institutions and politics in, 377–379
problem, 374–375
*Guardians of the Galaxy* series, 44
*Gwa'gwayems*, 97

**H**

Hamadryades, 40
Hapū (subtribe), 275*n*2
Harbour seals, 101
Hardwood tree species, 153
*Harry Potter* series of eight motion pictures, 44
Hawke's Bay, 267
Hawkes Bay Regional Planning Committee Act (2015), 277*n*21
Hazel, 153
Healthy kelp forests, 10
HELCOM, *see* Baltic Marine Environment Protection Commission
Heritage Lottery Fund, 150
*Hethla'tusla*, 98
*Hidden Life of Trees, The* (Wohlleben), 50
Hierarchy of Rights, 325
Highland biting midge, 158
Highland glens, 143, 149
Highlands of Scotland, 144
Highland Township Ordinance, 340–342
Himalayan black bear, 126
Hogweed, 153
Home Rule Charter, 340–342
Human exceptionalism, theory of, 8
Human factor in Llanos and Pantanal, 224–225
Humanity, 59, 90
Human landscape, 70
Human rights laws, 14
Human systems, natural systems with, 60
Human-wildlife conflict, 126–128
Human Wildlife Conflict Management Endowment Fund, 128
Human Wildlife Conflict SAFE Strategy, 128
Hydraulic fracturing, 323
Hydrocarbon exploitation, 280
Hydropower, 124

**I**

*Idea of Justice, The*, 17
Illusory effects, 319
Inanimate objects, 333
India
Ganga and Yamuna rivers as persons, 372–379
Indian legal system, 370
National Green Tribunal, 369–372
precursors of Nature's Rights in, 367–368
Rights of Nature approach in, 366–367
Indigenous conservation
culture, 178–180
laws, 180–182
Individualism, 95
Industrialisation, 8
Industrialised modern society, 79
Infiltration, 219
Infinite injury, 30
*'Infinitos'*, 215
Injustices, 29–31
Inorganic heavy metal pollution, 4
*Instinct* (film), 46
Instrumental rationality, 30
*Intelligence of the Substance, The* (Dulcan), 50–51

Intelligent robots, 53
Intergovernmental Panel on Climate Change (IPCC), 80
International Charter of Nature's Rights, 53
International Right Livelihood Award, 65
International Union for Conservation of Nature (IUCN), 15, 249
Introduced grasses, 219
  in Llanos and Pantanal, 222
IPCC, *see* Intergovernmental Panel on Climate Change
*Ishmael* (novel), 46
IUCN, *see* International Union for Conservation of Nature
Iwi (tribe), 275*n*2

J

Jigme Singye Wangchuck National Park (JSWNP), 127
Joint Nature Conservation Committee, 144
Jomolhari Mountain range from Chelala, 114
J. Paul Getty Conservation Leadership Award, 125
*Juglan regia, see* Walnut
Juju priests at River Ethiope Source Umuaja, 175
*Jungle Book* (film), 41–42
Justice, 17, 30
  system, 16

K

'Keeping it living' concept, 99–100
Keystone species, 100
Killer whales, 101
Killing, 95
Kiribati, 79
  Island Nation of, 79–80
  Makin, Kiribati, 81
  map of Kiribati relation to Pacific States, 80
  Mexico Village, Makin, Kiribati, 84
  response to climate change, 82–85
*Kodama*, 48
Kwakwaka'wakw, 90, 104
  clans and chiefs, 94
  mountains, 96–98
  Orca and Cedar, 101–104
  perspectives, 90
  relationships with natural world, 92–96
  salmon, eulachon, clams and plants, 98–101
  territory, 91
  wolves, 96–98
*Kwaxsistalla*, 108*n*4
*Kwelth'esta*, 98
Kyoto Protocol, 286

L

Lag periods of system, 404
Lake Erie, Bill of Rights, 338–341
Land, 388–390
  community, 205–206
  land-based pollution, 246
Land and Agrarian Development Decree, 225
Land use zoning system, 160
Large-leaved dense-canopied species, 153
Law creating conditions, 321–323
Lawmaking, 335
  contracts clause, 336
  corporate constitutional rights, 335–336
  Dillon's Rule, 336
  dormant commerce clause, 336
  Nature as property, 33
  pre-emption as ceiling on rights protections, 336
Least Developed Country, 121
Legacy of Godavari Marble 1 case, 302–303
Legal framework, 279
Legal realism, 322
Legal systems, 289
Liberty of contract, 334–335
Lincoln County, Oregon, ordinance against aerial spraying, 342
  map of Salish Sea and surrounding basin, 347
  Middle Siletz River watershed, Lincoln County, Oregon, 342–346
  Rights for Salish Sea, Washington State and British Columbia, 346–349
Litigation, 90
Livestock management practices, 223
Livestock production, 206
Living beings, 93
Living ethics for living earth, 18–19
Llanero descendants, 224–225
Llanos, 206–208
  climatic factors, 208
  fauna, 210–211
  fire, 209–210
  soils, 208–209
  sustainable management of, 219–225
Local democracy, 334–337
Local Governance Sustainable Development Program, 119
Locality, 391
Locaphilia, 23
*Lochner* Era, 335

*Lochner v. New York*, 335
*Lord of the Rings* (film), 41, 44
*L'ours* (film), 41
Love, 23–24, 28
*Luxiwey* in Kwak'wala, 100

M

Macrocosm, 321
*Macropius rufogriseus, see* Bennett's wallaby
Mahogany, 173
*Malus fusca, see* Crabapple
Management Council in Bering Sea, 246
Manas Wildlife Sanctuary, 115
Mangroves case, 286–287
Māori customary law, 263
Māori Land Wars, 275*n*5
Māori property rights, 269–270
Marine Coastal Act (2011), 277*n*33
Marine Strategy Framework Directive, 242
Materialism, 50
Medical fungus problem, 128–130
Medicinal cypress (*Cupressus corneyana*), 118
Mehta I, 369
Mehta II, 369
Mercury, 4
*Mesosetum* spp, 222
*Michelia champaca, see* Champ
Microcosm, 321
Middle Income Country, 121
Middle Siletz River watershed, Lincoln County, Oregon, 342–346
Migrants, 176
'Millennium Forest for Scotland' project, 150
Mines and Minerals Act, 302
Mites, 149
Modern Western resource management, 106
'Modulos de Apure', 221
*Mohd Salim v. State of Uttarakhand and Others*, 377
Monarch of Kingdom of Bhutan, 116
Montane scrub, 142
Montane willow shrubs, 152
Morality, 22
Mother Earth, *see* Pachamama
Moths, 149
Mountain ash (*Eucalyptus regnans*), 388
Mountains, 96–98
*Muppet Show, The*, 45
Mutual enhancement, 326–327
Mutual prey species, 101
Mycelial network, 74
Mycorrhizal partnership, 149

N

Namami Gange project, 377
Namche Bazaar, 298
Nanotechnology, 320
Narphung valley under samdrupjongkar dzonkha, 132
National Biodiversity Centre, 121
National Center for Ecological Analysis and Synthesis (NCEAS), 238
National Environment Commission (NEC), 121, 132–133
National Environment Protection, 133
National Forest Policy, 120
National Ganga River Rights Act, 372
*National Geographic*, 50, 53
National Green Tribunal, 366, 369, 372
  provisions close to Rights of Nature, 370–371
  standing and other procedural matters, 371–372
  Supreme Court, 369–370
National Oceanic and Atmospheric Administration (NOAA), 242, 249
National Parks Act, 182
National planning and institutional frameworks, 121–123
National sovereignty, uncertainties for, 85–86
National Trust for Scotland (NTS), 155
National Voluntary Environmental Service programme, 159
Native grass, 217–219
*Native Pinewoods of Scotland, The* (Steven), 144
Natural capital, 319–320
Natural disturbance regimes, 143
Natural ecological succession, 148
Natural law, 322
Natural regeneration, 157
Natural Resources Development Corporation, 122
Nature, 30, 52
  applying self-regulation and accept feedback, 69
  catching energy, 68
  creatively using and responding to change, 73
  designing from pattern to details
  using diversity, 72
  using edges, 73
  flourishing, 6
  inherent order, 67
  integration rather than segregation, 72
  legal protection, 5
  model, 59

no waste producing, 69–70
observing and interact, 67–68
obtaining yield, 69
principles of, 59, 66
as property, 33
using renewable resources and services, 69
resilience, 4
using small and slow solutions, 72
storing energy, 68
valuing diversity, 72
valuing marginal, 73
valuing renewable resources and services, 69
Von Kármán vortices in clouds downwind of Isla Socorro, 71
Nature Conservancy Council, 144
Nature Conservation Act, 119, 128
Nature's intelligence
continuous presence in popular culture, 39–40
portraying in cinema, 40–41
Nature's laws of reciprocity, 402
all relationships as inclusive and productive of outcome, 402
all relationships as self-reinforcing feedback loops, 403
all relationships having one or more trade-offs, 403
all relationships like irreversible process, 403
all relationships result in energy transfer, 402–403
all systems based on composition, structure and function, 403
all systems defining by function, 402
all systems having cumulative effects, lag periods and thresholds, 404
all systems in cyclical process, but none perfect circles, 404
change as process of eternal becoming, 403
dynamic disequilibrium rules all systems, 404
everything as relationship, 402
only true investment in global ecosystem, 402
restoration, 401
systemic change based on self-organized criticality, 404
Nature's Rights, *see* Rights of Nature
*Nawalux*, 95
NCEAS, *see* National Center for Ecological Analysis and Synthesis
NEC, *see* National Environment Commission
Neoliberalism, 318–323
Nepal, 296–297
environmental federalism in, 301–302
environmental problems in, 297–298
Godavari Marble 2 case, 303–305
legacy of Godavari Marble 1 case, 302–303
map, 296
Namche Bazaar, 298
prefederal environmental protection regime in, 298–301
Rights of Nature, 295
underpinnings of Rights of Nature in case, 305–306
Nereids, 40
*New Scientist*, 50, 53
New Zealand, 259
Ngai Tuhoe, 271
Ngati Whare Claims Settlement Act (2012), 277*n*19
Nga Wai o Maniapoto Act (2012), 277*n*19
NGOs, *see* Nongovernmental organisations
Niger Delta Ecosystem, 168
dredging, 169
gas flaring, 169
inadequate clean-up, 169–170
seismic surveys and construction of roads and pipelines, 169
wastewater, 168–169
Niger Delta indigenous peoples and sacred sites, 171
Adibe and Esiribi Lakes, 171–172
Boupere Lake, 172
Obi pond, 174
Ode Evil Forest, 174–175
*Okpagha* and *Ogriki* Trees, 172–173
*Ovughere* Shrine, 173
stewardship ethic of, 182–184
Umuaja Shrine, 175–176
*Usede* pond, 174
Niger Delta region of Nigeria, 167–168
indigenous conservation laws *vs.* statutory conservation laws, 180–182
preserving stewardship ethic of indigenous peoples, 182–184
rainforests and marshes, 176
Niger Delta States and Oil Fields (2012), 170
Nigerian National Policy, 181
NIMBY phenomenon, *see* Not in my backyard phenomenon
NOAA, *see* National Oceanic and Atmospheric Administration
Non-native grass, 217–219
Nongovernmental organisations (NGOs), 169
Nonmarket transactions, 319
Northern Rocky Mountains region, 10
Not in my backyard phenomenon (NIMBY phenomenon), 198

NTS, *see* National Trust for Scotland
Nuclear power plants, 192
Nutrient cycling, 143
Nymphs, 7, 39–40

O

Oak, 153
Obi Pond, 174
in Ethiope East Local Government Area, 173
Oceanides, 40
Ocean Rights
European Union waters and Baltic Sea, 239–240
fisheries policies, 245–246
Legal and Institutional Frameworks of European Union and Baltic Sea Region, 240–243
movement, 248–249
Rights of Nature affecting criteria for decision-making, 246–248
systemic solution, 243–245
World Ocean Health, 237–238
Ode Evil Forest, 174–175
Official Land Registry, 225
*Ogriki* Trees, 172–173
Oil exploration activities in the Niger Delta region, 177–178
Oil industry, 177
Oil spills in Niger Delta region, 168
*Okpagha* trees, 172–173
Orca, 101–104
Organisation for Economic Cooperation and Development, 321
Orinoco River basin, 206
Osprey, 144
Overexploitation, 96
Overgrazing in Llanos and Pantanal, 221–222
*Ovughere* Shrine, 173
Ovwian community river, 172
Ownership, 322
Oxygen-loving fish, 70

P

Pachamama, 13, 205–206, 281
Pacific Ocean, 342
Pacific Rising Project, 84
Pacific silverweed, 100
Palaeolithic cave art, 6
*Panchabhut, see Pancha Tatva*
*Pancha Tatva*, 305–306
Pandora, 47
*Panpsychism*, 51
*Pan's Labyrinth* (film), 41, 47
Pantanal, 211
biology and fauna, 213
climatic factors, 211–212
fire, 212–213
soils, 212
sustainable management of, 219–225
Pantaneiros, 212
descendants, 224–225
Paraguay River basin, 206
*Pa'sa* in Kwak'wala language, 93
Peat hags, 149
Permaculture, 59
care of Earth, 64–65
care of people, 66
discipline of, 64
ethics, 63
flower, 63
principles, 66
principles of nature, 59, 66–73
reproduction and redistribute surplus, 66
and Rights of Nature, 73–74
setting limits to consumption and reproduction, 66
teachings, 63
Permanent-finance model, 124
Personhood, 90
of deities, 375
religious doctrines of, 375–377
Petroleum-based fertilisers, 61
Philosophy, 29
Pine marten preys, 153
Pipiriki, 277*n*27
Pisang Valley, 300
Planetary crisis, 295
Planetary sustainability, 61
Planet Earth, 401
Plant/animal recovery, 60
Planting, 157
Plants, 98–101
common and scientific names of, 111
Plough-back mechanism, 124
Political rights, 284
Pollinating insects, 72
'Polluter pays' principle, 299, 302, 304
Potlatch, 93–94
Power, 29, 168
Pre-emption as ceiling on rights protections, 336
Precautionary principle, 317–318
Predator-prey dynamics, 143
Predictability of law, 322

Prefederal environmental protection regime in Nepal, 298–301
Principal South American Savanna Ecosystems
  Llanos, 207–211
  Pantanal, 211–213
Private ownership of land, 322
Private property, 323
Produced water, 168
Property, 332
  Nature as, 33
  ownership, 105, 263
  rights, 320
Provincial Court of Justice of Loja, 180
Public interest litigation, 367–368
Public policy domain, 279
Punaka dzong, 123

## Q

*Qawadillikala*, 96, 98
Quaternary extinctions, 213–214
*Qwak'qwala'owkw*, *see* 'Keeping it living' concept

## R

Ratification, 266
Rational anthropomorphism, 106
Rebound effect, 319
Reciprocity principle, 282
Reconciliation ecology, 24–25
Recruitment effects, 246
Recycle ethic, 66
Red-green coalition, 195
Red deer, 153
Redeem, 122
Red squirrels, 152
Regulatory phase in environmental law, 314
Relational liberty, 28
Relational love, 24
Relationship between indigenous peoples and Nature, 167–168
*Relato* methodology, 15, 20–21
Religious doctrines of personhood, 375–377
Renewable resources and services, 69
Republic of Kiribati, 79
Resilience, 327
Resource Management Act (1991), 272
Resources Development Cooperation Limited, 123
Resource stewardship, 94–96
Restoration, 283, 288, 401
*Restrisiko*, 193
Reverence for trees, 40
Rhyzosphere, 61
Rights-based approach, 14
Rights-based law, 394
Rights of Nature, 178–180, 243, 245, 247, 295, 326, 331, 401
  approach in India, 366–367
  in Australia, 386–388
  beginnings, 3–6
  binding jurisprudence on constitutional Rights of Nature language, 290
  biodigesters case, 285–286
  case law, 285
  constitutional case law relating to, 284
  content, 283–284
  in Ecuador, 279
  framework, 250
  in Godavari Marble case, 305–306
  grassroots system changing, 334–337
  language effectiveness, 279
  through lawmaking in United States, 331
  in legal system, 332–334
  local democracy *vs.* corporate state, 334–337
  mangroves case, 286–287
  Marine iguanas, 289
  Model, 324
  natural evolution of justice, 25–27
  origins, 281–282
  philosophy, 31
  precursors in India, 367–368
  and sacred lands tradition in Western culture, 6–8
  selected cases, 285
  selection criteria, 284–285
  unauthorised mining case, 287–289
  Vilcabamba River case, 289–290
  watershed protection through ecosystem rights, 337–338
  Yasuni National Park, 280
Rights of Nature case law, 285
Right to environmental protection, 299, 302
Right to restoration, 283
Rio Branco-Rupununi savannas in Guyana, 206
Rio Declaration, 302
Risk
  society, 193
  technologies, 200
River Ethiope watershed, 176
River pollution, 298
Roads and pipelines construction in Nigeria, 169
Romano-Germanic legal systems, 281
Root cuttings, 151
Rooted cosmopolitanism approach to law and life, 19–23

Royal Society for Protection of Birds (RSPB), 144, 155
Roy Dennis Foundation for Wildlife, 152
RSPB, *see* Royal Society for Protection of Birds
*Russula emetica*, 149

S

Sacred lands tradition in Western culture, 6–8
Sacred places, 98
  for Native cultures, 289
Sacred trees, 40
Salish Sea
  map and surrounding basin, 347
  Rights for, 346–349
Salmon, 98–101
*Sand County Almanac, A*, 19
SANParks, *see* South African National Parks
*Sapere aude!*, 27–29
Savanna Formation Theories, 211
Savannas, 206, 218
*Scientific American*, 50
Scots pine, 141, 144, 147–149
Scottish Natural Heritage, 155
Sea Fisheries Act, 182
Sea otters, 10
Seasonally flooded savannas of South America, 205–207
  European conquest of new world, 215–219
  principal South American savanna ecosystems, 207–213
  quaternary extinctions, 213–214
  solutions to creating sustainable savannas, 225–228
  sustainable management of Llanos and Pantanal, 219–225
Seismic surveys in Nigeria, 169
Self-love, 24
Self-organized criticality, systemic change based on, 404
Self-regulation, 69
Self-reinforcing feedback loops, all relationships as, 403
Sessile oak, 153
Settlement components, 266–267
7th Environmental Action Programme, 311–314, 319
Shallow ecology, 64
Shingkhar-Gorgan road, 131–133
*Shinrin-yoku*, *see* Forest bathing
Shorea robusta, 122
Silvanus, 40
Single-species crops, 62
Slash-and-burn technique, 60
Small and slow solutions, 72
Social-Democratic Party (SPD), 193
Social disharmony, 30
Social harmony, 30
Socialistic-Democratic Party under Chancellor Gerhard Schröder (SPD), 194
Social transformation process, 199
Sociophilia, 23
Soil(s)
  fertility, 60
  in Llanos, 208–209
  in Pantanal, 212
  protection, 219
*Solaris* (film), 41, 49–50
South African National Parks (SANParks), 20
Sovereignty, 322
Sozialadäquate Lasten, 193
Spanish Parliament's Environmental Committee, 178–179
SPD, *see* Social-Democratic Party; Socialistic-Democratic Party under Chancellor Gerhard Schröder
Spiders, 149
Spillover, 246
Spirit, 401
Springbank clover, 100
Squirrels, 68, 152
Statutes of the Global Alliance for the Rights of Nature, 15
Statutory conservation laws, 180–182
Stewardship ethic of indigenous peoples, 182–184
Stockholm Declaration, 370
Stratum function, 209
*Stromeinspeisungsgesetz*, 194
Substitution, 319
*Sumak Kawsay*, 18–19, 281–282, 292*n*13
*Suo motu*
  cognisance of pollution, 370
  powers, 372
Support wildlife, 226
Sustainability
  models, 324
  projects in United States, 8–11
Sustainable conservation, 124, 126
  human-wildlife conflict, 126–128
  problem of medical fungus, 128–130
  waste management and sustainable landscape, 130
Sustainable development, 317–318
  goals, 326
  model, 280

Sustainable Development Agenda, 314
Sustainable management
cattle, 223–224
of freshwater ecosystems, 184
grazing and overgrazing, 221–222
human factor, 224–225
introduced grasses, 222
of Llanos and Pantanal, 219–220
water, 220–221
Swidden agriculture, 60
*Sylvicultura oeconomica*, 193
Symbiotic relationship, 95
Sympathy, 28
Symptomatic fixes of local ecological devastation, 3
Synthetic biology, 320
Systemic analysis, 385–386
Systemic change based on self-organized criticality, 404

**T**

Taquari River, 220
Teak (*Tectona grandis*), 122
*Tectona grandis, see* Teak
Tectonic plate, movement of, 79
*Tekilakw*, 100
Temporal patterns, 71
Te Pou Tupua, 272–273, 275*n*1
Termites, 219
Te Urewera Act (2014), 277*n*34
Theory of justice, 26–27
Third Circuit, 341
Top-down rule of law, 322
Tourism, 113
Traditional anthropocentric approach, 283
Traditional knowledge-holders, 95
Traditional resource management, 100
Treaty of Waitangi Act, 261–263, 268
*Tree of Immortality*, 40
*Tree of Life*, 40
Tree planting, 157
Trees for Life, 144–147, 150–152, 155
Tree windbreaks, 70
Tribal cultures, 40
Truth and Reconciliation Commission, 30
Twinflower, 152

**U**

*Ubuntu*, 16, 24–25
Ugyen Wangchuck Institute for Conservation and Environment (UWICE), 129
UK Biodiversity Action Plan, 150
Umuaja Shrine, 175–176
Unauthorised mining case, 287–289
UNESCO, *see* United Nations Educational, Scientific and Cultural Organization
United National Conference on Environment and Development, 191
United National Environment Programme Ozone Ambassador, 125
United Nations Development Programme, 119, 130
United Nations Educational, Scientific and Cultural Organization (UNESCO), 303
United Nations Environment Programme, 125, 133
United Nations Framework Convention on Climate Change, 120
United Nations Ocean Conference, 250
United States (US)
sustainability projects in, 8–11
Nature's rights though lawmaking in, 331
Universal Declaration of Rights of Mother Earth, 15
Universal principles, 326
Universe, 393–394
Urban ecological gardening, 72
*Usede* pond, 174
U.S. Environmental Protection Agency, 133
UWICE, *see* Ugyen Wangchuck Institute for Conservation and Environment

**V**

Vegetation recovery, 149
Vegetative propagation of aspen, 151
Viable ecosystems, 401
Vilcabamba River case, 289–290
Vilcabamba ruling, 290
*Vindication of the Rights of Women, A*, 31
Vishishta Advaita, 375
Vision-seeking, 6–7

**W**

WAI 168 Whanganui River Claim Outline for Settlement, 277*n*35
Waikato-Tainui Raupatu Claims Settlement Act, 276*n*12
Waitangi Tribunal, 263–264
Walnut (*Juglan regia*), 122
Walt Disney Animation Studios, 42
Wangchuck Centennial National Park, 115

Wangchuck Institute for Conservation of Environment, 124
Washington State, Rights for, 346–349
Waste management and sustainable landscape, 130
Waste Prevention and Management Act, 130
Wastewater in Nigeria, 168–169
Water, 61, 379
  pollution, 285
Water Act (1974), 370–371, 373
*Water Horse-Legend of the Deep, The* (film), 44
Waterlogged soils, 159
Water management in Llanos and Pantanal, 220–221
Watershed protection through ecosystem rights, 337
  Bill of rights for Lake Erie, 338–340
  Highland Township Ordinance, 340–342
  Home Rule Charter, 340–342
  Lincoln County, Oregon, ordinance against aerial spraying, 342–349
Western culture, rights of nature and sacred lands tradition in, 6–8
Western industrial capitalism, 105
Western legal history, 90
Western philosophical and religious traditions, 93
Westminster Parliamentary system, 79
Wet desert, 144
Wetlands, 120
Whanganui River, 260–261
  claims settlement process, 264–266
  components of settlement, 266–267
  map, 260
  settlement, 268–271
  solution, 271–274
  Treaty of Waitangi, 261–263
  Waitangi Tribunal, 263–264
Whanganui River Claims Settlement Act, 272
Wholeness, 326
Wild boar, 126
Wildlife enforcement, 120
Wild zone, 74
Win-win for culture and nature, 182–184
*Wizard of Oz, The* (film), 43
Wolves, 96–98
*Wonderful Wizard of Oz, The* (Baum), 43
Wollestonecraft, 31
Woolly willow, 152
Work weeks, 155
World ocean health, 237–238
World Trade Organisation, 322
Wych elm, 153

## Y

Yamuna river, 370
  environmental conservation and protection, 373–374
  overlapping institutions and politics in 'guardianship', 377–379
  as persons, 372
  problem of guardianship, 374–375
  religious doctrines of personhood, 375–377
Yarra River Protection Act, 395
Yartsa goenbub (*Ophiocordyceps sinensis*), 115, 128
Yellowstone National Park, 9–10
Yggdrasil, 40

## Z

Zebu cattle, 226

www.ingramcontent.com/pod-product-compliance
Ingram Content Group UK Ltd.
Pitfield, Milton Keynes, MK11 3LW, UK
UKHW050757180926
475701UK00007B/1092
*9781032280684*